PHOTOABSORPTION,
PHOTOIONIZATION,
AND
PHOTOELECTRON SPECTROSCOPY

Photoabsorption, Photoionization, and Photoelectron Spectroscopy

Joseph Berkowitz

Physics Division
Argonne National Laboratory
Argonne, Illinois

ACADEMIC PRESS New York San Francisco London 1979

A Subsidiary of Harcourt Brace Jovanovich, Publishers

ACADEMIC PRESS, INC.
111 Fifth Avenue, New York, New York 10003

United Kingdom Edition published by
ACADEMIC PRESS, INC. (LONDON) LTD.
24/28 Oval Road, London NW1 7DX

Library of Congress Cataloging in Publication Data

Berkowitz, Joseph, Date
Photoabsorption, photoionization, and photoelectron spectroscopy.

Includes bibliographical references.
1. Photoelectron spectroscopy. 2. Photoabsorption. 3. Photoionization. I. Title.
QD96.P5B47 543'.085 78–67872
ISBN 0–12–091650–9

PRINTED IN THE UNITED STATES OF AMERICA

79 80 81 82 9 8 7 6 5 4 3 2 1

וַיֹּאמֶר אֱלֹהִים יְהִי אוֹר וַיְהִי־אוֹר׃

וַיַּרְא אֱלֹהִים אֶת־הָאוֹר כִּי־טוֹב

And God said: ''Let there be light.'' And there was light.
And God saw the light, that it was good.

Genesis 1:3–4.

CONTENTS

PREFACE

Although several recent books have concerned themselves with photoelectron spectroscopy, their emphasis has been primarily on spectroscopic rather than dynamic phenomena. While limited surveys exist on absolute photoabsorption cross sections in certain spectral domains, a comprehensive analysis encompassing the entire electromagnetic spectrum and treating many molecules in a systematic fashion is currently lacking. The field of photoionization mass spectrometry, after more than two decades of intense activity, has reached a stage where a detailed review seems warranted. In this monograph I have attempted a synthesis of these areas in order to present a panoramic description of the processes and reactions resulting from the interaction of electromagnetic radiation with isolated molecules.

In the literature the vacuum ultraviolet and x-ray domains are often treated separately because different tools and investigators are involved. Similarly, the techniques for measurement of total and partial cross sections are quite distinct, and research articles tend to focus on particular areas. The photoionization of molecules leads to some processes not encountered with atoms. As a consequence, while surveys of atomic photoionization processes are available, corresponding overviews of molecular photoionization processes are almost nonexistent. The decomposition of molecules and molecular ions is also fragmented in the literature. Workers performing experiments and developing theories for neutral molecules are often unaware of related activity with molecular ions, and vice versa. Molecules mildly excited (valence shell excitation) and highly excited (inner shell excitation) are rarely treated together.

A prime goal of this volume is to present an eclectic synthesis of these sometimes diverse fields of investigation.

The first four chapters and portions of Chapters VI and VII provide a general theoretical background, which should be useful in pedagogic situations at the graduate student level. The material in Chapters V–VII, which treats in detail certain classes of atoms and molecules, should be useful for research workers in various fields: i.e., radiation chemistry and physics and mass spectrometry. Chapter VIII on laboratory techniques is intended for prospective experimentalists who may be inspired to embark upon one of the avenues emanating from this broad domain.

Because of the time span required to write a monograph, parts of the manuscript had to be brought up to date, and new information as of December 1978 has been incorporated. A fair amount of previously unpublished material from my laboratory has been included here (particularly in Chapter VI on photoionization mass spectrometry) in order to achieve a more complete presentation.

ACKNOWLEDGMENTS

The synthesis of various fields which I have tried to achieve in this monograph harks back to my early scientific training as a kineticist and mass spectroscopist and to my subsequent awareness of the significance of the electric-dipole matrix element as a unifying feature, joining the fields involving electromagnetic radiation and those involving the effects of other ionizing radiation on matter. I owe much to the late Professor Robert L. Platzman for his kind and patient tutelage during our years of overlap at Argonne National Laboratory. He initially broadened my horizons and taught me the significance of the dispersion relations and the various sum rules. His comments and criticisms of this manuscript would have been invaluable.

I should like to express my gratitude to Dr. T. E. H. Walker for a brief period of intense interaction and collaboration when I formulated the nucleus of this monograph while preparing a graduate course at Northwestern University. Some of the proofs in the early chapters owe much to his insight. His departure from the scientific scene is sorely felt.

I have had the benefit of considerable discussion and interaction with Dr. Yong-Ki Kim during the course of this effort, and he has kindly read this monograph from his vantage point as a theoretical atomic physicist. His encouragement and assistance are greatly appreciated.

Finally, I should like to express my gratitude to Dr. Gordon L. Goodman, who has proofread this monograph from the rather different perspective of a theoretical chemist.

Needless to say, any errors of commission or omission are my own.

UNITS AND CONVERSION FACTORS*

A. Units of Absorption Probability in a Continuum

The basic unit of cross section (σ) is the megabarn (Mb), which is equal to 10^{-18} cm^2 and is defined by

$$-dI/I = \sigma n l,$$

where n is the molecular density in molecules per cubic centimeter, l the path length in centimeters, and I the light intensity in photons per second.

Sometimes the experiment is normalized to n_0 at STP. Then n_0 is the Loschmidt number, $2.686754(84) \times 10^{19}$ molecules/cm^3, and σn_0 is given the nomenclature k, in units of reciprocal centimeters:

$$k\,(\mathrm{cm}^{-1}) = \sigma\,(\mathrm{Mb}) \times 26.86754(84).$$

In some studies (particularly x-ray measurements), it is customary to report data as attenuation coefficients, defined as μ or μ/ρ, in units of square centimeters per gram. Then

$$\mu\left(\frac{\mathrm{cm}^2}{\mathrm{gm}}\right) = \frac{\sigma\,(\mathrm{Mb}) \times 6.022045(31) \times 10^5}{A},$$

where A is gram atomic weight. (Usually this unit is used for atomic constituents separately rather than for molecules. The A is then on the physical atomic weight scale.)

* Physical constants are taken from Cohen and Taylor.[1]

A unit used by solution chemists and occasionally finding its way into gas phase absorption studies is ε, which may be referred to in the literature as optical density. Consider the absorption equation in the form

$$\ln(I/I_0) = -ln\sigma.$$

Now if n is expressed in moles per liter (on the chemical scale), which we shall call n_{chm}, then

$$\sigma n = \varepsilon n_{\text{chm}}$$

or

$$\varepsilon\left(\frac{\text{liter}}{\text{mole cm}}\right) = \sigma\,(\text{Mb}) \times \frac{6.022045(31) \times 10^{23}}{1.000275}\,\frac{\text{molecules}}{\text{mole}} \times \frac{1\text{ liter}}{10^3\text{ cm}^2}$$

and

$$\varepsilon\left(\frac{\text{liter}}{\text{mole cm}}\right) = \sigma\,(\text{Mb}) \times 6.020389 \times 10^2.$$

Since the measurements are sometimes performed on an apparatus that directly measures $\log_{10}(I/I_0)$, a modified ε_{10} may be encountered whereby

$$\varepsilon_{10} = 0.434294\varepsilon.$$

Finally, we relate to the units of differential oscillator strength in a continuum, df/dE. From Fano and Cooper[2]

$$\begin{aligned}
\sigma(\omega) &= \frac{\pi e^2 h}{m_e c}\frac{df}{dE} \\
&= \frac{e^2}{\hbar c}\hbar^2\frac{2\pi^2}{m_e}\frac{df}{dE} \\
&= \frac{(1.0545887(57) \times 10^{-27}\text{ erg sec})^2 \times 19.7392088}{137.03604(11) \times 9.109534(47) \times 10^{-28}\text{ gm}}\frac{df}{dE} \\
&= 1.7585917 \times 10^{-28}\text{ cm}^2\text{ erg}\frac{df}{dE} \\
&= 1.097618 \times 10^{-16}\text{ cm}^2\text{ eV}\frac{df}{dE},
\end{aligned}$$

where df/dE is the differential oscillator strength per electron volt or

$$\sigma(\omega) = 8.0672784 \times 10^{-18}\text{ Ry cm}^2\frac{df}{dE},$$

where df/dE is the differential oscillator strength per rydberg.

B. Wavelength and Energy Units

The basic wavelength unit for this monograph is

$$
\begin{aligned}
1 \quad \text{Å} &= 10^{-8} \quad \text{cm}, \\
1 \quad \text{nM} &= 10^{-7} \quad \text{cm} \quad \text{(new units)}, \\
1 \quad \text{m}\mu &= 10^{-7} \quad \text{cm} \quad \text{(old units)}.
\end{aligned}
$$

Wavelengths of spectral lines above 2000 Å are usually given as measured in air, those below 2000 Å as measured in vacuum. In this monograph, we are dealing primarily with $\lambda < 2000$ Å. (To convert wavelengths measured in air to corresponding wavelengths in vacuum, $\lambda_{\text{vac}} = n\lambda_{\text{air}}$, where n is the index of refraction of air at that wavelength, or equivalently,

$$\lambda_{\text{vac}} = \lambda_{\text{air}} + (n - 1)\,\lambda_{\text{air}}.)$$

In the x-ray region, a slight variation on the angstrom unit is employed, the kilo x-ray unit (kxu), such that

$$\frac{\text{kxu}}{\text{angstrom}} = 1.0020772(54).$$

Since $hv = hc\omega$ is in energy units, ω is directly proportional to energy and is often used as an energy unit, in cm^{-1}. (Note that both absorption probability and energy can be expressed in units of cm^{-1}, but obviously have different meanings.)

Now

$$\lambda_{\text{vac}}\,(\text{Å}) = 10^8/\omega\,(\text{cm}^{-1}).$$

The unit cm^{-1} in the current context may be referred to as the wavenumber or kayser.

A common energy unit is the electron volt. The most recent voltage–wavelength product is

$$\lambda_{\text{vac}}\,(\text{Å}) \times E\,(\text{eV}) = 12398.520(32)\,\text{eV Å}.$$

Although λ_{vac} and ω may be fairly reliable in older literature, the voltage–wavelength product has changed slightly with time, as the relevant constants have been more accurately established.

Finally, we relate the above units, encountered mostly in experimental results, with the energy units preferred by calculators.

The Rydberg constant is 109,737.3177(83) cm^{-1} or 13.605803 eV. The Hartree unit has twice this value. Usually, calculational results expressed in atomic units (a.u.) refer to Hartree units, although they may occasionally refer to Rydberg units.

REFERENCES

1. E. R. Cohen and B. N. Taylor, *J. Phys. Chem. Ref. Data* **2**, 663 (1973).
2. U. Fano and J. W. Cooper, *Rev. Mod. Phys.* **40**, 441 (1968).

PHOTOABSORPTION, PHOTOIONIZATION, AND PHOTOELECTRON SPECTROSCOPY

INTRODUCTION

This monograph largely confines itself to the study of photoabsorption processes involving individual, isolated molecules in the wavelength or photon energy range from the ionization thresholds of molecules (usually in the vacuum ultraviolet region) through the soft and hard x-ray region and beyond the "K edge." It does not cover absorption processes in the liquid or solid phase. By molecule we mean here diatomic or larger species. Atoms are discussed when illustrative of molecular behavior in general. Reactions between excited or ionized molecules and other gases, while active areas of research, are logically separable from the primary interaction of electromagnetic radiation with molecules, and hence are not considered here.

Many of the techniques necessary for investigation in the vacuum ultraviolet have been developed since World War II, and particularly in the 1960s. The subject matter we are considering is still quite new; no single text or reference covers a significantly large portion of the material we hope to include. Indeed it is the aim of this monograph, which is based on a graduate course given by the author at Northwestern University in 1973, to temporarily fill this void. Some useful references are, however, listed at the conclusion of this chapter.

What do we learn from these various measurements? We shall demonstrate shortly that the absorption of electromagnetic radiation by molecules is the dominant process, generally much stronger than various scattering processes. Hence, if we can determine the absorption coefficient from an attenuation experiment, we know the probability for a molecule to be excited to some energetic state, E_{exc} where $E_0 + h\nu = E_{exc}$. If this excited state has an energy greater than the first ionization potential, then ionization may result, but need not do so 100% of the time. It is possible for such a state of a molecule to dissociate and even conceivable for it to reradiate its excitation energy. Therefore, an independent measurement of the ionization cross section, when combined with an absorption cross section, determines the quantum yield of ionization, commonly called the ionization yield. There are various theorems, collectively referred to as sum rules, which must be satisfied by the distribution of absorption probability throughout the electromagnetic spectrum. They in turn are related to other physical properties of the molecular system, such as electric dipole polarizability, refractive index, and the cross section for ionization by fast, charged particles.

After exploring this distribution of absorption probability (often referred to as oscillator strength distribution) and its relationship to other physical properties, we can ask more specific questions, such as, "What is the probability of formation of the ion in a particular electronic, vibrational, and rotational state?" This probability is a kind of partial cross section, and this type of information can be obtained from photoelectron spectroscopy. This type of study involves the measurement of kinetic energies of electrons produced when monochromatic electromagnetic radiation, having sufficient energy, interacts with molecules. From the conservation of energy criterion, we can write the equation $h\nu = (\mathrm{IP})_n + (\mathrm{KE})_n$ where $h\nu$ is the energy of the incident radiation, $(\mathrm{IP})_n$ corresponds to one of the n ionization potentials of the molecules, and $(\mathrm{KE})_n$ is the kinetic energy of electrons formed in that particular process. The $(\mathrm{IP})_n$ may include various vibrational and rotational, as well as electronic, states of the ion. For a complete understanding of partial cross sections we need to know the photoelectron spectrum as a function of $h\nu$.

Finally, we can ask, "What happens to these ions, in particular states, after they are formed?"

Do they remain "parent ions," or do they subsequently dissociate? Do particular states fragment in specific ways?

These questions can be answered partly by photoionization mass spectrometry, partly by charge exchange mass spectrometry, and more definitively by photoion–photoelectron coincidence spectrometry. If we can amass sufficient information, these results can then be related to the older field of electron impact mass spectrometry. A large portion of this monograph consists of an assemblage of such information for those molecules for which sufficient data are available to make a meaningful analysis.

The angular distribution of photoelectrons is currently an active field for experimentalists and theorists and is related to accurate measurements of partial cross sections. It is covered in a later chapter.

The various measurement techniques are also described in the last chapter. This may serve as a point of departure for experimentalists choosing to direct their research into one of the channels encompassed by this monograph. However, the major thrust of the book is not to answer the question "How to do it?" but rather "What does it mean?"

It can be seen from this discussion that the area covered is relevant to radiation chemists, radiation physicists, photochemists, mass spectrometrists, and perhaps radiation biologists. It is assumed throughout that the reader has had a basic course in quantum mechanics, with application to molecules. The structure of the monograph, with theoretical background followed by application to selected molecules, has been deliberately chosen so that it may serve either as a text for a graduate course or as a reference volume for researchers in relevant fields.

Despite the fact that an immense quantity of research in the past two decades has been focused on the interaction of electromagnetic radiation with gases, my experience has been that university training in this area has not kept pace. If this monograph were to inspire chemistry departments to incorporate such material into the graduate curriculum, it will have achieved much of its intended purpose.

SUGGESTED REFERENCES

Experimental Techniques

1. J. A. R. Samson, "Techniques of Vacuum Ultraviolet Spectroscopy." Wiley, 1967. Good for discussion of techniques for measuring total photoabsorption cross section and photoionization yield.
2. D. W. Turner, C. Baker, A. D. Baker, and C. R. Brundle, "Handbook of Molecular Photoelectron Spectroscopy." Wiley (Interscience), New York, 1970. Describes some of the methods of photoelectron spectroscopy. It is primarily a compilation of data obtained by the authors.
3. K. Siegbahn *et al.*, "ESCA—Atomic, Molecular and Solid State Structure Studied by Means of Electron Spectroscopy." Almquist & Wiksells, Uppsala, Sweden, 1967. (Often referred to in this text as ESCA 1.)
4. K. Siegbahn *et al.*, "ESCA Applied to Free Molecules." North-Holland, Amsterdam, 1969. (Often referred to in this text as ESCA 2.) A discussion of the methods of x-ray photoelectron spectroscopy, again useful as a concentrated source of data.

Compilations of Data

5. R. D. Hudson and L. J. Kieffer, Compilation of atomic ultraviolet photoabsorption cross sections for wavelengths between 3000–10 Å, *Atomic Data* **2**, 205 (1971).
6. R. D. Hudson, Critical review of ultraviolet photoabsorption cross sections for molecules of astrophysical and aeronomic interest, *Rev. Geophysics Space Science* **9**, 305–406 (1971). See also the National Bureau of Standards Reference Data Series, NSRDS–NBS 38 (1971).

Theory and Interpretation

7. U. Fano and J. W. Cooper, Spectral distribution of atomic oscillator strengths, *Rev. Mod. Phys.* **40**, 441 (1968). Best one currently available for this energy region, but restricted to atoms.

Other References

8. G. V. Marr, "Photoionization Processes in Gases." Academic Press, New York, 1967.
9. J. A. R. Samson, The measurement of the photoionization cross sections of the atomic gases, *in* "Methods of Experimental Physics," Vol. 7A, (B. Bederson and W. L. Fite, eds.), Chapt. 2.2, Academic Press, New York, 1968.

Chapter I

SURVEY OF ELECTROMAGNETIC INTERACTIONS WITH GASES

In the hierarchy of interactions of electromagnetic radiation with gaseous matter, one frequently sees allusions to first- and second-order processes. First-order processes refer in this context to those interactions which are deduced from first-order perturbation theory, and correspondingly, second-order processes are inferred from second-order perturbation theory. Although the preponderant emphasis in this monograph will be on first-order processes, we shall begin with a brief review of second-order processes. No attempt is made to derive the equations presented, although references are provided for the interested reader. Rather, the intent is to use the equations to estimate the order-of-magnitude intensities of the various processes, and hence their relative importance across the electromagnetic spectrum. When we discuss the first-order process, it will become apparent that this interaction dominates in the electromagnetic energy domain considered here. The succeeding chapters will thereby maintain continuity with the latter part of this one.

A. The Second-Order Processes

(1) Rayleigh scattering is a process whereby incident radiation is deflected in angle but unchanged in wavelength.

The differential cross section $\sigma(\theta)\,d\Omega$ for scattering of unpolarized light through an angle θ into the element of solid angle $d\Omega$ is[1]

$$\sigma(\theta)\,d\Omega = (2\pi/\lambda)^4|\alpha(\lambda)|^2\,\tfrac{1}{2}(1 + \cos^2\theta)\,d\Omega, \tag{1}$$

where λ is the wavelength of the incident radiation and $\alpha(\lambda)$ the wavelength-dependent polarizability.

The λ^{-4} wavelength dependence of $\sigma(\theta)\,d\Omega$ is quite strong, causing deflection of blue light much more than red light. This preferential scattering of sunlight by the molecules of our atmosphere makes the sky appear blue.

For an order-of-magnitude estimate, let $\lambda = 1000$ Å $= 10^{-5}$ cm, $\alpha(\lambda) \cong (1\ \text{Å})^3 = 10^{-24}\ \text{cm}^3$, then the integrated cross section (4π solid angle) is about $10^{-24}\ \text{cm}^2$, which is called 1 barn. The differential cross section will be of order 0.01–0.1 b.

We shall have occasion in a subsequent chapter to show that in the classical theory of radiation

$$\alpha(\lambda) \equiv \alpha(\omega) = \frac{Ne^2/m}{\omega_s^{\,2} - \omega^2 + ig_s\omega}, \tag{2}$$

where $\omega = 2\pi c/\lambda$ (c is the velocity of light), ω_s is one of the characteristic absorption frequencies of the molecule, g_s a damping factor, N the number of "oscillators," i.e., electrons in the atom or molecule, and e and m are the charge and mass of the electron, respectively.

When λ becomes very small, which means ω increases far beyond the characteristic absorption frequencies (ω_s), the polarizability α becomes proportional to ω^{-2} or λ^2. At such high energies, the wavelength dependence of the $|\alpha|^2$ term cancels the λ^{-4} term and the scattering cross section becomes wavelength independent. This domain is called the Thomson scattering region.

(2) The Thomson scattering cross section is defined as

$$\sigma_0 = \tfrac{8}{3}\pi r_0^{\,2}, \tag{3}$$

where $r_0 = e^2/mc^2$ is the "classical electron radius." Hence, we can write

$$\sigma_0 = \tfrac{8}{3}\pi(e^4/m^2c^4). \tag{4}$$

Note that our Rayleigh cross section at high energies can be written

$$\begin{aligned}\sigma(\theta)\,d\Omega &= \left(\frac{2\pi}{\lambda}\right)^4 \left(\frac{Ne^2\lambda^2}{4\pi^2c^2m}\right)^2 \times \text{angular factor}\\ &= (e^4/m^2c^4) \times \text{angular factor}.\end{aligned} \tag{5}$$

Hence, the Rayleigh cross section gradually becomes wavelength independent as λ approaches and becomes smaller than the dimensions of the molecule. At this point it is called the Thomson scattering cross section. (From the basic constants, $r_0 = 2.8 \times 10^{-13}$ cm, and hence $\sigma_0 = 0.66$ b.)

As the radiation becomes still harder ($\lambda \lesssim 2$ Å), the cross section begins to decrease with increasing photon energy. This is the Compton scattering region where the incoming photon can "see" one electron at a time. The usual theoretical treatment considers the electrons in an atom as unbound, although more sophisticated treatments take into account the effect of this binding. In the Compton process, the outgoing radiation has a different wavelength than the incoming radiation, as well as a different angle. Hence, this is a kind of inelastic scattering.

(3) The Compton cross section is given by the formula

$$\frac{\sigma}{\sigma_0} = \frac{3}{4}\left\{\frac{1+\gamma}{\gamma^3}\left[\frac{2\gamma(1+\gamma)}{1+2\gamma} - \log(1+2\gamma)\right] + \frac{1}{2\gamma}\log(1+2\gamma) - \frac{1+3\gamma}{(1+2\gamma)^2}\right\}, \tag{6}$$

where σ_0 is the aforementioned Thomson cross section and

$$\gamma = h\nu/mc^2.$$

In differential form, this is called the Klein–Nishina formula (see Heitler,[2] Chapter V).

In the nonrelativistic case ($\gamma \ll 1$), this may be expanded to yield

$$\frac{\sigma}{\sigma_0} = 1 - 2\gamma + \tfrac{26}{5}\gamma^2 - \tfrac{133}{10}\gamma^3 + \cdots. \tag{7}$$

In this form, we can readily see the falloff from the Thomson cross section with increasing energy, when $\gamma \approx 0.1$ or $\lambda = 0.25$ Å. Experiments fit the theoretical expression quite well (see graph in ref. 2, Chapter V, p. 223).

In the extreme relativistic limit ($\gamma \gg 1$), the Klein–Nishina formula reduces to

$$\frac{\sigma}{\sigma_0} = \frac{3}{8\gamma}\left\{\frac{1}{2} + \log(2\gamma) - \frac{2\log(2\gamma)}{\gamma}\right\}. \tag{8}$$

The strongest dependence in this expression is $1/\gamma$. Thus, for very high energies, the number of scattered quanta again decreases with the energy of the primary radiation. However, in this region another process, electron–positron pair production, begins to take over.

(4) Electron–positron pair production has a threshold at

$$2(mc^2) \cong 1 \text{ MeV} \cong 0.01 \text{ Å}.$$

The equations describing this process in different materials are rather complex, and need not concern us here since they are well beyond our energy range of interest. We merely note here some relative magnitudes. The cross sections in this region are expressed in units of $\bar{Q} = Z^2 {r_0}^2/137$, where Z is the atomic number. Since ${r_0}^2/137 = 6 \times 10^{-4}$ b, $\bar{Q} \cong 0.5$ b for $Z = 30$. For Cu, this is the value at $\gamma \approx 5$, and the cross section increases with γ.

(5) Inelastic scattering reconsidered. The Raman effect. We have previously described Rayleigh and Thomson scattering as elastic (no change in λ) and Compton scattering which occurs at higher energy as inelastic. Actually there exists an inelastic process at lower photon energy as well, which can satisfy the quantum restrictions and the conservation laws. One may view Rayleigh scattering as the initial absorption of the incoming photon by the atom or molecule, raising it to some virtual state which then reemits the same wavelength of light (in some characteristic angular distribution) and returns to its ground state. In this framework, one can allow for the possibility of a transition from the virtual state to a level other than the ground state, with the accompanying emission of radiation that differs in energy from the incoming radiation by an amount corresponding to the excitation induced in the atom or molecule. This shift of emitted radiation from the main Rayleigh line by quantized magnitudes is called Raman scattering. It has been extensively used to elucidate the vibrational levels of molecules. If the incident energy is high enough, the Raman effect can result as well in *electronic* excitation of the target atom or molecule. One can anticipate a smooth transition from the electronic (excitation) Raman effect to the Compton effect for bound electrons, which is a kind of ionization. Both are second-order effects, and differ from photoabsorption and the photoeffect, to be described below. This relationship between the electronic Raman effect and the Compton effect for bound electrons has been considered (see, e.g., Gavrila[3]), but to this author's knowledge the analytical union of the two effects has not yet been achieved.

We shall dwell on the Raman effect a bit longer in order to develop a feeling for its intensity. In Rayleigh scattering we noted that the cross section was proportional to the square of the dynamic polarizability $\alpha(\omega)$. We can think of the incident light wave of frequency ω as inducing an oscillating dipole moment μ in the molecule, proportional to the polarizability. Thus,

$$\mu = \alpha \mathscr{E}_0 \cos(\omega t). \tag{9}$$

If the molecule is vibrating, α is itself a function of time and can be expressed as

$$\alpha = \alpha_0 + \left(\frac{\partial \alpha}{\partial Q_k}\right) \Delta Q_k \cos(\omega_k t) \tag{10}$$

or

$$\mu = \left[\alpha_0 + \left(\frac{\partial \alpha}{\partial Q_k}\right) \Delta Q_k \cos(\omega_k t)\right] \mathscr{E}_0 \cos(\omega t). \tag{11}$$

The induced electric moment can therefore be regarded as the superposition of three periodically changing moments having frequencies ω,

$\omega + \omega_k$, and $\omega - \omega_k$. The reradiated light corresponds to the Rayleigh line, the anti-Stokes–Raman line, and the Stokes–Raman line.

The ratio of intensities of the kth fundamental Raman band to that of the Rayleigh line is expressed by[4,5]

$$\frac{I_k}{I_0} = \left(\frac{\omega_0 - \omega_k}{\omega_0}\right)^4 \left(\frac{\partial\alpha/\partial Q_k}{\alpha}\right)^2 \frac{\hbar}{2\omega_k}\left[\frac{1}{1 - \exp(-h\omega_k/kT)}\right]. \tag{12}$$

The most significant term in this expression is the ratio of the polarizability derivative to the polarizability, which reduces the Raman intensity by approximately one to two orders of magnitude from the Rayleigh line intensity.

B. The First-Order Process, Photoabsorption

We introduce the photoabsorption process in the same spirit with which we have discussed the second-order processes. It is not our purpose to rederive the quantum theory of radiation, but rather to present the essential equations in the framework of more familiar concepts.

Consider an attenuation experiment in which electromagnetic radiation having a bandwidth between ω and $\omega + d\omega$, and intensity $I(\omega)$ is permitted to pass normal to the surface of a slab of gas of thickness dx. The radiation exiting from this slab will have been attenuated by $dI(\omega)$, given by

$$-dI(\omega) = I(\omega)N_0\sigma_{on}(\omega)\,dx, \tag{13}$$

where N_0 is the number density of molecules of the gas in some lower (ground) state and the equation serves to define an absorption coefficient $\sigma_{on}(\omega)$ corresponding to a transition between the lower state 0 and upper state n.

The incident radiation induces transitions from state 0 to state n, thereby reducing the number density N_0. This is described by the equation

$$-\frac{dN_0}{dt} = N_0 \int B_{on}(\omega)\rho(\omega)\,d\omega, \tag{14}$$

where $\rho(\omega) = I(\omega)/c$ is the energy density of the electromagnetic radiation and $B_{on}(\omega)$ the Einstein coefficient for induced absorption [also equal to $B_{no}(\omega)$, the coefficient for induced emission].

$$B_{on}(\omega) = \frac{4\pi^2}{3h^2}\left|\left\langle \psi_n \middle| e\sum_j \mathbf{r}_j \middle| \psi_0 \right\rangle\right|^2, \tag{15}$$

where the quantity in bra-kets is the electric dipole matrix element. $I(\omega)$, which is in units of (say) ergs/cm^2 sec frequency interval can be written as

$N(\omega) \cdot \hbar\omega$, where $N(\omega)$ has units of number of quanta/cm^2 sec frequency interval.

Then we can write the probability/sec, which is $(-dN_0/N_0)/dt$, as

$$\frac{\text{Prob}}{\text{sec}} = \frac{4\pi^2}{3h^2}\left|\left\langle \psi_n \middle| e\sum_j \mathbf{r}_j \middle| \psi_0 \right\rangle\right|^2 \frac{N(\omega)\hbar\omega}{c}. \tag{16}$$

We can also write

$$\text{Prob/sec} = \int N(\omega)\sigma(\omega)\,d\omega. \tag{17}$$

Now for the moment, we restrict ourselves to a very narrow absorption line, such that $N(\omega)$ does not change substantially over this wavelength interval. Then

$$\frac{\text{Prob}}{\text{sec}} = N(\omega)\int \sigma(\omega)\,d\omega = \frac{4\pi^2}{3}\frac{e^2}{\hbar c}\left|\left\langle \psi_n \middle| \sum_j \mathbf{r}_j \middle| \psi_0 \right\rangle\right|^2 \omega N(\omega). \tag{18}$$

We now introduce a quantity

$$f_n = \frac{2m}{3\hbar}\omega\left|\left\langle \psi \middle| \sum_j \mathbf{r}_j \middle| \psi_0 \right\rangle\right|^2. \tag{19}$$

This quantity f_n is called the oscillator strength. It was first introduced in the classical theory of radiation. We shall have an opportunity to develop a feel for this quantity in a subsequent chapter. For the present, we merely note from the above equations that

$$\sigma(\omega) = \frac{2\pi^2 e^2}{mc} f_n. \tag{20}$$

In the domain of continuous absorption (dissociation or ionization continuum) this expression becomes

$$\sigma(\omega) = \frac{2\pi^2 e^2}{mc}\frac{df}{d\omega} = \frac{\pi e^2 h}{mc}\frac{df}{dE} \tag{21}$$

(cf. Fano and Cooper,[6] eq. 2.14). Now $df/d\omega$ is of order

$$\frac{2m}{3\hbar}\left|\left\langle \psi_n \middle| \sum_j \mathbf{r}_j \middle| \psi_0 \right\rangle\right|^2 \tag{22}$$

or

$$\sigma(\omega) \sim \frac{2\pi^2 e^2}{mc}\frac{2m}{3\hbar}\left|\left\langle \psi_n \middle| \sum_j \mathbf{r}_j \middle| \psi_0 \right\rangle\right|^2 \tag{23}$$

$$\sim \frac{4\pi^2}{3}\frac{e^2}{\hbar c}\left|\left\langle \psi_n \middle| \sum_j \mathbf{r}_j \middle| \psi_0 \right\rangle\right|^2. \tag{24}$$

Since $e^2/\hbar c$ = fine structure constant $\approx 1/137$ (pure number) and the r_j will be of the magnitude of 1 Å $= 10^{-8}$ cm and therefore

$$\left|\left\langle \psi_n \middle| \sum_j \mathbf{r}_j \middle| \psi_0 \right\rangle\right|^2 \approx 10^{-16}\ \text{cm}^2,$$

we obtain as an order-of-magnitude estimate of $\sigma(\omega)$ for the continuum (for allowed electric–dipole transitions)

$$\sigma(\omega) \sim \frac{4\pi^2}{3}(10^{-2} \times 10^{-16}) \approx 10 \times 10^{-18}\ \text{cm}^2 = 10\ \text{Mb}.$$

We shall see that the values of absorption cross sections for molecules in the continuum are indeed in the range of one to many tens of megabarns.

In summary, we note that the expression for photoabsorption can be derived from the quantum theory of radiation in first-order perturbation theory, and hence the relevant matrix element has the form expected from the first-order perturbation theory, i.e., $\langle \psi_n|\mathbf{r}|\psi_0\rangle$.

The other processes we have discussed can be derived from second-order perturbation theory, and will have characteristic terms of the form

$$\frac{\langle \psi_b|\mathbf{r}_1|\psi_c\rangle\langle \psi_c|\mathbf{r}_2|\psi_a\rangle}{E_a - E_c - \hbar\omega}$$

and coefficients of order $(e^2/mc^2)^2$. This accounts for the large difference in the magnitude (a factor of 10^6) of photoabsorption and second-order processes.

Multiphoton processes, although actively being pursued in many laboratories, are beyond the scope of our present interests.

C. The Hierarchy of Photoabsorption Processes

We have made passing reference to allowed, electric-dipole photoabsorption processes. It turns out that there are other types, and to examine these some additional mathematics is necessary.

A plane monochromatic wave may be represented by a vector potential $\mathscr{A}$, where

$$\mathscr{A} = A_0 \cdot e\{\mathscr{K} e^{-i\omega t} + \mathscr{K}^* e^{i\omega t}\} \tag{25}$$

and

$$\mathscr{K} = e^{i/\hbar c(\mathbf{k} \cdot \mathbf{r})}.$$

$\mathbf{k}$ is a vector describing the photon; its direction is the propagation direction, and its magnitude represents the energy ($\hbar\omega$) of the photon.

The interaction between the electrons in a molecule and the external field may be written

$$\mathscr{H}_{\text{int}} = \frac{e\hbar}{mci}\sum_j \int A_0(\omega)\{\mathscr{K}_j e^{-i\omega t} + \mathscr{K}_j^* e^{i\omega t}\}(e\nabla_j)\, d\omega. \tag{26}$$

We are interested in the value of the matrix elements corresponding to this interaction in some small frequency range $d\omega$.

$$\langle\psi_n|\mathscr{H}_{\text{int}}|\psi_0\rangle \sim \left\langle \psi_n \left| \sum_j \mathscr{K}_j(e\nabla_j) \right| \psi_0 \right\rangle. \tag{27}$$

Now $\nabla_j = \partial/\partial r_j$ and it can be shown by manipulation that

$$\left\langle \psi_n \left| \frac{\partial}{\partial r_j} \right| \psi_0 \right\rangle = \left(\frac{m\omega_n}{\hbar}\right)\langle\psi_n|\mathbf{r}|\psi_0\rangle. \tag{28}$$

By expanding $e^{i/\hbar c(\mathbf{k}\cdot\mathbf{r})}$ about $r = 0$, we obtain

$$\langle\psi_n|\mathscr{H}_{\text{int}}|\psi_0\rangle \sim \left\langle \psi_n \left| \sum_j \left(1 + \frac{i}{\hbar c}(\mathbf{k}_\lambda\cdot\mathbf{r}_j) - \frac{(\mathbf{k}_\lambda\cdot\mathbf{r}_j)}{(\hbar c)^2} + \cdots\right) r_j \right| \psi_0 \right\rangle. \tag{29}$$

The first term on the right-hand side is of the form

$$\left\langle \psi_n \left| \sum_j \mathbf{r}_j \right| \psi_0 \right\rangle.$$

which we recognize and identify as the electric-dipole term. The second term is of the form

$$\left\langle \psi_n \left| \sum_j \left[\frac{i}{\hbar c}(\mathbf{k}_\lambda\cdot\mathbf{r}_j)r_j\right] \right| \psi_0 \right\rangle.$$

The ratio of second term to first term is of the order

$$\frac{(i/\hbar c)(\mathbf{k}\cdot\mathbf{r})}{1}.$$

Since $\mathbf{k}$ has magnitude $\hbar\omega = \hbar c/\lambda$,

$$(i/\hbar c)(\mathbf{k}\cdot\mathbf{r}) = 2\pi i(r/\lambda). \tag{30}$$

If r is of the order of atomic dimensions (~ 1 Å) and $\lambda \cong 1000$ Å, then the second term is $\sim 10^{-3}$ of the first term. The corresponding probabilities go as the squares of these matrix elements, and hence differ by $\sim 10^6$ at 1000 Å.

The second term contains contributions that are referred to as electric-quadrupole and magnetic-dipole transitions. These two are of the same order. They become important when

(a) the electric-dipole term is zero because of selection rules or because of cancellation of positive and negative contributions to the electric-dipole term, or

(b) the incident radiation has λ of the order r.

This latter condition, which is of some interest to us, occurs for $\lambda \sim 1$ Å $\cong$ 12 keV.

REFERENCES

1. See, e.g., W. Kauzmann, "Quantum Chemistry," p. 585. Academic Press, New York, 1957.
2. W. Heitler, "The Quantum Theory of Radiation," 3rd Ed. Oxford Univ. Press (Clarendon), London and New York, 1954.
3. M. Gavrila, *Phys. Rev. A* **6**, 1348 (1972).
4. G. Placzek, *in* "Handbuch der Radiologie" 2nd ed. (E. Marx, ed.), Vol. VI, p. 276. Akademische Verlagsgesellschaft m.b.h., Leipzig, 1934.
5. T. Yoshino and H. J. Bernstein, *Spectrochim. Acta* **14**, 127 (1959).
6. U. Fano and J. W. Cooper, *Rev. Mod. Phys.* **40**, 441 (1968).

Chapter II

PHOTOABSORPTION BELOW THE FIRST IONIZATION LIMIT

From the analysis presented in the previous chapter, we can conclude that the dominant interaction of electromagnetic radiation with gaseous matter is electric-dipole allowed photoabsorption. At energies below the onset of dissociation and ionization, such absorption occurs to discrete states. Low-lying states are usually referred to as valence states, higher energy ones as Rydberg states. The distinction between these classes is often blurred, and is a favorite subject of discussion in some circles. Here we take the pragmatic viewpoint that a Rydberg state is one which (at least approximately) follows the Rydberg formula

$$E_n = \mathrm{IP} - R/(n - \delta)^2, \tag{1}$$

where E_n is the energy of the nth level, R the Rydberg constant = 109,737 cm^{-1}, n the quantum number of the level considered, and δ the quantum defect. In principle, the value of R varies slightly with the reduced mass of the electron-core system, and the modified value appropriate to the particular mass should be used.

For atomic hydrogen, $\delta = 0$ and the first Rydberg member occurs at 10.2 eV, precisely three-quarters of the ionization potential. In general, the first Rydberg member for a molecule is to be expected at roughly this fraction of its ionization potential. For many of the molecular systems we shall be examining in this monograph, the first ionization potentials are in the 10–15 eV region, and hence the first Rydberg member occurs at 7–10 eV.

Let us now consider the changes that occur as $\hbar\omega$ approaches the first ionization potential. It has been known since the early days of quantum theory (and is clear from the Rydberg formula) that an infinite number of electronic energy levels exist below the ionization potential of an atom or molecule. If one knows the value of E_n and $(n - \delta)$ for several values of n, one can compute the ionization potential. In practice, the value of δ fluctuates when n is small, becoming more or less constant as n exceeds ~ 10. However, it becomes more difficult to identify high n states, partly because the intensity of transitions to higher n states falls off as n^{-3}, and also because of small perturbations frequently present and the difficulty of resolving adjacent high n states. Hence, although the extrapolation of Rydberg levels E_n remains a powerful and accurate means of determining an ionization potential, it has not been as frequently applicable as the direct measurement of the onset of ionization, i.e., the minimum energy necessary to liberate an electron and leave behind a free ion. A prime function of photoionization studies is to determine this onset, as one varies the frequency of incident radiation across the threshold value.

While on the subject of these Rydberg states, it is useful to consider one further property that characterizes them. One can readily show from solutions to hydrogen-like atoms that the average radius of the "Rydberg electron" is

$$r_n = n^2\hbar^2/me^2Z. \tag{2}$$

Hence, the spatial extent of a Rydberg state of an atom or molecule can become quite large, approaching 100 Å when $n \to 14$. Such large physical entities can be expected to have correspondingly large cross sections for collisional reactions, such as

(1) collisional ionization

$$A^{**} + M \to A^+ + e + M, \tag{3}$$

(2) charge transfer

$$A^{**} + B \to A^+ + B^-, \tag{4}$$

(3) chemi-ionization

$$A^{**} + B \to AB^+ + e. \tag{5}$$

Now let us consider some properties characteristic of a molecular system. If the incident radiation is less energetic than the first ionization potential, the resultant molecular system may either be (a) elevated to a discrete state or (b) dissociated into separate entities, which may be neutral or charged, i.e.,

$$\hbar\omega + AB \to A + B \tag{6}$$

$$\to A^+ + B^-. \tag{7}$$

Thus, it is possible to cause ionization in a molecular species even if $\hbar\omega < \mathrm{IP}(1)$, provided that

$$\mathrm{IP(A)} - \mathrm{EA(B)} + \mathrm{D_0(AB)} < \mathrm{IP(AB)} \equiv \mathrm{IP}(1), \tag{8}$$

where IP, EA, and D_0 are ionization potential, electron affinity, and dissociation energy, respectively.

Process (6) is called direct dissociation and process (7) direct ion-pair formation.

What becomes of the molecular system undergoing process (a)? Let us consider some possibilities

$$\text{Process (a):} \quad \hbar\omega + \mathrm{AB} \rightarrow (\mathrm{AB})^* \tag{9}$$

$$\text{Reradiation:} \quad \mathrm{AB}^* \rightarrow \mathrm{AB} + \hbar\omega \tag{10}$$

$$\text{Predissociation:} \quad \mathrm{AB}^* \rightarrow \mathrm{A} + \mathrm{B} \tag{11}$$

$$\text{Predissociation into ion pairs:} \quad \mathrm{AB}^* \rightarrow \mathrm{A}^+ - \mathrm{B}^-. \tag{12}$$

Processes (11) and (12) can occur because of curve crossing, a mechanism first discussed by Landau[1] and Zener[2] (see Fig. 1). In brief, it comes about because the state AB* we have formed is described as a "mixture" of two or more "unperturbed wave functions," degenerate in energy. The excited system can undergo a "radiationless transition" or curve crossing from the bound portion to the unbound one, resulting in predissociation. The pre-

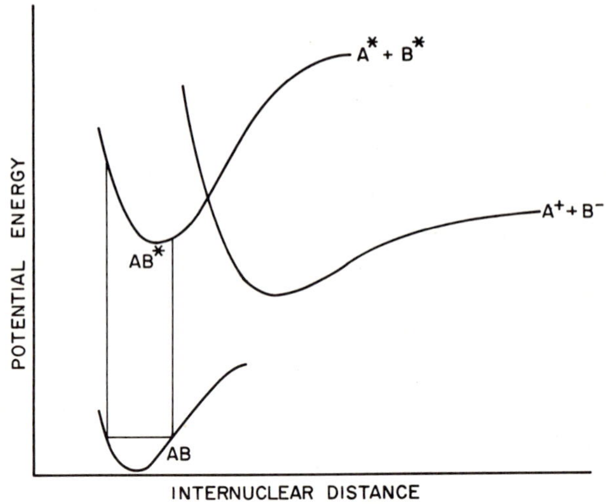

Fig. 1 Illustration of predissociation into ion pairs. This is the most commonly encountered mechanism when ion-pair formation is observed, rather than direct transition to the ion-pair state.

dissociation can result in neutral or ionic products, corresponding to (11) or (12). Indeed, the primary mechanism for ion-pair formation in most systems investigated to date has been the predissociative one (12), rather than (7). Examples of ion-pair formation below IP(1) which have been investigated include F_2, Br_2, I_2, HF, HCl, HCN, CH_3Cl, CH_3Br, and TlX (X = halogen atom).

Process (12) is most likely to occur below IP(1) when EA(B) is large [see Eq. (8)], which for practical purposes frequently restricts B to be a halogen atom, as can be seen in the above examples.

How do we distinguish between direct dissociation and predissociation? Very often it is the width of the peak corresponding to the type of dissociation which is its earmark. We note from Fig. 1 that predissociation is most probable over a narrow energy region where the curves cross. Direct dissociation can be viewed as a transition to a repulsive curve, which will generally result in a broad dissociation band. Curve crossing can also result in reradiation of a different frequency.[3]

Process (10) can occur if there are no states in the energy range of the excited state which interact with it. It can be shown from the Einstein A coefficient for spontaneous emission that typical lifetimes for such states are 10^{-6}–10^{-9} sec. From the uncertainty principle, this corresponds to an energy width of 10^{-9}–10^{-6} eV or 10^{-4}–10^{-7} Å. This is called the "natural width" of the absorption line. In practice, the Doppler effect and collisions will broaden this line somewhat. Nevertheless, it is very difficult to obtain experimental resolution widths in photoabsorption experiments smaller than the inherent line widths for small molecules. As a consequence, the determination of oscillator strengths f_n for absorption to discrete states are hazardous for "thick targets," which are commonly employed in photoattentuation experiments. This problem can be minimized for thin targets, which are common in electron scattering and photoionization mass spectrometric measurements. Peculiar features have sometimes been reported in photoabsorption studies which failed to recognize this. It is a consequence of the fact that absorption by sharp intense features can obey the Beer–Lambert law [the integrated form of Eq. (13), Chapter I] when the incident radiation is not too strongly absorbed (thin target), but fails when the target thickens.

In order to pursue our study of absorption processes to higher energy and ultimately (for this chapter) the ionization continuum, we make the following observations. If the oscillator strength f_n varies monotonically, then

$$\frac{\Delta f_n}{\Delta n} = \frac{f_{n+1} - f_n}{(n+1) - n} = \frac{\bar{f}_n}{1} \rightarrow f_n = \frac{df}{dn} \qquad (13)$$

for large n.

Now differentiating (1), we obtain

$$\frac{dE_n}{dn} = \frac{2R}{(n-\delta)^3} \approx \frac{2R}{n^3}. \tag{14}$$

Hence, we can write

$$\frac{df}{dE} \cong \frac{n^3 f_n}{2R}. \tag{15}$$

If $(n - \delta)$ is replaced by an effective quantum number n^* (which is generally nonintegral) then we can write (15) more precisely as

$$\frac{df}{dE} = \frac{(n^*)^3 f_n}{2R}.$$

Equation (15) is an indication of the manner in which discrete absorption features eventually merge with the continuous absorption at the ionization threshold. If it can be shown that $f_n \propto n^{-3}$, then the merger should be smooth and continuous. Sommerfeld[4] had already speculated about this smooth merger. Hartree[5] showed that for an atomic system $f_n \propto n^{-3}$.

In the Appendix, we examine the properties of the Laguerre polynomials which are the radial wave functions for the hydrogen atom. If high Rydberg states are hydrogenic and their wave functions can be approximated as hydrogenic wave functions, then it follows from the analysis in the Appendix that $f_n \propto n^{-3}$, and hence that the average oscillator strength in the discrete spectrum merges smoothly into the continuum.

The quantum defect method provides a detailed formulation of the qualitative arguments indicated above (see Fano and Cooper[6]; Seaton[7,8]).

Appendix

We would like to show that the photoabsorption cross section to successive Rydberg states falls off as n^{-3}. We assume that the wave function for a Rydberg electron (which spends most of its time relatively far from the positively charged center) may be approximated by the wave function for such a Rydberg electron for atomic hydrogen. The radial wave functions for the hydrogen atom are

$$R(r) = C\chi^l e^{-\chi/2} L_{n+l}^{2l+1}(\chi), \tag{16}$$

where

$$\chi = \frac{8\pi^2 \mu e^2}{nh^2} \tag{17}$$

and $L_{n+1}^{2l+1}(\chi)$ is the set of associated Laguerre polynomials. Their general formula is

$$L_\alpha^{\beta}(\chi) = (-1)^\alpha \frac{\alpha!}{(\alpha-\beta)!}\left\{\chi^{\alpha-\beta} - \frac{\alpha(\alpha-\beta)}{1!}\chi^{\alpha-\beta-1} + \frac{\alpha(\alpha-1)(\alpha-\beta)(\alpha-\beta-1)}{2!}\chi^{\alpha-\beta-2} + \cdots\right\}. \tag{18}$$

These Laguerre polynomials are not orthogonal functions, but the functions

$$e^{-\chi/2}L_\alpha(\chi)$$

form an orthogonal set. In addition, they have the normalization property

$$\int_0^\infty e^{-\chi}\chi^{\beta+1}L_\alpha^{\beta}L_\alpha^{\beta}\,d\chi = \frac{(2\alpha-\beta+1)(\alpha!)^3}{(\alpha-\beta)!}. \tag{19}$$

The normalization condition is

$$\int_0^\infty [R(r)]^2 r^2\,dr = 1 \tag{20}$$

$$\int_0^\infty C^2\chi^{2l}e^{-\chi}L_{n+l}^{2l+1}(\chi)r^2\,dr = 1. \tag{21}$$

From the definition of χ and of

$$r = \frac{n^2h^2}{4\pi^2\mu e^2} \qquad \text{[see (2)]}$$

and

$$a_0 = \frac{h^2}{4\pi^2\mu e^2}$$

(which is the Bohr radius), we can take

$$r = \frac{na_0}{2}\chi \quad \text{and} \quad r^2\,dr = \left(\frac{na_0}{2}\right)^3\chi^2\,d\chi.$$

This leaves us with

$$\int_0^\infty [R(r)]^2 r^2\,dr = C^2\left(\frac{na_0}{2}\right)^3\int_0^\infty \chi^{2l+2}e^{-\chi}[L_{n+l}^{2l+1}(\chi)]^2\,d\chi. \tag{22}$$

The integral on the right-hand side is given by (19), and is

$$\frac{[2(n+l)-(2l+1)+1][(n+l)!]^3}{[n+l-(2l+1)]!} = \frac{2n[(n+l)!]^3}{(n-l-1)!}. \tag{23}$$

Therefore,

$$C^2\left(\frac{na_0}{2}\right)^3 = \frac{(n-l-1)!}{2n[(n+l)!]^3}$$

and

$$C = \left\{\left(\frac{2}{na_0}\right)^3 \frac{(n-l-1)!}{2n[(n+l)!]^3}\right\}^{1/2}. \tag{24}$$

The n-dependence of $R_{n,l}(r)$ is of the form

$$R \propto \left[\frac{(n-l-1)!}{n^4[(n+l)!]^3}\right]^{1/2} \frac{1}{n^l} L_{n+l}^{2l+1}\left(\frac{2\rho}{n}\right), \tag{25}$$

where we take $\rho = (Z/a_0)r$, and $e^{-\rho/n} \approx 1$ for $\rho/n \ll 1$. The general formula for $L_\alpha{}^\beta(\chi)$ has been given in descending powers of $\chi = 2r/na_0 = (2/Z)(\rho/n)$. For $\rho/n \ll 1$, the most important term in the general formula for L is the one with the smallest exponent of χ, which is the last term. This last term will have a coefficient (apart from $\pm$ sign) of

$$\frac{\alpha!}{(\alpha-\beta)!}\frac{\alpha!}{\beta!} = \frac{(n+l)!(n+l)!}{(n-l-1)!(2l+1)!}.$$

Our current interest is not the l-dependence, although this expression with the above qualifications is true for all l. For the n-dependence, we can write

$$R \propto \left[\frac{(n-l-1)!}{n^4[(n+l)!]^3}\right]^{1/2} \frac{1}{n^l} \frac{(n+l)!(n+l)!}{(n-l-1)!}.$$

Cancelling some terms, we rewrite

$$R \propto \frac{1}{n^l}\left[\frac{(n+l)!}{n^4(n-l-1)!}\right]^{1/2}.$$

Now consider

$$\left[\frac{(n+l)!}{(n-l-1)!}\right]^{1/2} = \left[\frac{(n+l)(n+l-1)\cdots(n-l)(n-l-1)!}{(n-l-1)!}\right]^{1/2}$$
$$= [(n+l)(n+l-1)(n+l-2)\cdots(n-l)]^{1/2}. \tag{26}$$

There are $2l+1$ of the terms under the last square root, and their mean value is n. We approximate this expression by

$$(n^{2l+1})^{1/2}.$$

Hence,

$$R \propto \frac{1}{n^l}\left(\frac{n^{2l+1}}{n^4}\right)^{1/2} = \frac{1}{n^{3/2}}. \tag{27}$$

The oscillator strength f_n was defined in Chapter I, Eq. (19), and is seen to be proportional to the *square* of the dipole integral connecting the ground state and the excited state. Since the n-dependence occurs only in the wave function of the excited state,

$$f_n \propto \left(\frac{1}{n^{3/2}}\right)^2 \propto \frac{1}{n^3}.$$

We shall see in Chapter III that the normalization conditions which led us to the n-dependence of the cross section for formation of Rydberg states also explains the n-dependence of other properties of Rydberg states, such as their autoionization probability.

REFERENCES

1. L. Landau, *Phys. Z. Sowjetunion* **2**, 46 (1932).
2. C. Zener, *Proc. R. Soc. London, Ser. A* **137**, 696 (1932); **140**, 660 (1933).
3. A. E. Douglas, *J. Chem. Phys.* **45**, 1007 (1966).
4. A. J. W. Sommerfeld, "Atombau und Spektrallinien." Vieweg, Braunschweig, 1924.
5. D. R. Hartree, *Proc. Cambridge Philos. Soc.* **24**, 426 (1928).
6. U. Fano and J. W. Cooper, *Rev. Mod. Phys.* **40**, 441 (1968).
7. M. J. Seaton, *Proc. Phys. Soc. London* **88**, 801, 815 (1966).
8. M. J. Seaton, *Mon. Not. R. Astron. Soc.* **118, 504** (1958).

Chapter III
QUASI-DISCRETE STATES ABOVE THE FIRST IONIZATION POTENTIAL

A. Types of "Discrete" Excitation above the First Ionization Potential

Since the primary emphasis in this monograph is the description of photoabsorptive processes that lead to ionization in molecules, those Rydberg states with large n that correspond to excitation of an electron from the valence shell really define the energy range where our study begins. For all systems more complicated (i.e., having more electrons) than a hydrogen atom, there exist states whose energy of excitation is greater than that corresponding to the first ionization potential. In atomic systems, these states can generally be classified as two-electron excitations and inner-shell excitations. An example of the former is the transition $He(1s)^2 + \hbar\omega \rightarrow He(2s2p)$ (Fig. 1) which occurs at $\sim$60.1 eV, whereas the first ionization potential of He is 24.6 eV. The latter may be exemplified by the transition in argon,

$$Ar(1s)^2(2s)^2(2p)^6(3s)^2(3p)^6 + \hbar\omega \rightarrow (1s)^2(2s)^2(2p)^6 3s(3p)^6 np,$$

which occurs in the energy region 26.5–29.24 eV ($\lambda \cong$ 470–424 Å) for $n = 4 \rightarrow \infty$. The first ionization potential of Ar is at 15.76 eV. Hence, one can anticipate a Rydberg series superposed upon an ionization continuum and converging to the 29.24 eV limit. Such a series, characterized by a diminution of intensity rather than an enhancement, has been observed in the absorption spectrum of argon corresponding to this energy (see Fig. 2). When electron spin is taken into account, then even the excitation of an

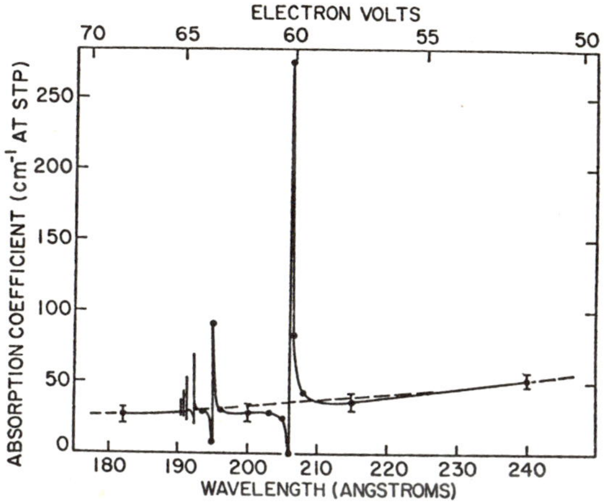

Fig. 1 Autoionization of two-electron excited states in helium converging on the double ionization potential. The data were obtained in an absorption experiment by Madden and Codling.[1]

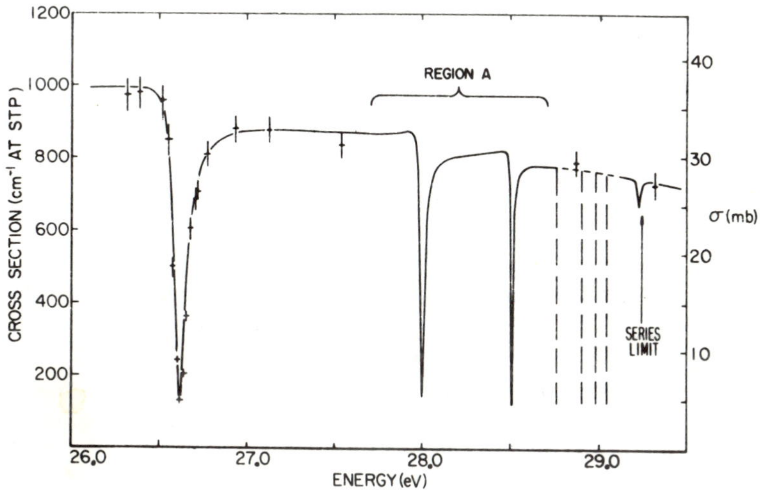

Fig. 2 Autoionization resulting from excitation of an electron from the (inner) 3s orbital of argon to np levels. The autoionization manifests itself in this case as "window resonances" converging upon a state of Ar^+ in which there is a hole in the 3s shell. (From Madden *et al.*[1a])

outermost shell electron into Rydberg levels can lead to states of excitation of the neutral atom above the first IP. In the aforementioned case of Ar, the outermost orbital is completely filled with p electrons. The electric-dipole selection rules permit ns and nd states to be reached for transitions emanating from this outer orbital. As n approaches ∞, the residual ionic system is left with $l = 1$ (the missing p electron) and $s = \pm 1/2$, which corresponds to the two states of Ar^+, $^2P_{3/2}$ and $^2P_{1/2}$, separated in energy by 0.177 eV. Since there exists a set of Rydberg levels converging to each ionization limit of an atomic or molecular system, there must exist Rydberg levels that converge to the excited $Ar^+(^2P_{1/2})$, as well as the ionic ground state $Ar^+(^2P_{3/2})$. Thus, there is a range of Rydberg levels converging to the excited state which also lie above the first ionization potential. (See Fig. 3.)

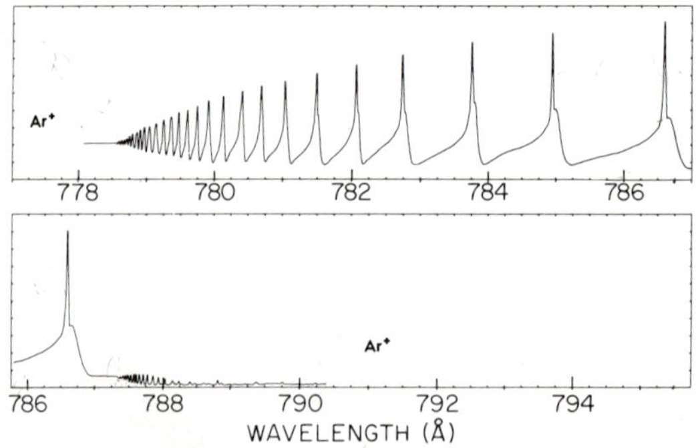

Fig. 3 Autoionization resulting from excitation of an electron from the (outer) 3p orbital of argon to ns and nd Rydberg levels which converge upon the higher energy member ($^2P_{1/2}$) of a spin-orbit pair and decay to the lower energy member ($^2P_{3/2}$).

Diatomic or More Complex Molecules

It is by now well established that the orbital picture of atomic structure can be extended to systems having two or more nuclear centers. These molecular orbitals, in the valence levels, tend to have trajectories (or better, probability densities) that encompass the several nuclear centers. When one focuses attention progressively to inner shells, these orbitals begin to look more and more like atomic (one center) orbitals, only slightly perturbed by the neighboring nuclei.

In the present context, all of the mechanisms for producing "discrete" states above the first ionization potential for atomic systems (two-electron excitation, inner-shell excitation, and spin-orbit splitting) can also exist for multicenter systems.

However, molecular systems have additional mechanisms for generating such excited states. The two forms of energy which are characteristic of molecular, but not atomic, systems are their vibrational and rotational modes. Recognizing this fact, it is possible for us to construct states of a molecular system whose total energy (electronic + vibrational + rotational) exceeds the first ionization potential, but in which the electronic component alone is insufficient for ionization. In real cases, the electronic component must be rather close to the ionization limit, since vibrational excitation imparted to a molecular system by photoabsorption is infrequently much larger than 1 eV, the rotational energy component still smaller, and the first ionization is generally in the neighborhood 10–15 eV. Hence, the electronic component will usually imply a high Rydberg state. Several examples are presently known (H_2, HF, F_2, HOF, F_2O) in which vibrational excitation plus electronic excitation leads to "discrete" states above the first ionization potential, but only in one case (H_2) has the rotational mechanism been observed.

How can one identify and distinguish these various mechanisms for producing "superexcited states"[1b] in molecules? An obvious method for recognizing their existence is the observation of discrete structure in the photoabsorption spectrum at photon energies above the first ionization potential, where that is known. Frequently, overlapping structure is found due to several transitions occurring in the same energy region. This fact, together with the perturbations that result from near coincidence of energies and line broadening due to interactions† with the underlying continuum, make spectroscopic analysis and identification above the first ionization potential very difficult. In such analyses, two patterns that one seeks are vibrational progressions

$$E_{\mathrm{v}} - E_0 = \hbar[(\mathrm{v})\omega_e - \mathrm{v}(\mathrm{v} + 1)\omega_e x_e],$$

where ω_e is the vibrational frequency, $\omega_e x_e$ the anharmonic constant, and v the vibrational quantum number and Rydberg progressions [Chapter II, Eq. (1)]. If the value of IP were known, the Rydberg type of analysis could obviously be simplified. Similarly, if ω_e were approximately known, the vibrational progressions could be more easily identified.

Since we are dealing here primarily with systems in which one electron (the "Rydberg electron") has been excited to an orbital far from the core, the vibrational motion of the core (and hence ω_e) approaches ever more closely the corresponding behavior of the ionic state. Therefore, if we knew ω_e for that particular ionic state, we would be in a better position to deduce the vibrational progression of the Rydberg states converging to that ionic level. The determination of both IP and ω_e has been greatly aided in recent years by the development of the field of photoelectron spectroscopy.

† This topic will be thoroughly discussed in Section C.

B. Photoelectron Spectroscopy

Heretofore, we have mentioned virtually nothing about the techniques employed to observe photoionization phenomena. Indeed, we have elected to defer all experimental methodology to the last chapter. In this section, however, it is necessary to describe phenomenologically the two major types of investigation pertinent to this study.

Visualize a device which selects a beam of photons, more or less monochromatic, and directs them into a chamber containing the molecules of interest. If ionization takes place, the ions formed are swept from this chamber by an electric field and directed into a mass spectrometer, which measures the number of ions per second being created, having a particular mass. If the wavelength of the incident photons is varied, then there exists some long wavelength limit beyond which no ionization occurs.† This long wavelength limit corresponds to the first ionization potential, i.e., formation of the ion in its ground state. As the wavelength of photons is decreased, a photon energy (hc/λ) will be attained which makes it energetically possible for the resulting ion to be found in an excited state, *as well as the ground state*. The mass spectrometer will measure an ion current, but this information does not enable one to distinguish whether the ions at this wavelength are in their ground state, excited state, or whether some are in the ground state and others in excited states. In other words, the measurement of ion intensity is related to a *total* cross section for ionization. A missing piece of information is the *partial* cross section for formation of individual states of the ion.

Now visualize a similar experiment at this shorter wavelength in which the mass analyzer is replaced with a device which detects the electrons concomitantly formed in the photoionization act. Furthermore, this device measures the kinetic energies of these electrons corresponding to the selected wavelength of photons. We have selected a wavelength such that two possible states of the ion, $E(0)$ and $E(1)$, are energetically possible. We can write two simple equations, based upon conservation of energy, that describe these possible processes,

$$h\nu + \mathrm{M} \rightarrow \mathrm{M}^{+}_{E(0)} + \mathrm{e}^{-}_{\mathrm{KE}(0)}$$

and (1)

$$h\nu + \mathrm{M} \rightarrow \mathrm{M}^{+}_{E(1)} + \mathrm{e}^{-}_{\mathrm{KE}(1)},$$

where KE(0) and KE(1) represent the kinetic energies of electrons liberated in the first or second process. Since these kinetic energies can be distinguished

† We neglect here the process of ion-pair formation, and other more subtle phenomena such as field ionization, chemi-ionization, collisional ionization, etc.

by our device and their relative intensities measured, we have a means of measuring the partial cross sections for formation of the ion in its ground and excited states. This is the basic philosophy of photoelectron spectroscopy.

From a careful measurement of these kinetic energies, one can determine the energetic thresholds for formation not only of the ionic ground state, but its various excited states up to the limit $h\nu$. These are referred to as the adiabatic ionization potentials of the molecule. In addition, it is frequently possible to observe fine structure corresponding to the excitation of a particular ionic state in different levels of vibrational excitation. From the spacings of these levels one can compute ω_e and $\omega_e x_e$ using the aforementioned vibrational progression formula. Thus, the photoelectron spectroscopic technique provides a first approximation to IP and ω_e which can be used to help analyze the superexcited states.

In addition to these energy parameters, two intensity criteria are frequently helpful in analyzing spectra. It has already been shown in Chapter II that the intensities of Rydberg members of the same series vary approximately as n^{-3}. For a molecular system, the intensity distribution of vibrational components of a Rydberg level will approach the corresponding distribution for the ionic state to which these Rydberg levels converge. This can be appreciated by noting that the potential energy curves of the high Rydberg levels, describing the interaction as a function of internuclear distance, approach the potential curve of the ion as the Rydberg electron moves ever farther away. Therefore, the vibrational wave functions of the Rydberg levels will have parameters approaching those of the corresponding ionic state and the vibrational transition probabilities, given by the overlap between neutral ground state vibrational wave function(s) and excited or ionic state vibrational wave functions, will be similar. Since the intensity distribution of vibrational levels corresponding to a given ionic state (which we shorten to vibrational envelope) can be measured by photoelectron spectroscopy, this intensity distribution can be utilized in the analysis of the superexcited Rydberg levels.

C. Discrete State–Continuum Interaction: Autoionization and Predissociation

The previous sections have outlined various mechanisms by which neutral molecules can be excited to states above the first ionization potential. Each of these states therefore has an energy which, in zero order, is the same as that corresponding to some state of the molecular ion plus the requisite kinetic energy of the departing electron. Since the first ionization potential of a molecule AB generally occurs at higher energy than the threshold for dissociating the molecule, these superexcited states are also degenerate in energy with some states of the system A + B, where either A or B can have internal and/or kinetic energy.

When such degenerate states in zeroth order appear in the discrete spectrum, they are usually split by some higher order interaction. For example, the well-known accidental degeneracy of the Coulomb field leaves the 2s and 2p states of atomic hydrogen identical in energy, but higher-order interactions described by quantum electrodynamics and measured by the Lamb shift, split these two levels. The P level is split into $^2P_{3/2}$ and $^2P_{1/2}$ states by the interaction of the orbital angular momentum and spin angular momentum of the electrons. Each of these states is in turn split by the interaction of the nuclear spin angular momentum with the resultant of the electronic interaction (known as hyperfine structure). The Zeeman and Stark effects are examples of external perturbations that enable zeroth order degeneracies to be split.

In the molecular domain, there are a number of other internal interactions that belong to this category. Linear molecules may be characterized by the component of orbital angular momentum Λ along the molecular axis, which is quantized. States are characterized as Σ, Π, Δ, etc., when this component has quantum number 0, 1, 2, etc. The molecule has another form of angular momentum due to its overall rotation. For states having $\Lambda > 0$, the molecular rotation can be clockwise or anticlockwise, and hence the vectorial addition of electronic and nuclear motions splits these Λ states into sublevels, a phenomenon known as λ-doubling. The Jahn–Teller effect is an internal interaction that causes a polyatomic molecule of high symmetry, formed in a degenerate state in zeroth order to split into states of lower symmetry.

The above phenomena can be formally treated by first-order perturbation theory, which may be written in the form

$$V_{km} = \langle \psi_m | \mathscr{V} | \psi_k \rangle, \; \Psi_k = - \sum_{m \neq k} \frac{V_{km}}{E_m - E_k} \psi_m,$$

where $\mathscr{V}$ is the perturbation operator and acts on the unperturbed wave functions ψ_k.

This approach can be extended to the case where degeneracy exists not between two discrete states, but between a discrete state and a state in the continuum. (Indeed, it may be extended to the case of many discrete states and more than one set of continuum states. For an experimental illustration and discussion, see Chapter VI, Section D.3.)

The first theoretical treatment of this general problem was given by Rice.[2] He considered the case where a potential curve, corresponding to a bound state, is intersected by a repulsive curve, these being the unperturbed states. His treatment showed that absorption from the molecular ground state to the aforementioned "bound" state gave rise to rotational fine structure lines that were not only broadened, but shifted in position and distorted in shape. In his analysis he divided the continuum energy states into two groups, those very near to the unperturbed "bound" state and others relatively farther away. The nearby states were quantized by the particle in a box model, the transition being made to a very large box, thus making the separation between states arbitrarily small.

In 1935, Fano[3] first interpreted some discrete structure in the absorption spectra of Ar and Kr, above the ionization potential,[4] as being due to the interaction of discrete states and an underlying ionization continuum. This treatment was later[5] made more general by using the Dirac delta function method instead of the particle in a box quantization. His derivation will not be repeated in detail, but rather roughly sketched below together with the pertinent results.

Let us consider a discrete state ϕ which is embedded in a continuum, and one member state of this continuum $\psi_{E'}$ which is nearly degenerate with ϕ in zeroth order. Then the total wave function at this energy is the linear combination

$$\Psi_{E'} = C_1\psi_{E'} + C_2\phi \tag{2}$$

and from first-order perturbation theory

$$C_2 = \frac{V_{E'}}{E' - E_\phi}. \tag{3}$$

Now the ratio of transition probabilities from the ground state ψ_0 to the compound state $\Psi_{E'}$ and to the continuum state $\psi_{E'}$ is given by

$$\frac{\langle\psi_0|r|\Psi_{E'}\rangle^2}{\langle\psi_0|r|\psi_{E'}\rangle^2}. \tag{4}$$

Substituting from (2), we obtain for this ratio

$$\left[C_1 + C_2\frac{\langle\psi_0|r|\phi\rangle}{\langle\psi_0|r|\psi_{E'}\rangle}\right]^2. \tag{5}$$

If $C_2 = 0$, then from (3) $V_{E'} = 0$, which means there is no interaction between the discrete state ϕ and the continuum state $\psi_{E'}$. Then the transition probability to the compound state is simply the sum of the individual transition probabilities. If the transition probability to the continuum is constant or slowly varying, the transition probability to the discrete state will appear as a symmetric peak superposed upon the continuum.

If $C_2 \neq 0$, then from (3) C_2 changes sign as the incident energy, E' varies from $<E_\phi$ to $>E_\phi$.

As C_2 changes sign, the transition probability expression (5) abruptly changes intensity in the vicinity $E' = E_\phi$, and hence the absorption in the peak region around E_ϕ will be asymmetric.

Fano's mathematically rigorous formulation arrives at an expression analogous to (5), e.g.,

$$(q + \varepsilon)^2/(1 + \varepsilon^2), \tag{6}$$

where

$$q = \frac{\langle \psi_0 | r | \Phi \rangle}{\Pi V_{E'} \langle \psi_0 | r | \psi_{E'} \rangle} \tag{7}$$

and ε is the varying incident energy measured in units of peak width. A large value of q corresponds to our previous weak interaction ($C_2 \sim 0$), an intermediate value of q gives rise to the asymmetric peak shape discussed earlier (for $C_2 \neq 0$). A value of $q = 0$ corresponds to an extremely weak transition probability to the discrete state, and a strong interaction between the discrete state and the nearby continuum. This interaction "repels" the nearby continuum states and results in the "window resonance" we noticed in Fig. 2.

Fano[5] has studied the effect of this interaction as one progresses through a Rydberg series embedded in a continuum. The value of q changes very slightly in going from one Rydberg member to another in the upper part of the series. From (7) we note that the numerator will have an $n^{-3/2}$ dependence, as shown in the Appendix to Chapter II. But the perturbation matrix element $V_{E'}$ also contains the function ϕ, and thus has the same $n^{-3/2}$ dependence. Hence, q, which is proportional to this ratio, is essentially independent of n. The width of each resonance will be proportional to $V_{E'}^2$, and hence to n^{-3}. The spacing between levels [Chapter II, Eq. (14)] is also proportional to n^{-3}. Therefore, as one progresses to higher members of the Rydberg series, the separation between peaks diminishes but so do the peak widths, enabling one in principle to continue resolving adjacent peaks to very high Rydberg members.

The profiles of the perturbed lines are frequently referred to as Beutler–Fano line shapes, and have been observed in predissociation as well as autoionization. (See Chapter VI, partial cross section of H_2.) Now let us summarize the salient features of this "configuration interaction."

(1) The complete wave function for the system is a linear combination of discrete and continuum states. Therefore, the molecular system, when once excited to the "discrete" level, has a certain probability of subsequently finding itself in its other form, i.e., as a molecular ion plus an electron or two neutral entities, in the case of a dissociation continuum. The two cases are conventionally referred to as autoionization and predissociation. Because of

this interaction and decay, we shall frequently refer to these superexcited states as quasi-discrete states.

(2) The probability for this decay into the continuum is governed by the perturbation matrix element, which is analyzed in more detail below. What would have been a sharp line is thereby broadened by this interaction, due to the uncertainty principle.

(3) The energy of the discrete peak is shifted by the interaction, just as in the case of two interacting discrete states.

(4) The shape of the absorption peak can take various forms, depending upon the interaction. This shape factor has been neatly parametrized by Fano.[5]

(5) The intensities of the various absorption peaks can be strongly influenced by the strength of the interaction between discrete state and continuum. This factor can confuse the identification of states based solely upon the aforementioned n^{-3} rule and vibrational envelopes.

(6) The angular distribution of the photoelectrons (see Chapter VII) produced by the autoionization can vary dramatically in the neighborhood of these absorption peaks.[6–8]

D. The Nature of the Perturbation Matrix Elements in Autoionization

Case (1). Excitation of an Inner-Shell Electron to a Rydberg Level

This mechanism can be visualized as the migration of energy from the Rydberg electron to a valence shell electron as the Rydberg electron re-occupies its inner shell hole, or a valence shell electron dropping into the inner shell hole and releasing its energy to the Rydberg electron. (This is obviously intended as a pictorial description. The indistinguishability of electrons must be recognized.) In either event, the interaction is of the form e^2/r_{jk}, where r_{jk} is the distance between the Rydberg electron and some core electron, and the transition probability takes the form

$$\int \phi_e^f \chi_v^f \frac{e^2}{r_{jk}} \phi_e^i \chi_v^i \, d\tau, \tag{8}$$

where ϕ_e^f includes the wave function of the departing electron.

Within the framework of the Born–Oppenheimer approximation, we can recast this matrix element in the form

$$\int \phi_e^f \frac{e^2}{r_{jk}} \phi_e^i \, d\tau_e \int \chi_v^f \chi_v^i \, d\tau_v. \tag{9}$$

The second integral is a vibrational overlap and is like a Franck–Condon factor, but refers to the radiationless transition between the quasi-discrete state and the ionization continuum. If the vibrational distribution predicted by this integral could be observed, it would constitute good evidence for this mechanism. Since the quasi-discrete state is frequently not well characterized, the problem is often inverted to determine the properties of the quasi-discrete state from the observed vibrational distribution.[9]

Case (2). Valence Shell Excitation Involving Spin Flip

This behavior has been observed primarily in the noble gas atoms Ne, Ar, Kr, and Xe, but should also be applicable to such molecules as HCl, HBr, and HI. For the noble gases, the ionic ground state is $^2P_{3/2}$ and the first excited state is $^2P_{1/2}$. Some members of the Rydberg series converging to the $^2P_{1/2}$ limit lie above the $^2P_{3/2}$ threshold. Autoionization can result from a spin flip in the core, releasing the Rydberg electron, or from an exchange between Rydberg and core electrons, culminating in a $^2P_{3/2}$ core. At this time, a quantitative theoretical description of the detailed interaction which can account for the differences among the noble gases is not available. It appears, however, as if the matrix element which most likely determines the width of both the $p \rightarrow d$ and $p \rightarrow s$ resonances is of the form

$$V = \left\langle \psi_n \left| \frac{e^2}{r_{jk}} \right| \psi_i \right\rangle, \tag{10}$$

where ψ_n represents the total wave function of the neutral Rydberg state and ψ_i represents the total wave function of final ion state and continuum electron. We can expand the operator $1/r_{12}$ (restricting ourselves for the moment to a two-electron problem) as

$$\frac{1}{r_{12}} = \sum_{l=0}^{\infty} \frac{r_<^{\,l}}{r_>^{\,l+1}} Y_{lm}(\mathbf{r}_1) Y_{l-m}(\mathbf{r}_2),$$

where $r_<$ refers to a core electron, $r_>$ to a Rydberg electron and $Y(\mathbf{r})$ are the Legendre functions.

Both ψ_n and ψ_i can be written as products of core functions and outer electron functions. The core functions of ψ_n and ψ_i have angular momenta $j = \frac{1}{2}$ and $j = \frac{3}{2}$, respectively. In the expansion of $1/r_{12}$, only two terms can couple $j = \frac{1}{2}$ and $j = \frac{3}{2}$, and they involve $l = 1$ or $l = 2$. Since the parity of initial and final states (ψ_n and ψ_i) is the same, the matrix element will vanish for $l = 1$ (odd parity). Hence, the expansion of $1/r_{12}$ reduces to a single term

$$\frac{1}{r_{12}} = \frac{r_<^{\,2}}{r_>^{\,3}} Y_{2m}(\mathbf{r}_1) Y_{2-m}(\mathbf{r}_2).$$

This in turn restricts the autoionization of s-like Rydberg states into d departing waves, and d-like Rydberg states into s and d waves. One general feature of this class of autoionizations is that the d-like Rydberg states generate broad resonances, whereas the s-like ones are characterized by sharp resonances. The radial overlap integrals connecting Rydberg and outgoing wave electron wave functions must be responsible for this different behavior. The above analysis permits d → d transitions, but forbids s → s transitions. Since high-lying Rydberg states and low energy continuum electron states of the same l can be expected to have a similar spatial distribution, the d → d radial overlap integral may be the dominant factor in describing the greater width of the d-like resonances. Other systematic features of these resonances, including some anomalous behavior, are summarized in a recent study.[10]

Since the perturbation involves electron coordinates, the corresponding molecular transition integral in the HX case should be factorable into an electronic and a vibrational term, the latter again being of the Franck–Condon form.

Case (3). Simultaneous Two-Electron Excitation

In this case, the excitation energy of each electron, by itself, is insufficient to cause ionization, but the sum of their excitations exceeds the ionization threshold. To account for the initial excitation, one must go beyond the independent particle approximation and include either configuration interaction or correlation.

Hence, when the latter effects are important (as in some alkaline earth metal atoms), we may anticipate such a type of autoionization to be prominent. The matrix element will again be of the form (10).

Case (4). Excitation of a Valence Electron to a Rydberg Level and Concomitant Excitation of Vibrational Energy

This problem has been treated rather exhaustively by Berry and collaborators[11–14] and Bardsley,[15] particularly for the case of H_2. The vibrational energy of the core must somehow be transmitted to the rather distant Rydberg electron. This transfer involves the interaction of vibrational and electron coordinates, and implies a breakdown of the Born–Oppenheimer separation. Two significant conclusions of the analyses by Berry and collaborators[11–14] and Bardsley[15] are appropriate for summary here.

(a) As the Rydberg level n increases, the average distance of the Rydberg electron from the nuclear core increases, and hence the interaction should be expected to decrease. Hence, the autoionization transition rate should decrease, and the lines should become narrower with larger *n*.

(b) The major perturbation element is of the form

$$\int \chi_v{}^f \frac{\partial}{\partial R} \chi_v{}^i \, d\tau_v,$$

where *R* is the internuclear distance of the diatomic molecule. This operator $(\partial/\partial R)$, when acting on a Hermite polynomial, has the effect of raising or lowering it by one. Since the vibrational wave functions (the Hermite polynomials) are an orthonormal set, the only nonvanishing contribution to the integral is one where v_f and v_i differ by one unit, i.e., $\Delta v = \pm 1$. This is similar to the selection rule for infrared absorption. Higher order terms will affect this conclusion, but the first-order result is seen to differ drastically from the conclusion of cases (1) and (2), which yield a broad vibrational envelope as a consequence of the autoionization whereas the vibrational autoionization can be expected to yield predominantly one vibrational member.[9]

Case (5). Rotational Autoionization

This mechanism has thus far been identified only in the case of H_2. Consequently, discussion of this point is postponed to Chapter VI, in the section dealing with processes in H_2.

REFERENCES

1. R. P. Madden and K. Codling, *Astrophys. J.* **141**, 364 (1965).
1a. R. P. Madden, D. L. Ederer, and K. Codling, *Phys. Rev.* **177**, 136 (1969).
1b. R. L. Platzman, *J. Phys. Radium* **21**, 853 (1960); *Vortex* **23**, 372 (1962); *Radiat. Res.* **17**, 419 (1962); *J. Chem. Phys.* **38**, 2775 (1963).
2. O. K. Rice, *J. Chem. Phys.* **1**, 375 (1933).
3. U. Fano, *Nuovo Cimento* **12**, 156 (1935).
4. H. Beutler, *Z. Phys.* **93**, 177 (1935).
5. U. Fano, *Phys. Rev.* **124**, 1866 (1961).
6. D. Dill, *Phys. Rev. A* **7**, 1976 (1973).
7. J. A. R. Samson and J. L. Gardner, *Phys. Rev. Lett.* **31**, 1327 (1973).
8. N. M. Kabachnik and I. P. Sazhina, *J. Phys. B* **9**, 1681 (1976).
9. J. Berkowitz and W. A. Chupka, *J. Chem. Phys.* **51**, 2341 (1969).
10. K. Radler and J. Berkowitz, *J. Chem. Phys.* **70**, 216 (1979).
11. R. S. Berry, *J. Chem. Phys.* **45**, 1228 (1966).
12. S. E. Nielsen and R. S. Berry, *Chem. Phys. Lett.* **2**, 503 (1968).
13. R. S. Berry and S. E. Nielsen, *Phys. Rev. A* **1**, 383, 395 (1970).
14. G. B. Shaw and R. S. Berry, *J. Chem. Phys.* **56**, 5808 (1972).
15. J. N. Bardsley, *Chem. Phys. Lett.* **1**, 229 (1967).

Chapter IV
THE IONIZATION CONTINUUM

The discussion thus far has implied an ionization continuum of nebulous properties. In this section, we take a closer look at this continuum and concern ourselves with (a) the variation of cross section in the vicinity of the ionization threshold, or energy region where ejection of an electron from an outer orbital becomes possible; (b) the variation of cross section with energy in the energy region where ejection of an electron from an inner orbital becomes possible; (c) deep minima in the absorption cross section (so-called Cooper minima); (d) the dependence of absorption cross section upon the type of orbital (s,p,d, etc.) from which the electron is being ejected; (e) molecular "barriers" to electron escape in certain molecules; (f) multiple electron excitations and ionizations; and (g) the overall variation of cross section from ionization threshold in the vacuum ultraviolet region to the x-ray region, sometimes referred to as the oscillator strength distribution in the continuum, and its dependence upon the size or complexity of the molecule.

A. The Variation of Cross Section in the Vicinity of the Ionization Threshold

Let us first consider the simplest system, atomic hydrogen. The energy levels in the discrete spectrum are given by

$$E_n = \mathrm{IP} - \frac{R}{n^2}.$$

The probability of photon absorption to the respective E_n levels goes as n^{-3}. As $n \rightarrow \infty$, the density of levels piles up near the ionization potential, and the probability of absorption to any one level goes to zero. However, the average absorption probability per unit energy remains a smooth function up to and beyond the ionization threshold, as we have seen in Chapter II.

The calculation of the probability of photon absorption beyond threshold for H atom (the ionization cross section) has been carried out by several authors (see Samson [1]). In Fig. 1, this cross section has been plotted as a function of photon energy. (Except for a single measurement at 14.58 eV, this cross sectional behavior has not been verified experimentally.) One sees an abrupt rise at threshold, followed by a monotonic (more or less exponential) decline to higher energies. The post-threshold decline for hydrogenic systems has been calculated[2] to be $E^{-8/3}$, asymptotically approaching $E^{-7/2}$ at high energy. Wigner[3] has calculated the post-threshold cross sectional behavior of the photoionization process in the limit of zero excess photon energy above threshold. In this limit, the details of the absorption probability characteristic of individual atomic and molecular entities does not enter, and the cross section is determined by the long range interaction between the departing particles, in this case a positive ion and electron. The coulombic attraction then leads to an abrupt onset to a finite cross section, with a post-threshold behavior independent of excess photon energy (so-called step-function behavior). It is clear that even in the case of atomic hydrogen, this behavior is not fulfilled for any finite energy beyond threshold. For helium, which has been studied both theoretically and experimentally, the situation is described in Fig. 2. Again, there is an abrupt rise to a finite cross section at threshold, immediately followed by a monotonic decline.

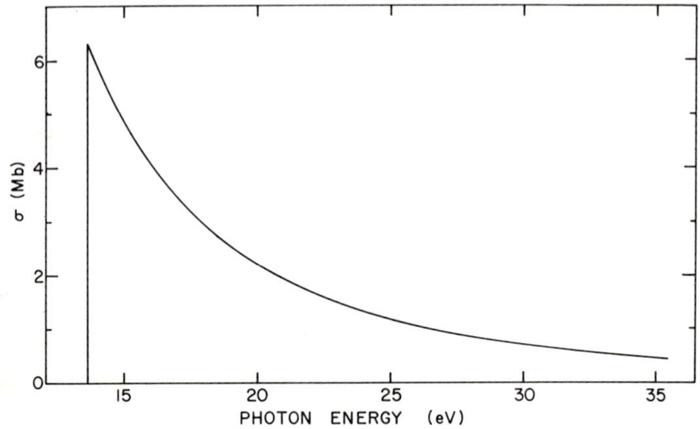

Fig. 1 The calculated photoionization cross section of atomic hydrogen.

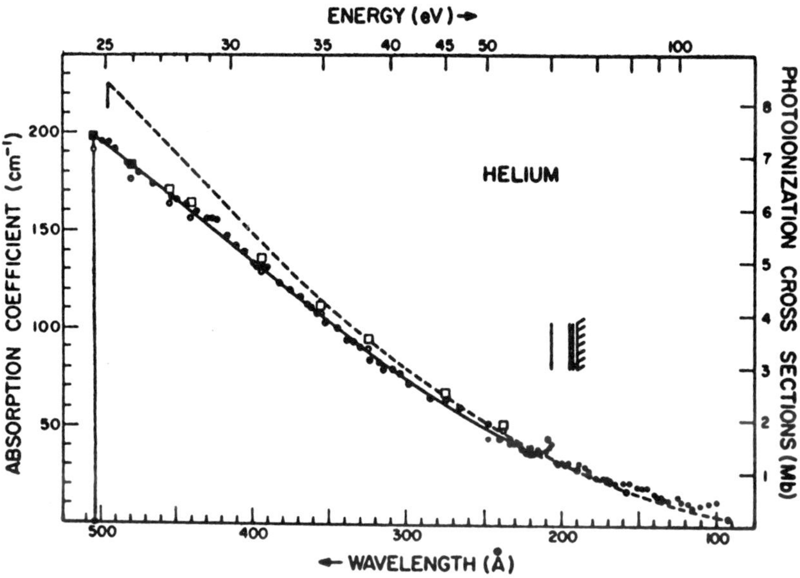

Fig. 2 The experimental photoabsorption (or photoionization) cross section of helium. The dotted line is from a calculation. (From Samson,[1] p. 185.)

Hence, for the simplest systems, a step-function description of threshold behavior for any finite energy beyond threshold would seem to be inappropriate. For the heavier noble gases, autoionization fine structure due to spin-orbit interaction (see Chapter III, Section D, Case (2) and Chapter 3, Fig. 3) manifests itself immediately beyond threshold, and prevents an examination of the isolated continuum. This point, together with (c) and (d) below, thwart an analysis of the unperturbed continuum in the alkali metals and other simple atomic systems.

For molecular systems, the situation is even more complicated. In addition to autoionization phenomena, one must be prepared for variation of ionization probabilities with vibrational state (Frank–Condon effects), with electronic transition moments which are a function of internuclear distance, with the effects of molecular rotation, and with the fact that there may be another competing channel (dissociation) for the disposal of the absorbed energy beyond threshold. Even the abrupt rise at threshold is not always apparent. In the simplest molecules, such as H_2, O_2, and N_2, the behavior of the continuum is largely obscured by very strong autoionization features. The molecule NO had for many years been considered the best example of step-function behavior, but recent results[4] indicate that autoionization is also prominent here. H_2O[5] and CO_2[5] exhibit a more or less steep rise at threshold, with opposite post-threshold behavior. COS[6] and CS_2[6] do not even have an abrupt rise at threshold. For more complex molecules, the abrupt rise is even less frequently encountered. Hence, one is led to the conclusion that step-function behavior, while a convenient idealization, has little correspondence with reality. For a finite region beyond threshold, the *dynamics* of molecular absorption that characterize a *particular* species immediately manifest themselves, and negate any generalized approach to post-threshold behavior.

B. The Energy Region for Inner-Shell Excitation

When the incident photon energy is high enough, an inner-shell electron can be excited to empty orbitals which eventually take on the characteristics of a Rydberg series as the electron is excited to ever larger orbits and "sees" the residual ion core more as a point charge than as a complex potential of the system. The high members of this series will, in general, be degenerate with the continuum corresponding to the system (ionic ground state + electron), and hence interaction and autoionization will ensue. The resulting spectrum will display autoionization features which themselves converge upon this higher ionization potential, but the average absorption *per unit energy* will go smoothly through the new ionization onset as seen in Fig. 3. As a consequence of this phenomenon, there is no abrupt rise in absorption intensity or ionization probability at the threshold for formation of an excited ionic state. Even the deepest shells (in x-ray terminology, the K and L shells) should exhibit this behavior and not give rise to "absorption edges." However, when absorption probability is plotted on a wavelength scale, the foreshortening in energy that occurs to low λ will "squeeze" the Rydberg levels in a small $\Delta\lambda$ which can be interpreted as a sharp edge when insufficient resolution is available. The detailed structure of these threshold regions is only now being investigated with continuum sources in the extreme ultraviolet.

Recently, such phenomena have begun to be observed in molecules. The Rydberg series due to excitation of K shell electrons in N_2 to unoccupied orbitals, and ultimately to the ionization limit, has been reported by Nakamura *et al.*[7] in the 30–31 Å range. Their analysis leads to the conclusion that the lower unoccupied levels to which the K electron is promoted have energy spacings very much like that of NO. In other words, the missing electron in the K shell appears to the valence electrons as an additional positive charge in the nucleus. (This analogy has been utilized by Jolly[8] in his thermochemical interpretation of x-ray photoelectron chemical shifts.) Such behavior has also been reported for O_2[9,10] and CO.[9] In the latter case, transitions from the K shells of C and of O have been observed, both apparently going to the same unoccupied valence orbitals. A more complex system recently investigated, SF_6, provides evidence for transitions from inner shells to unoccupied valence orbitals and Rydberg orbitals, but this system displays other complicating features which will be deferred to a later section.

Due to the complex structure in the vicinity of a higher ionization potential, measurements of ionization cross section as a function of photon energy do not readily lend themselves to the accurate determination of the corresponding ionization threshold. For this purpose, the complementary photoelectron spectroscopic technique provides this information more directly, as already noted.

The normal oscillator strength distribution of Rydberg members approaching the threshold for ionization from an inner shell may be dramatically modulated by centrifugal and other molecular potential barriers (see Sections D and E). Modulation has been observed well beyond (50–1000 eV) the absorption edge in molecules, but not in monatomic gases. (See Fig. 4 for comparison of Kr and Br_2.) Hence, the presence of more than a single scattering center appears to be a necessary condition for this postabsorption modulation, which has been dubbed EXAFS (extended x-ray absorption fine structure).†

When an electron is ejected from an inner shell, the radial wave function for the outgoing electron is approximately a sine function (free particle wave function) with a phase shift determined by the initial orbital angular momentum and the internal dynamics of the atom. If the excited atomic center is surrounded by other atomic centers, as in a solid or molecule, the outgoing wave scatters from the surrounding atoms, producing backscattered waves which can interfere with the primary wave from the initial scattering center. This interference causes an oscillatory behavior in the matrix element connecting the bound and continuum states as the incident photon energy is varied, since the photoelectron wavelength (and hence the phase between outgoing and incoming waves) changes.

The physical description of this process is a kind of internal electron diffraction, where the initial source of photoelectrons can be chosen by selecting $h\nu$ such that an inner shell of a particular type of atom is excited. The period of the oscillations should therefore bear some relation to the internuclear distances of atoms surrounding the primary photoelectron source. Indeed, the primary focus of both theorists and experimentalists toward this phenomenon has been to infer local geometry from the quasi-periodic structure. Most investigations to date have been in condensed phases, and the analyses have involved Fourier transformation of the oscillatory portion, combined with semiempirical treatment of the data, or comparison of *ab initio* calculations with experiment. At first[11] it was thought

† Interest in this type of study has been stimulated by the availability of intense continuum x-ray sources from high energy synchrotron storage rings.

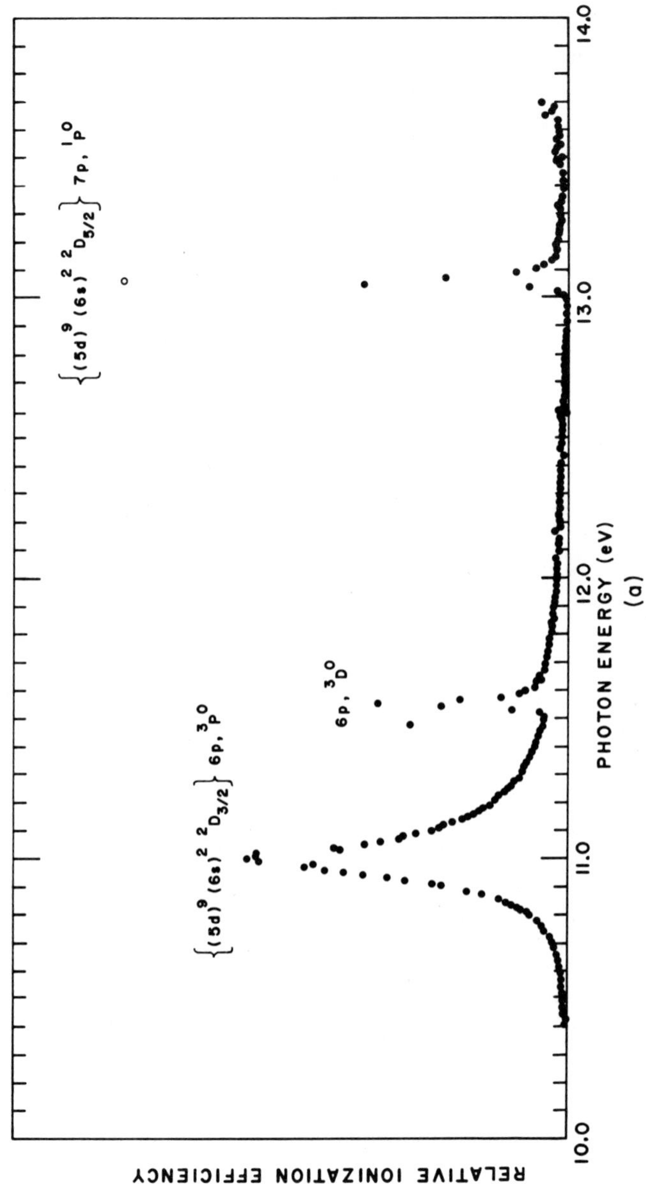
{(5d)9 (6s)2 2D5/2} 7p, 1P0
{(5d)9 (6s)2 2D3/2} 6p, 3P0
6p, 3D0
10.0
11.0
12.0
13.0
14.0
PHOTON ENERGY (eV)
(a)
RELATIVE IONIZATION EFFICIENCY

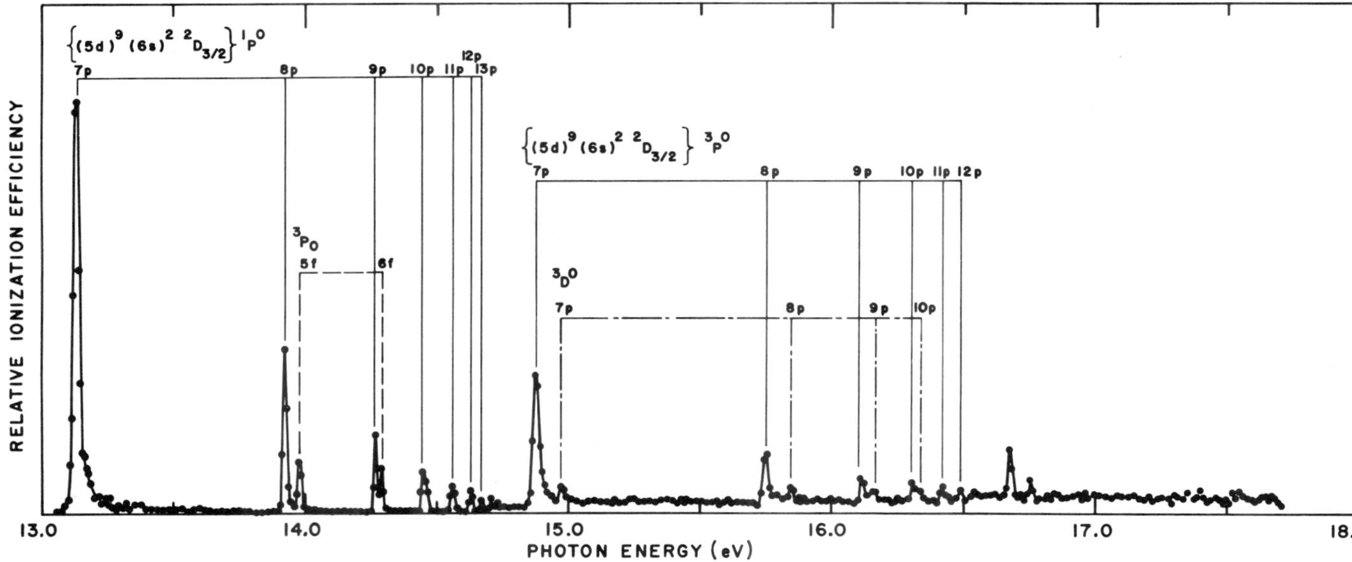

(b)

Fig. 3 Photoionization of atomic mercury. The prominent resonances involve excitation of an electron from the (inner) 5d orbital to np and nf states which then autoionize. The Rydberg states converge on two limits, at 14.841 and 16.705 eV, corresponding to the $^2D_{5/2}$ and $^2D_{3/2}$ states of Hg^+. Note that there is no abrupt increase in ionization probability at either limit. (From Berkowitz and Lifshitz.[6a])

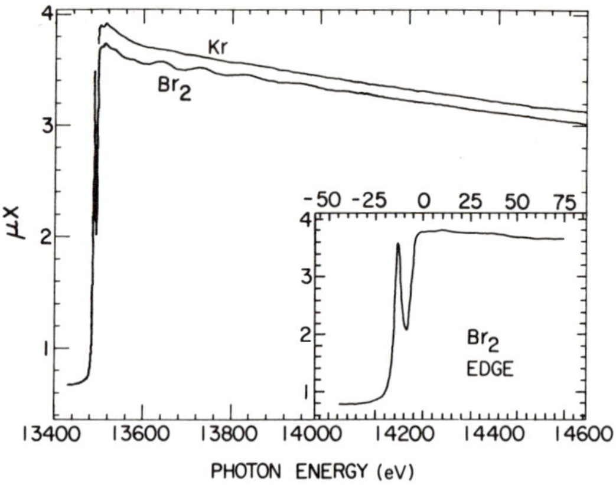

Fig. 4 K edge x-ray absorption spectra for Kr and Br_2 gases. The Kr energy scale has been shifted and the curve displaced for comparison with Br_2. The inset is an expanded threshold region for Br_2. (From Kincaid and Eisenberger.[14])

that long-range order was necessary for the manifestation of the undulations (implying crystalline structure) but it is now clear that it occurs also in free molecules (Fig. 4), and hence short-range order apparently makes the primary contribution. (In a crystal, other than next nearest neighbors may modify this effect.[11–13])

Since our interest in this monograph is confined to isolated molecules, we shall restrict our comments to the limited studies in this area to date.

It has been common in these investigations to subtract the monotonically varying post-threshold absorption cross section (usually post K shell and monotonically declining) and focus attention on the oscillatory component. The corresponding theoretical form of the matrix elements for photoabsorption is[12]

$$\chi = \frac{|\langle f_0 + f_{sc}|H_i|f_i\rangle|^2 - |\langle f_0|H_i|f_i\rangle|^2}{|\langle f_0|H_i|f_i\rangle|^2}. \tag{1}$$

Here, f_i is the wave function of the initial state, f_0 that of the final state in the isolated atom, $f_0 + f_{sc}$ the final state in the molecular system, with H_i being the dipole-approximation interaction Hamiltonian and χ the undulating portion of the photoabsorption.

With rather simplistic assumptions (plane waves for initially produced photoelectron and for backscattered wave, no inelastic processes, the molecular nature of the final and initial states being treated as atomic wave functions) the above formula (when restricted to K-shell ejection) leads to[13]

$$\chi(k) = -\frac{|f(k,\pi)|}{kR^2}\exp(-2\sigma^2 k^2)\sin[2kR + \alpha(k)],$$

where $f(k, \pi)$ is the backscattering amplitude, k the wave vector of the outgoing electron, R the internuclear distance (only one is considered), σ^2 the mean square fluctuation in R, and $\alpha(k) = 2\eta_1(k) + \arg f(k, \pi)$ is the phase shift, with $\eta_1(k)$ the outgoing p-wave phase shift of the photoelectron.

The partial wave formula is used to calculate $f(k, \pi)$.

$$f(k, \pi) = \frac{1}{k} \sum_{l=0}^{\infty} (-1)^l (2l + 1) \exp(i\delta_l) \sin \delta_l,$$

where δ_l are the phase shifts of the corresponding waves with angular momentum l.

Kincaid and Eisenberger[14] have used this prescription to compute oscillatory behavior above the K-shell edge in Br_2 and above the Ge K-shell edge in $GeCl_4$, and compared the computed structure with their experimental undulations (Fig. 5). The comparison clearly indicates that the essence of the physical situation has been incorporated in the calculations. Kincaid and Eisenberger note that the oscillations predicted by theory are approximately

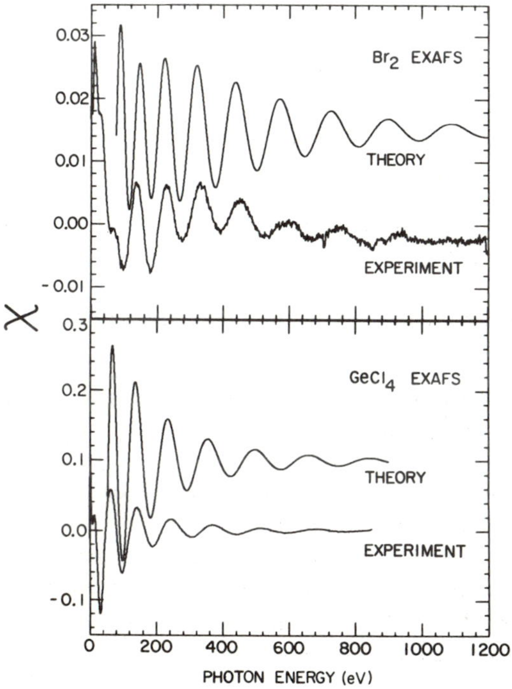

Fig. 5 Comparison of theory and experiment for the extended x-ray absorption fine structure of Br_2 and $GeCl_4$. The theoretical curves have been displaced for clarity. (From Kincaid and Eisenberger.[14])

twice those of experiment, and peak positions (after adjustment) still differ by $\sim 5\%$.

As better (molecular) wave functions are utilized, as well as continuum functions more sophisticated than plane waves, more satisfactory agreement with experiment can be expected. Inelastic (shake-up) processes may also influence the final result. In the present context, we note that the undulatory component of the post-threshold cross section amounts to ~ 1–5%, and results in a mild redistribution of the oscillator strength from the shell being excited.

C. "Cooper Minima"

From the definition for f_n [see Chapter I, Eq. (19)] and the expression for $\sigma(\nu)$ as a function of $(df/d\nu)$ in Eq. (21), Chapter I, we can deduce the cross section for absorption of a beam of photons of energy $h\nu$ greater than the first ionization potential of an atom or molecule in the dipole approximation

$$\sigma(\nu) = \tfrac{4}{3}\pi^2 \frac{e^2}{\hbar c}\left|\left\langle \psi_{E'} \middle| \sum_j \mathbf{r}_j \middle| \psi_0 \right\rangle\right|^2. \tag{2}$$

The term

$$\left|\left\langle \psi_{E'} \middle| \sum_j \mathbf{r}_j \middle| \psi_0 \right\rangle\right|^2$$

is the dipole moment strength per unit energy. For a molecule, the functions ψ can be taken as products of electron wave functions ϕ and nuclear functions χ.

To illustrate the essential feature of this section, we make the following approximations:

(1) The Born–Oppenheimer approximation, factoring the integral into electronic and nuclear integrals. Here we take the nuclear integral as some constant.

(2) The electronic wave functions ψ_0 and ψ_{f} are taken as antisymmetrized products of one-electron wave functions.

(3) Of the n one-electron wave functions comprising both ψ_0 and ψ_{f}, $n-1$ are identical for initial and final states. This assumption neglects the relaxation of the ion core during the ionization act.

The integral above has thus been reduced to

$$\int \phi_0(r_1)\mathbf{r}_1\phi_{\mathrm{f}}^*(r_1)\,d\tau_1 \tag{3}$$

an integral over two one-electron wave functions.

(4) The one-electron wave functions are separable in radial and angular coordinates (spin being ignored). The $\phi(r)$ above are then to be interpreted as only the radial wave functions.

Cooper[15] has shown how the nature of these $\phi(r)$ can cause dynamic variations in the absorption cross section. In Fig. 6 the radial wave functions of Ne(2p), Ar(3p), and Kr(4p) are shown, together with the final state d waves corresponding to each of these cases, when the *departing electron* has *zero* kinetic energy, i.e., $\varepsilon = 0$. (In the dipole approximation, the allowed transitions are $\Delta l \pm 1$, i.e., p electrons can be excited to s or d waves. The transition probability to d waves is an order of magnitude greater than that to s waves over most of the energy range, and in general, $I(l+1) \gg I(l-1)$ (see Bethe and Salpeter,[16] Sects. 61 and 62).

The above dipole integral in Eq. (3) can be roughly regarded as an overlap between the initial p function and the d wave. In the case of neon, there is only a small overlap for the condition $\varepsilon = 0$. As ε increases, the wavelength of the free electron wave decreases and the first cycle draws in closer to the nucleus. Hence, the overlap improves, and the cross section increases with increasing incident $h\nu$.

For argon, the wave functions of ϕ_0 and ϕ_f have opposite sign in the overlap region, and hence the dipole integral in Eq. (3) is negative at $\varepsilon = 0$. As ε increases, the free electron wave again draws closer to the nucleus and decreases in wavelength, so that at some point the negative lobe of the d wave contributes as much to the integral as the positive lobe.

At this point, the total integral vanishes. Since the absorption probability is proportional to the square of this integral, $\sigma(\nu) \to 0$ at some value of incident energy $h\nu$. Beyond this energy, there will again be a mismatch between positive and negative components to the integral, and the absorption coefficient will again increase. This phenomenon, which can also be visualized for the case of Kr in Fig. 6, is often referred to as the Cooper minimum.

In Fig. 7, the value of the integral in Eq. (3) as a function of ε is shown for Ne, Ar, and Kr. For the alkalis (except Li), this minimum occurs very near the first ionization threshold. For Li, it actually occurs in the discrete spectrum below the IP (Seaton had noted this behavior earlier than Cooper, for the alkalis). Cooper has drawn the generalization that the ground state photoionization cross sections from atomic subshells whose radial wave functions are nodeless have a different spectral shape than those from subshells whose radial wave functions have nodes. Thus, the cross sections from subshells 1s, 2p, 3d, etc., should not display Cooper minima, while all others should. (Number of nodes $= n - l - 1$.)

Recently, an apparently anomalous intensity ratio in the photoelectron spectra of spin-orbit partners has been interpreted as an extension of the

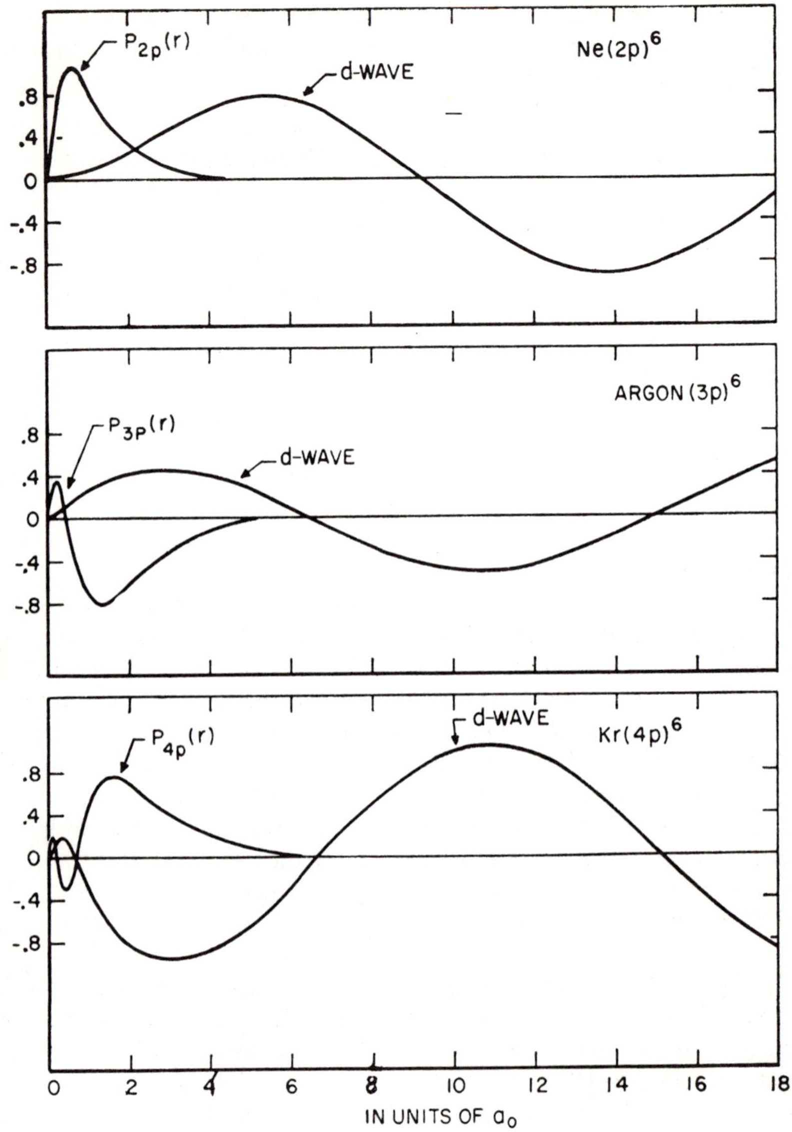

Fig. 6 Outer subshell radial wave functions and outgoing d waves with zero kinetic energy for Ne, Ar, and Kr. (From Cooper.[15])

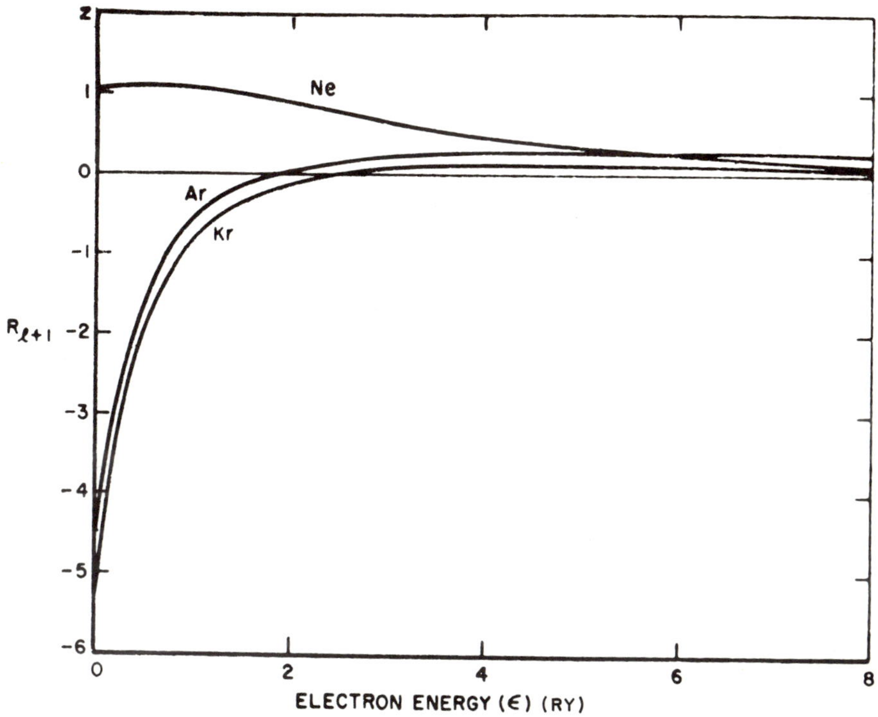

Fig. 7 Matrix elements for p → d transitions in Ne, Ar, and Kr. (From Cooper.[15])

"Cooper minimum" picture. Removal of the outermost (p) electron in the heavier noble gases results in the two spin-orbit split states $^2P_{3/2}$ and $^2P_{1/2}$ (see Chapter III, Section A). The statistical weight ratio of these two states is 2:1, yet the measurements made at 584 Å invariably yielded a smaller ratio. More recently, the set of atoms Zn, Cd, and Hg were studied. Their valence shells have the common configuration $\cdots nd^{10}(n+1)s^2$. With 584 Å radiation, it is possible to eject an electron from the d orbital and form $^2D_{5/2}$ and $^2D_{3/2}$ ionic states. Their statistical ratio should be 3:2, but was found to be larger. The explanation for both sets of observations begins by noting that spin-orbit splitting is a relativistic effect which separates an l shell into $l + \frac{1}{2}$ and $l - \frac{1}{2}$ subshells. The $l - \frac{1}{2}$ wave function is drawn closer to the nucleus, the $l + \frac{1}{2}$ wave function is slightly repelled. Hence, the $l + \frac{1}{2}$ subshell has an overlap with the continuum wave function in the ionization process which precedes the corresponding overlap with the $l - \frac{1}{2}$ subshell. If the partial cross section for ionization is rising, the $l + \frac{1}{2}$ portion will begin to rise before the $l - \frac{1}{2}$ subshell. If the partial cross section is falling because of a cancellation of positive and negative matrix elements, the $l + \frac{1}{2}$

subshell will also experience this decline first (i.e., at a lower photon energy). From this qualitative description, it may readily be seen that when the partial cross section for ionization from a filled l shell is falling, the spin-orbit intensity ratio will be less than statistical (the case of the noble gases at 584 Å), and when it is rising, the intensity ratio will be greater than statistical (Zn, Cd and Hg at 584 Å). We have recently[17] examined Hg over a range of wavelengths encompassing a rising and falling cross section (from 584 to 256 Å) and observed a variation in intensity ratio from ~2.0 to 1.3, in good agreement with the theoretical prediction.

The above illustration has been confined to atomic systems (central field) but the pictorial representation in Fig. 6 can obviously be extended to molecular systems, particularly when the molecular orbital has primary contribution from one atomic orbital, as is frequent in inner shells of molecules.

D. Dependence of Absorption Cross Section upon the Type of Orbital (s, p, d, . . .) from Which Ejection Takes Place

The radial wave functions $\phi_0(r)$ and $\phi_f(r)$ appearing in Eq. (3) are (for central fields) solutions of the radial Schrödinger equation (in Hartree units)

$$\left(-\frac{d^2}{dr^2} - 2V(r) - 2E + \frac{l(l+1)}{r^2}\right)\phi(r) = 0$$

with the following boundary conditions on $V(r)$:

$$V(r) \xrightarrow[r\to 0]{} Z/r \qquad \text{and} \qquad V(r) \xrightarrow[r\to\infty]{} 1/r.$$

The terms $[-2V(r) + l(l+1)/r^2]$, which combine the coulombic attraction and centrifugal repulsion, can be regarded as an effective potential $2\mathscr{V}$. Examples of this potential which would be appropriate for the computation of $\phi(r)$ for f waves ($l = 3$) are shown in Fig. 8. Except for Kr, these curves have the form of a "potential well" inside the atom, a small repulsive barrier (shown in the inset), and finally a broad well which is the same for all cases in the region where $V(r)$ has reached its asymptotic form $1/r$.

Each potential has a deep inner well, produced by the strong attraction of the nucleus to an electron, the high positive charge of the nucleus being only partially screened. At somewhat larger distances, screening is more effective, leaving only one or two units of residual nuclear charge. In this region, the repulsive centrifugal term dominates over the weakened nuclear attraction term. For a given value of l, the centrifugal term remains the same for a series of atoms, but the screening varies as one proceeds from a compact electron distribution (such as the closed shell Xe structure) to the more diffuse La

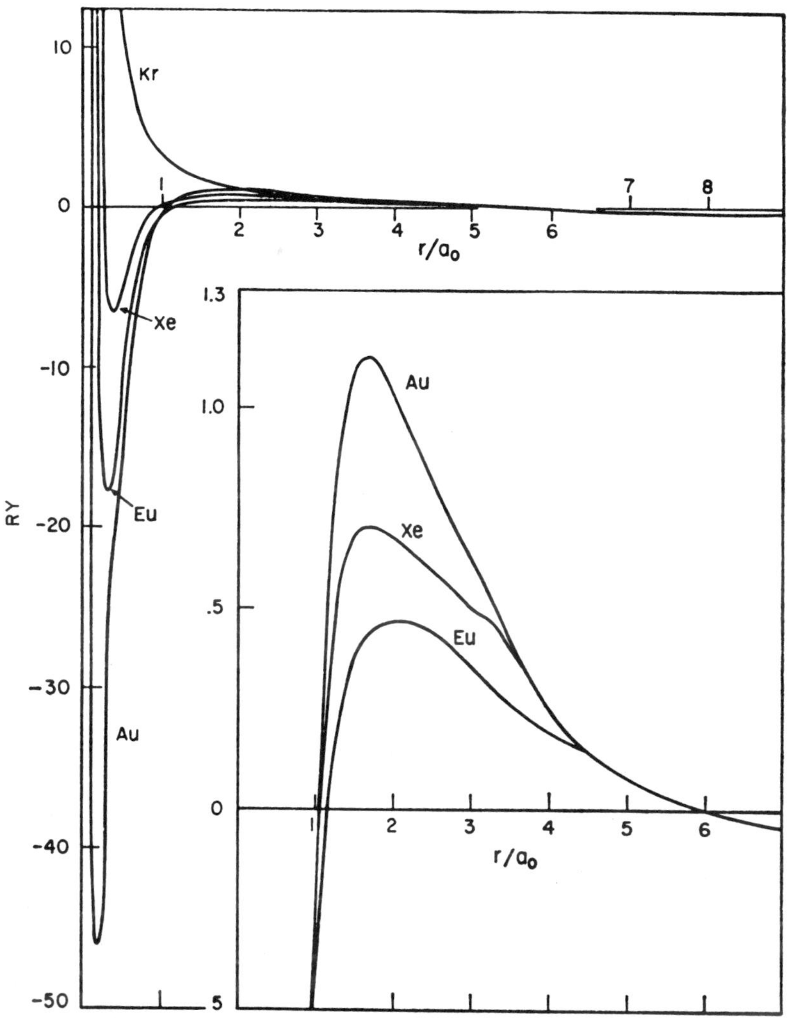

Fig. 8 The potential curves for f electrons in Kr, Xe, Eu, and Au. (From Manson and Cooper.[17a])

structure which has 6s and 5d orbitals outside the Xe configuration. The centrifugal term, and hence the repulsive barrier, will consequently be more pronounced in Xe than in La. As the 4f subshell fills, the lanthanides contract and at thulium the electron distribution becomes more compact than that of Xe, due to the nearly complete filling of the $n = 4$ shell.

At larger distances, the longer range $1/r$ attractive potential begins to overcome the repulsive centrifugal $1/r^2$ term. Beyond $r \sim 5$ a.u. (1 a.u. =

0.52917706 Å = 1 Bohr radius), the effective potential is given by

$$2\mathscr{V} = l(l+1)/r^2 - 2/r.$$

The outer well is in this region, and consequently we can give its coordinates analytically

$$r_{\min} = l(l+1) \qquad \text{in a.u.}$$

$$\mathscr{V}(r_{\min}) = -\frac{1}{2l(l+1)} \qquad \text{in a.u.}$$

For $l = 3$, the outer well reaches a minimum of -0.0417 a.u. at $r = 12$ a.u. (-1.13 eV at $r = 7.1$ Å).

The influence of the barrier is most pronounced in states with total energy E which fall below the barrier maximum, i.e., the valence, Rydberg, and low energy continuum states. States in this energy range tend to be eigenfunctions of either the inner well *or* the outer well. Examples of this behavior are given in Figs. 9 and 10. In Fig. 9, the 4f wave function of Ce is seen to be an eigenfunction of the inner well, while 4f(Xe) is an eigenfunction of the outer well. In Fig. 10, the 5f(Ce) is seen to be an eigenfunction of the outer well, although the 4f(Ce) is localized in the inner well. The 4d radial wave functions are also shown as a visual aid for the subsequent discussion.

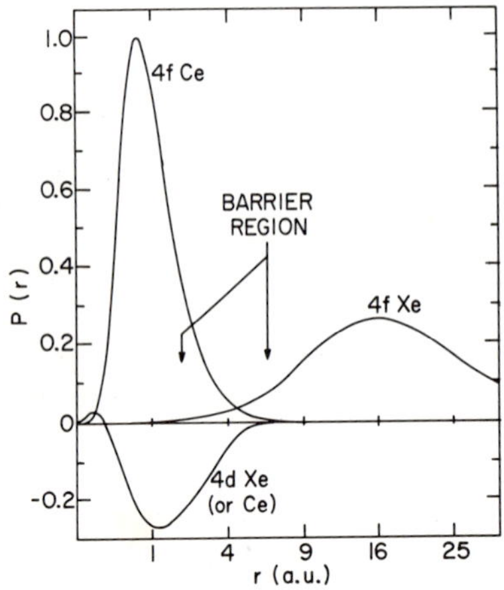

Fig. 9

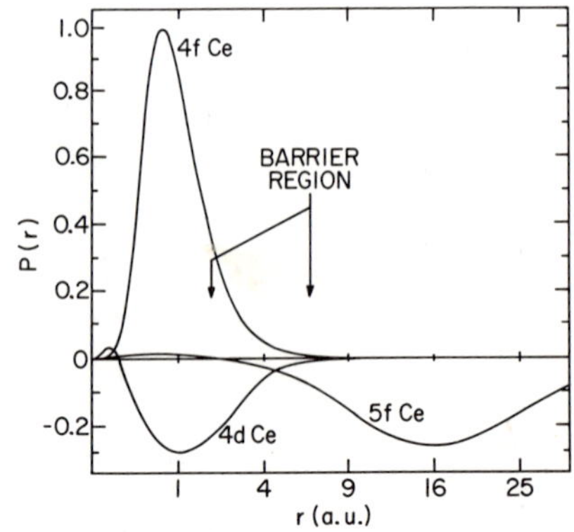

Fig. 10

Fig. 9 Wave function for 4d and 4f electrons of Ce and Xe. Three are eigenfunctions of an inner well potential, while the Xe (4f) is an eigenfunction of an outer well potential.

Fig. 10 The Ce 4d and 4f wave functions shown as eigenfunctions of an inner well potential, while Ce (5f) is an eigenfunction of an outer well potential.

Using these wave functions $\phi_f(r)$, the dipole integrals in Eq. (3) can be calculated for the transitions 4d → 4f and 4d → εf for Xe and Ce. The square of this integral which is proportional to the transition probability is plotted in Fig. 11. For Xe, the transition probability has a delayed onset, corresponding to the delayed penetration of the first lobe of the final state wave function into the region of the inner well, which occurs only with larger ε. The cross section then decreases toward a Cooper minimum at $\varepsilon = 2$ a.u., caused by cancellation in the dipole integral when the first node of the continuum f-wave function overlaps with the 4d wave function.

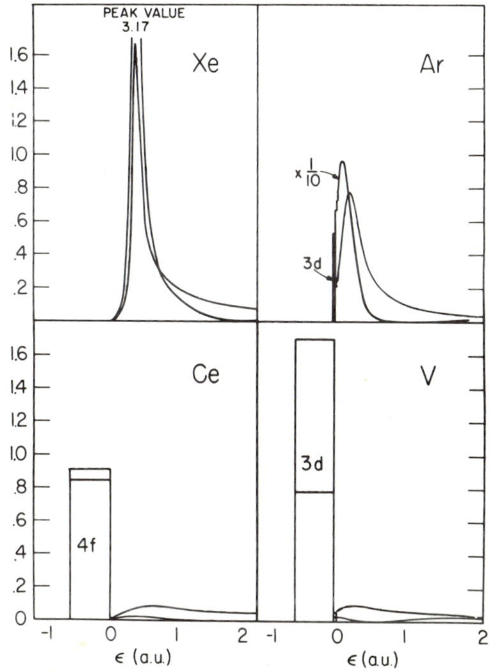

Fig. 11 Squares of matrix elements for the transitions 3p → d waves for Ar and V, and 4d → f waves for Xe and Ce, using the independent electron model. (From Dehmer *et al.*[17b])

The experimental absorption spectrum of xenon in the energy range of interest, is shown in Fig. 12. The delayed onset of ionization from the 4d shell, reaching a maximum ~30 eV above threshold, is clearly evident and interpretable in terms of the dynamics of the two-well potential described above.

The calculated transition probability for cerium in Fig. 11 clearly shows the dominant role played by the 4f state. Because of the large amplitude of

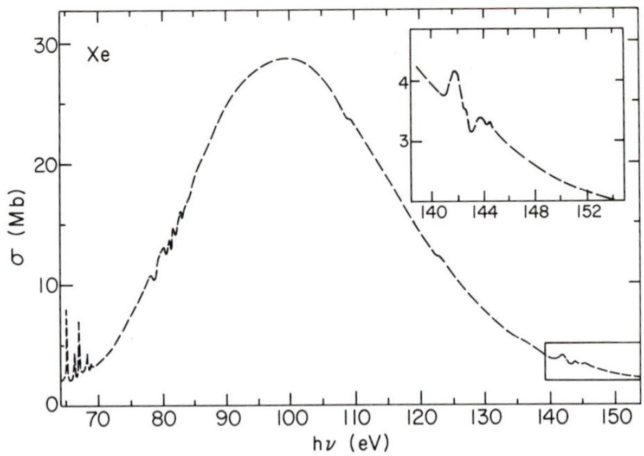

Fig. 12 Photoabsorption cross section of gaseous xenon in the region where ionization from the 4d shell is prominent. (From Haensel *et al.*,[17c] Fig. 3.)

the 4f wave function in the inner well, the 4d → 4f line dominates the spectrum. The nonpenetrating Rydberg and low energy continuum states have near-zero transition probability, and this increases only slowly at higher energy.

For cerium and the other rare earths, the 3d → 4f transition does indeed dominate the oscillator strength contribution from the 3d shell. However, the 4d → nf transitions introduce additional complications.

In brief, the evaluation of the final state ϕ_f in these cases requires the consideration of an exchange integral between the photoelectron and the hole in the 4d subshell. The effect of this exchange interaction on the 4d → εf spectrum of xenon has been calculated by Starace.[18] Its effect is to broaden the 4d absorption in the independent particle model, and to shift the maximum 30 eV above threshold, as observed experimentally.

For La, Ce, and several of the other rare earths, the major contributor to the exchange interaction is the intrashell exchange integral ⟨4d, 4f: 4f, 4d⟩. This term introduces an energy spread in the transition probability of about 20 eV, somewhat larger than experimentally observed. Components of these configurations will be raised into the continuum above the 4d ionization threshold, thus making the 4d absorptions qualitatively different from the 3d absorptions, which are dominated by the 3d → 4f transition below threshold.

Although the effective potential $\mathscr{V}$ generally does not have a positive "barrier" (as shown in Fig. 8) for $l < 3$, there will nevertheless be "inner" and "outer" valleys for large Z, since $V(r)$ is the same for all l. The formation of new subshells may thus be considered as a less drastic "moving in" of a particular nl radial eigenfunction with increasing Z, with a resulting increase in binding energy.

In summary, when $\phi(r)$ is determined by $\mathscr{V}$ for $l < 3$, we can expect transitions from the initial state to unoccupied valence and Rydberg states smoothly joining into continuum states in a "normal" fashion. When $\phi(r)$ corresponds to $l \geq 3$, the effect of the two-well potential will generally cause a delayed onset. For the rare earths, there are two additional effects; the $3d \rightarrow 4f$ transition below threshold is strongly enhanced and contains the bulk of the transition probability for this subshell, and the $4d \rightarrow nf$ transition is broadened and partially shifted above the threshold by exchange effects.

E. Molecular "Barriers" to Electron Escape

This phenomenon was uncovered as a result of anomalies existing in the K- and L-shell absorption spectra of sulfur in various sulfur-containing molecules. The K-shell sulfur absorption in H_2S (Fig. 13), and the L-shell sulfur absorptions in both H_2S and CS_2 (Fig. 14), display "normal" Rydberg progressions blending smoothly into the respective absorption edges. However, in SO_2, SO_4^{2-}, and SF_6, strong absorption peaks exist in the region above the edge (see Fig. 14). SO_2 still shows Rydberg series in addition to the post-threshold peaks, but in SO_4^{2-} and SF_6 the Rydberg levels are absent.

Dehmer[19] has explained these observations by noting that the oxygen atoms in SO_2 and SO_4^{2-} and the fluorine atoms in SF_6 contribute to the

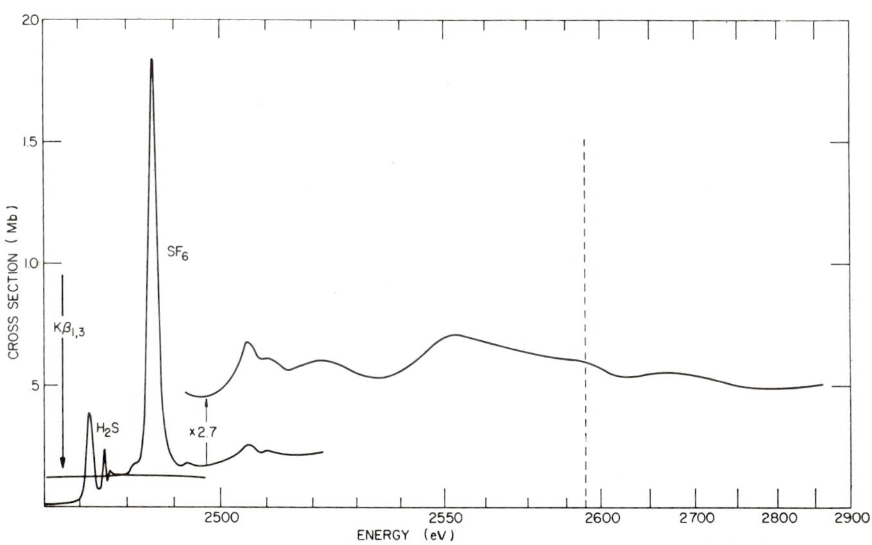

Fig. 13 Photoabsorption spectra of H_2S and SF_6 above the sulfur K threshold. (From LaVilla and Deslattes.[18a])

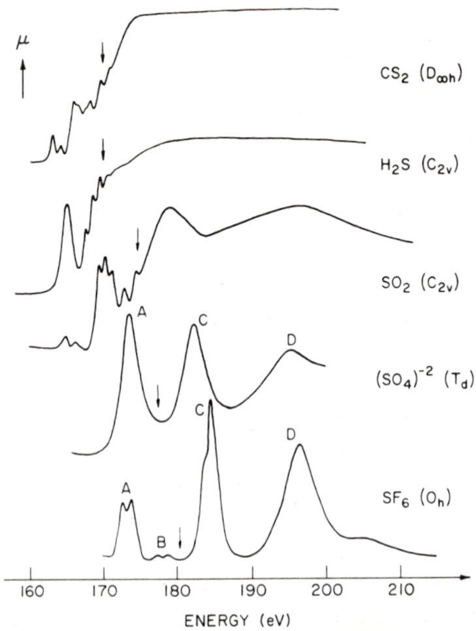

Fig. 14 The $L_{II,III}$ photoabsorption spectra of sulfur in CS_2, H_2S, SO_2, SO_4^{-2}, and SF_6. The arrows denote the ionization energies of S(2p) in the different molecular environments. (From data of Zimkina *et al.*[18b] The figure is taken from Dehmer.[19])

occupied upper molecular orbitals of their respective molecules in a specific fashion depending upon the symmetry of the molecule. For example, in the bent SO_2 molecule, one class of the upper occupied molecular orbitals containing large oxygen contribution exist in the molecular plane. Transitions from K- or L-shell sulfur to the unoccupied molecular orbitals will exhibit interference if the unoccupied molecular orbitals are in the molecular plane because of the large negative charge (and hence potential barrier) contributed by the filled oxygenlike orbitals. Other transitions to unoccupied orbitals out of the molecular plane will not experience such interference, and presumably these are the observed Rydberg transitions in the sulfur L-shell of SO_2.

The interference of the in-plane oxygenlike orbitals can be interpreted as a potential barrier that separates an "inner well" (attraction of the electron to the sulfur nucleus) and an "outer well" due to the distant $1/r$ coulomb potential. Transitions to inner well states will be strong due to better overlap with the initial state ϕ_0. If the barrier is high (10–20 eV), quasi-stationary states above the K- or L-shell edge can exist. Whether or not transition from

ϕ_0 to these states can occur depends also upon the relative symmetry of ϕ_0 and the excited state, i.e., they must have opposite parity for nonvanishing dipole transitions.

Dehmer[19] has examined the sulfur K and L absorption and the fluorine K absorption in SF_6. The sulfur K shell in SF_6 has molecular symmetry $1a_{1g}$. The first allowed transition is to $6t_{1u}$, and is identified with the prominent peak ~3 eV below the K edge in this spectrum. One of the sulfur L-shell orbitals in octahedral symmetry is $2t_{1u}$, and hence it makes transitions to gerade states, of which the lowest unoccupied one is $6a_{1g}$, then $2t_{2g}$, and $4e_g$. The large peaks A(A′), C(C′), and D in the sulfur L spectrum are identified with these three unoccupied orbitals. Hence, the K- and L-shell absorptions probe different symmetries of the unoccupied orbitals.

The boron K shell in BF_3 has been interpreted in a similar manner.[20] This behavior has also been seen in NF_3 recently.[21]

F. Multiple Electron Excitations and Ionizations

If one had at hand the exact wave functions of initial and final states for the photoabsorption process, the transition matrix elements would presumably give the correct measure of transition probabilities to all channels whether they involve single or multiple electron excitation. However, since even very accurate wave functions are available for only a few systems, and since single-electron processes usually dominate, most calculations use as a starting point the independent particle model. Various corrections are then applied to these basis functions, as required to explain particular phenomena. In photoelectron spectroscopy, one may observe satellite peaks or peaks forbidden in the independent particle model. In photoionization mass spectrometry, double or multiple ionization can be observed as a separate process. These and other effects which are manifestations of electron correlation are generally treated by three categories of corrections to the wave functions, described as core relaxation, initial state configuration interaction, and final state configuration interaction. The relative importance of these corrections varies with the atom or molecule under study, with the region of ionization (inner or outer shells) and with the photon energy. For a recent review see Åberg[22a]; see also Carlson *et al.*[22b] and earlier papers from this group, where the terms "monopole excitation" and "monopole ionization" are extensively used. Our purpose here is to acquaint the reader with this terminology, although it may be misleading.

The core-relaxation extension of the independent particle model (sometimes referred to as "shakeup" or "shakeoff") is important when a two-electron process involves electrons from different principal shells.

As an example, consider the process in which a 1s electron is ejected from Ne, and concomitantly a 2p electron is excited to 3p (or more generally, *n*p). The process can be viewed as an initial, rapid ejection of the 1s electron, followed by a relaxation of the other electrons adjusting to the new potential. By invoking the "sudden approximation," this latter process can be calculated as a "monopole" overlap integral between wave functions describing the initial state and those calculated in a potential which takes into account the weaker screening resulting from a hole in an inner shell, such as Ne 1s. Since the wave functions are not calculated on the same potential, the one-electron wave functions are no longer orthogonal. By treating this as a "monopole" transition, however, the angular components of the wave functions are still orthogonal. Hence, the second step in the process involves excitation of a second electron to a higher principal quantum number, but with the same orbital angular momentum quantum number.[22a,b]

The initial state configuration interaction correction is significant when electrons of the same subshell are involved. Examples include double ionization of the atomic gases treated in this monograph and the simultaneous ionization and excitation of otherwise forbidden states in Zn, Cd, and Hg. The methodology simply involves inclusion of other configurations to the ground state, in addition to the primary one, subject to the condition that the symmetries of such configurations can mix with the main configuration.

The final state configuration interaction is involved when configurations of the same symmetry are closely spaced, as in the presence of partially filled or empty subshells. It may manifest itself in the form of interchannel coupling in partial photoionization cross sections.

These various corrections are presumably included in a less transparent manner in the various many-body perturbation theory approaches.

G. Oscillator Strength Distribution in the Continuum

In the classical theory describing the interaction of electromagnetic radiation with individual molecular systems, the atom (or molecule) has its own complement of N electrons, each of which has a characteristic binding force. When electromagnetic radiation of a specific frequency is incident upon the system, such that one of these N electrons is set into forced oscillation by the action of the electric field on the electron, absorption of radiation will take place at that characteristic frequency.

If the number of characteristic frequencies s is less than the number of electrons (i.e., two or more electrons have the same frequency) we can subdivide the N electrons into groups $f_s(\nu)$, each value of f describing the number of electrons responding to the frequency ν. From such a description, it is obvious that

$$\sum_s f_s = N,$$

where the symbol f is referred to as the oscillator strength.

Modern concepts reject the identification of a characteristic frequency with a given electron in a system. Instead, the characteristic frequencies correspond to differences in eigenvalues of the entire N electron system, i.e.,

$$E_j - E_0 = h\nu_j = \hbar\omega_j.$$

The number of such frequencies is much greater than the number of electrons. Hence, if we wish to retain the concept of oscillator strength, f will generally be less than one for a particular absorption line.

Let us return for a moment to the classical model. A particular electron may experience a displacement $x(t)$ from its equilibrium position under the influence of an oscillating external electric field $\mathscr{E}_0 e^{-i\omega t}$ and its own elastic restoring force $(-m\omega_s^2 x)$ and a radiative damping force $[-mg_s(dx/dt)]$.

From Newton's equation of motion,

$$e\mathscr{E}_0 e^{-i\omega t} - m\omega_s^2 x - mg_s\left(\frac{dx}{dt}\right) = m\left(\frac{d^2x}{dt^2}\right), \tag{4}$$

for which a solution is

$$x = \frac{e}{m}\left(\frac{\mathscr{E}_0 e^{-i\omega t}}{\omega_s^2 - \omega^2 - ig_s\omega}\right). \tag{5}$$

The polarizability α of a system is defined as the dipole moment induced by unit electric field strength, i.e., α per electron $= \alpha_1$ is given by

$$\alpha_1 = \frac{ex}{\mathscr{E}_0 e^{-i\omega t}} = \frac{e^2}{m}\left(\frac{1}{\omega_s^2 - \omega^2 - ig_s\omega}\right). \tag{6}$$

If, as before, we have f_s oscillators of frequency ω_s, then the polarizability per atom (or molecule) is

$$\alpha_N = \frac{e^2}{m}\sum_s\left(\frac{f_s}{\omega_s^2 - \omega^2 - ig_s\omega}\right). \tag{7}$$

Quantum mechanically, one must consider the mean value of the displacement $\bar{x}$ (or equivalently, the mean value of the induced dipole moment $e\bar{x}$) when an external field is imposed upon an atomic or molecular system. The derivation of this quantity is a rather lengthy one, and can be found in various standard texts (e.g., Kauzmann,[23]). We shall not repeat it here, but merely accept the result,

$$e\bar{x} = \frac{4\pi}{h} \sum_j \frac{\omega_{j0} |eM_{j0}|^2}{\omega_{j0}^2 - \omega^2} \mathscr{E}_0 \cos \omega t, \tag{8}$$

where

$$\omega_{j0} = \frac{E_j - E_0}{\hbar}, \ M_{j0} = \int \phi_j{}^* x \phi_0 \, d\tau$$

is the one-dimensional equivalent of Eq. (3) and has been referred to as the electric-dipole integral, and $\mathscr{E}_0 \cos \omega t$ is the applied external electric field. The polarizability is given, as before, by

$$\alpha_N = \frac{e\bar{x}}{\mathscr{E}_0 \cos \omega t} = \frac{4\pi}{h} \sum_j \frac{\omega_{j0} |eM_{j0}|^2}{\omega_{j0}^2 - \omega^2}. \tag{9}$$

When generalized to three dimensions,

$$\alpha_N = \frac{4\pi e^2}{3h} \sum_j \frac{\omega_{j0} |\bar{M}_{j0}|^2}{\omega_{j0}^2 - \omega^2}, \tag{10}$$

where

$$\bar{M}_{j0} = \int \phi_j{}^* r \phi_0 \, d\tau.$$

Comparing Eq. (10) with the classical result (Eq. (7)), we can make the identification

$$f_{j0} = f_s = \frac{4\pi m \omega_{j0} |\bar{M}_{j0}|^2}{3h} = \frac{8\pi^2 m v_{j0} |\bar{M}_{j0}|^2}{3h}. \tag{11}$$

(The damping term g_s is somewhat more complicated to introduce into the quantum mechanical expression, and has been neglected here. See footnote in Fano and Cooper,[23a] p. 444.)

Furthermore, it can be shown by quantum mechanical considerations (see Kauzmann,[23] pp. 651–653) that

$$\sum \omega_{j0} |\bar{M}_{j0}|^2 = \frac{3Nh}{4\pi m}, \tag{12}$$

and hence

$$\sum_j f_{j0} = N.$$

This is the same as the classical result, and is known as the Thomas–Kuhn–Reiche sum rule. As mentioned earlier, it does not have the same interpretation as the classical description of individual electron oscillations, but rather as a sum of transitions involving the entire molecular system. A derivation of the Thomas–Kuhn–Reiche sum rule is given in the Appendixes to this chapter, together with two other theorems which we shall have occasion to use in Chapter V.

Note that $\sum_j \omega_{j0}|\bar{M}_{j0}|^2$ is related by a simple proportionality constant to $\sum_\nu \sigma(\nu)$. Hence, if one plots the absorption coefficient as a function of wave number or energy, the area under the curve is proportional to the number of electrons in the atomic or molecular system.

Note also that α_N, which is a macroscopic, measurable property called the static dipole polarizability, is of the form

$$\sum_{j\,\text{discrete}} \frac{f_{j0}}{\omega_{j0}^2} + \int_{\omega=\text{IP}}^{\infty} \frac{f}{\omega^2}\,d\omega.$$

Since $\omega = 2\pi c/\lambda$, the integral takes the form

$$\int_{\lambda\equiv\text{IP}}^{0} f\,d\lambda \propto \int_{\lambda\equiv\text{IP}}^{0} \sigma(\lambda)\,d\lambda.$$

Hence, if one plots the absorption coefficient $\sigma(\lambda)$ as a function of λ, the area under the curve is proportional to the continuum contribution to the dipole polarizability. The experimental polarizability should be larger than this integral due to the contribution of the discrete terms.

To evaluate the constants, we begin with (10)

$$\alpha_N = \frac{4\pi e^2}{3h}\sum_j \frac{\omega_{j0}|\bar{M}_{j0}|^2}{\omega_{j0}^2 - \omega^2} = \frac{2e^2}{3h}\sum_j \frac{\nu_{j0}|\bar{M}_{j0}|^2}{\nu_{j0}^2 - \nu^2}.$$

From Chapter I, Eq. (21),

$$\sigma(\nu)\,d\nu = \frac{\pi e^2}{mc}\left(\frac{df}{d\nu}\right)d\nu$$

and from Chapter I, Eq. (22),

$$\left(\frac{df}{d\nu}\right) = \frac{8\pi^2 m}{3h}\left|\left\langle \psi_j{}^* \middle| \sum_k \mathbf{r}_k \middle| \psi_0 \right\rangle\right|^2 = \frac{8\pi^2}{m}|\bar{M}_{j0}|^2.$$

Therefore,

$$\sigma(\nu)\,d\nu = \frac{8\pi^3 e^2}{3hc}|\bar{M}_{j0}|^2\,d\nu$$

and

$$\int \frac{\sigma(\nu)}{\nu^2}\,d\nu = \frac{8\pi^3 e^2}{3hc}\int \frac{|\bar{M}_{j0}|^2}{\nu^2}\,d\nu.$$

The right-hand side integral is of the form required to evaluate the continuum contribution to α_N. Comparing the constants before the $\sum$ in α_N and before the integral in the last expression, we deduce

$$\int \frac{\sigma(\nu)}{\nu^2}\,d\nu = \frac{4\pi^3}{c}\,\alpha_N.$$

Since

$$\frac{d\nu}{\nu^2} = -\frac{d\lambda}{c} \quad \text{and} \quad \sigma(\nu) = \sigma(\lambda), \qquad \alpha_N = \frac{1}{4\pi^3}\int \sigma(\lambda)\,d\lambda.$$

The limits of integration are from $\lambda = 0$ to $\lambda = \mathrm{IP}$. The polarizability α_N will come out in units of cubic centimeters if $\sigma(\lambda)$ is expressed in squared centimeters and λ in centimeters.

It is convenient to present a generalized formula for the various sum rules. For the continuum portions, the general expression (with energy expressed in Ry units) is

$$S(p) = \frac{(4\pi a_0/\alpha_0)^p}{\pi {\alpha_0}^2 a_0} \int_0^{\lambda_0} \sigma(\lambda)\lambda^{-(p+2)}\,d\lambda, \tag{13}$$

where a_0 is the Bohr radius and α_0 the fine structure constant, $e^2/\hbar c$. For $p = -2$ we have

$$S(-2) = \frac{(4\pi a_0/\alpha_0)^{-2}}{\pi {\alpha_0}^2 a_0} \int_0^{\lambda_0} \sigma(\lambda)\,d\lambda$$

or

$$S(-2) = \frac{1}{16\pi^3 {a_0}^3} \int_0^{\lambda_0} \sigma(\lambda)\,d\lambda. \tag{14}$$

But we had previously found that

$$\alpha_N = \frac{1}{4\pi^3} \int_0^{\lambda_0} \sigma(\lambda)\,d\lambda.$$

Therefore,

$$\alpha_N = 4{a_0}^3 S(-2). \tag{15}$$

To determine the contribution of the discrete absorptions to α_N, we divide the individual f_s by ${E_s}^2$, expressed in Rydberg units (13.605804 eV = 1 Ry), then multiply by $4{a_0}^3 = 0.592738 \times 10^{-24}\ \mathrm{cm}^3$.

The Thomas–Reiche–Kuhn rule is the $S(0)$ sum. From the general formula,

$$S(0) = \frac{1}{\pi \alpha_0{}^2 a_0} \int_0^{\lambda_0} \frac{\sigma(\lambda)}{\lambda^2} d\lambda. \tag{16}$$

Substituting

$$a_0 = 0.529177 \times 10^{-8} \text{ cm}$$

and

$$1/\alpha_0 = 137.03604,$$

we obtain

$$S(0) = 112.95846 \int_0^{\lambda_0} \frac{\sigma(\lambda)}{\lambda^2} d\lambda, \tag{17}$$

where $\sigma(\lambda)$ is in megabarns (10^{-18} cm^2) and λ is in angströms. The discrete portion is simply $\sum_s f_s$.

The above sum rules are just two of a set of sum rules which can be generalized in the expression $S(k) = \sum_s f_s E_s{}^k$.

We have considered the cases $k = 0, -2$. In Appendix 3, we make a brief digression to charged particle ionization, where we shall encounter the case $k = -1$. The cases $k = 1$, 2, and $-2r$ are discussed in Fano and Cooper,[22] but are not as pertinent to our current interest.

Of the $S(0)$ sum, the contribution of the discrete portion is characteristically of the order of one for most atoms and molecules, rarely outside the region 0.2–2. For most molecules, the continuum absorption cross section rises rapidly from threshold to a relatively large value (corresponding to ejection of the valence electrons), then slowly declines as λ decreases, and finally steeply declines as inner shell absorption dominates. In the latter region, the absorption cross section begins to approach hydrogenic behavior (particularly for the K shell) and declines asymptotically[24] as $E^{-l-7/2}$.

Even in the ultraviolet and soft x-ray region, it has been observed that the partial cross section for ejection of electrons from p-like orbitals declines more rapidly with excess energy than ejection from s-like orbitals. This behavior is in qualitative agreement with the Rau–Fano prediction.[24] Price *et al.*[25] have given a pictorial description of the change in matrix elements for s and p ionization as a function of the kinetic energy of the ejected electrons.

Since the sum of oscillator strengths increases with the number of electrons in the molecular system, it is reasonable to expect that average absorption cross sections will increase as the molecular system is enlarged. A graphic illustration of this behavior is provided in Fig. 15, where absorption cross sections of several paraffin molecules are shown on the same plot.

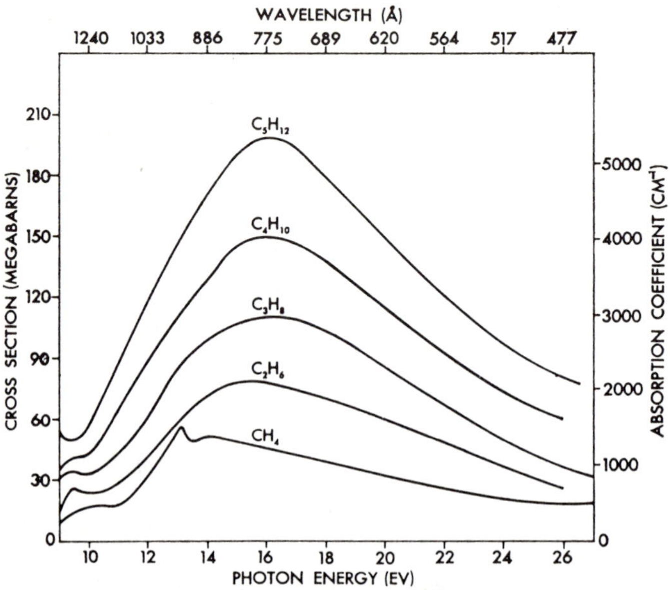

Fig. 15 Photoabsorption cross sections in the alkane series, qualitatively demonstrating the increase in oscillator strength as the number of electrons is increased. (From Schoen.[25a])

H. Dissociation in the Ionization Continuum

Typical energies required to remove the least bound electron in a molecular system are 10 eV. Typical molecular bond energies are 5 eV. Obviously, at all energies above the ionization threshold, energy is also available for molecular bond rupture.

However, electrons generally have greater velocity than nuclei. An electron provided with 0.1 eV above the minimum necessary to liberate it from a molecule will leave with a velocity of $>10^7$ cm/sec. Nuclear vibrational motion, typical of dissociating atoms, might be of order $\sim 10^5$ cm/sec. Hence, making the naive assumption that both events have competing rates based on these velocities, ionization would always predominate.

There are circumstances, however, in which electron departure is delayed. We have already had occasion to discuss this phenomenon, autoionization, in Chapter III. Absorption takes place to a quasi-discrete state, and the time of electron ejection is dependent upon the interaction rate between this state and the ionization continuum. During this time, the quasi-discrete state can also interact with states of the neutral system corresponding to the dissociation continuum, i.e., repulsive states. (Attractive states are also possible, but the energy contained in the system must be dissipated eventually, and reemission of the radiation, either directly or from a subsequent state, is usually slow (10^{-8} sec) compared to dissociation and ionization (10^{-11}–10^{-15} sec), unless the latter are severely restricted by selection rules). Competition between ionization and dissociation can occur from such quasi-discrete states. A quantitative illustration of this competition has been worked out[26] for a portion of the N_2 absorption spectrum, and more recent, higher resolution data for this system will be presented in Chapter VI.

For larger molecules, these quasi-discrete states may not be well resolved, and may indeed form a pseudocontinuum of their own. The competition between ionization and dissociation is still possible here, and can be investigated by measuring the absorption and ionization cross sections individually. Such observations were first reviewed by Weissler.[27] We have already noted (Section III.A) that Platzman denoted these accessible states above the ionization threshold which did not give rise to ionization, "superexcited states." Several workers (Schoen, Samson, Cook & Metzger, Watanabe, Person) have pursued such studies in recent years. This type of investigation is frequently summarized in terms of an ionization yield (defined as the ratio of ionization to absorption cross sections) plotted as a function of wavelength. Such measurements are almost nonexistent below 584 Å.

It has been speculated that for simple diatomic systems this yield is unity below 584 Å, and a few measurements bear this out, but it may not be so for polyatomic molecules. In Chapter V, we present the available information on ionization yields, and photoabsorption cross sections, both as functions of wavelength for those molecular systems about which enough data are available to permit a meaningful analysis to be made.

Summary of Useful Sum Rules

(1) $S(0) = N$, where N is the total number of electrons in the atom or molecule

$$S(0) = \sum_s f_s + 112.95846 \int_0^{\lambda \equiv \mathrm{IP}} \frac{\sigma_a(\lambda)}{\lambda^2} d\lambda.$$

(2) $S(-1)$: The portion of this sum above the first ionization threshold can be determined from the absolute cross section for ionization of atoms by fast charged particles. Also, by direct substitution of $a_0 = 0.529177 \times 10^{-8}$ cm and $1/\alpha_0 = 137.03604$,

$$S_i(-1) = 0.12396 \int_0^{\mathrm{IP}} \frac{\sigma_a(\lambda)}{\lambda} d\lambda,$$

where $S_i(-1)$, in units of $(\mathrm{Rydbergs})^{-1}$, is the ionization component of $S(-1)$. If we are talking about molecules rather than atoms, some of the absorption can lead to nonionizing processes, e.g., dissociation processes. The identification of the $S_i(-1)$ ionization component with the photoabsorption $S(-1)$ integral can be accomplished by replacing σ_a (the absorption cross section) with σ_i (the ionization cross section) or alternatively, by $\eta\sigma_a$ where η is an ionization quantum yield per absorption.

(3) $S(-2)$: The electric-dipole polarizability averaged over all directions of space α_N is a quantity which can be deduced by various measurements, including the Stark effect in microwave spectroscopy, Stark effect in radio-frequency spectroscopy, electric Stern–Gerlach experiments, dielectric constant, and molar refraction determinations. It is related to $S(-2)$ by the expression

$$\alpha_N = 4a_0{}^3 S(-2) = 0.59274 \times 10^{-24} S(-2).$$

The continuum component of $S(-2)$ can be obtained from photoabsorption measurements by the expression

$$S(-2) = \frac{1}{16\pi^3 a_0{}^3} \int_0^{\lambda \equiv \mathrm{IP}} \sigma(\lambda)\, d\lambda, \qquad \text{with } \sigma \text{ in cm}^2, \qquad \lambda \text{ in cm.}$$

or alternatively the contribution of the continuum to α_N can be directly computed from

$$\alpha_N = \frac{10^{-26}}{4\pi^3} \int_0^{\lambda \equiv \mathrm{IP}} \sigma(\lambda)\, d\lambda.$$

The contribution of the discrete portion of the spectrum to α_N requires attention to units, as discussed earlier. In the above expressions, the constants have been computed assuming σ to be expressed in megabarns, λ in angströms, unless otherwise noted.

Appendix 1. Quantum Mechanical Derivation of Thomas–Reiche–Kuhn Sum Rule

We wish to prove that $\sum_j f_j = N$, where N is the number of electrons in the atom or molecule, and

$$f_j = \frac{8\pi^2 m}{3h} v_{j0} |m_{j0}|^2,$$

where $|m_{j0}|$ is the rth matrix element, not x, y, or z. Let us first take the xth element and evaluate the expression $v_{j0}|m_{xj0}|$. Since $v_{j0} = (E_j - E_0)/h$ and $|m_{xj0}| = \int \psi_j{}^* x_n \psi_0 \, d\tau$, we can rewrite this expression as

$$v_{j0}|m_{xj0}| = \frac{1}{h} \sum_{n=0}^{N} (E_j - E_0) \int \psi_j{}^* x_n \psi_0 \, d\tau.$$

But

$$E_j\psi_j^* = \mathscr{H}\psi_j^* = (T + V)\psi_j^*$$
$$E_0\psi_0 = \mathscr{H}\psi_0 = (T + V)\psi_0$$

$$\therefore\ v_{j0}|m_{xj0}| = \frac{1}{h}\sum_{n=0}^{N}\left[\int(\mathscr{H}\psi_j^*)(x_n\psi_0)\,d\tau - \int(\psi_j^* x_n)(\mathscr{H}\psi_0)\,d\tau\right]$$
$$= \frac{1}{h}\sum_{n=0}^{N}\left[\int(T + V)\psi_j^*(x_n\psi_0)\,d\tau - \int(\psi_j^* x_n)(T + V)\psi_0\,d\tau\right].$$

The operator V is an ordinary function of the coordinates, and hence commutes with all the coordinates. Therefore, we can extract a component of the first integral and one from the second integral which cancel one another, i.e.,

$$\int\psi_j^* V x_n\psi_0\,d\tau - \int\psi_j^* x_n V\psi_0\,d\tau = 0,$$

leaving us with the terms

$$\int(T\psi_j^*)(x_n\psi_0)\,d\tau - \int(\psi_j^* x_n)(T\psi_0)\,d\tau.$$

Now

$$T = -\frac{h^2}{8\pi^2 m}\nabla^2 = -\frac{h^2}{8\pi^2 m}\left(\frac{\partial^2}{\partial x^2} + \frac{\partial^2}{\partial y^2} + \frac{\partial^2}{\partial z^2}\right).$$

Since T (and also $\mathscr{H}$) is both Hermitean and real, we can interchange $T\psi_j^*$ and we are left with the commutator relation

$$[T, x_n] = [\nabla^2, x_n] = \nabla^2 x_n\psi_0 - x_n\nabla^2\psi_0.$$

If ∇^2 concerns all the other electrons except the nth, we immediately obtain

$$\nabla^2 x_n\psi_0 = x_n\nabla^2\psi_0.$$

If the ∇^2 is the derivative with respect to the nth electron, we have for the xth component,

$$\frac{\partial}{\partial x_n}\left(\frac{\partial}{\partial x_n}\right)x_n\psi_0 = \frac{\partial}{\partial x_n}\left(\psi_0 + x_n\frac{\partial\psi_0}{\partial x_n}\right)$$
$$= 2\left(\frac{\partial\psi_0}{\partial x_n}\right) + x_n\left(\frac{\partial^2\psi_0}{\partial x_n{}^2}\right)$$

or, including the y and z components,

$$\nabla^2 x_n\psi_0 = 2\left(\frac{\partial\psi_0}{\partial x_n}\right) + x_n\nabla_n{}^2\psi_0.$$

Hence, we have a term from the first integral $(x_n \nabla_n{}^2\psi_0)$, for all n, which is identical and of opposite sign to the term of the second integral. We are left with

$$\nu_{j0}|m_{xj0}| = -\frac{h^2}{8\pi^2 m}\frac{2}{h}\sum_{n=1}^{N}\int \psi_j{}^* \frac{\partial\psi_0}{\partial x_n}\,d\tau.$$

Recalling that the momentum operator p_{xn} may be written $(h/2\pi i)/(\partial/\partial x_n)$, we can rewrite this

$$\nu_{j0}|m_{xj0}| = -\frac{h}{4\pi^2 m}\frac{2\pi i}{h}\sum_{n=1}^{N}\int \psi_j{}^* p_{xn}\psi_0\,d\tau.$$

Let us now multiply by

$$|m_{xj0}| = \int \psi_0{}^* x_n \psi_j\,d\tau.$$

Then

$$\nu_{j0}|m_{xj0}|^2 = -\frac{i}{2\pi m}\sum_{n=1}^{N}\langle\psi_0{}^*|x_n|\psi_j\rangle\langle\psi_j{}^*|p_{xn}|\psi_0\rangle$$

and now sum over all states j

$$\sum_j \nu_{j0}|m_{xj0}|^2 = -\frac{i}{2\pi m}\sum_j\sum_{n=1}^{N}\langle\psi_0{}^*|x_n|\psi_j\rangle\langle\psi_j{}^*|p_{xn}|\psi_0\rangle.$$

Now let us consider the operators x_n and p_{xn} of the nth term of this summation. We can expand any one function in terms of a complete set of functions. Therefore,

$$p_{xn}\psi_0 = \sum_j a_j\psi_j.$$

Multiplying by $\psi_j{}^*$ and integrating

$$\int \psi_j{}^* p_{xn}\psi_0\,d\tau = \sum_j a_j \int \psi_j{}^*\psi_j\,d\tau$$

or

$$a_j = \int \psi_j{}^* p_{xn}\psi_0\,d\tau.$$

Now multiply by x_n, then by $\psi_0{}^*$ and integrate

$$\begin{aligned}\int \psi_0{}^* x_n p_{xn}\psi_0\,d\tau &= \int \psi_0{}^* x_n \left[\sum_j \int \psi_j{}^* p_{xn}\psi_0\,d\tau\right]\psi_j\,d\tau\\ &= \sum_j \langle\psi_0{}^*|x_n|\psi_j\rangle\langle\psi_j{}^*|p_{xn}|\psi_0\rangle.\end{aligned}$$

Therefore,

$$\sum_j v_{j0}|m_{xj0}|^2 = -\frac{i}{2\pi m}\sum_{n=1}^{N} \langle\psi_0{}^*|x_n p_{xn}|\psi_0\rangle.$$

Alternatively, we could have taken the complex conjugate

$$v_{j0}|m_{xj0}| = \frac{i}{2\pi m}\sum_{n=1}^{N} \int \psi_j p_{xn}^* \psi_0{}^*\, d\tau$$

$$= \frac{i}{2\pi m}\sum_{n=1}^{N} \int \psi_0{}^* p_{xn}\psi_0\, d\tau,$$

by its Hermitean property. We now multiply by $|m_{xj0}|$ to obtain

$$v_{j0}|m_{xj0}|^2 = \frac{i}{2\pi m}\sum_{n=1}^{N} \langle\psi_0{}^*|p_{xn}|\psi_j\rangle\langle\psi_j{}^*|x_n|\psi_0\rangle.$$

Once again summing over all j transitions and replacing

$$\sum_j \langle\psi_0{}^*|p_{xn}|\psi_j\rangle\langle\psi_j{}^*|x_n|\psi_0\rangle$$

by

$$\langle\psi_0{}^*|p_{xn}x_n|\psi_0\rangle,$$

we obtain another expression for $\sum_j v_{j0}|m_{xj0}|^2$.

$$\sum_j v_{j0}|m_{xj0}|^2 = \frac{i}{2\pi m}\sum_{n=1}^{N} \langle\psi_0{}^*|p_{xn}x_n|\psi_0\rangle.$$

Summing these two expressions for $\sum_j v_{j0}|m_{xj0}|^2$,

$$2\sum_j v_{j0}|m_{xj0}|^2 = \frac{i}{2\pi m}\sum_{n=1}^{N} \langle\psi_0{}^*|p_{xn}x_n - x_n p_{xn}|\psi_0\rangle.$$

But

$$p_{xn}x_n\psi_0 = \frac{h}{2\pi i}\frac{\partial}{\partial x_n}(x_n\psi_0) = \frac{h}{2\pi i}\left(\psi_0 + x_n\frac{\partial\psi_0}{\partial x_n}\right),$$

$$x_n p_{xn}\psi_0 = x_n\frac{h}{2\pi i}\frac{\partial\psi_0}{2x_n}$$

$$\therefore \langle\psi_0{}^*|p_{xn}x_n - x_n p_{xn}|\psi_0\rangle = \frac{h}{2\pi \mathrm{i}} \qquad \text{(famous Heisenberg expression)}$$

and

$$2\sum_j v_{j0}|m_{xj0}|^2 = \frac{i}{2\pi m}\frac{h}{2\pi i}\sum_{n=1}^{N} = \frac{h}{4\pi^2 m}N.$$

Similarly,

$$2\sum_j \nu_{j0}|m_{yj0}|^2 = 2\sum_j \nu_{j0}|m_{zj0}|^2 = \frac{h}{4\pi^2 m} N.$$

Recalling that

$$f_j = \frac{8\pi^2 m}{3h} \nu_{j0}|m_{j0}|^2 = \frac{8\pi^2 m}{3h} \nu_{j0}[|m_{xj0}|^2 + |m_{yj0}|^2 + |m_{zj0}|^2],$$

we obtain

$$\sum_j f_j = \frac{8\pi^2 m}{3h} 3 \frac{h}{8\pi^2 m} N = N.$$

Appendix 2. Relationship of the Electric-Dipole Matrix Elements to the Ground State Wave Function

We wish to prove that

$$\int \psi_0{}^* r^2 \psi_0 \, d\tau = \sum_{n=1}^{\infty} \left| \int \psi_n{}^* \mathbf{r} \psi_0 \, d\tau \right|^2 = \sum_{n=1}^{\infty} |r_n|^2.$$

In one dimension and for a one-electron system, we can write the left-hand side

$$\int (\psi_0{}^* x)(x\psi_0) \, d\tau.$$

Now expand $(x\psi_0) = \sum_{n=0}^{\infty} a_n \psi_n$. Multiplying on the left by $\psi_n{}^*$ and integrating,

$$\int \psi_{n'}^* x \psi_0 \, d\tau = \int \psi_{n'}^* \sum_{n=0}^{\infty} a_n \psi_n \, d\tau.$$

The left-hand side is one of the desired matrix elements which we label $|x_{n'}|$. The right-hand side is zero unless $n' = n$, when it reduces to $a_{n'}$. Hence, $|x_{n'}| = a_{n'}$ for all n. We now perform the same operation on the complex conjugate $(\psi_0{}^* x)$. We are left with

$$\int (\psi_0{}^* x)(x\psi_0) \, d\tau = \int \sum_{n'=0}^{\infty} |x_{n'}^*| \psi_{n'}^* \sum_{n'=0}^{\infty} |x_{n'}| \psi_{n'} \, d\tau = \sum_{n'=0}^{\infty} \langle x_{n'} \rangle^2.$$

The derivation can be generalized to three dimensions and to many-electron systems. In the latter case, cross terms of the form $|r_i \cdot r_j|$ enter, where i and j refer to different electrons.

Appendix 3. Interaction of Fast Charged Particles with Atoms and Molecules—A Digression

In the limit of the first Born approximation, the interaction of fast charged particles with atoms was first derived by Bethe.[28] We shall not attempt to rederive his expression here, since it is an involved development. We shall instead refer to a concise summary in a recent review article by Inokuti.[29] The differential cross section $d\sigma_n$ for excitation of the atom to the state N as a result of the interaction between the fast charged particle and the atom is given by

$$d\sigma_n = \frac{M^2}{(2\pi)^2(\hbar)^4}\left(\frac{k'}{k}\right)\left|\int e^{iKr} u_n{}^*(r_1, \ldots, r_z) V u_0(r_1 \cdots r_z)\, dr_1 \cdots dr_z\, dr\right|^2 d\omega,$$

where

$$M = \frac{M_1 M_2}{M_1 + M_2}\cdot$$

is the reduced mass of the colliding system, r the position of the projectile relative to the center of the atom, $\hbar k$ the momentum of the projectile before collision, $\hbar k'$ the momentum of the projectile after collision, $\hbar K = \hbar(k - k')$ the momentum transfer and the u's are the eigenfunctions of the atomic system, whose electron coordinates are r_j and the total number of electrons in the atom is Z. The $d\omega$ refers to a solid angle interval of measurement. If the interaction is coulombic, V is written as

$$V = -\sum_{j=1}^{Z} \frac{ze^2}{|r - r_j|} + \frac{zZ_n e^2}{r},$$

where ze is the charge of the projectile, and $Z_n e$ is the charge on the atomic nucleus. Bethe showed how this integration can be performed and leads to

$$d\sigma_n = 4z^2\left(\frac{Me^2}{\hbar^2}\right)\left(\frac{k'}{k}\right)\frac{1}{K^4}|\varepsilon_n(K)|^2\, d\omega,$$

where

$$\varepsilon_n(K) = \int u_n{}^* \sum_{j=1}^{z} e^{i\mathbf{K}\cdot\mathbf{r}_j} u_0\, dr_1 \cdots dr_z.$$

When the momentum transfer K is large, $d\sigma_n$ will be small for two reasons: (1) the $1/K^4$ term outside the integral and (2) $e^{i\mathbf{K}\cdot\mathbf{r}}$ will oscillate very rapidly over a distance r where the wave functions u_n^* and u_0 do not. Hence, the integrand will tend toward zero. When the momentum transfer K is small, we can expand the exponential for each electron in the atom

$$e^{i\mathbf{K}\cdot\mathbf{r}} = 1 + i\mathbf{K}\cdot\mathbf{r} - \frac{(\mathbf{K}\cdot\mathbf{r})^2}{2!} - \cdots.$$

The first term, when inserted into the expression for $\varepsilon_n(K)$ will make the integrand nonvanishing only when $u_n^* = u_0^*$, which means only elastic scattering has taken place. The second term gives rise to the dominant inelastic scattering cross section, when permitted. It is the familiar dipole moment matrix element. Thus,

$$\varepsilon_n(K) \Rightarrow \int u_n^* \sum_{j=1}^{z} (\mathbf{K}\cdot\mathbf{r}_j)u_0\, dr_1 \cdots dr_z$$

and

$$d\sigma_n \propto \left| \int u_n^* \sum_j \mathbf{r}_j u_0\, dr_1 \cdots dr_z \right|^2.$$

We had previously derived an expression for the total oscillator strength, designated $S(0)$, which was proportional to

$$v \left\langle \psi_n^* \middle| \sum_j \mathbf{r}_j \middle| \psi_0 \right\rangle^2.$$

The dipole polarizability was shown to be proportional to $S(-2)$, which was of the form f/v^2 or $\langle \psi_n^* | \sum_j \mathbf{r}_j | \psi_0 \rangle^2 / v$.

The Bethe expression for the excitation or ionization of an atom (or molecule) by a fast charged particle is proportional to $\langle \psi_n^* | \sum_j \mathbf{r}_j | \psi_0 \rangle^2$, and hence is related to $S(-1)$.

If we restrict ourselves to ionization phenomena, the set of ψ_n are states of the ionization continuum. One can formally write the Bethe equation in this case as

$$\sigma = 4\pi(\hbar/mc)^2(M_i^2 x_1 + C x_2),$$

where x_1 and x_2 are functions of the velocity of the incoming charged particle, C is a constant characteristic of the atom or molecule, and M_i^2 is the square of the total dipole matrix element for ionization. From the slope of a plot of σ versus kinetic energy of the incident charged particle, it is possible to extract M_i^2, which is in fact the ionization component of $S(-1)$ designated $S_i(-1)$.

From the generalized definition of $S(p)$ we obtain

$$S_{\mathrm{i}}(-1) = \frac{1}{4\pi^2 a_0{}^2 \alpha_0} \int_0^{\lambda_0} \frac{\sigma_{\mathrm{i}}(\lambda)}{\lambda} d\lambda$$

which is the contribution to the $S(-1)$ sum from some λ_0 (say the ionization threshold) to very large energy.

This value, obtained solely from photoabsorption (or photoionization) studies can be compared with an independent value of $S_{\mathrm{i}}(-1)$ which can be deduced from the absolute cross section for charged particle excitation and/or ionization of the atom or molecule under study (see also Fano[30]).

REFERENCES

1. J. A. R. Samson, *Adv. At. Mol. Phys.* **2**, 237 (1966).
2. H. A. Bethe and R. Jackiw, "Intermediate Quantum Mechanics," 2nd Ed., p. 236. Benjamin, New York, 1968.
3. E. P. Wigner, *Phys. Rev.* **73**, 1002 (1948).
4. P. C. Killgoar, Jr., G. E. Leroi, J. Berkowitz, and W. A. Chupka, *J. Chem. Phys.* **58**, 803 (1973); E. Miescher, Y. T. Lee, and P. Gürtler, *J. Chem. Phys.* **68**, 2753 (1978).
5. J. H. D. Eland and J. Berkowitz, to be published; J. Berkowitz, unpublished data.
6. V. H. Dibeler and J. A. Walker, *J. Opt. Soc. Am.* **57**, 1007 (1967).

6a. J. Berkowitz and C. Lifshitz, *J. Phys. B* **1**, 438 (1968).

7. M. Nakamura, M. Sasanuma, S. Sato, M. Watanabe, H. Yamashita, Y. Iguchi, A. Ejiri, S. Yamaguchi, T. Sagawa, Y. Nakai, and T. Oshio, *Phys. Rev.* **178**, 80 (1969); see also ref. 21.
8. W. L. Jolly, *J. Am. Chem. Soc.* **92**, 3260 (1970); W. L. Jolly, *in* "Electron Spectroscopy" (D. A. Shirley, ed.), p. 629. North-Holland Publ., Amsterdam, 1972.
9. M. Nakamura, Y. Morioka, T. Hayaishi, E. Ishiguro, and M. Sasanuma, *Int. Conf. VUV Radiat. Phys., 3rd, Tokyo* Pap. 1pA1-6 (1971).
10. S. Bodeur, C. Senemaud, C. Bonnelle, and J. P. Connerade, *Proc. Int. Conf. VUV Radiat. Phys., 4th*, Hamburg. p. 94 (1974).
11. E. A. Stern, *Phys. Rev. B* **10**, 3027 (1974).
12. C. A. Ashley and S. Doniach, *Phys. Rev. B* **11**, 1279 (1975).
13. P. A. Lee and J. B. Pendry, *Phys. Rev. B* **11**, 2795 (1975).
14. B. M. Kincaid and P. Eisenberger, *Phys. Rev. Lett.* **34**, 1361 (1975).
15. J. W. Cooper, *Phys. Rev.* **128**, 681 (1962).
16. H. A. Bethe and E. E. Salpeter, "Quantum Mechanics of One- and Two-Electron Atoms." Springer-Verlag, Berlin and New York, 1957.
17. J. L. Dehmer and J. Berkowitz, *Phys. Rev. A* **10**, 484 (1974).

17a. S. T. Manson and J. W. Cooper, *Phys. Rev.* **165**, 126 (1968).

17b. J. L. Dehmer, A. F. Starace, U. Fano, J. Sugar, and J. W. Cooper, *Phys. Rev. Lett.* **26**, 1521 (1971).

17c. R. Haensel, G. Keitel, P. Schreiber, and C. Kunz, *Phys. Rev.* **188**, 1375 (1969).

18. A. F. Starace, *Phys. Rev. A* **2**, 118 (1970).

18a. R. E. LaVilla and R. D. Deslattes, *J. Chem. Phys.* **44**, 4399 (1966).

18b. T. M. Zimkina and V. A. Fomichev, *Sov. Phys.-Dokl.* **11**, 726 (1966); A. S. Vinogradov, T. M. Zimkina, and V. A. Fomichev; Engl. transl., *Russian J. Structural Chem.* **12**, 823 (1971).

19. J. L. Dehmer, *J. Chem. Phys.* **56**, 4496 (1972).
20. B. Cadioli, U. Pincelli, E. Tosatti, U. Fano, and J. L. Dehmer, *Chem. Phys. Lett.* **17**, 15 (1972); L. N. Mazalov, F. K. Gel'muskhanov, and V. M. Chermoschentsev, *Zh. Strukt. Khim.* **15**, 1099 (1974); Engl. transl., *J. Struct. Chem.* (USSR) **15**, 975 (1974).
21. A. S. Vinogradov, T. M. Zimkina, V. N. Akimov, and B. Shlarbaum, *Bull. Acad. Sci. USSR, Phys. Ser.* **38**, 69 (1974).
22a. T. Åberg, *in* "Proceedings Int'l. Conf. on Inner Shell Ionization Phenomena, Atlanta (1972) (R. W. Fink *et al.*, eds.), p. 1509. USAEC Technical Information Center, Oak Ridge, Tennessee (1973, CONF-720404).
22b. T. A. Carlson, M. O. Krause, and W. E. Moddeman, *J. Phys.* (*Paris*) **32**, C4–76 (1971).
23. W. Kauzmann, "Quantum Chemistry." Academic Press, New York, 1957.
23a. U. Fano and J. W. Cooper, *Rev. Mod. Phys.* **40**, 441 (1968).
24. A. R. P. Rau and U. Fano, *Phys. Rev.* **162**, 68 (1967).
25. W. C. Price, A. W. Potts, and D. G. Streets, *in* "Electron Spectroscopy" (D. A. Shirley, ed.), p. 187. North-Holland Publ., Amsterdam, 1972.
25a. R. I. Schoen, *J. Chem. Phys.* **37**, 2032 (1962).
26. V. L. Carter and J. Berkowitz, *J. Chem. Phys.* **59**, 4573 (1973).
27. G. L. Weissler, *in* "Encyclopedia of Physics. Vol. XXI: Electron-Emission, Gas Discharges I" (S. Flugge, ed.), pp. 304–382. Springer-Verlag, Berlin and New York, 1956.
28. H. Bethe, *Ann. Phys.* (*Leipzig*) **5**, 325 (1930).
29. M. Inokuti, *Rev. Mod. Phys.* **43**, 297 (1971).
30. U. Fano, *Phys. Rev.* **95**, 1198 (1954).

Chapter V

TOTAL PHOTOABSORPTION AND PHOTOIONIZATION CROSS SECTIONS FOR SELECTED MOLECULES

We shall now embark upon an investigation of specific molecules to see how well the principles discussed in the preceding chapters apply. It is convenient for this purpose to distinguish total cross sections and partial cross sections. In this chapter, we shall examine experimental photoabsorption cross sections (supplemented where necessary with high energy electron impact studies) across the entire electromagnetic spectrum. The oscillator strength distribution will be tested against the sum rules $S(0) = N$, and $S(-2) = \alpha_N/4a_0{}^3$ and where possible, with $S_i(-1)$ and $|M_i|^2$. In some cases (He and H_2) where accurate wave functions are available for some states, theoretical evaluations of oscillator strengths for some transitions will be incorporated.

In order to present a panoramic view of the entire ionization continuum (from ionization threshold through the "K edge"), we have replotted existing data on a $\log\sigma$(Mb) versus λ(Å) scale. It will be seen that this has some advantages in expanding the energy scale where experimental resolution is generally highest, and in seeing whether the x-ray (<100 Å), extreme ultraviolet (100–500 Å), and vacuum ultraviolet (~500 Å–ionization threshold) data, often obtained by groups with quite different experimental techniques, join smoothly. It is also instructive to examine partial contributions of various wavelength regions to the $S(0)$, $S(-1)$, and $S(-2)$ sum rules, since they are sensitive to different portions of the spectrum. While these sum rules cannot distinguish minor discrepancies, they do serve to detect gross inaccuracies.

Also located on the individual graphs are the adiabatic ionization potentials which roughly correspond to the energies required to remove single electrons from successively deeper orbitals. In most cases, this information has been obtained directly from photoelectron spectroscopy. Theoretical calculations have been incorporated to supplement experimental data on ionization potentials, where the latter are unavailable. References used to construct the graphs are explicitly given. The integrations have been performed by computer, using a cubic spline fitting of the data.

Photoionization yields are presented on a separate graph. In the atomic gases, one might expect the yields to be unity, but even here exceptions are possible. Recent studies of atomic oxygen[1,1a] have revealed that some absorption peaks above the ionization potential only partially autoionize, due to unfavorable selection rules. Our knowledge of molecular photoionization yields below ~584 Å is, at best, fragmentary at the moment. For monatomic and diatomic gases (and largely due to ignorance, for polyatomic gases), it is frequently assumed to be unity at the higher energies.

The gases analyzed in this chapter (and also Chapter VI) have been selected for several reasons.

(a) In most cases, sufficient information exists on photoabsorption cross sections across the entire electromagnetic spectrum, and ionization yields in important wavelength regions have been measured.

(b) The data required for partial cross section analysis (Chapter VI) also exist.

(c) They represent an attempt to include prototypes of different classes of compounds, i.e., some atoms (He, Ne, Ar, Kr, Xe, Hg), diatomic (H_2, N_2, O_2, CO, NO) and triatomic (H_2O, CO_2, N_2O) molecules, NH_3, alkynes (C_2H_2), alkenes (C_2H_4) and normal alkanes (CH_4, C_2H_6), normal alcohols (CH_3OH), benzene (a representative of aromatics) and SF_6 (an example of a larger, but highly symmetric inorganic molecule).

Major omissions, which are indicative of the areas where much more investigation is required, include gases containing elements below the second row of the periodic chart and inorganic molecules, most of which are involatile at room temperature and hence require different experimental techniques. Some photoionization mass spectrometry has been performed on these systems, and also some photoelectron spectroscopy, but photoabsorption and ionization yield measurements in the vacuum ultraviolet and x-ray regions are almost nonexistent.

A. Monatomic Gases

1. Helium

Apart from atomic hydrogen (whose oscillator strength distribution in the continuum is known from theory, with little experimental verification), helium gas provides us with our best example for testing sum rules. A calculation of the ground state wave function for He was performed by Pekeris,[1b] with an accuracy which far surpasses our knowledge of any other system, except possibly molecular H_2.[2] We have seen how this knowledge enables us to compute $S(-1)$, for which Pekeris obtained 0.752_5. We have not discussed the fact that $S(+1)$ and $S(+2)$ can also be calculated from an accurate ground state wave function, because we shall not find these quantities generally useful, except for He and H_2. (For larger systems, these quantities are too sensitive to slight errors in the cross sections at high energy.) For He, Pekeris found $S(+1) = 8.167_5$ and $S(+2) = 121.34$.

In Fig. 1, the cross sections for He throughout the ionization continuum, as summarized by Samson,[3] are shown on a $\log \sigma$ versus λ plot. The data from threshold to 209 Å are from Samson's experiments using line sources, those from 209–101 Å are obtained by Lowry *et al.*[4] using synchrotron light. Samson's compilation has a gap between 100.9 and 44.6 Å, beyond which older x-ray values of Henke *et al.*[5] become available. Subsequently, Henke *et al.*[6] have proposed a new set of values in the 3–200 Å region, apparently based on measurements at four points and correlations with other systems, which are also plotted on the figure. They are uniformly lower than the earlier values. The cross sections tabulated by Samson below 1.0 Å appear to be clearly in error, since at this high energy we can expect the energy dependence[7] of σ to be $\sim E^{-7/2}$, and the tabulated values decline slower than E^{-1}. In the following analysis, we choose $\sigma = 3.7 \times 10^{-8}$ Mb at 1.0 Å, and $E^{-7/2}$ behavior to higher energy. We shall initially employ the older Henke data, then examine the effect of the newer Henke values.

From an examination of Fig. 1, we note that there is significant scatter below 157 Å $\equiv$ 79 eV, which is the threshold for double ionization of He. No structure can be expected for He above 79 eV, and hence this region has been smoothed for the purpose of integration. In particular, the three points between $\sim$100.9 and 106 Å are much too high. The large gap between $\sim$105 and 45 Å has also been joined by a smooth curve.

Some autoionization due to two-electron excitation does exist for He. The first member of this series is at $\sim$207 Å, and can be noted in the figure. The limit of this series is the aforementioned double ionization limit at $\sim$157 Å. The bulge in Fig. 1 below the onset of this series (245–285 Å) has also been removed by smoothing.

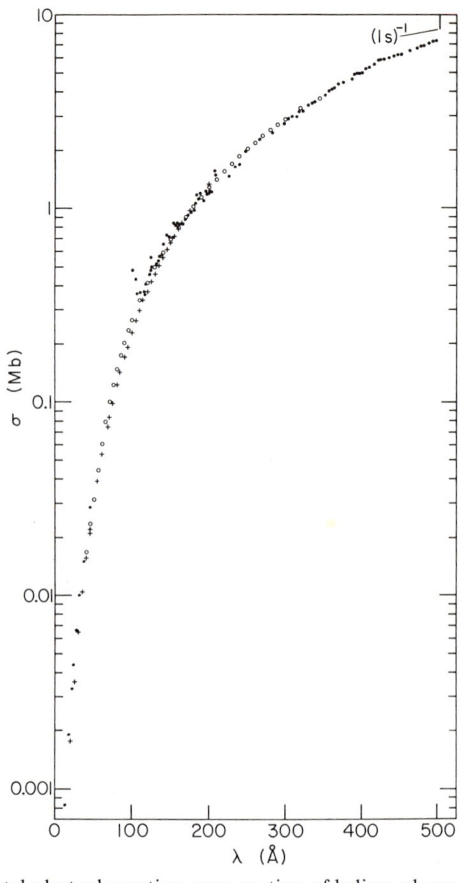

Fig. 1 Experimental photoabsorption cross section of helium above the ionization threshold. ● Samson[3]; + Henke *et al.*[6]; ○ West and Marr.[15]

After this doctoring procedure, the various weighted sums were obtained, and are summarized in Table 1. The values of f_n below the ionization limit have been obtained using the prescription of Inokuti and Kim,[8] the value for the first transition being the generally accepted one calculated by Schiff and Pekeris.[9] Inokuti and Kim extended their analytical form to $n \rightarrow \infty$, and obtained 0.2633 for the subionization portion of $S(-1)$, slightly higher than the truncated value (0.2621) obtained here. Dalgarno and Parkinson[10] have obtained 0.424 for the partial $S(0)$ below the ionization continuum, again in excellent agreement with the value in Table 1, 0.423.

The continuum contribution of $S(0)$ from Table 1 is 1.569, which is very close to the value (1.573) computed by Bell and Kingston.[11] The total $S(0)$,

TABLE 1 *The Oscillator Strength Distribution in He*

Energy range		$S(0)$	$S(-1)$	$S(-2)$	$S(+1)$	$S(+2)$
Discrete ($^1P_1$; eV)						
$n = 2$	21.2168	0.27615	0.17708[a]	0.11355	0.43063	0.67155
3	23.0858	0.07347	0.0433	0.02552	0.12467	0.21154
4	23.7408	0.03019	0.0173	0.00991	0.05268	0.09192
5	24.0445	0.01521	0.00861	0.00487	0.02688	0.04751
6	24.2097	0.00869	0.00488	0.00274	0.01546	0.02751
7	24.3094	0.00543	0.00304	0.00170	0.00970	0.01733
8	24.3742	0.00362	0.00202	0.00113	0.00648	0.01161
9	24.4186	0.00253	0.00141	0.00079	0.00455	0.00816
10	24.4504	0.00184	0.00102	0.00057	0.00331	0.00595
11	24.4739	0.00138	0.00077	0.00043	0.00248	0.00447
12	24.4918	0.00106	0.00059	0.00033	0.00191	0.00344
13	24.5058	0.00083	0.00046	0.00026	0.00150	0.00271
14	24.5168	0.00067	0.00037	0.00021	0.00120	0.00217
15	24.5258	0.00054	0.00030	0.00017	0.00098	0.00176
16	24.5331	0.00045	0.00025	0.00014	0.00081	0.00145
17	24.5391	0.00037	0.00021	0.00011	0.00067	0.00121
18	24.5442	0.00031	0.00017	0.00010	0.00057	0.00102
19	24.5485	0.00027	0.00015	0.00008	0.00048	0.00087
20	24.5522	0.00023	0.00013	0.00007	0.00041	0.00074
Partial sum		0.42324	0.26206	0.16268	0.68537	1.11292
Continuum (Å)						
209.3–503.0		1.015	0.3976	0.1643	2.753	7.953
44.6–209.3		0.5170	0.0761	0.0125	4.079	38.51
0.139–44.6		0.0369	0.0012	0.0000_4	1.555	169.7
Partial sum		1.5689	0.4749	0.1768_4	8.387	216.2
Total		1.992	0.737	0.3395	9.072	217.3

[a] See Schiff and Pekeris.[9]

1.992, is nearly the precise value needed to satisfy the Thomas–Reiche–Kuhn sum rule.

The contribution of the continuum to $S(-1)$, which we write $S_i(-1) \equiv |M_i|^2$ computed from Table 1 is 0.475. Kim and Inokuti[12] have obtained 0.489 for this quantity from theoretical considerations. Our total $(S-1) = 0.737$ is slightly lower than that required to satisfy Pekeris' value of 0.7525, the discrepancy evidently residing primarily in the continuum.

Rieke and Prepejchal[13] have measured the slope of the absolute ionization cross section for ionization of He by fast electrons. Under most circumstances, this slope is expected to be the quantity $S_i(-1)$. Rieke and Prepejchal obtain an experimental value of 0.752, which is almost identical to the *total* $S(-1)$ from Pekeris' calculation. The explanation that has been given for this anomaly lies in the nature of their experiment. In order to make the measurement, helium is mixed with another gas. The excited states of helium lie between 21.2 and 24.6 eV. At the pressures used in the Rieke–Prepejchal experiment, the helium atoms formed in excited states collide many times with the molecules of the other gas, and transfer their excitation energy. Since even the 21.2 eV state is well above the ionization potential of all other molecules except neon, ionization is recorded even for the discrete portion of the spectrum. Unfortunately, this explanation should not hold for the heavier noble gases, but as we shall see, the anomaly persists there.

The $S(-2)$ sum is almost equally divided between the continuum (0.1768) and discrete (0.1627) portions. From their sum and Eq. (15), we obtain $\alpha_N = 0.201 \times 10^{-24}\ \mathrm{cm}^3$. The recommended[14] polarizability of helium is $0.204956 \times 10^{-24}\ \mathrm{cm}^3$, and is based on a perturbation–variation procedure.

The $S(+1)$ sum contributed by the continuum alone is 8.38_7, already slightly larger than Pekeris' total $(S+1) = 8.1675$. Kim[14a] has computed the continuum $S(+1)$ contribution to be 7.078. Subtraction of our discrete portion of $S(+1)$ from Pekeris' total $S(+1)$ yields a continuum $S(+1)$ contribution of 7.48_2.

The $S(+2)$ sum, which we have already mentioned as being extremely sensitive in the higher energy region, is 216.2 in the continuum, much larger than Pekeris' total $S(+2)$ of 121.34. Substitution of the newer Henke *et al.*[6] values in the range 3 Å $< \lambda <$ 175 Å yields the following sums:

$S(0)$	$S(-1)$	$S(-2)$	$S(+1)$	$S(+2)$
1.93_1	0.731	0.339	7.93	93.0

As expected, the $S(-1)$ and $S(-2)$ sums are very slightly affected. Both the $S(+1)$ and $S(+2)$ sums are over corrected; the comparison suggests values in this energy range approximately midway between the two sets of Henke *et al.* The most serious effect is on $S(0)$, which acquires a $3\frac{1}{2}\%$ error. For this reason, either the older values or a mean set should be tentatively employed, pending further measurements.

Very recently, West and Marr[15] have remeasured the 40–340 Å region at the Daresbury synchrotron, and their data are also plotted in Fig. 1. In the region of overlap, their data do appear to lie between the two Henke sets. Their computed oscillator strength sums are given below:

$S(0)$	$S(-1)$	$S(-2)$	$S(+1)$	$S(+2)$
2.0	0.741	0.340	8.72	166

These values are in better agreement with theory, particularly for the sensitive $S(+1)$ and $S(+2)$ sums.

2. Neon

The data tabulated by Samson[3] are shown in Fig. 2 on a log σ versus λ plot. Samson's own experimental results, using line sources, encompass the region from ionization threshold to 40 Å. The region from threshold to 80 Å is also covered by Ederer and Tomboulian[16] using synchrotron radiation. Samson comments that the agreement between the two sets of data is good around threshold, but that below 450 Å the Ederer and Tomboulian data are about 10% too high. It may be that Samson's values are too low, as the ensuing sum-rule analysis and comparison with other types of experiments seem to

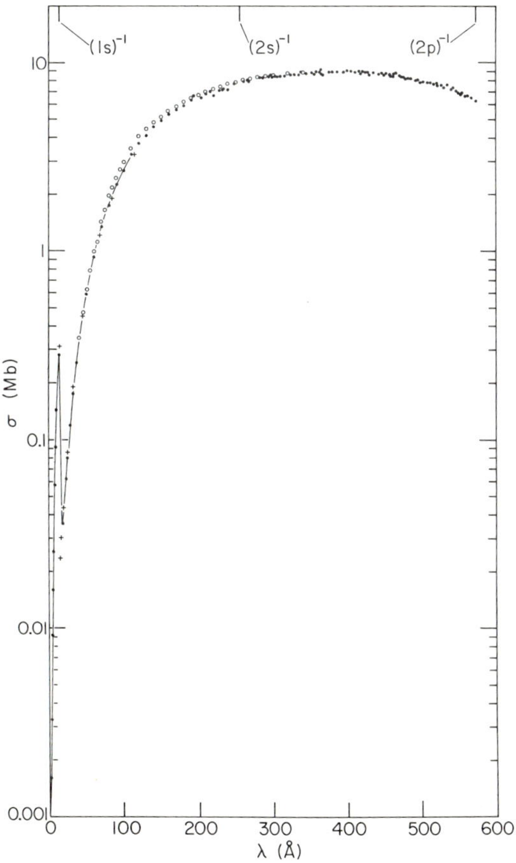

Fig. 2 Experimental photoabsorption cross section of neon above the ionization threshold. ● Samson[3]; + Henke[30]; ○ West and Marr.[15]

indicate. Below 44.6 Å the data of Henke *et al.*[5] seem reliable, but below 0.1 Å values of Victoreen[17] again appear to fall off with excess energy much too slowly.

Autoionization should manifest itself within 0.09 eV of the ionization threshold, as a consequence of the $^2P_{3/2}$–$^2P_{1/2}$ separation. A complex Beutler–Fano peak shape is observed for the autoionization resulting from excitation of the 2s electron to Rydberg series. Figure 2 shows some weak evidence for the first member of this series at ~272 Å.

The various contributions to the sum-rule analysis are listed in Table 2. The continuum contribution to $S(0)$ from Table 2 is 9.56. The contribution of the discrete portion has been taken from electron energy loss data obtained by Natali *et al.*[18] at the National Bureau of Standards and amounts to 0.277, or a total $S(0)$ of 9.83_7, ~1.6% low.

For $S(-1)$, the continuum provides 1.61, the discrete segment adds 0.208, for a total of 1.81_8. Naon *et al.*[19] have calculated $S(-1) = 1.91 \pm 0.02$, using a Bethe–Goldstone correlated wave function. The high energy electron ionization experiments of Rieke and Prepejchal[13] yield $|M_i|^2 = 2.02 \pm 0.05$, which is larger than $S(-1)$, let alone $S_i(-1)$. The explanation presumably follows the argument already given in the preceding section on helium.

TABLE 2 *The Oscillator Strength Distribution in Neon*

Energy region	$S(0)$	$S(-1)$	$S(-2)$
Discrete (eV)			
16.670	0.012	0.0098	0.0080
16.848	0.158	0.1275	0.1030
19.687	0.013	0.0090	0.0062
19.779	0.016	0.0110	0.0076
20.026	0.002	0.00136	0.0009
20.040	0.015	0.0102	0.0069
20.139	0.006	0.0041	0.0027
20.570	0.006	0.0040	0.0026
20.662	0.0043	0.0028	0.0019
20.702	0.0043	0.0028	0.0019
20.708	0.0042	0.0028	0.0018
20.805	0.0043	0.0028	0.0018
21.559	0.031	0.0196	0.0123
Partial sum	0.277	0.20776	0.1576
Continuum (Å)			
44.6–573.5	7.023	1.571	0.4859
0.10–44.6	2.537	0.0395	0.00096
Partial continuum sum	9.560	1.610_5	0.4869
Total	9.837	1.818	0.645

The $S(-2)$ contribution of the continuum, according to Table 2, is 0.487; that of the discrete portion, 0.158. Their sum implies $\alpha_N = 0.382 \times 10^{-24}\ \text{cm}^3$. The recommended[14] electric-dipole polarizability of neon, based on index of refraction data, is $0.3946 \times 10^{-24}\ \text{cm}^3$.

In all three cases, our computed oscillator sums are smaller than well-documented values. The $S(-1)$ sum is in poorer agreement than $S(-2)$. This suggests a "missing" contribution above the discrete portion but not too deep in the continuum. At this point, the 10% discrepancy between Samson's values and those of Ederer and Tomboulian below 450 Å come into question. Kim[20] has recently examined the secondary electron energy spectra reported by Opal *et al.*[21] resulting from ionization by high energy electrons. His comparison between their data and the photoabsorption results we have used is shown in Fig. 3. It can be seen that the secondary electron results imply additional oscillator strength between ~275 and 365 Å, in general agreement with the implications of our oscillator sum analysis.

West and Marr[15] have recently remeasured the absorption cross sections of He, Ne, Ar, and Kr between 40 and 340 Å using the Daresbury synchrotron. They have also made their own assessment of data in other wavelength ranges, and compared oscillator strength sums with corresponding values deduced from Hartree–Fock wave functions.

For neon, their *experimental* values alter the partial sums in the 44.6–573.5 Å region as follows: $S(0) = 7.37$, $S(-1) = 1.621$, and $S(-2) = 0.4939$. As a consequence, the total sums become: $S(0) = 10.18$, $S(-1) = 1.87$, $S(-2) = 0.652_5$, the latter implying $\alpha = 0.386 \times 10^{-24}\ \text{cm}^3$.

Both $S(-1)$ and $S(-2)$ are improved, but $S(0)$ has experienced an overshoot. The complete set of values selected by West and Marr yields a noticeably higher $S(-1)$ than is implied by experiment, suggesting that the f sum for the discrete transitions chosen by these authors may be too high (0.4, cf. 0.277 in this work).

3. Argon

The data compiled by Samson[3] are quite extensive for this atom, covering the region from ionization threshold through the K edge with no significant gap. Samson's own values encompass the region from approximate threshold to 200 Å, where they overlap the x-ray absorption measurements of Lukirskii and Zimkina.[22] Deeper x-ray values from Henke *et al.*[6] and Allen[23] appear reasonable, but those of Victoreen[17] must be rejected because of their erroneous energy dependence. As Samson notes, the Rayleigh scattering cross section begins to contribute significantly below 0.5 Å, and hence measured attenuation coefficients much below this region would distort the sum rule

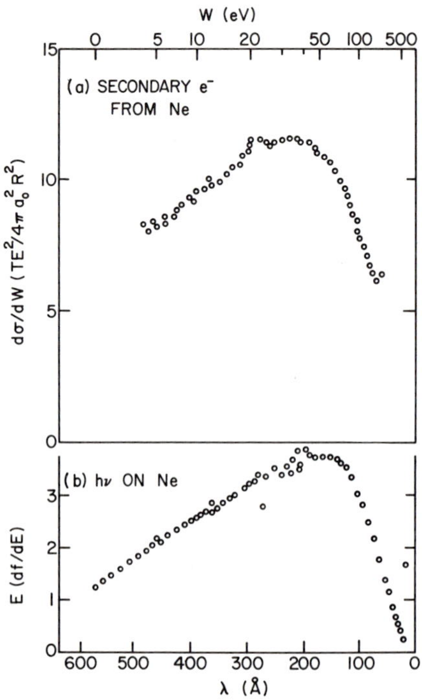

Fig. 3 A comparison between ionization produced by high energy electrons and photoionization, both performed on neon. In order that the experiments be meaningfully compared, the secondary electron energy spectra obtained in the electron impact experiment (Opal *et al.*[21]) are normalized in a manner given by the ordinate in (a) (see Appendix 3 to Chapter IV, and references therein). This presentation is sometimes referred to as a Platzman plot. Here $d\sigma/dW$ is the cross section for producing secondary electrons of energy W, T the kinetic energy of incident electrons, a_0 the Bohr radius, R the Rydberg constant, and E is the excitation energy. The ordinate in (b) is essentially the photoabsorption cross section $[\sigma = (\pi e^2 h/mc)(df/dE)]$ weighted by the excitation energy E. Kim[20] shows that for fast incident particles, the shape of the Platzman plot for slow secondary electrons is expected to resemble the energy weighted photoabsorption cross section.

analysis unless appropriate corrections were made. The panoramic log σ versus λ plot is shown in Fig. 4.

In argon, we begin to see for the first time some of the variations in the continuum which have been discussed in Chapters III and IV. The autoionization near threshold, between the $^2P_{3/2}$ and $^2P_{1/2}$ thresholds, has already been discussed (see Chapter III, Fig. 3). The vertical scale of Fig. 3 can be made absolute by taking the convergence limit of the oscillations to be 31 Mb, the value found by Samson near threshold. Beyond the $^2P_{1/2}$ threshold, the cross section remains relatively flat to ~500 Å, then begins to decrease rather

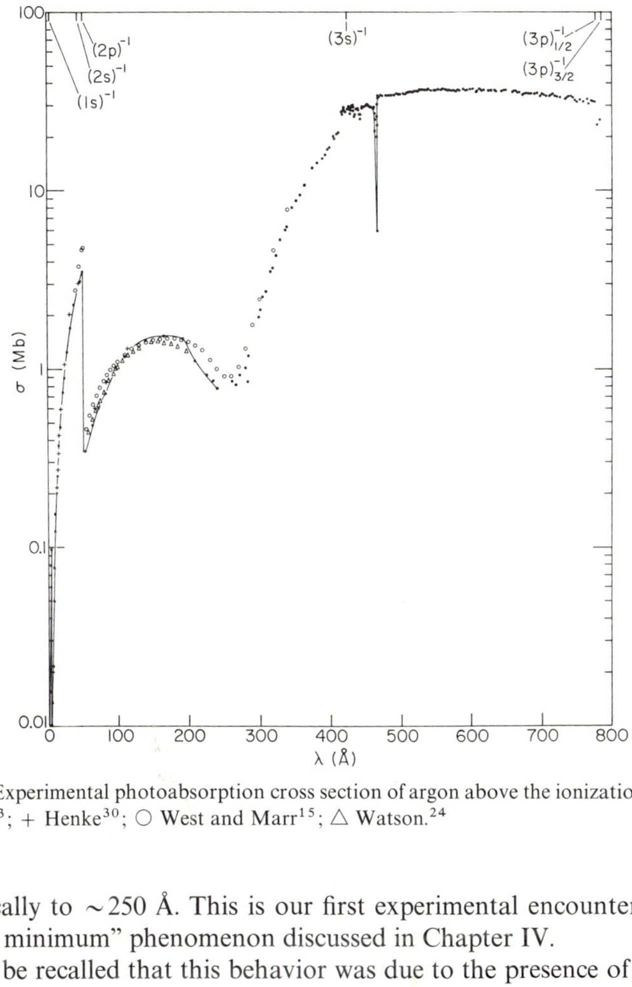

Fig. 4 Experimental photoabsorption cross section of argon above the ionization threshold. ● Samson[3]; + Henke[30]; ○ West and Marr[15]; △ Watson.[24]

dramatically to ~250 Å. This is our first experimental encounter with the "Cooper minimum" phenomenon discussed in Chapter IV.

It will be recalled that this behavior was due to the presence of a node in the bound state wave function. The argon 3p orbital has such a node, the neon 2p orbital does not. The overlap between the continuum wave function and the bound state wave function most nearly cancels the contributions from the positive and negative lobes of the bound state function at ~250 Å; the cross section rises somewhat at lower wavelengths, still primarily representing the 3p contribution to the ionization cross section.

At ~466 Å another kind of minimum appears. The first Rydberg member correspondency to excitation of the inner 3s orbital occurs here. The shape of this resonance corresponds to $q \approx 0$ (Chapter III, Section C) which implies a very weak transition probability from ground state to quasi-discrete state, and a strong interaction between this Rydberg state and the underlying continuum. The detailed behavior of these "window resonances" has already been seen earlier (Chapter III, Fig. 2).

Nakamura *et al.*[25] have photographically observed resonances preceding the $L_{II,III}(2p)$ edge, occurring between 50.8 and 49.9 Å. These studies, performed on the Tokyo synchrotron, clearly demonstrate that the abrupt increase in absorption cross section does not occur when ionization from a new shell commences, but rather when excitation from that shell to a vacant valence or Rydberg orbital can first occur. A Rydberg series should then be seen converging to the ejection threshold, and the average absorption cross section should be continuous through the "edge."

The contributions of the various regions to the $S(k)$ are summarized in Table 3.

TABLE 3 *The Oscillator Strength Distribution in Argon*

Energy region	$S(0)$	$S(-1)$	$S(-2)$
Discrete (eV)			
11.623	0.070	0.0819	0.0958
11.827	0.278	0.3197	0.3676
13.863	0.001	0.0010	0.0010
14.089	0.028	0.0270	0.0261
14.152	0.092	0.0884	0.0850
14.254	0.0124	0.0118	0.0113
14.303	0.110	0.1046	0.0995
14.710	0.004	0.0037	0.0034
14.847	0.0094	0.0086	0.0079
14.858	0.048	0.0439	0.0402
15.003	0.015	0.0136	0.0123
15.021	0.0224	0.0203	0.0184
15.117	0.0032	0.0029	0.0026
15.185	0.0139	0.0124	0.0111
15.189	0.0234	0.0210	0.0188
15.755	0.096	0.0829	0.0715
Partial sum	0.827	0.8437	0.8725
Continuum (Å)			
742.4–783.2	0.2456	0.2053	0.1717
209.3–742.4	5.173	2.890	1.701
50.6–209.3	1.431 (1.556-West)	0.1782 (0.187-West)	0.0252 (0.0258-West)
44.6–50.6	(1.278-West) 0.9919	(0.0667-West) 0.0518	(0.0035-West) 0.0027
3.871–44.6	6.950	0.1968	0.0066
0.1–3.869	1.743	0.0046	0.00001
Partial continuum sum	16.53	3.5267	1.907
Total	17.357	4.37	2.78

The f_n values in the subionization region are from electron energy loss experiments performed by Natali *et al.*[18] They provide a partial $S(0)$ sum of 0.827, which, when combined with the ionization portion (16.53), yields $S(0) = 17.357$.

We begin to see too low an $S(0)$, indicating insufficient data in the high energy region. Just as we found $S(+1)$ and $S(+2)$ not useful after helium, we shall find $S(0)$ only roughly helpful for high Z atoms and molecules.

The value of $S_i(-1)$ provided by the above table is 3.53, while the subionization contribution to $S(-1)$ is 0.84, or $S(-1) = 4.37$. The best correlated wave function yields a value[19] for $S(-1)$ of 4.3. Rieke and Prepejchal[13] report 4.22 for $|M_i|^2$, which is clearly much closer to $S(-1)_{total}$ than to $S_i(-1)$, a behavior already encountered for He and Ne.

The contributions to $S(-2)$ from the discrete (0.872_5) and continuum (1.907) regions yield $S(-2) = 2.78$, or $\alpha_N = 1.647 \times 10^{-24}$ cm^3. The currently recommended value[14] is $1.641 \pm 0.002 \times 10^{-24}$ cm^3, based on dielectric constant measurements.

Lee[26] has very recently computed f_n values for the discrete argon spectrum using a variational quantum defect method.

Although his individual f_n values differ by more than 10% from Natali *et al.*[18], in some cases, his partial sums are remarkably close. Lee's f_n provide 0.8198, 0.8399, and 0.8717 for the discrete contributions to $S(0)$, $S(-1)$, and $S(-2)$, respectively, which are seen to be within 1% of the values in Table 3.

The recently reported data of West and Marr[15] are a mixed blessing. Between 50.6 and 175 Å, they are in good agreement with values of Lukirskii and Zimkina,[22] and of Watson.[27] Between 40 and 50.6 Å, they provide a significant increase in $S(0)$—about 0.3 units—the net effect in the interval 40–175 Å being to raise $S(0)$ to 17.78. This is a significant improvement over the earlier result (17.36), relative to the required value (18.00). In this same wavelength interval, $S(-1)$ becomes 4.394, and $S(-2)$ is virtually unaffected.

However, the values of West and Marr[15] are substantially larger than the earlier results at $\lambda > 175$ Å, and here they begin to cause overshoot in all three sum rules. The earlier results were obtained with line sources, and are less likely to be affected by scattered light in the neighborhood of a Cooper minimum. Hence, we prefer the earlier values for $175 \text{ Å} < \lambda < 350$ Å. It may be significant that these authors have also chosen other values for this wavelength range in their compilation of best values.

4. Krypton

Our detailed knowledge begins to diminish significantly in the high energy region as we proceed to this 36-electron system. From threshold to 209 Å,

Samson's[3] results can be accepted to the same degree of reliability as they were for the lighter noble gases. The x-ray absorption data of Lukirskii *et al.*[28] effectively cover the region from ~250 to 23 Å. From ~138 to 31.0 Å, the DESY synchrotron results[29] complement those values and provide evidence for structure undetected in the x-ray line source experiments. Samson quotes values of Ederer and Tomboulian at 170 and 185 Å. Henke's recent[30] tabulation carries our knowledge from 31.6 to 8.34 Å with the same reliability. Below 8.34 Å, the data of Wuilleumier[31] have been utilized out to 1.5 Å. In order to estimate the $S(0)$ contribution of the 2s–2p shell, the calculated values of Veigele *et al.*[32] have been used between 0.866 and 7.40 Å. As will be noted, these shells contribute very slightly to $S(-1)$ and $S(-2)$, but the $S(0)$ contribution is roughly correct and makes the entire $S(0)$ sum more reasonable. The above considerations are summarized in Fig. 5.

As with argon, krypton also manifests autoionization structure between $^2P_{1/2}$ and $^2P_{3/2}$ thresholds near the onset of ionization. Since the spin-orbit splitting is significantly larger for krypton, it is easier to resolve the individual resonances, and in fact to observe sharp and diffuse components corresponding to s-like and d-like quasi-discrete states (see Chapter VI, Fig. 10 and Huffman *et al.*[33]). Below ~700 Å, the partial cross section due to ejection

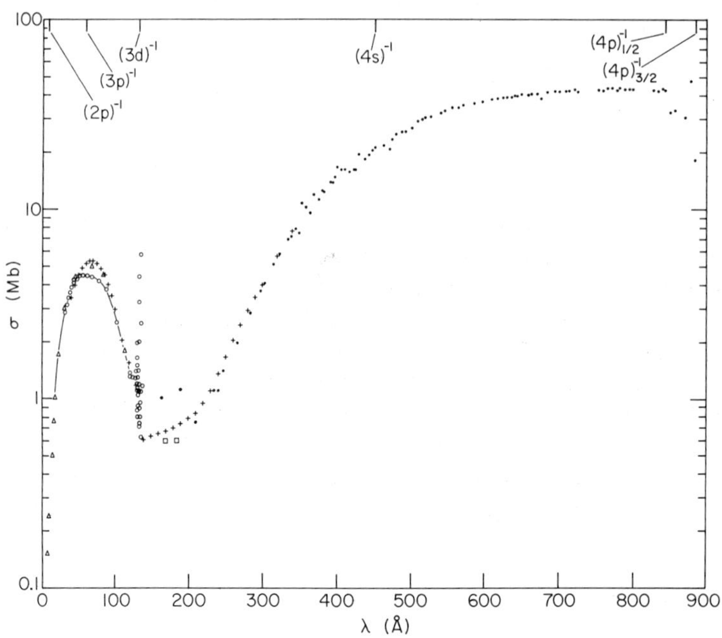

Fig. 5 Experimental photoabsorption cross section of krypton above the ionization threshold. ● Samson[3]; ○ Haensel *et al.*[29]; △ Henke[30]; □ Ederer and Tomboulian[16]; + West and Marr.[15]

of a 4p electron diminishes (as with the 3p of argon) and approaches a Cooper minimum at about 175 Å. The excitation of 4s electrons to Rydberg levels again gives rise to resonances having q parameters ~0, beginning at ~500 Å and extending to the threshold for 4s ejection at ~460 Å (see Samson[34]; Madden and Codling[35]).

At ~135 Å we encounter the first example in this experimental section of ionization from a d-like orbital. It will be recalled that the larger l value of such an orbital gives rise to an angular momentum barrier (see Chapter IV, Section D) separating inner well states and outer well states. The excitation to inner well states appears as resonances below the threshold for d ejection and these have been observed in the DESY experiments and are seen in Fig. 5. Because of the need for the wave corresponding to the departing electron to overcome the barrier, d-electron ejection is delayed and reaches its maximum probability at ~55 Å, about 100 eV above threshold.

The piecewise contributions to the $S(k)$ sums are tabulated in Table 4. The values of the individual f_n below the ionization potential are from

TABLE 4 *The Oscillator Strength Distribution in Krypton*

Energy region	$S(0)$	$S(-1)$	$S(-2)$
Discrete (eV)			
10.032	0.212	0.2874	0.3896
10.633	0.191	0.2443	0.3125
12.037	0.0044	0.0050	0.0056
12.354	0.0817	0.0899	0.0990
12.385	0.152	0.1669	0.1833
12.869	0.0138	0.0146	0.0154
13.004	0.0420	0.0439	0.0459
13.036	0.0056	0.0058	0.0061
13.099	0.119	0.1236	0.1283
13.114	0.048	0.0498	0.0516
13.350	0.0024	0.0024	0.0025
13.442	0.0295	0.0298	0.0302
13.436	0.0290	0.0294	0.0297
⋮			
13.996	0.168	0.1632	0.1586
Partial sum	1.10	1.26	1.458
Continuum (Å)			
209.3–883.6	5.897	3.647	2.442
137.8–209.3	0.4006	0.0703	0.0126
7.40–137.8	18.20	0.8158	0.0504
0.866–7.40	7.765	0.0395	0.0002
0.01–0.866	1.645	0.0009	0.000
Partial continuum sum	33.91	4.57	2.505
Total	35.01	5.83	3.963

electron energy loss measurements at the National Bureau of Standards (Natali *et al.*[18]).

The $S(0)$ sum for the continuum is 33.91 and for the discrete portions 1.10, or $S(0)_{\mathrm{total}} = 35.0$, which lacks one unit. The $S(-1)$ sum for the continuum is 4.57, significantly less than Rieke and Prepejchal's[13] experimental $|M_i|^2 = 6.09$. In fact, this latter number is larger than the total $S(-1)$ of this compilation, 5.83. It appears once more as if all of the discrete states are providing ionization paths in the electron ionization experiment. The $S(-2)$ partial sums from discrete (1.458) and continuum (2.505) contributions imply $\alpha_N = 2.35 \times 10^{-24}\ \mathrm{cm}^3$.

The recommended[14] experimental value for α_N based on refractive index data is $2.480 \times 10^{-24}\ \mathrm{cm}^3$. A 4–5% increase in the discrete portion would seem desirable, although the accuracy of the cross sections in the near continuum may also be questionable to this extent. Electron energy loss measurements by Geiger[36] result in still smaller values for the discrete portion.

The recent data of West and Marr,[15] also plotted in Fig. 5, provide a welcome addition to the otherwise sparsely studied region around the Cooper miminum (135–250 Å). Their values around 70 Å also alter the shape of the d absorption. The new data narrow the discrepancy in the $S(0)$ sum, but not the $S(-2)$ sum. Thus, utilizing the data of West and Marr[15] between 40 and 210 Å, we compute $S(0) = 35.41$, $S(-1) = 5.88$, and $S(-2)$ is virtually unchanged. The primary effect is to increase $S(0)$ by 0.41 units, leaving us 1.6% shy of the predicted value.

5. Xenon

Most of the experimental results for this 54-electron system have been reported in the last decade. In Fig. 6 and the ensuing sum rule analysis, Samson's[3] data have been used from the 5p ionization threshold to 209.3 Å. From 209.3 to ~19 Å, the values of Ederer[37] and Lukirskii *et al.*[28] have been substantially supplemented by the recent DESY synchrotron data.[29] A section from 18.37 to 15.69 Å has been extracted from Deslattes.[38] The x-ray data of Wuilleumier[31] have been adopted between 1 and 15 Å. The values quoted for $\lambda < 0.1$ Å are unreliable.

The detailed description of the absorption curve (Fig. 6) follows closely that of krypton; autoionization between $^2P_{3/2}$ and $^2P_{1/2}$ thresholds (see Huffman *et al.*[39] and Chapter VI, Fig. 14), followed by the characteristic decline of the 5p partial cross section toward a Cooper minimum at ~200 Å. The opening of the 5s excitation channel gives rise to window resonances commencing at ~596 Å and terminating in the 5s ejection threshold at ~530 Å. The 4d excitation to inner well states begins at 190.4 Å, followed by the 4d ejection threshold at ~183.6 Å and the peak of 4d ionization at ~125 Å. We have already encountered the detailed structure of this peak

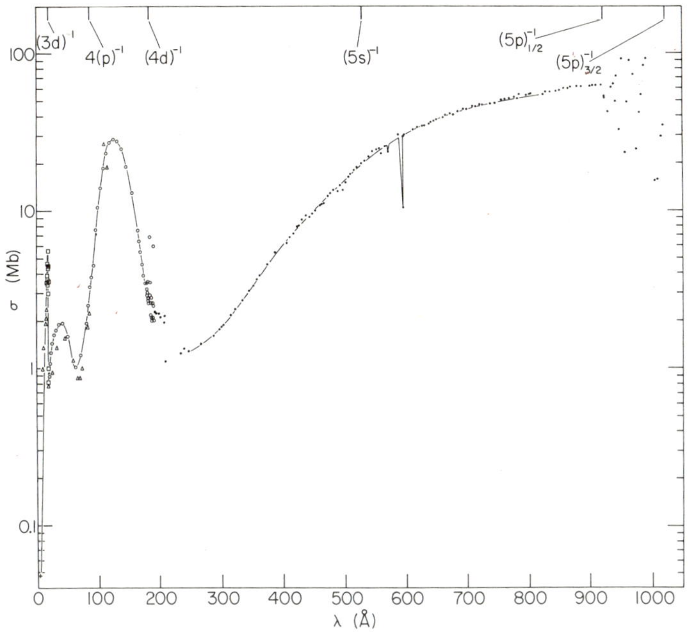

Fig. 6 Experimental photoabsorption cross section of xenon above the ionization threshold. ● Samson[3]; ○ Haensel *et al.*[29]; □ Deslattes[38]; △ Henke.[30]

in Fig. 12 of Chapter IV. The peak at ~40 Å is presumably a post–Cooper minimum rise, which may be due to 5p, 4d, or 4p. The 3d edge is also clearly seen. Table 5 summarizes the contributions of the various spectral regions to $S(k)$.

The f_n in the subionization region are from recent data by Geiger[36] and differ slightly from the NBS data.[18] The $S(0)$ sum for the discrete transitions (1.640) when added to the integrated continuum (51.08) yields 52.72, respectably close to the desired value of 54. As already noted, the deep x-ray region is quite uncertain. The continuum contribution to $S(-1)$, or $S_i(-1) = |M_i|^2 = 6.12$, while the discrete portion adds 2.13, or a total of 8.25. The Rieke–Prepejchal[13] electron ionization experiment reports 8.04, which is again much closer to $S(-1)$ total than to the computed $|M_i|^2$. It is more difficult to explain this behavior in xenon than in the light noble gases, because the discrete excitations in xenon do not provide enough energy to ionize many impurities. Some of the additional ionization may result from the chemi-ionization process

$$\mathrm{Xe}^* + \mathrm{Xe} \rightarrow \mathrm{Xe_2}^+ + \mathrm{e}.$$

The $S(-2)$ contribution of continuum (3.55) and subionization region (2.86) imply $\alpha_N = 3.78 \times 10^{-24}\ \mathrm{cm}^3$. The recommended[14] experimental value based on refractive index data is $\alpha_N = 4.044 \times 10^{-24}\ \mathrm{cm}^3$.

TABLE 5 *The Oscillator Strength Distribution in Xenon*

Energy region	$S(0)$	$S(-1)$	$S(-2)$
Discrete (eV)			
8.437	0.26	0.4193	0.6762
9.570	0.19	0.2701	0.3840
9.917	0.0095	0.0130	0.0179
10.401	0.395	0.5167	0.6759
10.593	0.0968	0.1243	0.1597
10.979	0.0025	0.0031	0.0038
11.163	0.0862	0.1051	0.1281
11.274	0.0236	0.0285	0.0344
11.423	0.0217	0.0258	0.0308
11.495	0.004	0.0047	0.0056
11.583	0.006	0.0070	0.0083
11.607	0.205	0.2403	0.2817
11.683	0.0096	0.0112	0.0130
11.740	0.123	0.1425	0.1652
11.752	0.0204	0.0236	0.0273
11.807	0.0042	0.0048	0.0056
11.840	0.0625	0.0718	0.0825
11.852	0.0025	0.0029	0.0033
11.878	0.029	0.0332	0.0381
11.892	0.0035	0.0040	0.0046
11.908	0.0386	0.0441	0.0504
11.918	0.005	0.0057	0.0065
11.955	0.025	0.0285	0.0324
11.989	0.016	0.0182	0.0206
Partial sum	1.640	2.15	2.86
Continuum (Å)			
209.3–1018.5	5.459	4.087	3.318
82.65–209.3	11.01	1.506	0.2123
2.59–82.65	26.90	0.5126	0.0163
0.359–2.59	6.18	0.0110	0.000022
0.001–0.359	1.53	0.0003	0.0000
Partial continuum sum	51.08	6.12	3.55
Total	52.72	8.27	6.41

These results, as with krypton, may be due to some missing oscillator strength in the low energy region (discrete or near continuum) or to a failure of the sum rules for high Z atoms.

6. Mercury

This is the most complex atomic gas we shall examine. The data available are far from complete, but it is a useful exercise to assemble the pieces, and we shall find the sum rules helpful in assessing them.

The subionization region has been the subject of three recent papers by Skerbele and Lassettre.[40] From these electron impact energy loss studies, a more or less complete set of f_n values for the discrete region is available. A large, broad autoionization peak appears just above threshold at 1126.6 Å. The absolute photoabsorption cross section of this peak has been reported by Lincke and Stredele,[41] and the integrated oscillator strength in this vicinity evidently agrees with the revised results of Skerbele and Lassettre[40] in 1972. Two photoionization mass spectrometric studies by Brehm[42] and Berkowitz and Lifshitz[43] report relative cross sections which include this 1126.6 Å peak. Hence, it is possible to place these relative cross sections on an absolute scale using the Lincke and Stredele data. Finally, Cairns *et al.*[44] have reported approximate absolute photoionization cross sections using line sources that extend the data from ~18 to 72 eV. The region from the ionization threshold to 72 eV is plotted in Fig. 7. This energy range encompasses ionization from $(6s)^2$ and $(5d)^{10}$ outer shells. The cross section for ionization of 6s electrons is relatively low, and the huge autoionization peaks correspond to excitation of 5d electrons to Rydberg levels. The 5d shell is split by spin-orbit interaction into $d_{5/2}$ and $d_{3/2}$ subshells; the angular momentum selection rule $\Delta l = \pm 1$

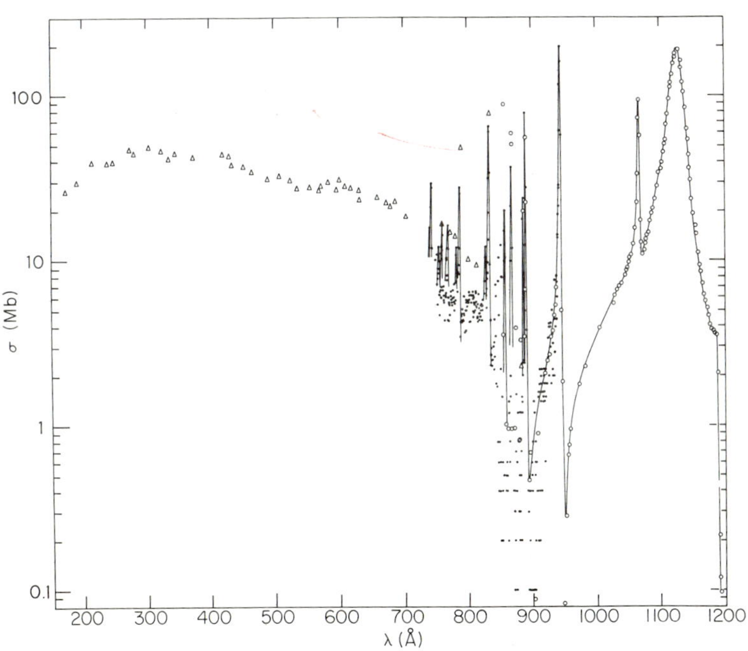

Fig. 7 Experimental photoabsorption cross section of atomic mercury above the ionization threshold. ○ Brehm[42]; ● Berkowitz and Lifshitz[43]; △ Cairns *et al.*[44]

permits transitions to p and f one-electron states. Hence the autoionizing series will consist of singlet and triplet states, further characterized by the coupling of p and f electrons, respectively, with a $(5d)^9$ core. The spectroscopic observations and assignments were first made by Beutler[45] and Beutler and Guggenheimer.[46] More recent assignments have been reported by Garton and Connerade[47] and by Mansfield.[48]

The strong one-electron transitions converge to an ionization potential corresponding to a $^2D_{5/2}$ ion at 14.840 eV ≡ 835.5 Å, and another corresponding to a $^2D_{3/2}$ ion at 16.704 eV ≡ 742.2 Å. To higher energy, direct ionization from the d shell dominates, exhibiting a slow rise from threshold characteristic of a delayed onset due to the centrifugal barrier (Chapter IV, Section D). Double ionization becomes possible in this region, and the cross section taken from Cairns *et al.*[44] includes both single and double ionization contributions.

The sum rule analysis of this limited region of the Hg spectrum is detailed in Table 6.

TABLE 6 *The Oscillator Strength Distribution of Hg*[a]

Energy region	S(0)	S(−1)	S(−2)
Discrete (eV)			
6^3P_1(4.886)	0.0285	0.0793	0.2208
6^1P_1(6.703)	1.18 ± 0.07	2.394	4.858
7^1S_0(7.926)	~0.004	0.0069	0.0118
7^1P_1(8.838)	0.049	0.0754	0.1160
8^1P_1(9.529)	0.021	0.0300	0.0428
$6p'^1P_1$(9.771)	0.24	0.334	0.465
9^1P(9.914)	0.115	0.1578	0.2164
10^1P(10.061)	0.029	0.0392	0.0530
11^1P(10.158)	0.012	0.016	0.0215
Partial sum	1.6785	3.1326	6.0053
Continuum (including autoionization) (Å)			
1073–1188.7	0.5147	0.6345	0.7824
949.0–1073	0.0906	0.1035	0.1184
894.2–949.0	0.0710	0.0733	0.0756
833.0–894.2	0.05014	0.0475	0.04506
788.0–833.0	0.05468	0.04893	0.04380
738.0–788.0	0.08326	0.06959	0.05819
Partial continuum sum to 738 Å	0.8644	0.9773	1.1234
(172.0–738.0	18.56	6.445	2.576)[b]

[a] To 172 Å.

[b] Based on Cairns *et al.*[44] Sum rule analysis indicates that these values should be reduced by approximately factor two.

One immediate and striking feature of this table is the large $S(0)$ sum. Our value to 172.0 Å is ~21, of which 18.56 involve solely the data of Cairns *et al.*[44] In this energy range, the only orbitals accessible are $(5d)^{10}$ and $(6s)^2$. While there is no rigorous sum rule that operates for each shell, the interaction between the next deeper shells and the ones considered appears slight, and our partial sum appears about a factor of two too large. There is indication from Fig. 7 that the higher energy data of Berkowitz and Lifshitz,[43] placed on an absolute scale by normalization to the Lincke and Stredele absolute values, lie about a factor of two lower than the results of Cairns *et al.*[44] in the region of overlap between ~750 and −825 Å, whereas the data of Brehm[42] and Berkowitz and Lifshitz[43] are more nearly consistent. Hence, the $S(0)$ sum rule strongly suggests reducing the high energy values ($\lambda < 750$ Å) by a factor of approximately two.

The partial $S_i(-1)$ value is 7.422, which may be compared with an electron impact ionization measurement (Riecke and Prepejchal[13]) of $|M_i|^2 = 5.69 \pm 0.94$.

If we halve the portion below 738 Å, as suggested by the $S(0)$ analysis, we obtain $S_i(-1) \cong 4.2$ which may be about right since there will undoubtedly be some significant contribution to $S(-1)$ for $\lambda < 172$ Å.

The partial $S(-2)$ sum has a large contribution from the subionization region, and is much less sensitive to the data below 750 Å. The calculated partial $S(-2) = 9.704$ implies $\alpha_N = 5.75 \times 10^{-24}$ cm^3. The most reliable experimental results[14] lie between 5.03 and 5.17×10^{-24} cm^3. If we again halve the contribution below 738 Å, we compute $\alpha_N = 4.99 \times 10^{-24}$ cm^3 which seems reasonable since it allows for a small contribution from $\lambda <$ 172 Å.

In summary, this synthesis of the mercury oscillator strength from diverse sources appears to be fairly reliable to λ ~750 Å, and the higher energy data of Cairns *et al.*[44] would be about right if they were uniformly reduced by approximately a factor of two.

B. Diatomic Molecules

1. H_2

The photoabsorption cross section from the approximate ionization threshold to shorter wavelength is shown in Fig. 8. The hydrogen molecule has no inner orbitals, but autoionization is nonetheless prominent in the near threshold region (15.4–16.7 eV) due to the vibrational autoionization mechanism (see Chapter III, Section D.3). The scale of Fig. 8 is too compressed to permit this display, which is given in considerable detail elsewhere.[49,50] We

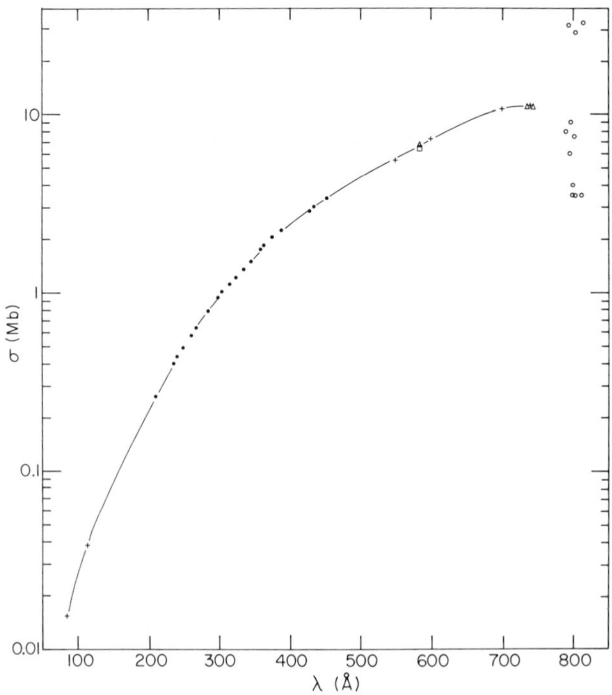

Fig. 8 Experimental photoabsorption cross section of molecular hydrogen. ● Samson and Cairns[57]; + Cook and Metzger[47a]; ○ Comes and Wellern[47b]; △ Rebbert and Ausloos[47c]; □ Bennett *et al.*[99]

shall find it convenient to separate the absorption cross section into three energy regions: (a) The "discrete" region below the ionization potential ($\lambda > 803.749$ Å); (b) The autoionization region from the ionization threshold to (somewhat arbitrarily) 743.7 Å ≡ 16.67 eV; and (c) The higher energy "direct ionization" region.

(a) The major contributors to oscillator strength below the ionization potential are the Lyman and Werner bands, corresponding to transitions to the B(2pσ) and C(2pπ) states, respectively. These can be thought of as the first Rydberg members of the σ and π series, both of which necessarily converge to the ionization potential. Various electron impact energy loss studies have reported oscillator strengths for these transitions,[51,52] but the most reliable values are probably the calculated ones of Allison and Dalgarno[53] based on the transition moments given by Wolniewicz.[54] These authors give 0.311 and 0.356 for the f values of the B and C states, respectively. The next Rydberg members, B′(3pσ) and D(3pπ) are much weaker. From Geiger and Schmoranzer,[52] we obtain $f(B') = 0.052$ and $f(D) = 0.071$.

(b) The autoionization region has many extremely narrow lines; accurate photoabsorption cross sections would require very high resolution, and electron impact energy loss measurements would give a blurred mean value at best. We have elected to assign an f value to this region which is the difference between $S(0) = 2$ and the sum of the f values from (a) and (c). We have apportioned this f value (which we find to be 0.173) to an ionizing component and a nonionizing component as follows:

(i) Dalgarno and Allison[55] have calculated contributions to a photodissociation continuum in this region due to transitions to the states $B'^1\Sigma_u^+$, $C^1\Pi$, and $B^1\Sigma_u^+$. Mentall and Gentieu[56] have provided experimental verification of the magnitude of this direct dissociation. We have computed $f =$ 0.0126 as the contribution of this dissociative continuum to the 15.4–16.67 eV region.

(ii) The remainder ($0.173 - 0.012_6$) is apportioned by utilizing the data of Chupka and Berkowitz.[49] In this photoionization study, one notes an underlying direct ionization continuum, upon which are superimposed autoionization peaks. The relative ionization yield of these peaks is also reported. We have directly integrated the continuum and the ionization peaks. Then, assuming the maximum ionization yield observed for these peaks is unity, an equivalent total absorption area has been computed from the individual ionization yields. The total area has been normalized to $f = 0.173 - 0.0126 =$ 0.1604. From the foregoing analysis, the fractional portion of this f-value leading to ionization is 0.722. (More detailed considerations regarding this region are presented in Chapter VI, Section C.1.b.)

(c) Experimental data in the energy region above 16.67 eV are not very extensive. Samson and Cairns[57] provide the major source between ~21 and 60 eV. Values at higher energy have been obtained from Henke[6,30] which check reasonably well with those of Denne.[58] The partial f-sum for the region 16.67 eV $\to \infty$ is found to be 1.037.

From the total $S(k)$ in Table 7, we are now in a position to test the oscillator strength distribution. We have already utilized the $S(0)$ condition to fix the contribution of the autoionization region. The $S(-1)$ sum can be tested by the theorem $S(-1) \equiv \int \psi_0{}^* r^2 \psi_0 \, d\tau$, (described in Chapter IV, Appendix 2) which a number of authors[13,59,60] have computed with highly accurate values of the ground state wave function of H_2. These values, averaged over internuclear distance, range from 1.532 to 1.568, very close to our computed value 1.556. The electric-dipole polarizability has been computed by Kolos and Wolniewicz[61] and found to be 0.805×10^{-24} cm^3. From $S(-2) = \alpha_N / 4a_0{}^3$ we obtain $S(-2) = 1.358$, within 1.5% of our computed value 1.377. The value for $S(+1)$ (which we shall not find very useful for other molecules because of its great sensitivity to the high energy spectrum) has been

TABLE 7 *The Oscillator Strength Distribution in H_2*

Energy region (eV)	$S(+1)$	$S(0)$	$S(-1)$	$S(-2)$
Lyman bands (12.1)	0.277	0.311	0.350	0.393
Werner bands (12.7–14.4)	0.332	0.356	0.381	0.408
B′, D bands (14.1–14.4)	0.128	0.123	0.117	0.111
(Higher Rydberg levels assumed too weak to contribute significantly below IP)				
Autoionization region (15.4–16.67)	0.205	(0.173)[a]	(0.146)	(0.123)
Ionization continuum (16.67–∞)	2.367	1.037	0.564	0.342
Total	3.309	2.000	1.558	1.377

[a] The autoionization contribution has been fixed (see text) to make $S(0) = 2.0$.

computed by various authors[59,60] to be 3.40 compared to our computed sum of 3.309.

A further useful comparison can be made with $|M_i|^2$ independently measured in high energy electron impact ionization measurements, which corresponds to that portion of $S(-1)$ which results in ionization. We have previously determined that 0.722 (0.1604) = 0.116 is the appropriate contribution from the autoionization region. The ionization continuum region should contribute fully to $|M_i|^2$, since the dissociation continua are weak and this energy region is outside the Franck–Condon region for vibrational autoionization (see Chapter III, Section B). Our result is 0.116 + 0.564 = 0.680, compared to the experimental value (Rieke and Prepejchal[13]) of 0.695 ± 0.015.

Gerhart[62] has modified the above oscillator strength distributions slightly, to enforce agreement not only with the sum rules discussed above, but with $S(k)$ sums for higher negative k. However, it can be appreciated that even the above procedure, which forces the fit only for $S(0)$ agrees within 2% with most criteria, and within 3% for $S(+1)$, constituting strong evidence for the general reliability of the distribution. A problem with the fitting procedure is that it is not unique.

2. N_2

The photoabsorption cross section from the vicinity of the first ionization threshold to successively shorter wavelengths below the K edge is given in Fig. 9. The near-threshold region (740–796.7 Å) is dominated by intense autoionization structure (see Chapter VI, Fig. 32), only a few points of which are shown on this compressed scale. This autoionization is of the electronic type (see Chapter III, Section D.1) and will be described in more detail in

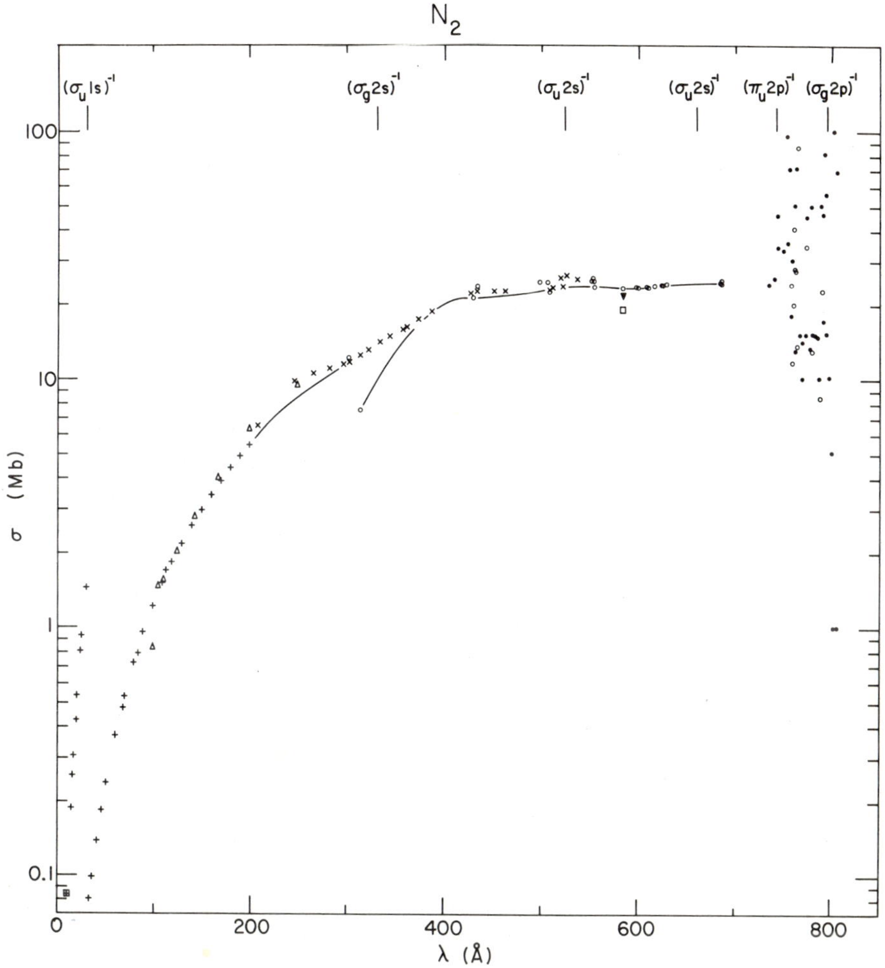

Fig. 9 Experimental photoabsorption cross section of molecular nitrogen above the ionization threshold. ● Carter[62a]; ○ Samson and Cairns[63]; × Samson and Cairns[57]; ——— Sasanuma *et al.*[65]; + Henke *et al.*[6]; △ de Reilhac and Damany[64]; □ Bennett *et al.*[99]; ▼ Brolley *et al.*[65a]

Chapter VI. Below 740 Å, the data of Samson and Cairns[57,63] are taken as most reliable to ~200 Å, where they merge rather smoothly with values from de Reilhac and Damany,[64] with direct measurements at some wavelengths by Henke *et al.*,[6] and with twice the atomic values taken from interpolated values of Henke.[30] The results of the Tokyo synchrotron group[65] deviate significantly from the selected data below 375 Å, possibly because of scattered light problems. The Stoughton synchrotron results[66] have quoted errors of $\pm 5\%$ in the 520–700 Å region, and hence are probably not quite as reliable as those of Samson and Cairns. The recent Glasgow synchrotron values[67] are quoted to $\pm 3\%$ accuracy in the 380–700 Å region, and are stated by the authors to be in excellent agreement with the Samson and Cairns measurements. Since the latter were taken with line sources, which largely avoid the order-sorting and scattered light problems that must be faced with synchrotron light sources, we have chosen to retain these older measurements for our subsequent analysis.

The adiabatic onsets for the ejection of single electrons from the shells $\sigma_g 2p$, $\pi_u 2p$, $\sigma_u 2s$, $\sigma_g 2s$, and the nearly degenerate $\sigma_u 1s$ and $\sigma_g 1s$ are marked at the top of the figure. These ionization potentials have been taken from ultraviolet photoelectron spectroscopy[68] and from x-ray photoelectron spectroscopy.[69]

We have previously noted (Chapter IV, Section B) that Rydberg series should manifest themselves in the energy region approaching each successive ionization limit. We can see evidence for this in the complicated structure of Fig. 9 approaching $(\pi_u 2p)^{-1}$, in the data of Lee *et al.*[66] for the region approaching $(\sigma_u 2s)^{-1}$, and the region approaching the double excitation state (C state of N_2^+) around 500 Å. Although this behavior has not yet been observed in the region approaching the $(\sigma_g 2s)^{-1}$ ionization, it has been detected near the K-shell edge $(\sigma_g 1s)^{-1}$, $(\sigma_u 1s)^{-1}$ by the Tokyo synchrotron group[70] and others.[71] Excitation of a 1s-like electron of N_2 to unoccupied and Rydberg levels results in a set of states analogous to NO*. The excited electron "sees" an inner core with one extra positive charge, due to the absent 1s electron, which is very similar to the NO inner core, where the extra charge occurs in the nucleus itself.

Note added in proof: This region has now been examined with photoelectric detection by Bianconi *et al.*[71a] The $(1s)^{-1}$ threshold is at 409.9 eV. Below this threshold (at 400.8 eV) a large, narrow peak is observed, having a maximum cross section of ~17.4 Mb. It corresponds to the first member of the Rydberg series having π_g upper states. Its dominant d-symmetric term (partial wave expansion—see Chapter VII) is coupled by the molecular field to the p-wave which is favored by photoabsorption of a 1s-like electron in an essentially atomic field. Evidence for some higher members of this series is also apparent in this spectrum. In addition, there is a broad "shape resonance" at ~420 eV, i.e. above the 1s threshold, which has the effect of redistributing the oscillator strength in the 1s continuum region.

The selected absorption cross sections for the region below 740 Å (which is essentially continuum with minor autoionization structure) are summarized in Table 8. For the sum-rule analysis, we shall again find it convenient to divide the spectrum into (a) subionization with $\lambda > 796.7$ Å, (b) autoionization with $796.7 > \lambda > 740$ Å, and (c) primarily continuum ionization with $0 < \lambda < 740$ Å.

TABLE 8 *Selected Values of Absorption Cross Sections for* N_2[a]

λ(Å)	σ(Mb)	λ(Å)	σ(Mb)	λ(Å)	σ(Mb)
740	25	463.7	22.6	200	5.40
735	24	452.2	22.6	190	4.88
685	24.4	434.3	22.4	180	4.37
629.7	24.2	387.4	18.6	170	3.86
625.8	24.0	374.4	17.3	160	3.40
617	23.7	362.9	16.1	150	2.98
610.8	23.3	358.5	15.7	140	2.56
608.4	23.4	345.1	14.8	130	2.18
599.6	23.4	335.1	14.0	120	1.83
597.8	23.4	323.6	13.1	110	1.50
584.3	23.1	314.9	12.4	100	1.21
555.3	24.8	303.1	11.6	90	0.95
554.5	24.6	290	10.7	80	0.73
554.0	25.3	280	10.1	70	0.53
537	25.2	270	9.55	60	0.37
525.8	26.2	260	9.00	50	0.24
522.2	23.6	250	8.40	40	0.138
519.6	25.8	240	7.75	35	0.099
512.1	23.2	230	7.20	30	1.43
508.2	22.8	220	6.70	25	0.926
		210	6.10	20	0.535
				15	0.256
				9.89	0.083

[a] From $\lambda = 0 - 740$ Å.

The primary sources of data for these three regions, together with parenthetical notes, are given below.

(a) From Lassettre and Skerbele[72] we note that the dominant contributions are primarily from the $b^1\pi_u$ state at ~12.85 eV, having $f = 0.682$, and the $b'^1\Sigma_u^+$ state at ~13.99 eV, having $f = 0.471$.

(b) The autoionization information, best garnered from Carter's high resolution photoabsorption data giving individual oscillator strengths for the various speaks, yields $f = 0.3299$. See also the more recent data of Gürtler *et al.*[72a] in this region.

(c) The integrated continuum region, based on the selected cross sections in Table 8, yields $f = 12.1833$.

The corresponding contributions to the other $S(k)$ sums are listed in Table 9.

The $S(0)$ sum is seen to be ~3% less than the 14.0 required. The $S(-2)$ sum implies $\alpha_N = 1.73 \times 10^{-24}$ cm^3, which is in excellent agreement with the experimental[73] value of 1.74×10^{-24} cm^3. This suggests that the missing oscillator strength contribution to $S(0)$ is at relatively high energy, where its

TABLE 9 *The Oscillator Strength Distribution in N_2*

Energy region	$S(0)$	$S(-1)$	$S(-2)$
Primarily $b^1\pi_u$ ($\sim$12.85 eV)	0.682	0.722	0.763
Primarily $b'\ ^1\Sigma_u^+$ ($\sim$13.99 eV)	0.471	0.458	0.445
Autoionization region (740–796.7 Å)	0.329_9	0.278_6	0.235
Ionization continuum ($\lambda < 740$ Å)	12.183_3	3.153_6	1.476
Total	13.67_2	4.61	2.919

effect on $S(-2)$ will be negligible. It is tempting to attribute this missing oscillator strength to the large peak at 400.8 eV recently observed.[71a] Its contribution to the $S(0)$ sum would be $\sim$0.2, which would reduce the discrepancy with theory to less than 1%. As surmised, the corresponding additions to $S(-1)$ of $\sim$0.007, and to $S(-2)$ of $\sim$0.0002, are negligible.

A highly accurate ground state wave function for N_2, of sufficient quality to test the $S(-1)$ sum by the theorem $S(-1) = \int \psi_0^*(\Sigma_i r_i^2)\psi_0\, d\tau$, is not yet available.

An overestimate of $|M_{\rm i}|^2$, which we designate as $|M_{\rm I}|^2$, can be obtained by evaluating the $S(-1)$ contributions from the first ionization threshold at 796.7 Å to shorter wavelength. From Table 9, this value is 3.432. In Fig. 10, the ionization yield of N_2 is plotted for the region between $300 < \lambda < 796.7$ Å. It is seen that for most of this region the yield is unity. The analysis of the autoionization peaks by Carter and Berkowitz[74] enables us to estimate the fraction of absorption in the near-threshold region which does not lead to ionization. In this way, we obtain a diminution from $|M_{\rm I}|^2 = 3.432$ to $|M_{\rm i}|^2 = 3.3$. This value is significantly lower than the directly measured

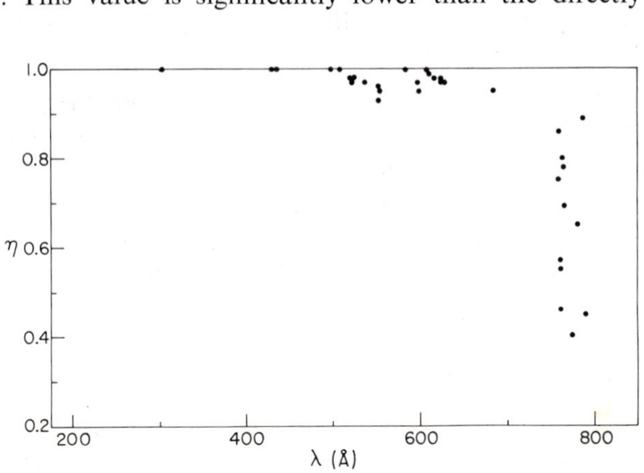

Fig. 10 The quantum yield of ionization (η) experimentally observed in the photoabsorption of molecular nitrogen. (Data are from Samson and Cairns.[63])

$|M_i|^2 = 3.74 \pm 0.14$ reported by Rieke and Prepejchal.[13] We have speculated[75] that the discrepancy may be due to chemi-ionization processes contributing to the ionization observed by Rieke and Prepejchal.

The general agreement with sum rules is quite good, the discrepancy for $S(0)$ and $S(-2)$ being $\leq 3\%$, which is perhaps more accurate than the available measurements can be expected to give.

3. O_2

Figure 11 summarizes photoabsorption data from selected sources from the vicinity of the ionization threshold to successively shorter wavelengths below the K edge. A rather large section of the ionization region beyond threshold ($\sim$500–1027 Å) is either dominated or influenced significantly by autoionization structure. As far as is currently known, this autoionization is of the electronic type (Chapter III, Section D.1) and is described further in the chapter on partial cross sections. Below $\sim$500 Å, autoionization structure ceases to be prominent.

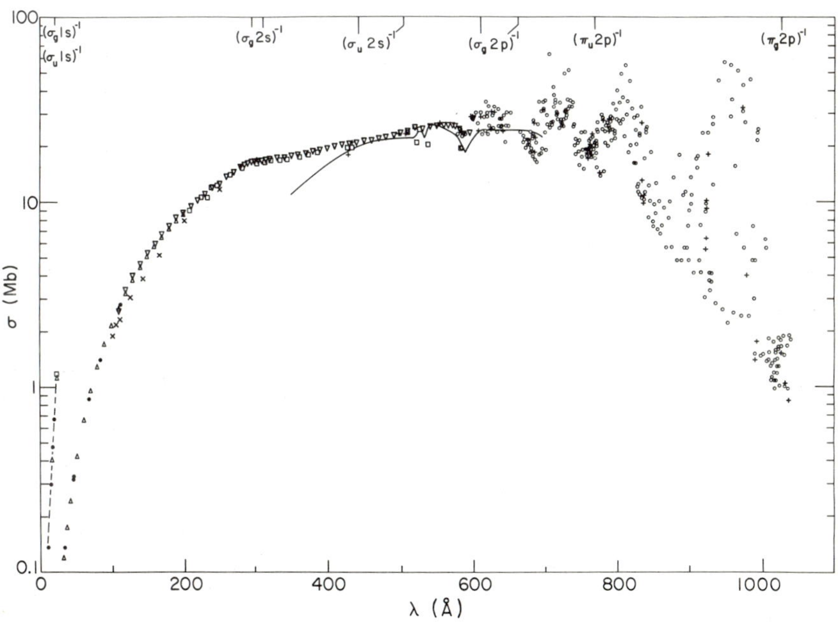

Fig. 11 Experimental photoabsorption cross section of molecular oxygen from the ionization threshold thru the K edge. + Samson and Cairns[63]; □ Samson and Cairns[57]; ○ Matsunaga and Watanabe[78]; ——— Sasanuma *et al.*[65]; △ Henke *et al.*[6]; × de Reilhac and Damany[64]; ▽ Samson *et al.*[57a]

The very recent data of Samson *et al.*[57a] supersede earlier results (Samson and Cairns[57]) and merge smoothly with x-ray data from Henke *et al.*,[6] both directly obtained and interpolated. Cross sections in the 350–400 Å region reported by the Tokyo synchrotron group[65] are significantly smaller, but the Stoughton synchrotron values[66] and those of the Glasgow synchrotron group[76] appear to be in very good agreement. The measurements of de Reilhac and Damany[64] in the 100–250 Å region tend to be a bit lower than these values.

Note added in proof: The more recent tabulation by de Reilhac and Damany[64a] leaves their earlier results[64] essentially unchanged. Mehlman *et al.*[76a] have reexamined the 55–350 Å region, using the NBS storage ring. Their values are in excellent agreement with the data of Samson *et al.*[57a] below ~275 Å, but are about 8% lower to higher λ. Their data exhibit a marked "change of slope" at ~275 Å. Bodeur[76b] provides new values between 9.89 and 17.57 Å which agree well with the earlier data of Henke *et al.*[6] Hence, these more recent experimental results will have a negligible effect on the sum rule analysis presented below.

The adiabatic onsets for the ejection of single electrons from the shells $\pi_g 2p$, $\pi_u 2p$, $\sigma_g 2p$, $\sigma_u 2s$, $\sigma_g 2s$, and the nearly degenerate $\sigma_u 1s$ and $\sigma_g 1s$ are marked at the top of the figure. Since O_2 has only two electrons in the outermost $\pi_g 2p$ orbital, giving rise to a triplet neutral ground state, ejection of a $\pi_g 2p$ electron leads to the formation of $X^2\Pi_g$, the ground state of $O_2{}^+$. Ejection of an electron from any of the inner orbitals leads to both doublet and quartet states, due to coupling of the ensuing "hole" with the angular momentum of the $\pi_g 2p$ electron.

Hence, two ionization potentials have been noted for each orbital, where measurements have distinguished the corresponding energies. The ionization potentials have been taken from Edqvist *et al.*[77] for the uv region and from Siegbahn *et al.*[69] for the x-ray region.

Again we note that the autoionization can, in general, be associated with Rydberg series terminating at the various ionization potentials. The enormous oscillations in cross section values given by Matsunaga and Watanabe[78] for 800 Å $< \lambda <$ 1027 Å appear to damp out at the onset for ejection of a $\pi_u 2p$ electron. A new set of oscillations then begins to appear, and terminates at the $(\sigma_g 2p)^{-1}$ onset. The oscillations converging on $(\sigma_u 2s)^{-1}$ are most clearly seen in the data of Watson *et al.*,[76] although some indication may also be observed in the curve of Lee *et al.*[66] The latter curve does not display autoionization structure in the vicinity of the $(\sigma_g 2s)^{-1}$ threshold, but it again becomes apparent in the region approaching $(\sigma_u 1s)^{-1}$, $(\sigma_g 1s)^{-1}$ as reported by the Tokyo synchrotron group and others.[79]

The photoabsorption cross sections we have selected for the sum-rule analysis are those of Matsunaga and Watanabe[78] for the region 600 $< \lambda <$

1027 Å, Samson *et al.*[57a] for 110 Å $< \lambda <$ 600 Å, and the interpolated values of Henke *et al.*[6] below 110 Å. We rather arbitrarily divide the spectral regions for purposes of analysis into (a) subionization; (b) "autoionization region" from 1027 to 750 Å; and (c) "continuum ionization" below 750 Å.

(a) Lassettre and Skerbele[72] found two major bands contributing to oscillator strength below the first ionization potential. The first, the Schumann–Runge bands at ~8.4 eV, are assigned a corrected oscillator strength of 0.179; the next, at ~9.9 eV, is reported to have $f = 0.020$.

(b) The integrated values of Matsunaga and Watanabe between 750 and 1027 Å yield $f = 0.7241$.

(c) The integrated continuum region, based on the selected cross sections in Table 10, gives $f = 15.784$.

The corresponding contributions to the other $S(k)$ sums are listed in Table 11.

The $S(0)$ sum is 4.4% larger than the desired 16.0. The $S(-2)$ sum implies $\alpha_N = 1.68 \times 10^{-24}$ cm^3, about 7% larger than the experimental value[73] of 1.57×10^{-24} cm^3. It is mildly disturbing to note the rather good agreement among several groups reporting recent results in the continuum region, and still obtain an $S(0)$ sum which is 4.4% too high. The data of Watson *et al.*[76] are, if anything, still higher. The $S(-2)$ sum deviates somewhat further from the direct measurement of polarizability. The implication is that we have somewhat too much oscillator strength either in the autoionization or subionization portion. In this instance, the enormous gyrations in the autoionization region may not have been mapped out sufficiently well, or the integration program may not have been able to handle these rapid oscillations.

A test of $S(-1)$ by means of $\langle \psi_0{}^*(\sum_i \mathbf{r}_i)^2 \psi_0 \rangle$ awaits the determination of an approximately accurate ground state wave function of O_2.

The value of $|M_\mathrm{I}|^2 = 4.473$ can be directly deduced from Table 11. However, we note from Fig. 12 that the ionization yield of O_2 departs from unity over a substantial portion of the measured region, stretching from 300 to 1027 Å. In particular, there is a large dip centering around 820 Å, which is associated with a strong dissociation continuum. By utilizing the photoionization cross sections (instead of photoabsorption cross sections), we can calculate contributions to $S_\mathrm{i}(-1)$, the portion of $S(-1)$ which leads to ionization. In this way, we find for $\lambda <$ 750 Å, $S_\mathrm{i}(-1) = 3.7115$, and for 750 Å $< \lambda <$ 1027 Å, $S_\mathrm{i}(-1) = 0.360_8$.

The sum of these two quantities, $|M_\mathrm{i}|^2 = 4.072$. This is *lower* than the electron impact ionization value[13] of 4.20 ± 0.18, despite the fact that $S(0)$ and $S(-2)$ were somewhat higher than expected. We are once again (as in the case of N_2) led to speculate that other ionization processes may be contributing to the measurements of Rieke and Prepejchal.[13]

TABLE 10 *Selected Values of Absorption Cross Sections for O_2*[a]

λ(Å)	σ(Mb)	λ(Å)	σ(Mb)	λ(Å)	σ(Mb)
1028.1	0.67	970.4	2.4	878.1	13.4
1025.7	1.64	965.5	51.0	875.2	5.6
1025.3	1.79	962.8	6.7	871.4	10.0
1024.6	1.00	961.9	15.6	870.0	8.2
1023.4	1.86	960.0	2.5	867.6	5.6
1022.4	1.30	957.0	35.3	864.6	9.3
1021.6	1.64	956.7	29.0	861.0	7.4
1021.1	1.30	955.9	54.7	859.2	6.7
1020.8	1.60	950.3	2.2	857.3	10.0
1020.4	1.19	947.7	56.2	855.0	7.1
1019.4	1.41	944.6	2.6	853.2	12.6
1018.8	1.08	939.3	45.0	851.8	8.6
1018.3	1.52	935.6	2.8	850.6	9.7
1017.8	0.97	932.4	28.6	849.2	7.4
1017.2	1.60	931.5	12.3	848.5	7.8
1016.9	1.08	930.6	26.0	847.6	7.4
1016.4	1.38	930.0	3.7	845.9	18.6
1016.0	1.19	929.1	4.1	844.6	10.0
1015.8	1.75	928.1	3.3	843.8	12.3
1013.9	1.12	927.6	4.1	842.1	8.2
1013.5	1.34	926.4	3.7	839.1	24.9
1012.3	1.08	924.5	23.4	838.9	23.4
1011.4	1.30	923.5	8.9	838.6	24.5
1009.4	1.52	923.1	9.7	837.8	11.2
1009.1	1.41	920.4	3.0	836.3	17.1
1007.9	1.82	917.2	23.4	835.4	10.0
1006.8	1.38	915.6	4.1	834.5	10.8
1004.6	6.30	914.7	7.4	834.1	10.4
1004.3	5.60	913.5	4.8	832.5	32.7
1004.0	6.30	910.5	17.5	831.0	10.8
1000.0	1.49	910.0	15.2	829.8	23.1
997.2	1.45	909.6	17.1	829.6	22.3
993.5	24.5	906.4	4.1	829.4	22.7
993.2	21.6	903.8	10.8	828.3	11.2
992.9	23.4	902.0	9.7	827.8	12.3
989.6	1.50	901.1	13.4	826.8	11.9
988.5	4.80	900.2	8.9	826.0	29.0
988.0	3.00	898.7	4.8	825.3	20.1
985.9	7.40	897.3	7.4	824.9	24.2
985.2	4.80	895.8	5.6	824.1	16.0
983.3	46.1	894.0	11.2	823.2	28.3
980.5	2.4	893.1	5.6	821.3	16.0
975.3	26.8	891.6	10.8	819.8	31.6
974.5	5.6	889.1	4.8	818.2	20.1
972.9	42.8	885.8	17.9	817.2	45.4
972.5	31.6	883.3	4.8	814.9	18.6

Table 10 (*continued*)

λ(Å)	σ(Mb)	λ(Å)	σ(Mb)	λ(Å)	σ(Mb)
813.7	35.0	767.3	20.1	721.8	27.5
812.5	29.4	766.7	19.0	721.3	26.8
811.8	55.0	765.4	22.3	720.4	34.2
809.3	22.3	764.6	17.5	719.4	24.9
808.2	36.4	763.2	21.9	717.6	29.8
807.0	27.9	762.5	19.3	717.2	27.5
805.1	49.1	762.1	19.7	716.5	30.5
803.5	26.0	761.5	19.3	716.0	29.0
802.6	34.2	760.7	21.2	714.7	35.0
801.6	27.2	759.4	17.1	714.0	33.1
799.5	39.8	758.4	19.3	712.9	37.9
798.1	26.8	758.0	17.5	711.9	32.4
797.7	31.6	756.2	19.3	711.0	41.7
796.0	21.9	756.0	19.0	709.7	25.7
795.2	23.4	755.0	23.4	709.3	26.8
795.0	23.1	752.9	14.9	708.9	25.7
794.1	33.8	751.6	19.0	707.9	26.4
792.8	24.2	751.0	17.5	706.6	24.9
792.4	27.9	750.0	23.8	705.3	63.2
791.3	21.2	748.0	15.6	703.6	26.0
790.0	28.3	747.0	21.2	703.1	29.0
789.0	26.8	746.4	18.6	701.6	19.0
788.6	27.5	745.0	20.5	700.8	35.0
788.0	24.2	743.2	17.9	699.6	23.4
787.6	25.7	742.2	21.6	698.3	34.6
786.4	19.7	741.2	20.1	697.7	31.6
786.0	26.8	740.0	25.7	697.3	30.5
784.8	23.8	739.3	24.9	696.0	19.3
784.4	25.7	737.5	34.2	695.2	30.5
783.2	20.5	737.2	32.4	694.0	18.2
782.9	20.8	735.3	35.3	692.4	34.6
781.5	14.5	733.3	32.7	691.3	17.9
780.0	27.9	732.5	51.3	690.1	26.0
778.8	24.5	731.8	31.6	689.1	16.4
778.1	29.4	731.1	35.3	688.7	20.8
775.1	13.8	729.8	29.0	687.9	16.7
773.1	26.8	729.4	29.8	686.6	23.8
772.4	23.8	729.0	27.9	686.1	16.7
771.6	24.2	728.3	30.5	684.9	27.2
770.8	19.7	727.3	30.1	683.8	17.5
770.5	20.5	726.4	49.1	682.8	21.9
770.2	15.2	725.0	25.3	682.3	18.6
769.6	21.2	724.2	27.5	681.6	22.3
769.2	17.9	723.4	25.3	681.4	21.9
768.8	20.8	722.9	26.0	681.1	22.3
768.4	17.1	722.6	25.7	680.7	19.3

Table 10 (*continued*)

λ(Å)	σ(Mb)	λ(Å)	σ(Mb)	λ(Å)	σ(Mb)
679.9	23.8	604.3	30.1	270	14.7
679.2	18.2	600	28.6	260	13.7
677.0	21.2	595	23.8	250	12.8
676.6	20.1	590	19.4	240	11.9
676.2	21.6	585	22.4	230	11.1
675.7	20.1	580	24.8	220	10.3
675.0	21.6	575	25.5	210	9.57
673.6	19.3	570	25.8	200	8.86
670.5	21.6	565	26.0	190	8.15
670.0	20.8	560	26.1	180	7.33
669.6	22.7	555	26.1	170	6.70
665.8	23.8	550	25.9	160	6.03
651.8	29.8	540	25.4	150	5.32
649.4	25.3	530	24.9	140	4.62
646.7	30.1	520	24.5	130	3.95
645.9	24.9	510	24.0	120	3.24
645.0	25.3	500	23.6	110	2.61
644.2	20.8	490	23.0	100	2.16
642.5	29.0	480	22.6	90	1.70
638.7	25.3	470	22.1	80	1.30
638.1	27.9	460	21.7	70	0.95
637.3	25.3	450	21.4	60	0.66
636.3	27.9	440	21.0	50	0.42
635.8	25.3	430	20.7	40	0.24
634.4	31.2	420	20.3	35	0.17
633.0	23.4	410	20.0	30	0.11_7
629.6	32.4	400	19.6	23	1.15
627.1	24.9	390	19.2	15	0.404
626.6	29.0	380	18.8	13.78	0.312
624.5	25.3	370	18.4	12.40	0.245
621.9	32.7	360	18.0	9.89	0.135
620.5	23.1	350	17.8	8.265	0.083_4
618.7	29.8	340	17.5	6.199	0.0372
618.2	28.3	330	17.3	4.133	0.0105
617.7	34.6	320	17.2	3.100	0.0049
616.3	24.2	310	17.0	2.480	0.0025
615.2	28.6	300	16.7	2.066	0.0014
613.1	28.3	295	16.5	1.55	0.0006
612.7	23.1	290	16.2	1.24	0.0004
610.8	30.1	285	15.9	0.827	0.0001
608.9	26.4	280	15.5	0.620	0.0000

[a] $0 < \lambda < 1028$ Å.

TABLE 11 *The Oscillator Strength Distribution in O_2*

Energy region	$S(0)$	$S(-1)$	$S(-2)$
(Schumann–Runge bands)			
B $^3\Sigma_u^-$ (7–9 eV)	0.179	0.290	0.469
Unidentified state			
(~9.9 eV)	0.020	0.027	0.038
Autoionization region			
(750–1027 Å)	0.724_1	0.682_8	0.649_3
Ionization continuum			
(λ < 750 Å)	15.784	3.790	1.684
Total	16.71	4.79	2.840

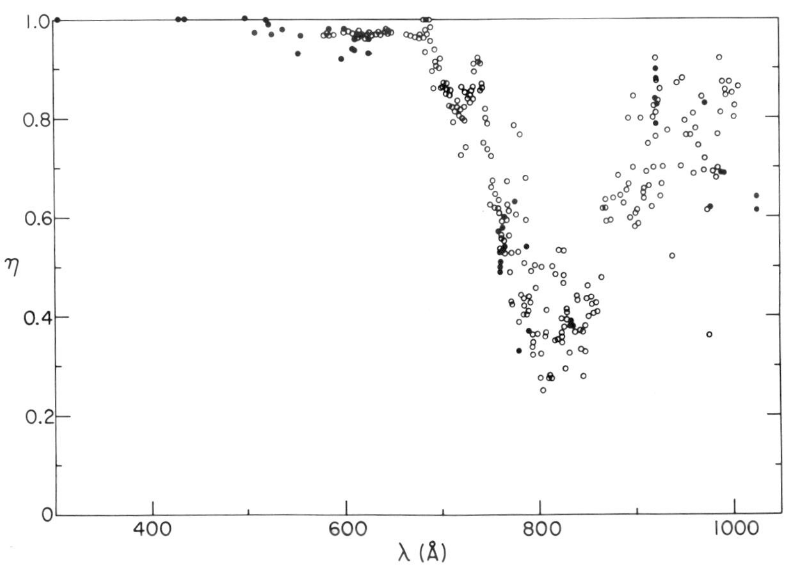

Fig. 12 The quantum yield of ionization (η) experimentally observed in the photoabsorption of molecular oxygen. ● Samson and Cairns[63]; ○ Matsunaga and Watanabe.[78]

4. CO

The photoabsorption curve for CO across its ionizing range is presented in Fig. 13. It can immediately be appreciated that there are more unsolved problems here than in the previously considered molecules, but this quality of data offers an opportunity for the sum rules to aid in making selections. The region from 600 Å to the ionization threshold (~885 Å) is again dominated by autoionization structure. The experimental data of two groups of

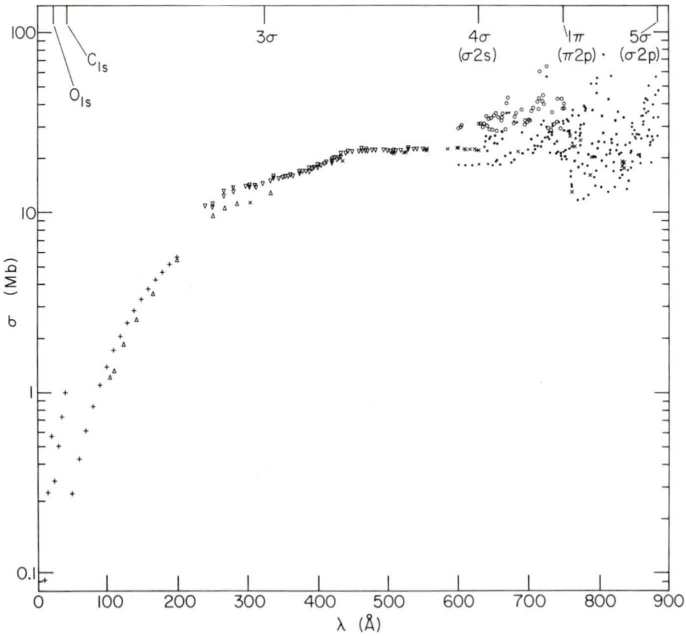

Fig. 13 Experimental photoabsorption cross section of carbon monoxide above the ionization threshold. × Cairns and Samson[82]; ▽ Cairns and Samson[83]; + Henke *et al.*[6]; ● Cook *et al.*[80]; △ de Reilhac and Damany[64]; ○ Huffman *et al.*[80a]

workers (Cook *et al.*[80]; Huffman *et al.*[81]) differ substantially in this region. The values of Cairns and Samson[82,83] appear to be a reasonably good selection from 625 to 350 Å, but at that point they do not merge well with atomic additivity (Henke *et al.*[6]) nor with the measurements of de Reilhac and Damany.[64,64a] The absorption cross sections of the Stoughton synchrotron group[66] (not plotted) are increasingly higher than both the atomic additivity and de Reilhac–Damany values as λ decreases from 250 Å, but they also fall somewhat below the values of Cairns and Samson in the 300–500 Å region. This behavior will be seen to be characteristic of many photoabsorption studies using synchrotron radiation. At the shorter wavelengths, scattered light which is composed of wavelengths that are absorbed more strongly (longer λ) give rise to artificially high cross sections. At the longer nominal wavelengths, the reverse may be true. The Tokyo synchrotron group[84] has observed absorption structure at energies just below the K edges of both C and O in CO, which they attribute to transitions from the respective 1s shells to unoccupied valence and Rydberg levels. These levels match up well with an NO-like molecule, as was found in the corresponding case of N_2 excitation near the K edge. This region has also been studied recently by the electron energy loss (pseudo-photoabsorption) technique.[84a] An intense Rydberg peak was observed below the carbon K edge, followed by weaker and higher Rydberg members, and a broad "shape resonance" beyond the K edge, rather similar to the behavior in N_2.

We once more divide the spectral region into (a) subionization, (b) autoionization region (600–885 Å), and (c) continuum ionization regions, and describe our selection of data below.

a. Subionization Region. Lassettre and Skerbele[85] have reported oscillator strength determinations for four states that contribute significantly below 14.0 eV. They include the $A^1\Pi$ state, over the energy range 8.03–9.95 eV, with $f = 0.1945$; the $B^1\Sigma$ state at ~ 10.78 eV, with $f = 0.017$; the $C^1\Sigma$ state at 11.4 eV, with $f = 0.170$; and the $E^1\Pi$ state at 11.52–11.77 eV, with $f = 0.101$. An additional contribution between 12 and 14 eV can be obtained either from Cook *et al.*[80] ($f = 0.8165$) or from Huffman *et al.*[81] ($f = 0.7921$). These values may be bandwidth dependent, but are helped somewhat by the expected predissociation in this region.

b. Autoionization Region (600–885 Å). The data of Cook *et al.*[80] yield $f = 1.479$ in this region; the corresponding values of Huffman *et al.*[81] give $f = 2.386$. We treat this region as an unknown and look to the sum rules as a guide to the proper selection.

c. Continuum Region ($\lambda < 600$ Å). Starting from the x-ray limit, we choose the atomic addivitity values below 50 Å and an average of the atomic additivity values, obtained from Henke, *et al.*,[6] and the measured values of de Reilhac and Damany,[64,76a] between 50 and 200 Å. These two sets of numbers are in fairly good agreement in this region, in any event. Between 200 and 350 Å we choose de Reilhac and Damany, which has the support of the Lee *et al.*[66] measurements and joins the Cairns and Samson[82,83] data at ~ 350 Å. The latter are accepted for the 350–600 Å region. The major question in this selection is the rejection of the Cairns and Samson values between 240 and 350 Å, which uniformly possess *higher* cross sections. The continuum contribution to f arrived at in this way is 10.672. The corresponding contributions to the other $S(k)$ sums are listed in Table 12.

The $S(0)$ sum of Cook *et al.* is clearly closer to the required sum of 14.0. The $S(-2)$ sums imply $\alpha_N = 2.18 \times 10^{-24}$ cm^3 (Cook *et al.*[80]) and 2.47×10^{-24} cm^3 (Huffman *et al.*[81]). The experimental result[73] is 1.97×10^{-24} cm^3, already significantly smaller than the Cook value and much smaller than the value using the Huffman cross sections. Both sum rules clearly favor the photoabsorption cross sections of Cook *et al.* in the autoionization region.

Note: After the completion of this analysis, detailed photoabsorption data covering the range 520–730 Å were reported by Lee *et al.*[66a] These new

TABLE 12 *The Oscillator Strength Distribution in CO*

Energy region	$S(0)$	$S(-1)$	$S(-2)$
Discrete			
A $^1\Pi$ (8.03–9.95 eV)	0.1945	0.2829	0.445
B $^1\Sigma$ (10.78 eV)	0.017	0.021	0.027
C $^1\Sigma$ (11.4 eV)	0.170	0.203	0.242
E $^1\Pi$ (11.52–11.77 eV)	0.101	0.118	0.141
Other states (~12–14 eV)	0.80	0.83_7	0.87_6
Autoionization region (600–885 Å)			
Cook *et al.*[80]	(1.479)	(1.192)	(0.972_3)
or			
Huffman *et al.*[81]	(2.386)	(1.887)	(1.511)
Continuum region			
($\lambda < 600$ Å)	11.226_5	2.482	0.9744
Total			
(Using Cook *et al.*[80])	13.99	4.14	3.68
(Using Huffman *et al.*[81])	14.90	5.83	4.22

values are about 5% lower than those of Cairns and Samson in the region of overlap (~520–660 Å) and clearly are in better agreement with the cross sections of Cook *et al.* than with the data of Huffman *et al.*, as indicated by our sum-rule analysis. The recent Lee data suffer from insufficient resolution, and hence do not reach the maximum peaks and minimum valleys in the region of strong autoionization structure (650–730 Å). Samson and Gardner[83a] have also recently offered modified values for total photoabsorption cross sections between 304 and 745 Å, using line sources. These values generally agree with the recent Lee data,[66a] and the point at 304 Å is considerably lower than that reported by Cairns and Samson[83] in 1966, justifying our rejection of the latter data.

The ionization yield measurements of Samson and Gardner[83a] between 600–750 Å (see Fig. 14) should be used in future analyses, since the values of η_i obtained by Cook *et al.* in this region appear more dubious. We may now note that a selection of the Cairns and Samson cross sections in the 240–350 Å region would have increased our calculated sums [particularly $S(0)$] still further, and hence these are suspect for another reason than those previously mentioned. In view of the general uncertainty of photoabsorption cross sections for CO in several wavelength regions, we have elected not to tabulate selected values. For specific purposes, the accuracy that the data warrants can be obtained from Fig. 13. In the 520–730 Å region additional information can be extracted from Lee *et al.*[66a] and Samson and Gardner.[83a]

The value of $|M_I|^2 = 3.674$ (Cook *et al.*[80]) or 4.369 (Huffman *et al.*[81]) may be obtained directly from Table 12. The photoionization yield between 885

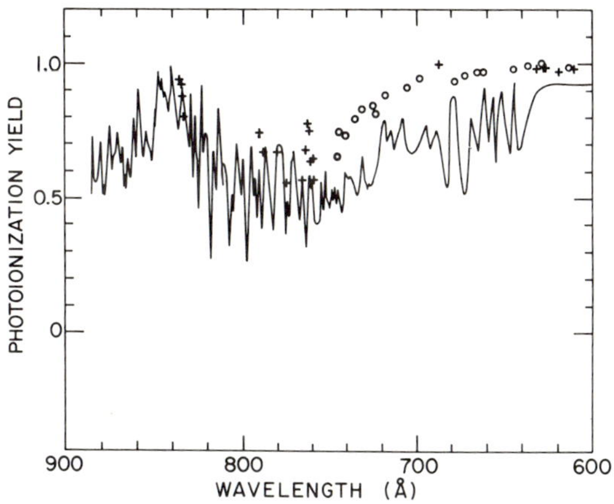

Fig. 14 The quantum yield of ionization experimentally observed in the photoabsorption of carbon monoxide. ——— Cook *et al.*[80]; + Cairns and Samson[82]; ○ Samson and Gardner.[83a]

and 600 Å is shown in Fig. 14. It, too, has a significant margin of error. Restricting ourselves for now to the measurements of Cook *et al.*, and focusing on that portion of absorption which leads to ionization in the autoionization region, we obtain $S_i(-1) = 0.858$ between 600 and 885 Å, or $|M_i|^2 = 3.340$. This is again substantially lower than the electron impact ionization value of (3.70 ± 0.15) of Rieke and Prepjchal.[13] However, in view of the lower reliability of the data as deduced from the other sum rules, it would be presumptuous to attribute additional ionization mechanisms in the electron impact experiments for CO, although our tentative analysis maintains the trend encountered previously.

5. NO

The rather extensive graph of photoabsorption cross section versus wavelength for NO from the vicinity of its ionization threshold to beyond the K edges of N and O is displayed in Fig. 15. In the region above 675 Å, the detailed measurements of Watanabe *et al.*[86] are preferred, and the data of Metzger *et al.*[87] are not shown for reasons of clarity. Between 584 and 675 Å, the recent data of Gardner *et al.*[88] using the Glasgow synchrotron, show more detail, and are preferred. However, a comparison of the Glasgow values in regions of nonresonance with those of Watanabe *et al.*, and (to shorter wavelengths) with the equally recent work of Lee *et al.*,[66] de Reilhac and Damany,[64] and with atomic additivity calculations, indicates that the Gardner data should perhaps be reduced uniformly by $\sim 8\%$ to bring them into line with the other measurements. (There appears to be a typesetting

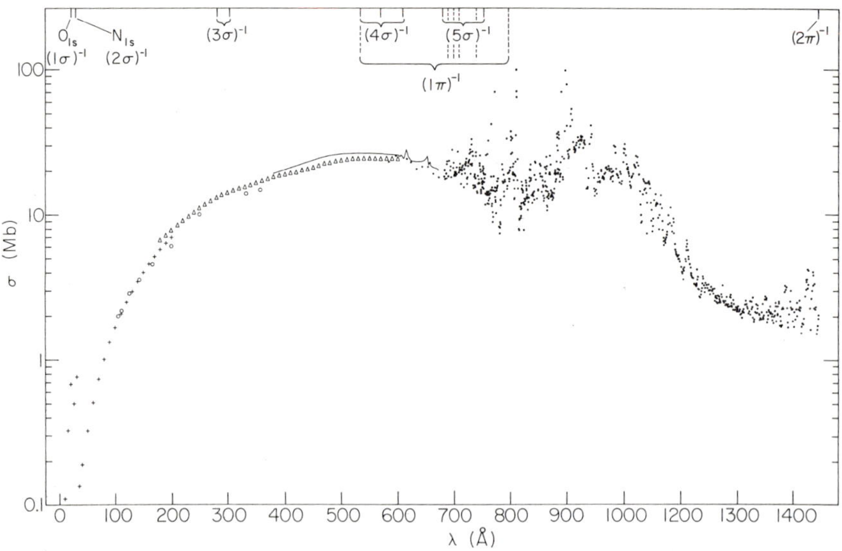

Fig. 15 Experimental photoabsorption cross section of nitric oxide above the ionization threshold. ● Watanabe *et al.*[86]; + Atomic additivity using data of Henke *et al.*[6]; ○ de Reilhac and Damany[64]; ——— Gardner *et al.*[88]; △ Lee *et al.*[66]

error in the more recent article by de Reilhac and Damany[64a].) We have made our sum-rule analysis with this kind of correction. The detailed selection of cross sections below 672.5 Å is listed in Table 13. For the detailed information above 672.5 Å, the reader is referred to the article of Watanabe *et al.*[86]

Nitric oxide has a single electron in its outermost occupied (2π) orbital, which is rather easily removed. As a consequence, the first IP of NO is well separated from the next higher ones. The next two deeper orbitals are (4σ) and (1π), and these are relatively closely spaced. Each is in turn further split, partially into different Λ states, and partially because removal of an inner electron enables the resulting hole to couple with the unpaired (2π) electron to form singlet and triplet states. Hence, a manifold of states is necessary to account for the ejection of a (4σ) or (1π) electron, as indicated at the top of Fig. 15. The 304 Å photoelectron spectrum of Edqvist *et al.*[89] and the theoretical analysis of Thulstrup and Öhrn[90] are the best current interpretations of these inner ionizations. The ionizations originating from 3σ and the K shells of N and O are obtained from x-ray photoelectron spectroscopy.[69] For the sum-rule analysis, we would like to divide the spectrum into a subionization portion, an autoionization region, and a continuum ionization region. At the present time, the oscillator strength distribution in the subionization region is poorly known, at best. Hence, we

TABLE 13 *Selected Values of Absorption Cross Sections for NO*[a]

λ(Å)	σ(Mb)	λ(Å)	σ(Mb)	λ(Å)	σ(Mb)
672.5	18.9	600.0	24.2	110.0	2.08
668.0	19.2	575.0	24.2	100.0	1.68
660.0	19.8	550.0	24.4	90.0	1.33
657.5	20.1	525.0	24.2	80.0	1.01
656.5	21.2	500.0	23.7	70.0	0.74
654.0	20.3	475.0	23.0	60.0	0.51
653.5	21.0	450.0	21.2	50.0	0.33
651.0	23.4	425.0	20.0	40.0	0.19
650.0	22.8	400.0	18.8	35.0	0.136
646.5	21.9	375.0	17.5	30.0	0.77
640.0	21.4	350.0	15.8	25.0	0.50
630.0	21.3	325.0	14.5	20.0	0.68
625.0	21.5	300.0	13.2	15.0	0.33
622.0	21.9	275.0	12.0	9.89	0.109
619.0	22.8	250.0	10.5	8.265	0.067
617.0	23.7	225.0	8.8	6.199	0.030
616.0	24.7	200.0	6.98	4.153	0.0091
615.0	25.9	190.0	6.35	3.10	0.0038
614.0	25.6	180.0	5.75	2.48	0.0020
613.0	23.7	170.0	5.15	2.066	0.0011
611.0	22.8	160.0	4.57	1.550	0.0005
610.0	22.5	150.0	4.01	1.240	0.0002
607.5	22.8	140.0	3.49	0.827	0.0001
606.5	23.6	130.0	2.98	0.62	0.0000
603.5	23.2	120.0	2.51		

[a] $0 < \lambda < 672.5$ Å.

are forced to concentrate our attention on two primary regions, the first dominated by autoionization structure (672.5–1339 Å) and involving the data of Watanabe *et al.*,[86] and the second, primarily smooth continuum ionization, below 672.5 Å and containing the selected data of Table 13.

The corresponding contributions to the $S(k)$ sums are given in Table 14.

TABLE 14 *The Oscillator Strength Distribution in NO*

Energy region	$S(0)$	$S(-1)$	$S(-2)$
Autoionization region (672.5–1399 Å)	1.413	1.331	1.281
Ionization continuum ($\lambda < 672.5$ Å)	12.759	3.005	1.259
Subtotal	14.17	4.34	2.54

The $S(0)$ sum implies an oscillator strength in the subionization region of 0.83. Lassettre *et al.*[91] present a spectrum of NO between ~5.5 and 9.5 eV, but the incident electron energy is only 50 eV, and no oscillator strength measurement is given. Marmo[92] has done photoabsorption measurements

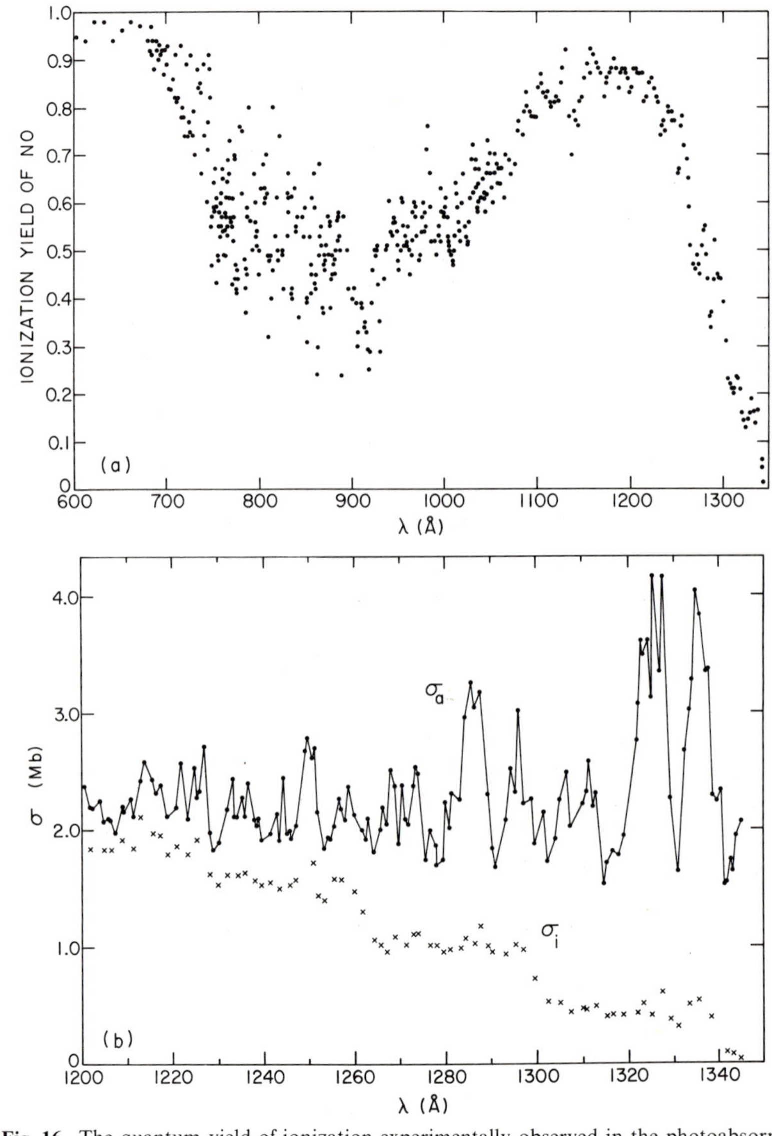

Fig. 16 The quantum yield of ionization experimentally observed in the photoabsorption of nitric oxide. (Data are from Watanabe *et al.*[86])

in the 1350–2350 Å region, but his resolving power is insufficient to take into account sharp rotational structure, and hence pressure and bandwidth effects make a quantitative determination of oscillator strength contributions from his data impractical.

The $S(-2)$ sum implies α_N (partial) $= 1.505 \times 10^{-24}$ cm^3. The directly measured[73] $\alpha_N = 1.75 \times 10^{-24}$ cm^3. At the state of present knowledge, the difference between these quantities, representing the subionization contribution, seems reasonable.

Figure 16a shows the ionization yield curve of NO from 600 Å to the threshold region. In Fig. 16b the region between 1200–1340 Å is displayed in the form of individual photoabsorption (σ_a) and photoionization (σ_i) cross sections. The latter figure clearly indicates that predissociation of vibrationally excited Rydberg states is more effective than autoionization in the threshold region. If we assume that the yield is essentially unity for $\lambda <$ 672.5 Å, and compute the $S_i(-1)$ contribution for $\lambda > 672.5$ Å, primarily from the photoionization cross section values given by Watanabe *et al.*,[86] we obtain 0.850 for $S_i(-1)$ above 672.5 Å, and hence $|M_i|^2 = 3.855$. Rieke and Prepejchal[13] have reported 4.31 ± 0.48 for the directly measured $|M_i|^2$ from high energy electron ionization. Our computed $|M_i|^2$ is within the range of the rather large error limit of the directly measured value, but the difference between these numbers is in the same direction as previously observed for three other diatomic molecules, and again suggests either a systematic error or additional ionization mechanisms operating in the electron impact ionization measurements.

C. Triatomic Molecules

1. H_2O

Photoabsorption cross sections of H_2O in the vacuum ultraviolet have been measured by several investigators, the most recent being Phillips *et al.*,[93] Katayama *et al.*[94] and Gürtler *et al.*[94a†] These publications provide references to earlier work. In Fig. 17, the values of Katayama *et al.* are selected for the region from the ionization threshold (~982.5 Å) to 584 Å. Cairns *et al.*[95] have measured the photoionization spectra producing H_2O^+, OH^+, and H^+ below 600 Å, have normalized their total ionization to that

† This photoabsorption spectrum has much better resolution (0.03 Å) than others, but the data are presented graphically on a rather compressed scale, making accurate numerical abstraction difficult. The prime focus of these authors was on spectroscopic assignments.

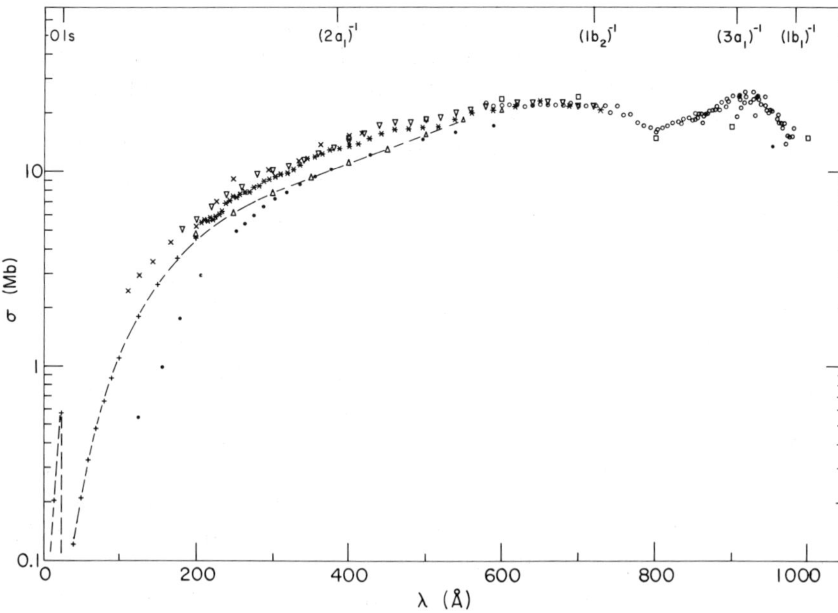

Fig. 17 Experimental photoabsorption cross section of water vapor from the ionization threshold thru the K edge. ○ Katayama *et al.*[94]; × de Reilhac and Damany[114]; □ Wainfan *et al.*[119]; + Henke *et al.*[6]; △ Cairns *et al.*[95]; ● Branton and Brion[94b], ▽ Phillips *et al.*[93]; ∗ Tan *et al.*[95a]

of Katayama *et al.* in the region of overlap, and thereby extended the data to 200 Å. On a photoabsorption plot, this extension assumes that the ionization yield is unity below 600 Å, which is implied by the asymptotic behavior of the photoionization yield reported by Katayama *et al.* (see Fig. 18). However, the directly measured photoabsorption cross sections in this region reported by Phillips *et al.*[93] and by de Reilhac and Damany[64] are significantly larger. The values calculated by atomic additivity[6] are in good agreement with Cairns *et al.* at 200 Å, but the slope of the atomic additivity cross sections suggests the selection of mean values between those of Cairns *et al.* on the one hand, and of Phillips *et al.* and de Reilhac–Damany, on the other. (It will be recalled that the Stoughton group's[66] results tend to be characteristically high for $\lambda < 250$ Å—see Chapter V, Section on CO.) This choice of intermediate values receives some corroboration from the (e, 2e) experiments of Tan *et al.*[95a] However, these authors determine their absolute absorption scale by making use of a truncated sum rule, so that use of such data to test sum rules is tantamount to circular reasoning. The more recent data of de Reilhac and Damany,[64a] which are presented in tabular rather than graphical form, are in excellent agreement

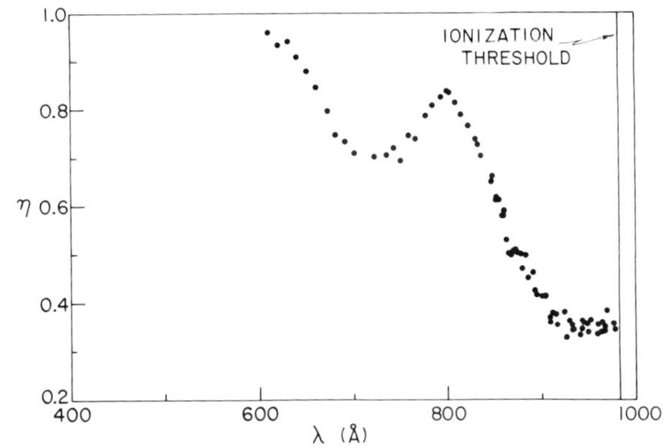

Fig. 18 The quantum yield of ionization (η) experimentally observed in the photoabsorption of water vapor. (Data are from Katayama *et al.*[94])

with atomic additivity below 175 Å and merge with their earlier values[64] above this wavelength. The selected photoabsorption cross sections are summarized in Table 15.

Although some significant structure can be noted in the photoabsorption curve (this shows up more clearly on a linear plot, e.g., Fig. 2 of Katayama *et al.*,[94]), particularly between 850 and 950 Å, this first example of a triatomic molecule displays less dramatic oscillations than the diatomics that we have considered. Even in the first few electron volts above threshold, a major fraction of the photoabsorption appears to lead to direct ionization and direct dissociation, or at least to highly diffuse structure. Hence, our earlier separation of the spectrum above the ionization threshold into "autoionization region" and "continuum region" is not necessary. We shall merely record subionization and supraionization contributions in Table 16. In the subionization region, we are able to present two alternate sources of oscillator strength information in this case, i.e., electron impact energy loss and photoabsorption data.

Photoabsorption cross section values between 1065 and 1855 Å were taken from measurements of Watanabe *et al.*[96,97] Photoabsorption data in the wavelength region between 1065 Å and the onset of ionization (982.5 Å) have been extracted from Fig. 2 of the paper by Katayama *et al.*[94]

The agreement between the electron energy loss data and the photoabsorption values below the ionization threshold is satisfactory. It would appear that less than one-third of a unit of oscillator strength lies in the subionization region. The total oscillator strength is in excellent agreement with the TRK sum rule. The $S(-2)$ sum implies $\alpha_N = 1.44_5 \times 10^{-24}$ cm^3 (EEL data) or $\alpha_N = 1.40 \times 10^{-24}$ cm^3 (based solely on photoabsorption

TABLE 15 *Selected Photoabsorption Cross Sections for H_2O*

λ(Å)	σ(Mb)	λ(Å)	σ(Mb)	λ(Å)	σ(Mb)
980.6	16.7	874.1	20.5	620.0	22.0
978.0	15.1	872.3	20.6	610.0	22.0
977.2	15.1	870.2	20.5	600.0	21.5
973.6	15.0	867.9	19.9	590.0	21.5
972.5	15.1	866.8	19.8	580.0	22.2
970.0	13.8	864.8	19.7	550.0	20.0
968.3	17.7	862.5	19.1	525.0	19.0
967.3	17.6	861.0	17.8	500.0	18.0
965.4	16.7	859.2	19.7	475.0	17.0
963.3	17.8	857.4	19.8	450.0	16.0
960.8	17.7	855.2	18.8	425.0	15.0
959.7	19.4	852.6	18.6	400.0	13.5
958.2	18.6	851.0	19.8	375.0	12.2
952.1	20.2	848.8	18.3	350.0	11.2
951.0	20.2	844.0	18.8	325.0	10.1
949.4	20.6	836.5	18.6	300.0	9.0
948.0	20.0	833.7	17.6	275.0	8.0
945.3	20.6	829.6	18.0	250.0	6.5
943.4	21.8	822.2	17.7	225.0	5.5
942.6	24.1	815.8	17.1	200.0	4.5
941.2	22.9	808.7	16.8	175.0	3.53
936.3	22.0	802.5	16.4	150.0	2.63
934.8	23.5	800.0	16.0	125.0	1.8
933.7	24.3	793.6	16.7	100.0	1.11
932.4	24.0	784.0	17.2	90.0	0.87
930.0	19.3	776.7	17.8	80.0	0.66
927.4	23.4	766.6	19.5	70.0	0.476
926.2	25.4	758.7	19.3	60.0	0.33
923.8	22.5	750.0	21.7	50.0	0.21
917.5	24.1	742.0	20.0	40.0	0.121
916.6	25.7	734.5	21.8	35.0	0.087
915.5	22.5	722.8	21.8	30.0	0.059
911.5	21.1	700.0	21.4	23.0	0.575
910.0	23.2	695.5	22.2	15.0	0.202
909.0	24.2	693.0	21.7	9.89	0.068
908.1	24.0	689.0	22.2	8.27	0.042
905.0	19.1	686.0	21.7	6.20	0.019
900.4	24.2	683.0	22.2	4.13	0.0057
894.4	22.8	680.0	21.6	3.10	0.0025
892.9	23.7	677.0	22.3	2.48	0.0012_5
891.4	21.6	670.0	21.9	2.07	0.00072
885.8	22.6	660.0	22.0	1.55	0.000296
883.3	20.5	650.0	22.0	1.24	0.000147
879.0	21.7	640.0	21.6	0.827	0.0000409
877.7	20.5	630.0	21.4	0.62	0.0000163

TABLE 16 *The Oscillator Strength Distribution in H_2O*

Energy region	$S(0)$	$S(-1)$	$S(-2)$
Subionization (EEL)[a] (eV)			
~7.37	0.061	0.1126	0.2077
~10.09	0.097	0.1307	0.1762
~11.02	0.067	0.0827	0.1020
11.64–12.62 (photoabsorption)	0.1016	0.1138	0.1277
Partial sum	0.3266	0.4398	0.6136
Subionization (photoabsorption) (Å)			
1363.5–1855.0	0.04517	0.08006	0.1425
1065.0–1361.5	0.1555	0.2039	0.2686
982.5–1065.0	0.1016	0.1138	0.1277
Partial sum	0.3023	0.3978	0.5388
Supraionization			
Partial sum, $\lambda < 982.5$ Å	9.6903	3.166	1.824
Total (using EEL[a] data)	10.02	3.61	2.44
Total (photoabsorption data only)	9.99	3.56	2.36

[a] Electron energy loss data from Lassettre and Skerbele.[72]

data). The experimental[73,98] values ($1.45 - 1.48 \times 10^{-24}$ cm^3) tend to favor the oscillator strengths deduced from the electron scattering experiments.

A direct assessment of $S(-1)$ does not seem feasible at the present time. However, $S_i(-1)$ can be computed with the aid of the ionization yield data of Katayama *et al.*, if the assumption is made that the yield is unity below 600 Å. Bennett *et al.*[99] report an ionization yield of unity at 584 Å. With the above qualifications, we compute $S_i(-1) = 2.77$. The electron ionization experiment of Rieke and Prepejchal[13] obtained $|M_i|^2 = 3.24 \pm 0.15$, a value slightly larger than our $S_I(-1)$. The direction and approximate magnitude of the discrepancy between these two types of study has been discussed previously in connection with the analyses of monatomic and diatomic gases and apparently is also present with triatomics.

2. CO_2

The spectral distribution of oscillator strength in CO_2 (Fig. 19) affords us another illustration of the utility of sum rules in evaluating data. Additivity of atomic cross sections (Henke[6]) has been employed to construct the curve from $0 < \lambda < 200$ Å, from where it joins fairly smoothly with the data of Cairns and Samson[82,83] between ~260 and 625 Å. From 600 Å to the ionization threshold at 900 Å, detailed photoabsorption data are available from Nakata *et al.*[100] and Cook *et al.*[101] Only the data of Nakata *et al.*

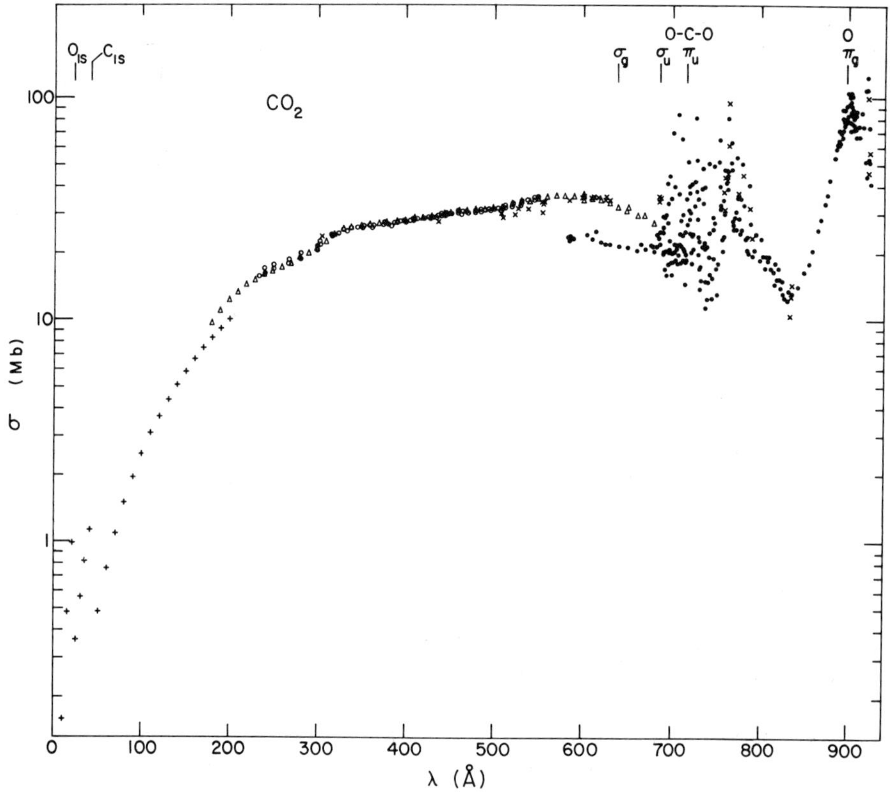

Fig. 19 Experimental photoabsorption cross section of carbon dioxide vapor above the ionization threshold. × Cairns and Samson[82]; ○ Cairns and Samson[83]; + Henke *et al.*[6]; ● Nakata *et al.*[100]; △ Lee *et al.*[66]

have been plotted in this region to avoid confusion. It is immediately striking that between 580 and 680 Å (a region relatively free of autoionization) the Cairns and Samson values are about 50% larger than those of Nakata *et al.* and Cook *et al.* The discrepancy is well outside the scatter of the respective experiments. There appears to be a systematic error, and one's first tendency is to reject the Cairns and Samson values, since the other two groups of investigators are in closer agreement. We attempted to establish the more correct set of values from a sum rule analysis, and at about the same time Samson *et al.*[102] reexamined the photoabsorption cross section at 584 Å. The early analysis of the oscillator distribution is summarized in Table 17.

In Table 17, the f_n values of the bands at 8.42 and 9.30 eV have been taken from Meyer and Lassettre.[103] These authors also provide f_n values for two intense transitions at 11.07 and 11.40 eV, which together amount to $\sim$0.25.

TABLE 17 *The Oscillator Strength Distribution in CO_2*[a]

Energy range	$S(0)$	$S(-1)$	$S(-2)$
Subionization			
8.42 eV band	0.0049	0.0079	0.0128
9.30 eV band	0.0072	0.0105	0.0154
10.78–12.40 eV (1000.0–1150.0 Å)	0.4712	0.5538	0.6520
920.0–1000.0 Å	0.5649	0.5922	0.6211
902.9–920.0 Å	0.1784	0.1784	0.1783
Partial sum	1.2266	1.3428	1.4796
Ionization continuum (Å)			
836.0–902.9	0.3670	0.3546	0.3427
(600.0–836.0)	(1.372)	(1.067)	(0.8357)
200.0–600.0	8.184	3.263	1.420
9.89–200.0	8.989	0.7028	0.0937
0.62–9.89	0.9656	0.0069_4	0.00006
Partial sum	19.8776	5.3943	2.69216
Total	21.10	6.74	4.17

[a] Preliminary.

However, from Fig. 1 of their paper, we note that a complex of other bands also occurs between ~11 eV and the IP at ~13.7 eV, and hence we have chosen to adopt the photoabsorption data of Nakata *et al.* in order to encompass all of the oscillator strength between 10.78 and 13.7 eV. In this preliminary analysis, the Nakata values are used further to 600 Å; for the shorter wavelengths, the selection of data is uniquely determined.

From the total sums, we immediately note that $S(0)$ is lacking ~0.9 units. The $S(-2)$ total implies $\alpha_N = 2.47 \times 10^{-24}$ cm^3 compared with experimental values of 2.91×10^{-24} cm^3 [98] and 2.71×10^{-24} cm^3.[73] The suggestion is clear that some of the "missing" oscillator strength is in the low energy region around the ionization limit. Returning now to Fig. 19, we can draw the plausible inference that the discrepancy between the Cairns and Nakata values between 580 and 680 Å is no longer apparent at 836 Å and longer wavelengths. A possible cause of this discrepancy is the onset of fluorescence from the $A^2\Pi_u$ and $B^2\Sigma_u^+$ states of CO_2^+ at 716 Å,[104] which could be viewed as *diminished* absorption in an attenuation experiment if the light detector was sensitive to the fluorescence. Cairns and Samson used an ionization detector, Nakata *et al.* and Cook *et al.* did not. Hence, this tentative interpretation supports the Cairns and Samson values between 580 and 680 Å. We can overcorrect the data of Nakata *et al.* by scaling their sums between 600.0 and 836.0 Å by $\frac{35}{23}$ (the approximate ratio of Cairns *et al.* to

Nakata *et al.*[100] in the region of discrepancy). This results in a total $S(0)$ of 21.82 and a total $S(-2)$ of 4.76, the latter yielding $\alpha_N = 2.82 \times 10^{-24}$ cm^3. Both of these adjusted sums would be in much better agreement with experiment, lending support to this hypothesis. The more recent reexamination of photoabsorption by Samson *et al.*[102] does indeed confirm the earlier Cairns–Samson result, arriving at a revised $\sigma(CO_2) = 33 \pm 1.5$ Mb at 584 Å. Finally, Lee *et al.*[66] have measured the photoabsorption cross section of CO_2 between ~230 and 730 Å, utilizing the Stoughton synchrotron. Their data are in good agreement with those of Cairns and Samson across the entire range of overlap. In particular, they obtain $\sigma = 36$ Mb at 580–590 Å. The ionization yield of CO_2 between ~300 and 900 Å (Fig. 20) indicates that below 600 Å the yield is very likely unity. Samson *et al.*[102] have recently reexamined this quantity at 584 Å and concluded that it is 0.99 ± 0.01. Hence, the contribution to $|M_i|^2$ between 0.62 and 600 Å can be taken as that listed in Table 17. Between 600 and 836 Å, which contains the troublesome portion of the Nakata data, we have chosen to increase the cross section by the aforementioned factor (35/23) but to use the Nakata *et al.* ionization yields. This results in a contribution of 1.353 to $S(-1)$ for this wavelength region. Straightforward application of the Nakata photoionization values to the 836–902.9 Å region adds 0.3546, or a total of 5.60 for $|M_i|^2$. This may be compared with the directly measured electron impact ionization value (Rieke

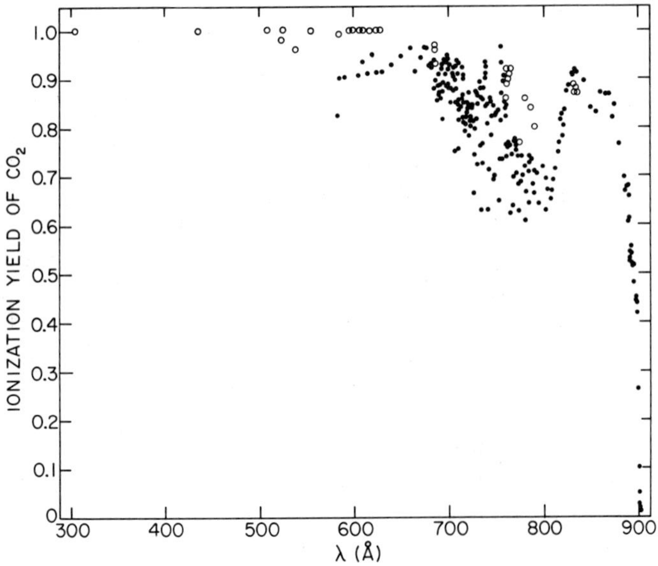

Fig. 20 The quantum yield of ionization experimentally observed in the photoabsorption of carbon dioxide. ○ Cairns and Samson[82]; ● Nakata *et al.*[100]

and Prepejchal[13]) of 5.75 ± 0.073. The agreement is probably better than is warranted by the uncertainty of the photoabsorption data.

3. N_2O

The photoabsorption studies on this asymmetric triatomic molecule have recently become sufficiently extensive to make a complete analysis possible. In Fig. 21, we note that Cook *et al.*[105] have studied the autoionization region from ionization threshold to 600 Å, and the recent investigation at the Stoughton synchrotron[66] has covered the "grazing incidence" region from 180 to 600 Å. Atomic additivity (Henke *et al.*[6]) has been used to estimate the molecular cross sections below 200 Å, receiving strong confirmation in the 100–200 Å region from the data of de Reilhac and Damany.[64,64a]

There is little room for controversy and selection, since the various experimental sources tend to cover different energy ranges. The synchrotron data appear to be ~10% high below 300 Å, but at this wavelength they appear to merge smoothly with the values of de Reilhac and Damany. Similarly, at their upper end (~600 Å), the values of Lee *et al.*[66] are ~10%

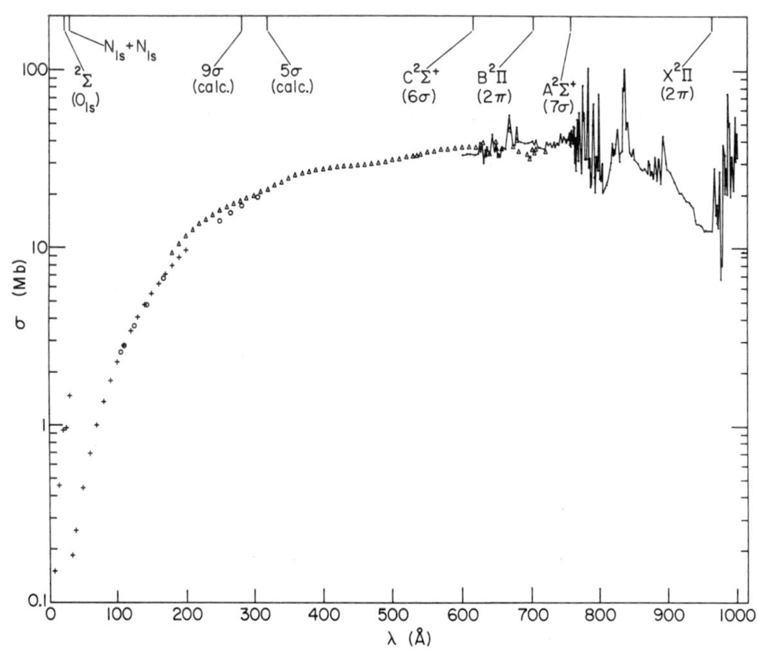

Fig. 21 Experimental photoabsorption cross section of nitrous oxide above the ionization threshold. ——— Cook *et al.*[105]; + Atomic additivity from Henke *et al.*[6]; ○ de Reilhac and Damany[64]; △ Lee *et al.*[66]

higher than those of Cook *et al.*,[105] but agree within closer limits at ~650 Å. In the following analysis, we have selected atomic additivity values to 200 Å, the de Reilhac–Damany cross sections between 200 and 320 Å, Lee *et al.* between 320 and 624 Å, and those of Cook *et al.* from 624 Å to the ionization threshold.

In this instance, we have also used photoabsorption data to deduce the subionization contribution, partly because fully analyzed electron energy loss values for f_n are not readily available, and partly because the structure in the absorption curve for this triatomic molecule is not so sharp as to be seriously affected by photon band width problems. Toward this end, we have utilized the photoabsorption cross sections of Zelikoff *et al.*[97,106] The results are summarized in Table 18.

The total $S(0)$ is 4% less than required to fulfill the Thomas–Reiche–Kuhn rule. However, the total $S(-2)$ leads to $\alpha_N = 2.908 \times 10^{-24}$ cm^3, in excellent agreement with the direct[73,98] experimental values (2.919×10^{-24} cm^3, 2.994×10^{-24} cm^3) derived from molar polarization data. (There is a gap in the photoabsorption data between 1000 and 1082 Å; by examining electron impact energy loss data[107] we have estimated that the oscillator strength contribution between 10.33 and 11.46 eV is very nearly equal to the oscillator strength contribution between 11.46 and 12.89 eV, and hence the values in parenthesis in Table 18 have been estimated in this way.) The inference is that the high energy region requires a small increase in photoabsorption cross section, perhaps suggesting that values intermediate between those of de Reilhac–Damany and Lee, *et al.* be chosen.

TABLE 18 *The Oscillator Strength Distribution in N_2O*

Energy range	$S(0)$	$S(-1)$	$S(-2)$
Discrete (eV)			
6.9	0.00143	0.00283	0.00564
7.80–9.00	0.0257	0.0410	0.0655
9.00–10.33	0.3729	0.5266	0.7438
10.33–11.46	0.3490	0.4350	0.5427
11.46–12.89	(0.3490)	(0.3897)	(0.4351)
Partial sum	1.0980	1.3951	1.7927
Continuum (Å)			
626.0–960.0	2.191	1.816	1.525
320.0–626.0	4.916	2.436	1.252
200.0–320.0	3.024	0.8568	0.2472
9.89–200.0	9.042	0.6730	0.0889
0.62–9.89	0.924	0.0066	0.00005
Partial sum	20.097	5.7884	3.1131
Total sums	21.20	7.18	4.91_8

An additional factor which should diminish the discrepancy between experimental values and the $S(0)$ sum rule, without affecting $S(-1)$ or $S(-2)$ significantly, is absorption to Rydberg levels below the respective K edges of N and O. Bianconi *et al.*[71a] report two peaks below the nitrogen K edge (corresponding to the nonequivalent N atoms) but their absorption scale is in arbitrary units. From their data on N_2, we can estimate $f \simeq 0.1$ for each of these peaks and also for oxygen, bringing the computed $S(0)$ to within 2% of the TRK rule.

The ionization yield measurements (Fig. 22) are inconsistent. The extensive data of Cook *et al.*[105] differ substantially from unity, even at the shortest wavelength (~600 Å). However, isolated measurements at ~740[108] and 584 Å[99] provide evidence for a much higher value of η_i. Since a directly measured value of $|M_i|^2$ is unavailable, and the implication of Fig. 22 is that the bulk of the ionization yield measurements are unreliable, we have made no further attempt to estimate $S_i(-1)$.

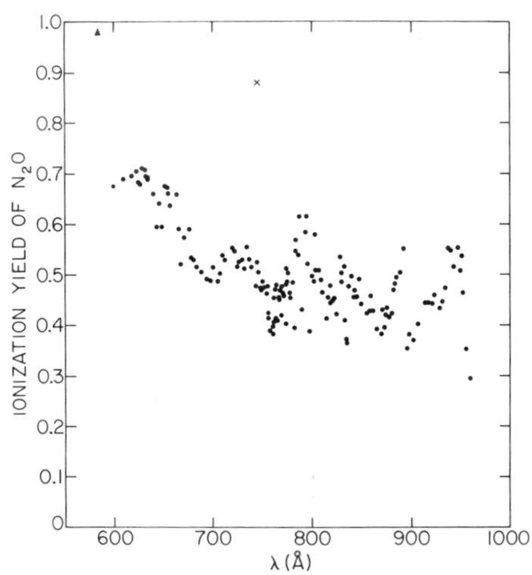

Fig. 22 The quantum yield of ionization experimentally observed in the photoabsorption of nitrous oxide. ○ Cook *et al.*[105]; × Rebbert and Ausloos[108]; ▲ Bennett *et al.*[99]

D. Polyatomics

1. NH_3

Much of our understanding of the photoabsorption spectrum of NH_3 is attributable to the experimental investigations of Watanabe and co-workers. In Fig. 23, we note the detailed curve between 580 and 1230 Å obtained from

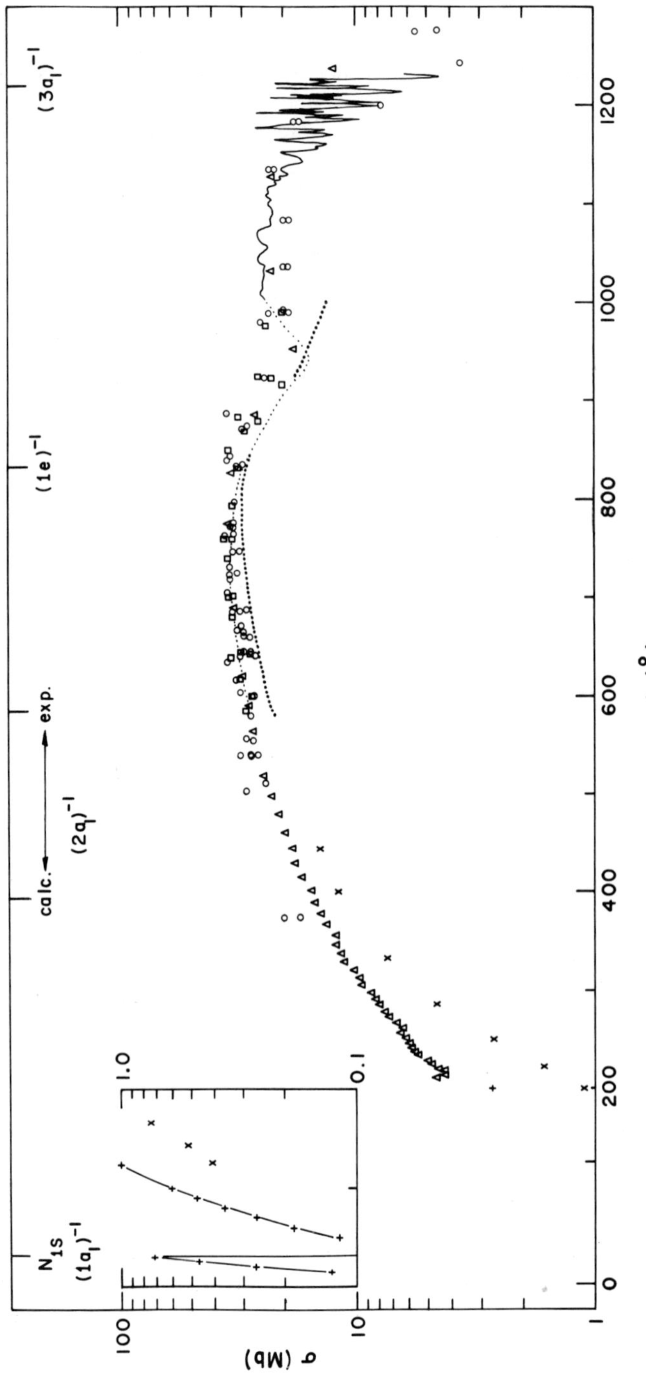

Fig. 23 Experimental photoabsorption cross section of ammonia vapor above the ionization threshold. △ Wight *et al.*[114a]; ●●● ——— Watanabe and Sood[109]; ● ● ● Metzger and Cook[112]; □ Walker and Weissler[109a]; × de Reilhac and Damany[114]; + Atomic additivity from Henke *et al.*[6]; ○ Sun and Weissler.[109b]

Watanabe and Sood,[109] which already takes us below the ionization threshold. Values at still longer wavelengths have been read from earlier work by Watanabe,[110] while the photoionization cross sections in the thresold region can be obtained from Watanabe and Mottl.[111] Unfortunately, all of these data have been presented in graphical form, and have had to be extracted.

The data of Metzger and Cook,[112] covering the 580–1000 Å range, deviate from the Watanabe and Sood values at their upper and lower extremities, and have not been used in the sum-rule analysis. Below 200 Å, atomic additivity and the values of Henke *et al.*[6] have been utilized. Structure in the photoabsorption spectrum in the 30–31 Å region (the vicinity of the K edge of nitrogen) can be inferred from the electron energy loss spectrum of Wight and Brion.[113] This is interpreted as being associated with the promotion of a K-shell electron to Rydberg orbitals which converge to the K-shell ionization threshold. Between 125 and 450 Å, photoabsorption cross sections have been reported by de Reilhac and Damany.[114] While they appear reasonably close to the interpolated curve between 300 and 450 Å, they appear much too low for $\lambda < 300$ Å, and hence have not been used in the analysis. (The more recent tabulation of de Reilhac and Damany[64a] has increased the cross sections below 300 Å, but they still appear to be ~30% too low.) A summary of this analysis is given in Table 19.

The $S(0)$ sum is seen to be ~6% too low, while the $S(-2)$ sum implies $\alpha_N = 2.07_9 \times 10^{-24}$ cm^3. Experimental data on molar polarization[98] can be used to derive values of α from 2.10 to 2.58×10^{-24} cm^3, the molar refraction at infinite λ yielding $2.2_0 \times 10^{-24}$ cm^3. Thus, $S(-2)$ would also

TABLE 19 *The Oscillator Strength Distribution in* NH_3

Energy region	$S(0)$	$S(-1)$	$S(-2)$
Subionization (Å)			
2038–2180	0.00759	0.01729	0.0394
1640–2038	0.07072	0.1460	0.3019
1444–1640	0.009384	0.01574	0.02642
1218–1444	0.1358	0.1956	0.2822
Partial sum	0.22349	0.37463	0.64992
Supraionization (Å)			
1185–1218	0.03469	0.04571	0.06025
1000–1185	0.3892	0.4615	0.5484
600–1000	2.194	1.805	1.514
200–600	3.265	1.466	0.7114
9.89–200	3.017	0.1984	0.02438
0.62–9.89	0.2517	0.001818	0.000015
Partial sum	9.15159	3.978428	2.85844
Total	9.38	4.35	3.51

appear to be ~5% too low. The low value of the $S(0)$ sum tends further to mitigate against the de Reilhac–Damany values, since they are substantially below our interpolated ones, and $S(0)$ has a significant contribution in the 200–600 Å region. The consistently low values of $S(0)$ and $S(-2)$ point to a systematic error in the Watanabe and Sood values, which appear to require a uniform increase of ~5%.

The ionization yield curve (Fig. 24) (from Watanabe and Sood,[109] Fig. 4) appears to approach unity at ~700 Å and remains there to 600 Å. We have assumed that this behavior persists to shorter wavelength. The yield has not been plotted as such for $\lambda > 1100$ Å, because the structure in the photoabsorption curve is not paralleled in the photoionization curve and the dips in the ionization yield that would result would be confusing. Instead, the photoabsorption and photoionization curves (reproduced from Watanabe and Sood,[109] Watanabe,[110] and Watanabe and Mottl[111]) have been individually plotted in Figs. 25–27. The inference from this behavior is that the

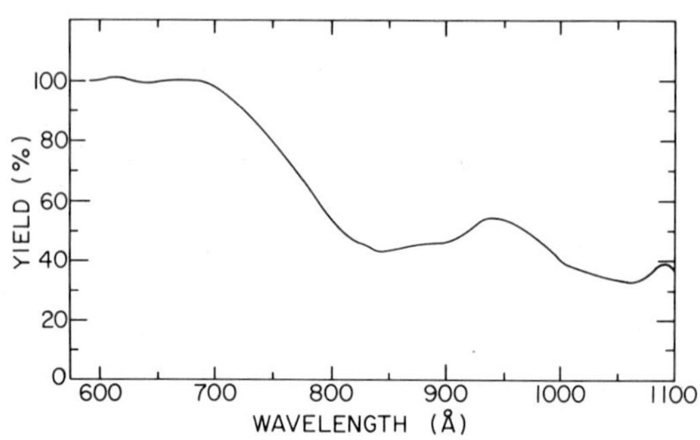

Fig. 24 The quantum yield of ionization experimentally observed in the photoabsorption of ammonia. (Data from Watanabe and Sood.[109])

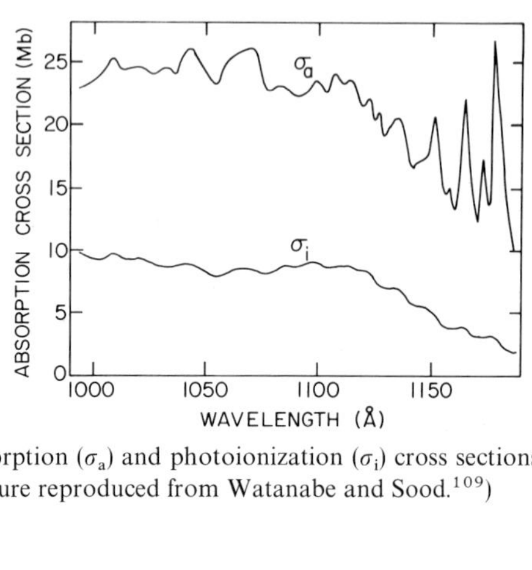

Fig. 25 Photoabsorption (σ_a) and photoionization (σ_i) cross sections of ammonia between 1000 and 1180 Å. (Figure reproduced from Watanabe and Sood.[109])

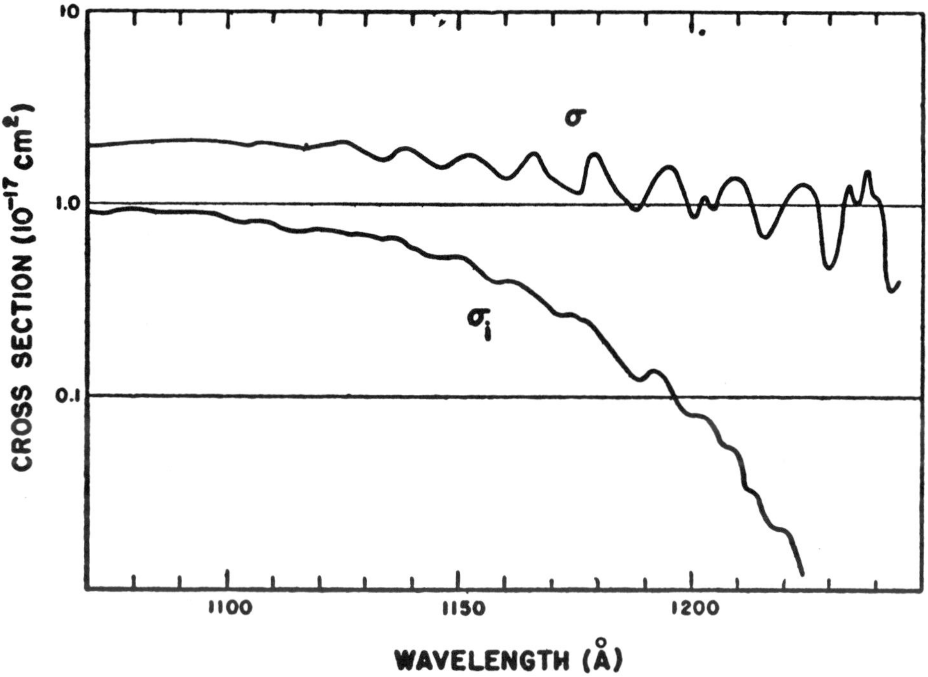

Fig. 26 Photoabsorption and photoionization cross sections of ammonia between 1070 and 1220 Å. (Figure reproduced from Watanabe.[110])

quasi-discrete levels in this region lead predominantly to predissociation, rather than autoionization.

From the photoionization cross sections shown in these figures, we compute contributions to $S_i(-1)$ of 1.359 (600–1000 Å), 0.1559 (1000–1185 Å), and 0.0011 (1185–1218 Å), for a total $S_i(-1)$ of 3.18_2. The value of $|M_i|^2$ reported by Rieke and Prepejchal[13] from electron ionization experiments is 3.58 ± 0.35. If one were to increase the value of Watanabe and Sood by ~5%, as suggested by the $S(0)$ and $S(-2)$ analysis, the Rieke–Prepejchal number would still be about 10% too high.

In view of the uncertainty in the photoabsorption data below 600 Å, we have chosen not to tabulate selected data for NH_3. For $\lambda > 600$ Å, a good approximation is to increase the σ values of Watanabe and Sood[109] by 5%. Below 600 Å, our interpolated line should also be increased by 6%. The recent electron impact energy loss (pseudophotoabsorption) data of Wight *et al.*[114a] merge nicely with the values of Watanabe and Sood at ~600 Å, but increasingly diverge from our interpolated values toward shorter wavelengths by as much as 30%. Although Wight *et al.* utilized a truncated form

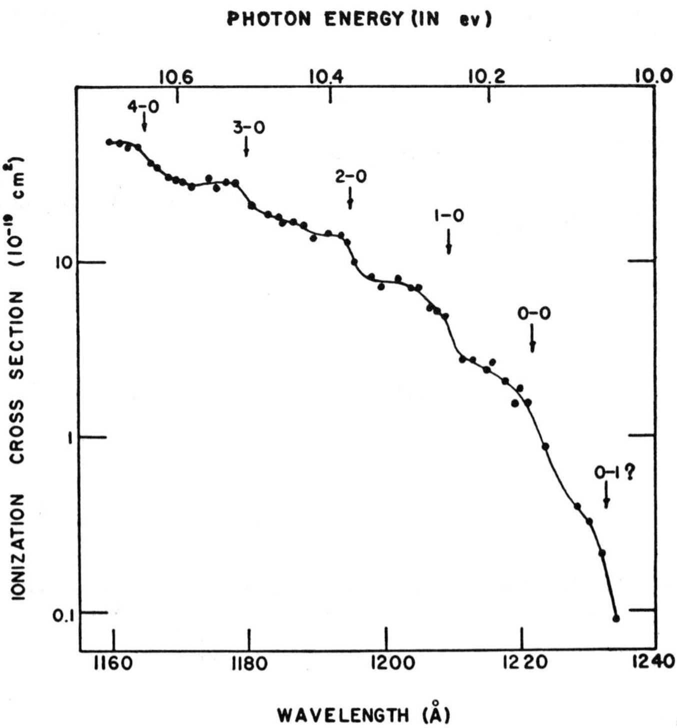

Fig. 27 Photoionization cross section of ammonia between 1160 and 1235 Å. (Figure reproduced from Watanabe and Mottl.[111])

of the TRK sum rule to normalize their data, our analysis would indicate too large a result for $S(0)$ if we were to adopt their numbers. The measurements of ionization yield by these authors, while not as detailed as those of Watanabe and Sood above 600 Å, do tend to confirm that η_i is essentially unity below 600 Å.

2. CH_4

A wealth of experimental data exists for CH_4 above the first ionization potential, as evidenced by Fig. 28. In the x-ray region, Lukirskii *et al.*[29] have reported values between 23.6 and 250.5 Å. These have been adopted in preference to those of Denne[58] which are about 30% lower, and would not merge well with the longer wavelength data of other workers nor with the sum-rule calculations, although they are very close to the values for atomic carbon alone (Henke *et al.*[6]). (In the region below 23 Å, the atomic carbon

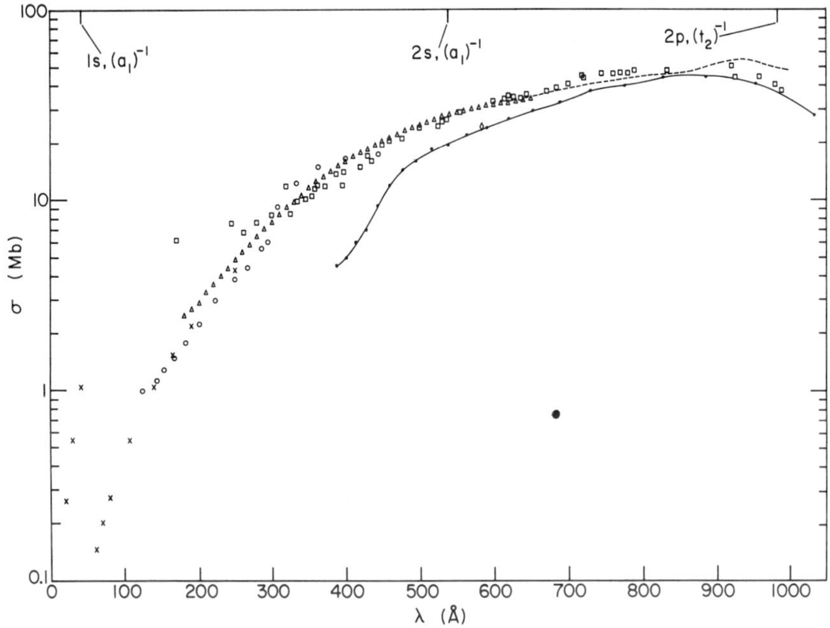

Fig. 28 Experimental photoabsorption cross section of methane above the ionization threshold. ○ de Reilhac and Damany[114]; × Lukirskii *et al.*[129]; Lukirskii and Zimkina[22]; □ Rustgi[115]; --- Metzger and Cook[112]; ●—● E. E. Koch and M. Skibowski[116]; △ Lee *et al.*[66]; ◇ Bennett *et al.*[99]

values have been used.) From 143 to 250 Å, the values of de Reilhac and Damany[114,64a] are in good agreement with those of Lukirskii[22,129] and merge with Rustgi's data[115] at ~300 Å. A smoothed set of values based primarily on the Rustgi data is taken from ~300 to 985 Å, the ionization threshold. A relative photoabsorption curve for CH_4 reported by members of the DESY synchrotron group[116] has been normalized to other data at its peak, and is seen to have a significantly different shape, and hence has been rejected for this analysis. Above 900 Å, the cross sections of Metzger and Cook[112] deviate somewhat from those of Rustgi. A choice between these is difficult to make, but in view of the satisfactory merging of Rustgi's values with the x-ray data, and the satisfactory agreement between Rustgi and the recent data from the Stoughton synchrotron (Lee *et al.*[66]) from ~450 to 650 Å, we have elected to remain with the (smoothed) Rustgi cross sections to the ionization threshold. The values of Lee *et al.* are about 20% larger than the x-ray values in the region of overlap. The selected values across the entire spectrum are summarized in Table 20. The contributions of the various regions to $S(k)$ are listed in Table 21.

TABLE 20 *Selected Values of the Photoabsorption Cross Section of CH_4* [a]

λ(Å)	σ(Mb)	λ(Å)	σ(Mb)
23.6	0.263	425.0	16.0
31.4	0.550	450.0	18.6
40.96	1.06	475.0	21.0
44.4	0.0617	500.0	23.2
64.35	0.148	525.0	25.2
72.20	0.205	550.0	28.0
81.98	0.276	575.0	30.3
108.65	0.550	600.0	33.0
139.50	1.049	625.0	34.5
164.60	1.532	650.0	36.0
190.3	2.19	675.0	38.0
250.5	4.31	700.0	40.0
		725.0	44.0
275.0	5.35	750.0	46.0
300.0	6.9	775.0	46.5
325.0	8.5	800.0	48.0
350.0	10.2	825.0	48.0
375.0	12.0	850.0	49.0
400.0	14.0	875.0	51.0
		900.0	51.0
		925.0	51.0
		950.0	47.5
		975.0	41.5
		985.0	39.0

[a] 0–985 Å.

Table 21 *The Oscillator Strength Distribution in CH_4*

Energy region	S(0)	S(−1)	S(−2)
Discrete (eV)			
8.55–10.95	0.28	0.381	0.518
10.95–12.6	0.45	0.521	0.603
Partial sum	0.73	0.902	1.121
Continuum (Å)			
625.0–985.0	2.913	2.518	2.213
400.0–625.0	2.321	1.306	0.7468
250.0–400.0	1.338	0.4822	0.1767
0.62–250.0	2.608	0.2480	0.0435
Partial sum	9.180	4.5542	3.1800
Total	9.91	5.456	4.30

The oscillator strength of the 8.55–10.95 eV band has been taken from the electron impact energy loss experiment of Harshbarger and Lassettre,[117] which is in good agreement with earlier photoabsorption data quoted therein. Most of the photoabsorption measurements do not adequately cover the energy range from 10.95 to 12.6 eV, which is beyond the LiF cutoff for some experiments and below the ionization potential (and hence of apparently less interest) for others. The value selected in Table 21 for f_n between 10.95 and 12.6 eV is primarily based on the spectrum reported by Harshbarger and Lassettre (Fig. 1 of their paper) and on equivalent data obtained at NBS and analyzed by Huebner.[118]

The total $S(0)$ sum (9.91) is fortuitously close, considering the uncertainty of some of the data. The total $S(-2)$ sum (4.37) implies $\alpha_N = 2.55 \times 10^{-24}\ \text{cm}^3$, in excellent agreement with the value ($2.56 \times 10^{-24}\ \text{cm}^3$) deduced from the molar polarization of methane.[98] The value of $S_I(-1)$ immediately available from Table 21 is 4.55. If the ionization yield (Fig. 29) between the ionization threshold and ~600 Å is taken into account, and the yield below 600 Å is assumed to be unity, one obtains $S_i(-1) = 3.77$. The corresponding quantity ($|M_i|^2$) arrived at by Rieke and Prepejchal[13] from electron impact ionization is 4.23 ± 0.13. This discrepancy is in the same direction and of about the same magnitude as has been previously encountered in a number of molecules when comparing photoabsorption values of $S_i(-1)$ with $|M_i|^2$ from the Rieke and Prepejchal measurements.

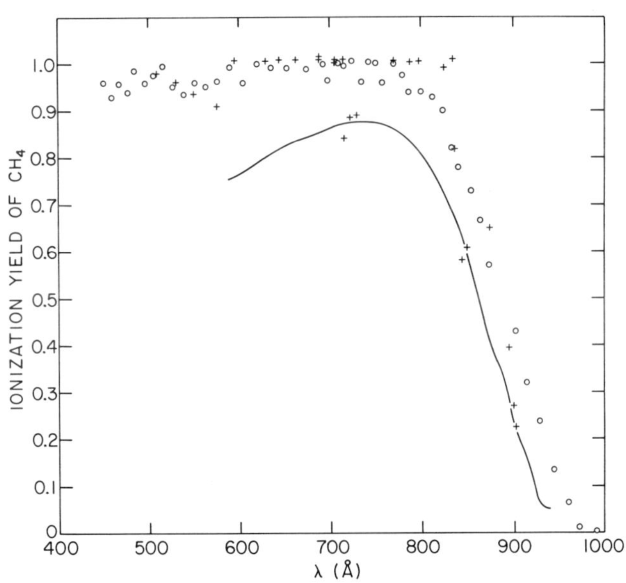

Fig. 29 The quantum yield of ionization experimentally observed in the photoabsorption of methane. ——— Metzger and Cook[112]; + Wainfan *et al.*[119]; ○ Backx *et al.*[120]

The ionization yield of CH_4 based solely on measurements with photons is not very satisfactory. The data of Metzger and Cook[112] indicate a decrease in η_i below 750 Å, to a value of ~0.75 at 584 Å. The earlier measurements of Wainfan *et al.*[119] do not indicate this behavior, but they are afflicted with poorer statistics. Quite recently, Backx *et al.*[120] have employed electron–electron coincidence measurements with their "pseudo-photoabsorption" technique (see Chapter VII.D.3). Their results for $\eta_i(CH_4)$ have less statistical uncertainty than those of Wainfan *et al.* and clearly demonstrate that $\eta_i \approx 1.0$ below 750 Å.

3. C_2H_2

The absolute photoabsorption cross section of acetylene requires more extensive study, as will be seen from a glance at Fig. 30 and from the ensuing sum-rule analysis. Nakayama and Watanabe[121] have examined the subionization region, and include a short portion of the ionization region (~1052–1088 Å). Person and Nicole[122] have also studied this latter region, and report somewhat higher cross sections. A gap exists in the available data between 1000 and 1052 Å, below which an important region (for sum rule contributions) from 580 to 1000 Å has been measured by Metzger and

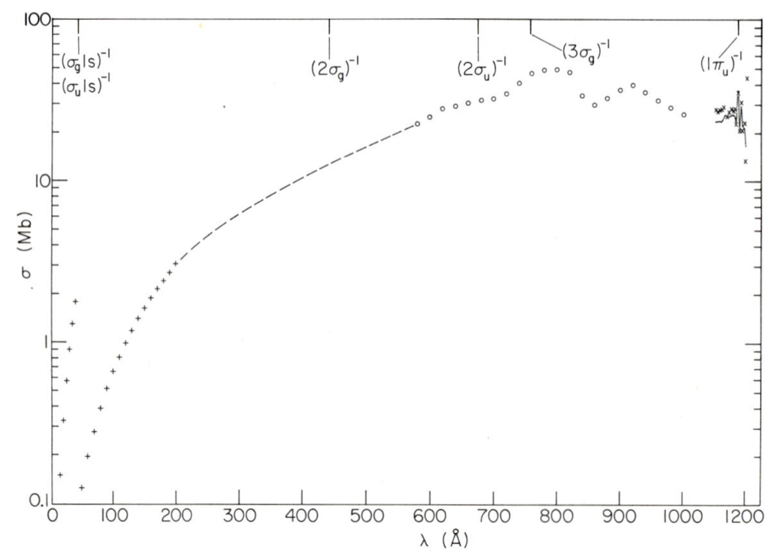

Fig. 30 Experimental photoabsorption cross section of acetylene above the ionization threshold. ●—● Nakayama and Watanabe[121]; ○ Metzger and Cook[112]; + Atomic additivity from Henke *et al.*[6]; × Person and Nicole.[122]

Cook.[112] Below 580 Å, no further data are available, although we have used atomic additivity (Henke *et al.*[6]) to estimate the region $0 < \lambda < 200$ Å.

The wavelength regions for which no data exist (200–580 Å; 1000–1052 Å) have been interpolated in the manner shown in Fig. 30. The results of the sum-rule analysis are summarized in Table 22.

TABLE 22 *The Oscillator Strength Distribution in C_2H_2*

Energy region	$S(0)$	$S(-1)$	$S(-2)$
Discrete (Å)			
1087.5–1272.5	0.3669	0.4735	0.6126
1272.5–1550.0	0.5894	0.8887	1.345
Partial sum	0.9563	1.3622	1.9576
Continuum (Å)			
1052.5–1087.5	0.09	0.105	0.125
1000–1052.5	0.1409	0.1586	0.1785
580–1000	2.784	2.330	1.991
200–580	3.180	1.358	0.6261
44.3–200	1.170	0.1631	0.02547
9.89–44.3	3.483	0.1108	0.003899
0.62–9.89	0.2685	0.001958	0.00001596
Partial sum	11.1164	4.227458	2.94998
Total	12.07	5.59	4.91

It can immediately be seen that the $S(0)$ sum is ~15% too low. The $S(-2)$ sum implies $\alpha_N = 2.909 \times 10^{-24}$ cm^3. Several experimental values of dipole polarizability are reported in the literature. Hirschfelder *et al.*[123] give $\alpha_{expt} = 3.33 \times 10^{-24}$ cm^3. From Maryott and Buckley[98] we can infer 3.95×10^{-24} cm^3 (based on dielectric constant measurements) and 3.40×10^{-24} cm^3 (based on molar refraction); and from Landolt-Bornstein,[73] 3.46×10^{-24} cm^3. A polarizability of $3.40 \pm 0.10 \times 10^{-24}$ cm^3 appears reasonable, and is about 15–20% higher than the quantity derived from photoabsorption data.

From the behavior of $S(0)$ and $S(-2)$, we can plausibly conclude that neither the subionization region nor the region below 200 Å is sufficiently sensitive to account for the simultaneous lack in both quantities. The most sensitive region is that between 580 and 1000 Å, and we infer that the cross sections measured in this region are too low. Of course, the absence of any data in large portions of the spectrum make even this conclusion tenuous.

From the measurements of Metzger and Cook on ionization yield between 580 and 1000 Å, and those of Person and Nicole between 1052 and 1088 Å (with interpolation between 1000 and 1052 Å) (see Fig. 31), we can estimate

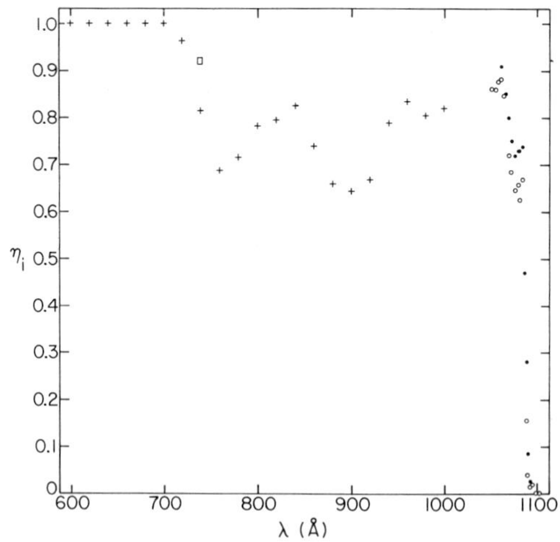

Fig. 31 The quantum yield of ionization (η_i) experimentally observed in the photoabsorption of acetylene. ○ Samson[123a]; ○ Person and Nicole[122]; + Metzger and Cook[112]; □ Rebbert and Ausloos.[108]

$S_i(-1)_{\text{total}} \equiv |M_i|^2$. The 580–1000 Å region contributes 1.958, the interpolated region, 0.142; and the 1052–1088 Å region, 0.082. According to Metzger and Cook, the yield below ~700 Å is unity, and hence we can use the $S(-1)$ entries in Table 22 below this wavelength. The value of $S_i(-1)_{\text{total}}$ that results is 3.815. Rieke and Prepejchal[13] have obtained a value of 5.21 ± 0.086 for $|M_i|^2$. The discrepancy of 25–30% is even larger than that already observed for $S(0)$ and $S(-2)$ and suggests the contribution of two effects:

(a) The photoabsorption cross section in the region 580–1088 Å is too low.

(b) The experiment of Rieke and Prepejchal is recording ionization in excess of that to be expected for $|M_i|^2$ alone.

4. C_2H_4

From Fig. 32 we note that experimental data exist over essentially all of the relevant electromagnetic spectrum. Some overlap exists between the experimental ranges of different investigators, with relatively minor discrepancies. Person and Nicole[124] present rather detailed information from the ionization threshold (~1180 Å) to the LiF cutoff (~1050 Å). Schoen[125] has covered a very large range, from below threshold to ~460 Å, but his line sources are sparse in some portions. The data of Metzger and Cook[112]

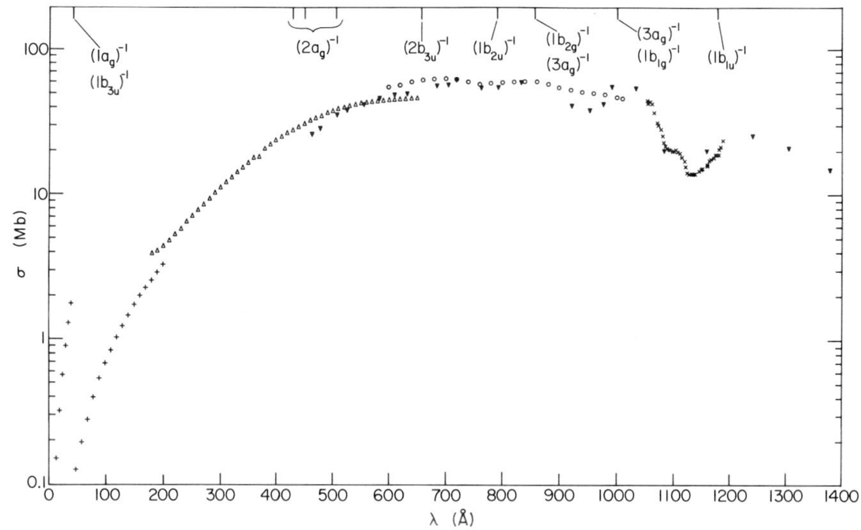

Fig. 32 Experimental photoabsorption cross section of ethylene above the ionization threshold. △ Lee *et al.*[66]; × Person and Nicole[124]; ○ Metzger and Cook[112]; ▽ Schoen[125]; + Atomic additivity from Henke *et al.*[6]

encompass the Hopfield continuum region (600–1010 Å), while Lee *et al.*[66] have reported values in the grazing incidence spectral range (180–650 Å) using the Stoughton synchrotron. The region below 200 Å has been estimated as before using atomic additivity and the atomic cross sections of Henke *et al.*[6]

The cross sections of Lee *et al.* appear too large in their short wavelength limit, presumably due to scattered light problems, and too small at their upper wavelength limit, perhaps due to the weakness of transmitted light in this region. This behavior of the Lee *et al.* data occurs with other molecules and seems to be characteristic of their experimental arrangement. In the sum-rule analysis, we utilize their values between 200 and 600 Å, accepting for the moment the abrupt jump at 200 Å between their results and those based on atomic additivity. Beyond 600 Å, where the Lee *et al.* data cross the Schoen values, we have utilized the cross sections of Schoen to 850 Å; between 850 and 1050 Å, values closer to those of Metzger and Cook have been used, since they merge more smoothly with those of Person and Nicole. Absorption coefficients in the subionization region have been extracted from the graphs of Zelikoff and Watanabe.[126] The results of the sum-rule analysis are given in Table 23.

The $S(0)$ sum of Table 23 is about 2.5% too high. From the $S(-2)$ sum we can deduce $\alpha_N = 3.99 \times 10^{-24}$ cm^3, whereas molar refraction data[98]

TABLE 23 *The Oscillator Strength Distribution in C_2H_4*

Energy region	S(0)	S(−1)	S(−2)
Subionization (Å)			
1500.0–1970.0	0.3042	0.5443	0.9756
1180–1500.0	0.3958	0.5674	0.8169
Partial sum	0.7000	1.1117	1.7925
Supraionization (Å)			
1050.0–1180	0.2656	0.3208	0.3879
600–1050.0	4.324	3.685	3.217
200.0–600.0	6.071	2.723	1.313
9.89–200.0	4.779	0.2861	0.03104
0.62–9.89	0.2685	0.001958	0.00001596
Partial sum	15.7081	7.016858	4.94896
Total	16.41	8.13	6.74

give $\alpha = 4.10 \times 10^{-24}$ cm^3. Hence, our $S(-2)$ sum is about 2.5% too low. This suggests that a tilt in the cross section curve of Lee et al, raising their values in the longer λ range and decreasing them in the shorter λ region, would tend to minimize the residual discrepancy, and would also permit a smoother merging of their data with those from the adjacent wavelength regions.

The ionization yield (Fig. 33) displays a rather dramatic minimum at ~1030 Å. The curve approaches unity at ~750 Å and we have assumed a

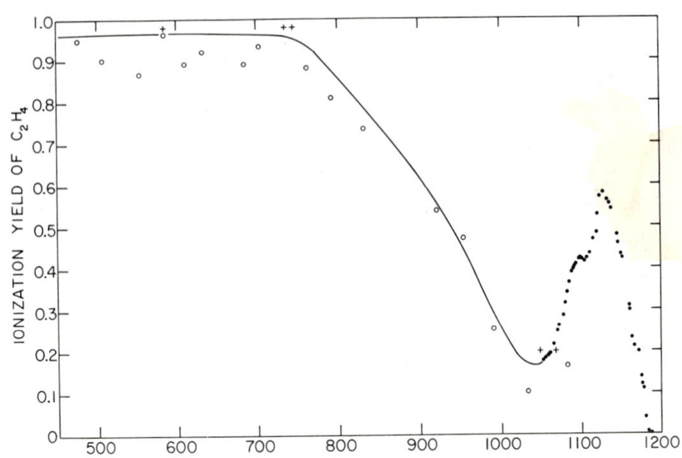

Fig. 33 The quantum yield of ionization experimentally observed in the photoabsorption of ethylene. ○ Schoen[125]; ● Person and Nicole[124]; + Rebbert and Ausloos.[108]

value of unity below 600 Å for the calculation of $S_i(-1)$. Between 600 and 1180 Å, we have combined the smoothed ionization yield curve with the selected photoabsorption curve to arrive at the best choice of photoionization cross sections. The contribution to $S_i(-1)$ between 600 and 1050 Å becomes 2.580, and between 1050 and 1180 Å, 0.103. The value of $S_i(-1)_{total}$ is thus 5.69_4, whereas Rieke and Prepejchal[13] report 6.75 ± 0.10 for $|M_i|^2$ from electron ionization experiments. The direction and magnitude (~15–20%) of the discrepancy have been noted previously in other comparisons with the Rieke–Prepejchal values.

5. C_2H_6

Direct experimental measurements of the photoabsorption cross section of ethane are now available which span the entire relevant electromagnetic spectrum. Raymonda and Simpson[127] and Lombos *et al.*[128] have reported results for the subionization region. Schoen[125] and Metzger and Cook[112] have presented both photoabsorption and photoionization values for the 600–1000 Å region. (Schoen's data covers a somewhat wider range, but uses line sources.) The DESY synchrotron group[116] have covered the 350–1100 Å region, while the Stoughton synchrotron workers[66] encompassed the 180–650 Å region. The soft x-ray domain (~9.9–200 Å) has been supplemented by atomic additivity with values from Henke *et al.*[6] The various measurements above the first IP are summarized in Fig. 34.

Additivity agrees very well with the directly measured values of Henke *et al.*[6] However, in the 180–200 Å region of overlap with Lee et al, the latter data are about 25% higher. This direction of discrepancy between the data of Lee *et al.* and other sources of data in the range $180 \text{ Å} < \lambda < 350 \text{ Å}$ has previously been observed in the cases of CO, NO, CO_2, N_2O, CH_4, and C_2H_4 (see also Phillips *et al.*,[93] H_2O).

The two sets of synchrotron measurements demonstrate wide discrepancy between 350 and 475 Å, the variation approaching a factor of two at 400 Å. At about 584 Å, five sets of investigations are in remarkable agreement, and from 600 to 1000 Å, the values of Schoen,[125] Metzger and Cook,[112] and Koch and Skibowski[116] generally agree, within their combined experimental errors.

The two subionization measurements are also in considerable discord with one another. The Raymonda and Simpson values are generally ~30% higher, and merge more smoothly with Schoen's values at the ionization potential.

This cursory review of the data implies that the sum-rule analysis may have a useful role to play in the selection of data from among the various experimental results. This analysis is detailed in Table 24.

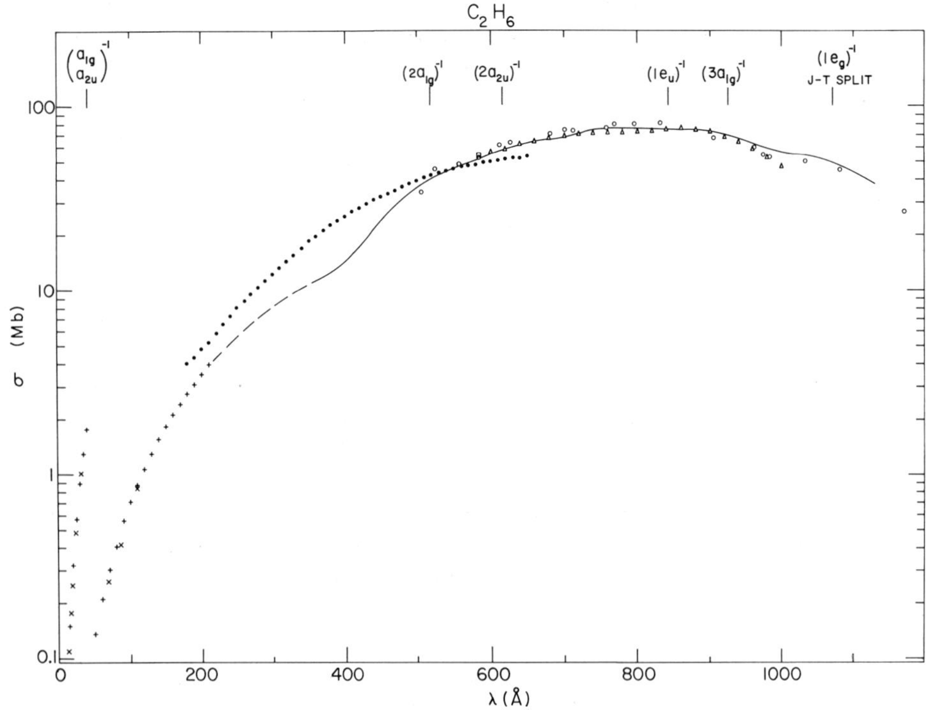

Fig. 34 Experimental photoabsorption cross section of ethane above the ionization threshold. ○ Schoen[125]; □ Bennett *et al.*[99]; + Atomic additivity from Henke *et al.*[6]; (calculation) and × (directly measured); ——— Koch and Skibowski[116]; △ Metzger and Cook[112]; ● Lee *et al.*[66]

TABLE 24 *The Oscillator Strength Distribution in C_2H_6*

Energy region	S(0)	S(−1)	S(−2)
Subionization			
1072.5–1539 Å	0.9505	1.2698	1.7072
(Raymonda and Simpson[127])			
Supraionization (Å)			
1000–1072.5	0.4050	0.4598	0.5222
58.4–1000	5.277	4.396	3.743
354–584	3.181	1.684	0.9077
210–354	1.502	0.4667	0.1481
(210–584 Lee *et al.*)	(6.169)	(2.736)	(1.296)
0.62–210	5.2135	0.3195	0.0376
Total	16.529	8.596	7.066
(Lee *et al.*[66a])	(18.02)	(9.18)	(7.31)

In order to simplify Table 24, the subionization values of Lombos *et al.*[128] have been omitted. It will readily be seen that the lower cross sections reported by these workers are in greater discord with the sum rules than the cross sections of Raymonda and Simpson.[127]

The significant difference between two possible $S(0)$ sums, given in Table 24, largely reflects the large discrepancy between the two synchrotron studies already alluded to. The cross sections of Lee *et al.* would lead to excellent agreement with the $S(0)$ sum rule, but they do not join smoothly with Henke's soft x-ray values. Some mean values between these latter two sets of data would appear to be a reasonable selection, on the basis of the $S(0)$ rule.

The $S(-2)$ sum implies $\alpha_N = 4.10 \times 10^{-24}$ cm^3, or $\alpha_N = 4.33 \times 10^{-24}$ cm^3, the latter utilizing the data of Lee *et al.*[66] The experimental value of α_N is given as 4.47×10^{-24} cm^3 from Hirschfelder *et al.*,[123] 4.39×10^{-24} cm^3 (based on R_∞ from Maryott and Buckley[98] and 4.49 from Landolt-Bornstein.[73] At this point, it will be noted that a significantly lower set of values in the subionization region (to which α_N is quite sensitive) as reported by Lombos *et al.* would widen the discrepancy with this sum rule. In addition, the $S(-2)$ analysis provides further support, though weak, for the results of Lee *et al.*,[66] rather than Koch and Skibowski,[116] in the region of major disagreement.

The $S(-1)$ values cannot be tested directly. However, we can estimate $S_i(-1)$ by making use of photoionization and ionization yield measurements. Although there is significant discrepancy between different groups on the ionization yield of ethane (see Fig. 35), the prevailing evidence (other than that of Metzger and Cook) suggests that the yield approaches unity at ~600 Å. We shall assume that it remains unity for $\lambda < 600$ Å. The actual photoionization cross section measurements between 584 and 1072 Å (see Fig. 36) are fortunately in better agreement with one another than the yields. Utilizing this latter figure, we have computed the $S_i(-1)$ contribution between 584 and 1072.5 Å to be 3.576. When combined with the $S(-1)$ contributions at shorter wavelengths, we obtain either $S_i(-1)_{\text{total}} = 6.047$ or $S_i(-1)_{\text{total}} = 6.632$, the latter based on the measurements of Lee *et al.* The electron impact ionization value of $|M_i|^2$ reported by Rieke and Prepejchal[13] is 6.80 ± 0.36.

It would be difficult to favor the values of Lee *et al.* over those of Koch and Skibowski on the basis of better agreement with $|M_i|^2$, however, since previous experience with other molecules has led us to expect the Rieke and Prepejchal[13] values to be large by ~10–20%.

The overall analysis has enabled us to choose between two sets of data in the subionization region, and strongly favors the values of Lee *et al.* over most of the wavelength range they covered, though perhaps they are too high in their short wavelength limit.

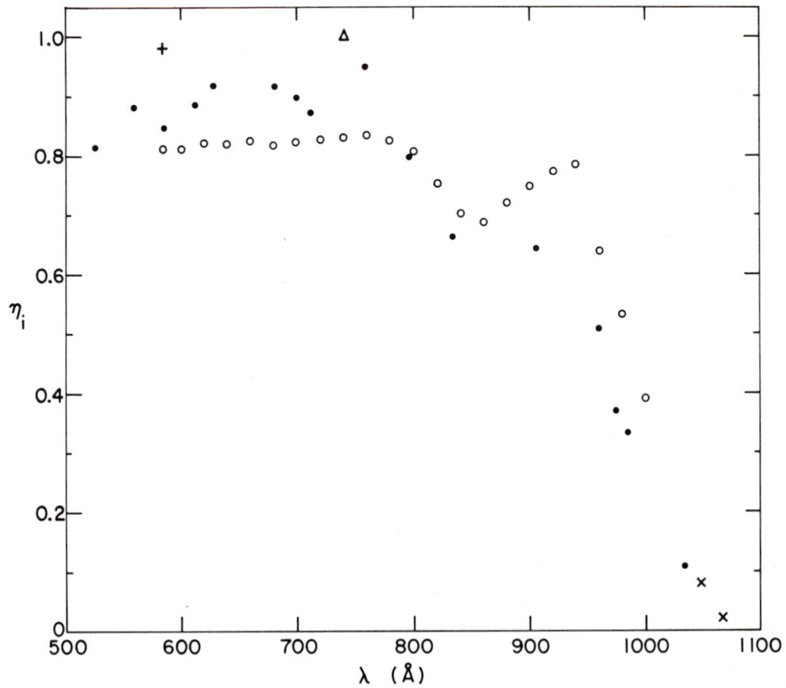

Fig. 35 The quantum yield of ionization experimentally observed in the photoabsorption of ethane. ● Schoen[125]; ○ Metzger and Cook[112]; + Bennett *et al.*[99]; △ Rebbert and Ausloos[108]; × Klots.[128a]

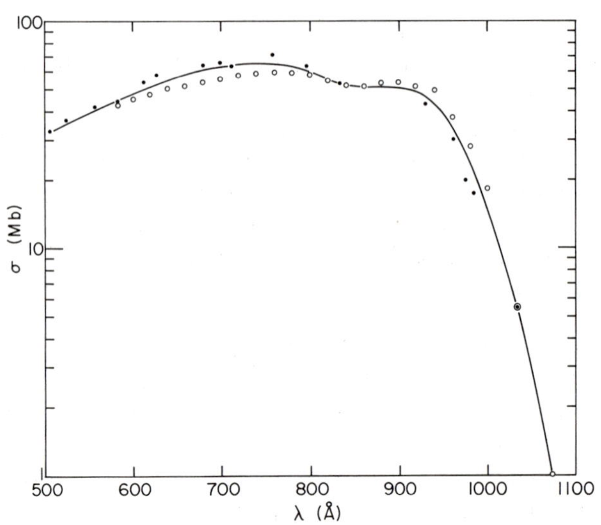

Fig. 36 Experimental photoionization cross section of ethane in the valence ionization region. ● R. I. Schoen,[125]; ○ P. H. Metzger and G. R. Cook.[112]

6. CH_3OH

The photoabsorption cross section of methanol in the ionization continuum can be mapped out rather completely, although some regions clearly need further study. Ogawa and Cook[129] have provided data from about 500 Å to the ionization threshold. The portion from the ionization onset at ~1148 Å to the LiF cutoff (~1050 Å) has been investigated in more detail by Person and Nicole[130] and preliminary data from these authors[131] has extended this work to ~580 Å. De Reilhac and Damany[64] cover a section between 200 and 325 Å, and the region below 200 Å has been approximated by atomic additivity. (This approximation can be tested for C_2H_5OH and the agreement with directly measured values is very good.) Sanson and Damany-Astoin[132] have obtained values photographically between ~180 and 1000 Å which are not unreasonable above ~550 Å, but which seem totally unacceptable below 550 Å (see Fig. 37).

The atomic additivity values join rather smoothly with the data of de Reilhac and Damany. For a preliminary sum-rule test, the gap between 325 and 500 Å has been filled by a smooth extrapolation between the values of

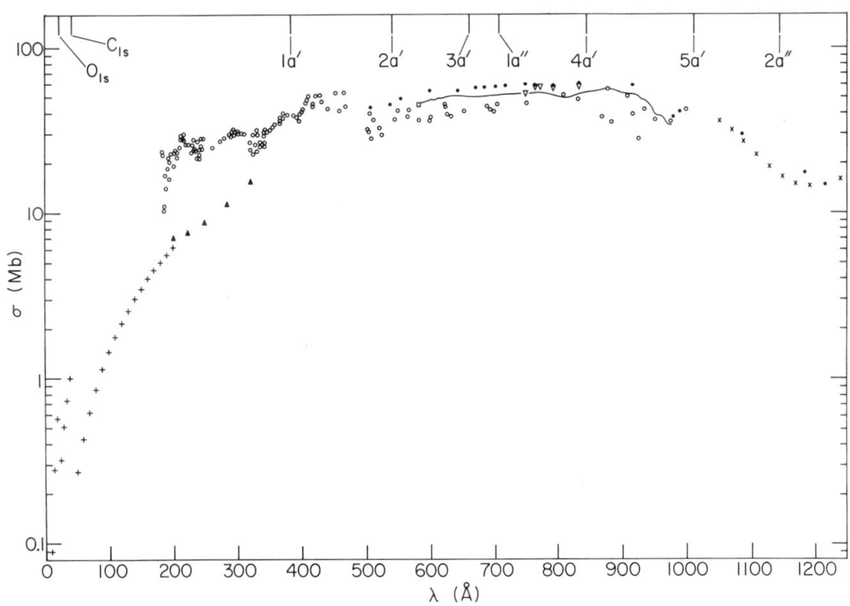

Fig. 37 Experimental photoabsorption cross section of methanol vapor above the ionization threshold. ● Ogawa and Cook[129]; ▽ Ogawa and Cook[129]; using grazing incidence; × Person and Nicole[130]; ○ Sanson and Damany-Astoin[132]; □ Bennett *et al.*[99]; + Atomic additivity from Henke *et al.*[6]; ——— Person and Nicole[131]; ▲ de Reilhac and Damany.[64]

de Reilhac and Damany and Ogawa–Cook. The sum-rule analysis is given in Table 25.

The subionization photoabsorption cross section has been reported by Salahub and Sandorfy,[133] but recently analyzed electron impact energy loss data (Huebner[118]) are believed to be more reliable for the purpose of estimating the oscillator strength contribution.

The preliminary analysis presented in Table 25 points to a slight overestimate ($\sim 1\%$) in $S(0)$. The $S(-2)$ sum implies $\alpha_N = 3.391 \times 10^{-24}$ cm^3, about 5% larger than the experimental value ($\alpha = 3.23 \times 10^{-24}$ cm^3, from Landolt-Bornstein[73] and Maryott and Buckley[98]). At this point, it is of interest to examine the preliminary data of Person and Nicole in the 500–990 Å region [which carries a large fraction of both $S(0)$ and $S(-2)$] to see its effect on the sum rules. The Person–Nicole data yield 5.205, 3.998, and 3.181 for $S(0)$, $S(-1)$, and $S(-2)$ in the 500–990 Å range. The effect is to reduce $S(0)_{\text{total}}$ to 17.424, and $S(-2)_{\text{total}}$ to 5.276. The $S(0)_{\text{total}}$ now becomes $\sim 3\%$ too low, and the polarizability calculated from $S(-2)_{\text{total}}$ (3.127×10^{-24} cm^3) is also about 3% too low.

Hence, we conclude that a set of values between those of Ogawa–Cook and Person–Nicole (which do not differ greatly, as can be noted in the figure) would give the best agreement with these sum rules. Note that a relatively small contribution is provided by the subionization region, even for $S(-2)$.

The ionization yield has been measured over a short region near threshold and at 584 Å (see Fig. 38), leaving a large gap. From Table 25, we note that this is just the region which makes the largest contribution to $S_i(-1)$,

TABLE 25 *The Oscillator Strength Distribution in CH_3OH*

Energy region	$S(0)$	$S(-1)$	$S(-2)$
Normalized Huebner[118] data (EELS) (Å)			
1815.2–2069.8	0.002566	0.005316	0.01102
1405.7–1815.2	0.06626	0.1105	0.1848
1148.0–1405.7	0.1999	0.2733	0.3748
Partial total	0.2687_3	0.3891_2	0.5706_2
Supraionization (Å)			
990–1148	0.4904	0.5668	0.6562
500–900	5.979	4.574	3.627
200–500	5.135	1.981	0.811
9.89–200	5.770	0.4359	0.05686
0.62–9.89	0.5545	0.00398	0.00003
Partial sum	17.9289	7.56168	5.1511
Total	18.20	7.95	5.72

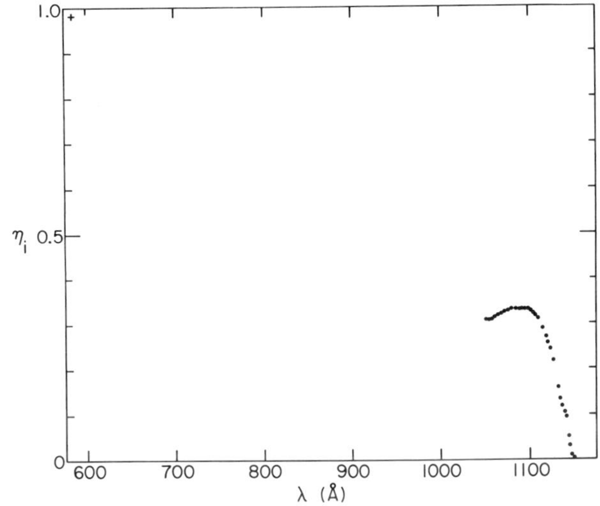

Fig. 38 The quantum yield of ionization (η_i) experimentally observed in the photoabsorption of methanol. ● Person and Nicole[130]; + Bennett *et al.*[99]

and hence $|M_i|^2$. An obvious overestimate is $|M_I|^2$, which we compute to be 7.56. The reduction to $|M_i|^2$ awaits further experiments on η_i between 600 and 950 Å. The value of $|M_i|^2$ directly determined by Rieke and Prepejchal[13] is 6.22 ± 0.18.

7. C_6H_6

Figure 39 is a superposition of photoabsorption data from a number of investigators in the range from the ionization threshold through the "K edge." Earlier work of Bunch *et al.*[134] covers a large energy range down to ~430 Å, but with substantial scatter. The region from ~1345 Å to the LiF cutoff (~1050 Å) has been examined by Person[135] and by Samson.[136] The wavelength range 600 Å ≤ λ ≤ 1150 Å has been investigated by Yoshino *et al.*[137] An investigation by Koch and Otto,[138] utilizing the DESY synchrotron, has encompassed a very broad range from ~350 to ~2280 Å, which covers the most significant subionization bands.

The region below 200 Å has been estimated by atomic additivity (Henke *et al.*[6]). A smooth interpolation has been employed between 200 and 350 Å in Fig. 39 (covering the region in which data is lacking).

The initial sum-rule test has been performed using the data of Koch and Otto from 2278 to 350 Å, together with the atomic additivity estimate (λ < 200 Å) and the interpolated region. The results of this analysis are presented in Table 26.

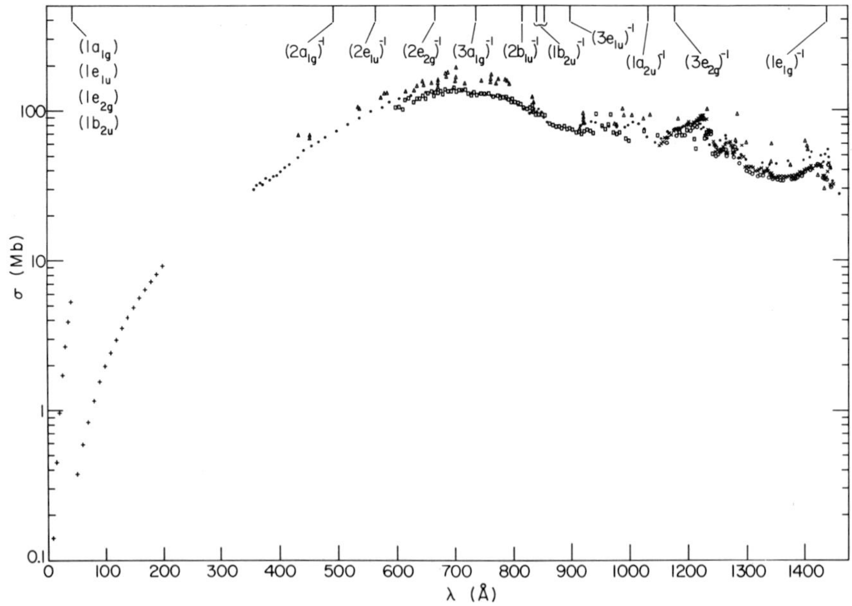

Fig. 39 Experimental photoabsorption cross section of benzene vapor above the ionization threshold. + Atomic additivity from Henke *et al.*[6]; ● Koch and Otto[138]; △ Bunch *et al.*[134]; □ Yoshino *et al.*[137]; ○ Person[135]; × Samson.[136]

TABLE 26 *The Oscillator Strength Distribution in* C_6H_6

Energy region	$S(0)$	$S(-1)$	$S(-2)$
Subionization			
1345–2278 Å	1.063	1.983	3.744
Supraionization (Å)			
1052–1345	1.358	1.745	2.253
600–1052	8.917	7.426	6.333
200–600	11.72	5.335	2.621
9.89–200	14.17	0.8297	0.0884
0.62–9.89	0.8054	0.00587	0.00005
Partial sum	36.9704	15.34157	11.29545
Total	38.0	17.32	15.04

The $S(0)$ sum is 38.0, about 9.5% below its required value. The $S(-2)$ sum is equivalent to $\alpha_N = 8.91 \times 10^{-24}$ cm^3, whereas the directly determined value of $\alpha_N = 9.99 \times 10^{-24}$ cm^3 (Landolt-Bornstein[73]), leaving us 10.8% too low for $S(-2)$.

Returning now to Fig. 39, we note that the photoabsorption cross sections of Koch and Otto are larger than those of Person and of Samson in the region of overlap. Since our preliminary sum-rule analysis already lacks oscillator strength, the alternative use of Person's or Samson's values clearly would not improve matters. The data of Yoshino *et al.* are quite close to those of Koch and Otto and would not significantly change our results. Only the values of Bunch *et al.* are significantly higher, and here the data are too scattered to be considered reliable to better than $\pm 15\%$.

From the ionization yield measurements of Person and Yoshino *et al.* (Fig. 40), it is possible to estimate $S_i(-1)$. In this way, we estimate a contribution to $S_i(-1)$ of 0.74 (1052–1345 Å) and 6.80 (600–1052 Å). Since the ionization yield appears to asymptotically reach unity above 600 Å, we have assumed a value of $\eta_i = 1$ below 600 Å. The total $S_i(-1)$ obtained by this procedure is 13.71.

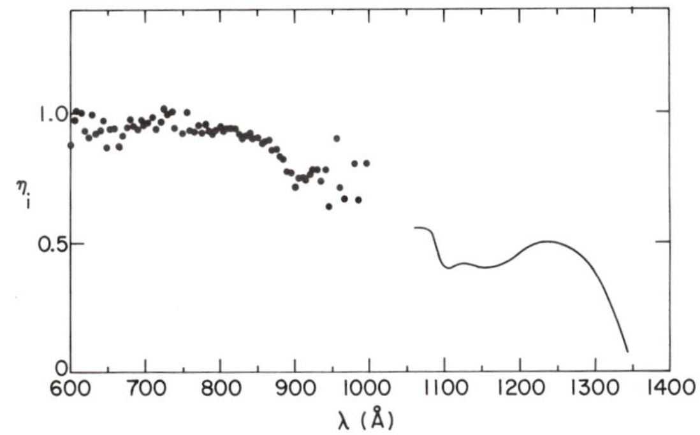

Fig. 40 The quantum yield of ionization (η_i) experimentally observed in the photoabsorption of benzene. ○ Yoshino *et al.*[137]; ——— Person.[135]

The value of $|M_i|^2$ which should be compared with $S_i(-1)$, deduced from high energy electron and positron ionization measurements[13] is 17.54 ± 0.37. The difference is 21.9%, rather larger than the discrepancy in $S(0)$ and $S(-2)$. However, we have already noted in almost all other comparisons of $S_i(-1)$ with $|M_i|^2$ that the latter tends to be $\sim 10\%$ too high. Hence, the general conclusion to be drawn from all three sum rules is that the available photoabsorption data are too low by $\sim 10\%$, with the discrepancy being roughly uniformly spread across the electromagnetic spectrum. If anything, the lower energy (more explored) region is lacking more than the higher energy region.

It is not clear at this time what the source of such a discrepancy might be, especially since several groups of investigators have obtained the lower values. Possible sources of error include pressure determinations of benzene, or fluorescence that might be registered by the photodetector as diminished absorption.

8. SF_6

The photoabsorption cross section from approximately the ionization threshold to the vicinity of the K edge of sulfur is displayed in Fig. 41. The most complete set of data from 300 to 800 Å are from the Tokyo synchrotron group (Sasanuma *et al.*[65,138a]). Their values are in reasonable agreement with a few isolated points given by Codling.[139] The DESY synchrotron results (Blechschmidt *et al.*[140]) and the recent data from the group using the Stoughton synchrotron[141] provide significantly lower photoabsorption cross sections. The latter join rather smoothly to the x-ray data of Vinogradov and Zimkina,[142] which continue on to ~7 Å. Measurements by Lavilla and Deslattes[143] extend the experimental data to 4.35 Å, beyond which some extrapolation is necessary. Vinogradov and Zimkina, who made their own sum-rule analysis, estimate 3.0 as the *S*(0) contribution for this hard x-ray region, and our estimate is in rough agreement. Interesting features of this spectrum, alluded to in Chapter IV, Section E, are the pre-

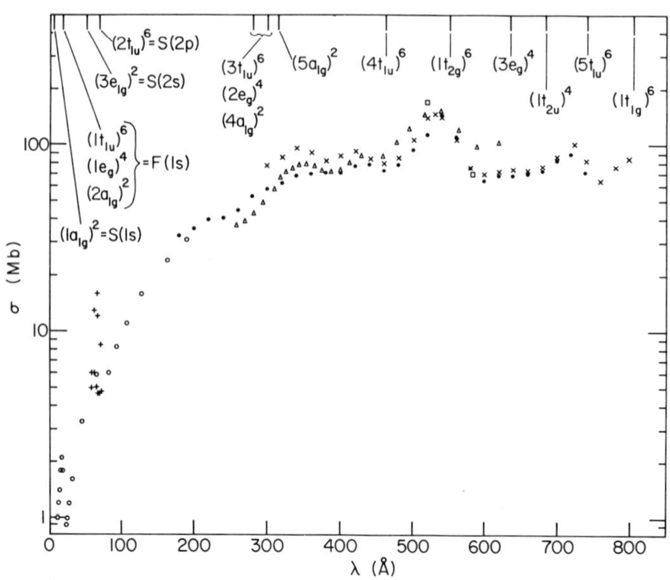

Fig. 41 Experimental photoabsorption cross section of sulfur hexafluoride above the ionization threshold. □ Codling[139]; + Zimkina and Fomichev[139a]; ○ Vinogradov and Zimkina[142]; △ Bleckschmidt *et al.*[140]; × Sasanuma *et al.*[65]; ● Lee *et al.*[141]

and post-threshold structures near the K and L edges of sulfur and the K edge of fluorine, which have been interpreted as resulting from "molecular barriers" to the escape of inner electrons. Hitchcock and Brion[143a] provide a useful and up-to-date set of references on this topic, as well as additional data. However, the recent articles have focused attention on spectroscopy and the line shapes, rather than absolute cross sections, and hence have had a negligible effect on the sum-rule analysis. The summary of the contributions of the various energy regions to the $S(k)$ are given in Table 27.

TABLE 27 *The Oscillator Strength Distribution in* SF_6

Energy region	$S(0)$		$S(-1)$		$S(-2)$	
Discrete (eV band)						
11.54	0.45		0.5303		0.6250	
13.30	0.10		0.1023		0.1046	
14.35	0.18		0.1706		0.1617	
14.90	~0.03		0.0274		0.0250	
Partial sum	0.76		0.8306		0.9163	
Continuum (Å)						
190.15–800.0	31.83	(27.30)	13.78	(12.00)	6.839	(6.081)
5.04–190.15	31.14		2.353		0.2958	
4.35–5.04	0.7535		0.00389		0.00002	
0–4.35	~3.0		0.007		0.00002	
Partial sum	66.72	(62.19)	16.144	(14.364)	7.1348	(6.3768)
Total	67.48	(62.95)	16.97	(15.19)	8.05	(7.29)

The contribution of the continuum to $S(0)$ is 66.46, rather close to the value (64 or 66.5, depending upon choices) obtained by Vinogradov and Zimkina. These authors presumably intended their results to include the discrete portion. Our explicit inclusion of the discrete region from the electron impact energy loss data of Simpson *et al.*[144] yields $S(0) = 67.2$, about 4% below the value necessary to satisfy the TRK sum rule.

The total $S(-2)$ implies $\alpha_N = 4.77 \times 10^{-24}$ cm^3. Two rather different experimental values exist—those based on dielectric constant measurements, and others based on molar refraction (refractive index) measurements. The former (Maryott and Buckley[98]; Nelson and Cole[145]) give $\alpha = 6.54 \times 10^{-24}$ cm^3, the latter (Maryott and Buckley[98]; Landolt-Bornstein[73]) give $\alpha = 4.48 \times 10^{-24}$ cm^3. This large difference is apparently due to the fact that the molar refraction measurements, performed at optical frequencies, probe only the electronic excitations, whereas the dielectric constant measurements include polarization due to atomic motions as well. The two types of measurement usually yield very similar values of α, but in this particular case they are grossly disparate.

Our sum-rule analysis has concentrated on electronic excitations, ignoring the low frequency vibrational contributions. Hence, our computed α is in rather good agreement with the value obtained from molar refraction.

If the recent data of Lee *et al.*[141] are used between 190 and 800 Å, one obtains $S(0) = 62.69$, $S(-1) = 15.19$, and $\alpha = 4.32$. The sum-rule analysis implies that these photoabsorption cross sections are too low, and the earlier ones are to be preferred over most of this wavelength range.

The discrete contribution is probably not an underestimate (although the 14.9–15.3 eV region is not accounted for) since the electron impact values which we have chosen are significantly higher than earlier photoabsorption results (see the table in Simpson *et al.*[144]).

Hence, in order to approach the higher experimental value of α, it is probably necessary to include very low frequency vibrational contributions, which have been omitted in our discussion.

The ionization yield values reported by the Tokyo synchrotron group[65,138a] are much less than unity ($\sim$0.30) even at 300 Å. A test of such low yields could come from electron ionization measurements of $|M_i|^2$, but unfortunately none are currently available.

REFERENCES

1. P. M. Dehmer, J. Berkowitz, and W. A. Chupka, *J. Chem. Phys.* **59**, 5777 (1973).
1a. J. A. R. Samson and V. E. Petrosky, *J. Electron Spectrosc. Relat. Phenom.* **3**, 461 (1974).
1b. C. L. Pekeris, *Phys. Rev.* **115**, 1216 (1959).
2. W. Kolos and L. Wolniewicz, *J. Chem. Phys.* **41**, 3663 (1964); **43**, 2429 (1965).
3. J. A. R. Samson, *Adv. At. Mol. Phys.* **2**, 177 (1966).
4. J. F. Lowry, D. H. Tomboulian, and D. L. Ederer, *Phys. Rev.* **137**, A1054 (1965).
5. B. L. Henke, R. White, and B. Lundberg, *J. Appl. Phys.* **28**, 98 (1957).
6. B. L. Henke, R. L. Elgin, R. E. Lent, and R. B. Ledingham, *Norelco Rep.* **14**, 112 (1967).
7. A. R. P. Rau and U. Fano, *Phys. Rev.* **162**, 68 (1967).
8. M. Inokuti and Y.-K. Kim, *Phys. Rev.* **186**, 100 (1969).
9. B. Schiff and C. L. Pekeris, *Phys. Rev.* **134**, A638 (1964).
10. A. Dalgarno and E. M. Parkinson, *Proc. R. Soc. London, Ser. A* **301**, 253 (1967).
11. K. L. Bell and A. E. Kingston, *Proc. Phys. Soc. London* **90**, 31 (1967).
12. Y.-K. Kim and M. Inokuti, *Phys. Rev. A* **3**, 665 (1971).
13. F. F. Rieke and W. Prepejchal, *Phys. Rev. A* **6**, 1507 (1972).
14. R. R. Teachout and R. T. Pack, *At. Data* **3**, 195 (1971).
14a. Y.-K. Kim, unpublished calculation (1969).
15. J. B. West and G. V. Marr, *Proc. R. Soc. London, Ser. A* **349**, 397 (1976).
16. D. L. Ederer and D. H. Tomboulian, *Phys. Rev.* **133**, A1525 (1964).
17. J. A. Victoreen, *J. Appl. Phys.* **20**, 1141 (1949).
18. S. Natali, C. E. Kuyatt, and S. R. Mielczarek, unpublished observations (1973).
19. M. Naon, M. Cornille, and Y.-K. Kim, *J. Phys. B* **8**, 864 (1975).
20. Y.-K. Kim, *Radiat. Res.* **61**, 21 (1975).
21. C. B. Opal, E. C. Beatty, and W. K. Peterson, *At. Data* **4**, 209 (1972).
22. A. P. Lukirskii and T. M. Zimkina, *Bull. Acad. Sci. USSR, Phys. Ser.* **27**, 808 (1963).

23. S. J. M. Allen, cited in A. H. Compton and S. K. Allison, "X-Rays in Theory and Experiment," p. 799. Van Nostrand, Princeton, New Jersey, 1935.
24. W. S. Watson, *J. Phys. B* **5**, 2292 (1972).
25. M. Nakamura, M. Sasanuma, S. Sato, M. Watanabe, H. Yamashita, Y. Iguchi, A. Ejiri, S. Nakai, S. Yamaguchi, T. Sagawa, Y. Nakai, and T. Oshio, *Phys. Rev. Lett.* **21**, 1303 (1968).
26. C. M. Lee and K. T. Lu, *Phys. Rev. A* **8**, 1241 (1973); C. M. Lee, *Phys. Rev. A* **10**, 584 (1974).
27. W. S. Watson, *J. Phys. B* **5**, 2292 (1972).
28. A. P. Lukirskii, I. A. Brytov, and T. M. Zimkina, *Opt. Spectrosc.* (*USSR*) **17**, 234 (1964).
29. R. Haensel, G. Keitel, P. Schreiber, and C. Kunz, *Phys. Rev.* **188**, 1375 (1969).
30. B. L. Henke, *Proc. Int. Conf. Inner Shell Ioniz. Phenom. Future Appl., USAEC Conf.-720404* **2**, 816 (1972).
31. F. Wuilleumier, *Phys. Rev. A* **6**, 2067 (1972).
32. W. J. Veigele, E. Briggs, L. Bates, E. M. Henry, and B. Bracewell, "X-Ray Cross Section Compilation from 0.1 KeV to 1 MeV," Def. Nucl. Ag. Doc. DNA 2433F, Vol. 1 (1971) (available from Kaman Sciences Corp., Colorado Springs, Colorado); Graphs available from J. H. Hubbell and W. J. Veigele, *Natl. Bur. Stand.* (*U.S.*), *Tech. Note No.* 901 (1976).
33. R. E. Huffman, Y. Tanaka, and J. C. Larrabee, *Appl. Opt.* **2**, 947 (1963).
34. J. A. R. Samson, *Phys. Lett.* **8**, 107 (1964).
35. R. P. Madden and K. Codling, *J. Opt. Soc. Am.* **54**, 268 (1964).
36. J. Geiger, *Z. Phys. A* **282**, 129 (1977).
37. D. L. Ederer, *Phys. Rev. Lett.* **13**, 760 (1964).
38. R. D. Deslattes, *Acta Crystallogr., Sect. A* **25**, 89 (1969).
39. R. E. Huffman, Y. Tanaka, and J. C. Larrabee, *J. Chem. Phys.* **39**, 902 (1963).
40. A. Skerbele, K. J. Ross, and E. N. Lassettre, *J. Chem. Phys.* **50**, 4486 (1969); A. Skerbele and E. N. Lassettre, *J. Chem. Phys.* **52**, 2708 (1970); A. Skerbele and E. N. Lassettre, *J. Chem. Phys.* **56**, 8405 (1972).
41. R. Lincke and B. Stredele, *Z. Phys.* **238**, 164 (1970).
42. B. Brehm, *Z. Naturforsch., Teil A* **21**, 196 (1966).
43. J. Berkowitz and C. Lifshitz, *J. Phys. B* **1**, 438 (1968).
44. R. B. Cairns, H. Harrison, and R. I. Schoen, *J. Chem. Phys.* **53**, 96 (1970).
45. H. Beutler, *Z. Phys.* **86**, 710 (1933); **87**, 19 (1933).
46. H. Beutler and K. Guggenheimer, *Z. Phys.* **87**, 176 (1933).
47. W. R. S. Garton and J. P. Connerade, *Astrophys. J.* **155**, 667 (1969).
47a. G. R. Cook and P. H. Metzger, *J. Opt. Soc. Am.* **54**, 968 (1964).
47b. F. J. Comes and H. O. Wellern, *Z. Naturforsch., Teil A* **23**, 881 (1968).
47c. R. E. Rebbert and P. Ausloos, *J. Res. Natl. Bur. Stand., Sect. A* **75**, 719 (1971).
48. M. W. D. Mansfield, *Astrophys. J.* **180**, 1011 (1973).
49. W. A. Chupka and J. Berkowitz, *J. Chem. Phys.* **51**, 4244 (1969).
50. P. M. Dehmer and W. A. Chupka, *J. Chem. Phys.* **65**, 2243 (1976).
51. J. Geiger and M. Topschowsky, *Z. Naturforsch., Teil A* **21**, 626 (1966).
52. J. Geiger and H. Schmoranzer, *J. Mol. Spectrosc.* **32**, 39 (1969).
53. A. C. Allison and A. Dalgarno, *Mol. Phys.* **19**, 567 (1970); *At. Data* **1**, 289 (1970).
54. L. Wolniewicz, *J. Chem. Phys.* **51**, 5005 (1969).
55. A. Dalgarno and A. C. Allison, *J. Geophys. Res.* **74**, 4178 (1969).
56. J. E. Mentall and E. P. Gentieu, *J. Chem. Phys.* **52**, 5641 (1970).
57. J. A. R. Samson and R. B. Cairns, *J. Opt. Soc. Am.* **55**, 1035 (1965).
57a. J. A. R. Samson, J. L. Gardner, and G. N. Haddad, *J. Electron Spectrosc. Relat. Phenom.* **12**, 281 (1977).
58. D. R. Denne, *J. Phys. D* **3**, 1392 (1970).

59. G. A. Victor and A. Dalgarno, *J. Chem. Phys.* **50**, 2535 (1969).
60. A. L. Ford and J. C. Browne, *Phys. Rev. A* **7**, 418 (1973).
61. W. Kolos and L. Wolniewicz, *J. Chem. Phys.* **46**, 1426 (1967).
62. D. E. Gerhart, *J. Chem. Phys.* **62**, 821 (1975).
62a. V. L. Carter, *J. Chem. Phys.* **56**, 4195 (1972).
63. J. A. R. Samson and R. B. Cairns, *J. Geophys. Res.* **69**, 4583 (1964).
64. L. de Reilhac and N. Damany, *J. Phys.* (*Paris*), *Colloq.* **32**(C4), 32 (1971).
64a. L. de Reilhac and N. Damany, *J. Quant Spectrosc. Radiat. Transfer* **18**, 121 (1977).
65. M. Sasanuma, E. Ishiguro, Y. Morioka, and M. Nakamura, *Int. Conf. VUV Radiat. Phys., 3rd, Tokyo* Pap. 1p A2-3 (1971).
65a. J. E. Brolley, L. E. Porter, R. H. Sherman, J. K. Theobald, and J. C. Fong, *J. Geophys. Res.* **78**, 1627 (1973).
66. L. C. Lee, R. W. Carlson, D. L. Judge, and M. Ogawa, *J. Quant. Spectrosc. Radiat. Transfer* **13**, 1023 (1973).
66a. L. C. Lee, R. W. Carlson, and D. L. Judge, *Mol. Phys.* **30**, 1941 (1975).
67. W. S. Watson, J. Lang, and D. T. Stewart, *J. Phys. B* **6**, L148 (1973).
68. D. W. Turner *et al.*, "Molecular Photoelectron Spectroscopy" (see Introduction for complete reference).
69. K. Siegbahn *et al.*, ESCA 2 (see Introduction for complete reference).
70. See Chapter IV, ref. 7.
71. See Chapter IV, ref. 21.
71a. A Bianconi, H. Petersen, F. C. Brown, and R. Z. Bachrach, *Phys. Rev* A **17**, 1907 (1978).
72. E. N. Lassettre and A. Skerbele, *in* "Molecular Physics," 2nd ed., Part B (D. Williams, ed.), Methods of Experimental Physics, Vol. 3, p. 868. Academic Press, New York, 1974.
72a. P. Gürtler, V. Saile, and E. E. Koch, *Chem. Phys. Lett.* **48**, 245 (1977).
73. "Landolt-Bornstein Zahlenwerte und Funktionen," 6th Ed., Vol. II, Part 8, "Optische Konstanten." Springer-Verlag, Berlin and New York, 1963.
74. See Chapter IV, ref. 26.
75. J. Berkowitz, M. Inokuti, and J. C. Person, *Int. Conf. Phys. Electron. At. Collisions, 7th, Belgrade* Abstr., p. 561 (1973).
76. W. S. Watson, J. Lang, and D. T. Stewart, *Phys. Lett. A* **44**, 293 (1973).
76a. G. Mehlman, D. L. Ederer, and E. B. Salomon, *J. Chem. Phys.* **68**, 1862 (1978).
76b. S. Bodeur, *J. Appl. Phys.* **47**, 4911 (1976).
77. O. Edqvist, E. Lindholm, L. E. Selin, and L. Åsbrink, *Phys. Scr.* **1**, 25 (1970).
78. F. M. Matsunaga and K. Watanabe, *Sci. Light* (*Tokyo*) **16**, 31 (1967).
79. See Chapter IV, refs. 7, 10, and 21.
80. G. R. Cook, P. H. Metzger, and M. Ogawa, *Can. J. Phys.* **43**, 1706 (1965).
80a. R. E. Huffman, J. C. Larrabee, and Y. Tanaka, *J. Chem. Phys.* **40**, 2261 (1964).
81. R. E. Huffman, J. C. Larrabee and Y. Tanaka, *J. Chem. Phys.* **40**, 2261 (1964).
82. R. B. Cairns and J. A. R. Samson, *J. Geophys. Res.* **70**, 99 (1965).
83. R. B. Cairns and J. A. R. Samson, *J. Opt. Soc. Am.* **56**, 526 (1966).
83a. J. A. R. Samson and J. L. Gardner, *J. Electron Spectrosc. Relat. Phenom.* **8**, 35 (1976).
84. See Chapter IV, ref. 9.
84a. R. B. Kay, Ph. E. Van der Leeuw, and M. J. Van der Wiel,*J . Phys. B* **10**, 2513 (1977).
85. E. N. Lassettre and A. Skerbele, *J. Chem. Phys.* **54**, 1597 (1971).
86. K. Watanabe, F. M. Matsunaga, and H. Sakai, *Appl. Opt.* **6**, 391 (1967).
87. P. H. Metzger, G. R. Cook, and M. Ogawa, *Can. J. Phys.* **45**, 203 (1967).
88. A. Gardner, M. Lynch, D. T. Stewart, and W. S. Watson, *J. Phys. B* **6**, L262 (1973).
89. O. Edqvist, L. Åsbrink, and E. Lindholm, *Z. Naturforsch., Teil A* **26**, 1407 (1971).
90. E. W. Thulstrup and Y. Öhrn, *J. Chem. Phys.* **57**, 3716 (1972).

91. E. N. Lassettre, A. Skerbele, M. Dillon, and K. Ross, *J. Chem. Phys.* **48**, 5066 (1968).
92. F. F. Marmo, *J. Opt. Soc. Am.* **43**, 1186 (1953).
93. E. Phillips, L. C. Lee, and D. L. Judge, *J. Quant. Spectros. Radiat. Transfer* **18**, 309 (1977).
94. D. H. Katayama, R. E. Huffman, and C. L. O'Bryan, *J. Chem. Phys.* **59**, 4309 (1973).
94a. P. Gürtler, V. Saile, and E. E. Koch, *Chem. Phys. Lett.* **51**, 386 (1977).
94b. G. R. Branton and C. E. Brion, *J. Electron Spectros. Relat. Phenom.* **3**, 129 (1974).
95. R. B. Cairns, H. Harrison, and R. I. Schoen, *J. Chem. Phys.* **55**, 4886 (1971).
95a. K. H. Tan, C. E. Brion, P. E. Van der Leeuw, and M. J. Van der Wiel, *Chem. Phys.* **29**, 299 (1978).
96. K. Watanabe and M. Zelikoff, *J. Opt. Soc. Am.* **43**, 753 (1953).
97. K. Watanabe, M. Zelikoff, and E. C. Y. Inn. Absorption coefficients of several atmospheric gases, Tech. Rep. 53-23. U.S. Air Force Cambridge Res. Cent. Cambridge, Massachusetts (1953).
98. A. A. Maryott and F. Buckley, Tables of Dielectric Constants and Electric Constants and Electric Dipole Moments of Substances in the Gas Phase. *Natl. Bur. Stand.* (*U.S.*), *Circ.* No. 537 (1953).
99. S. W. Bennett, J. B. Tellinghuisen, and L. F. Phillips, *J. Phys. Chem.* **75**, 719 (1971).
100. R. S. Nakata, K. Watanabe, and F. M. Matsunaga, *Sci. Light* (*Tokyo*) **14**, 54 (1965).
101. G. R. Cook, P. H. Metzger, and M. Ogawa, *J. Chem. Phys.* **44**, 2935 (1966).
102. J. A. R. Samson, J. L. Gardner, and J. E. Mentall, *J. Geophys. Res.* **77**, 5560 (1972).
103. V. D. Meyer and E. N. Lassettre, *J. Chem. Phys.* **42**, 3436 (1965).
104. J. A. R. Samson and J. L. Gardner, *J. Geophys. Res.* **78**, 3663 (1973).
105. G. R. Cook, P. H. Metzger, and M. Ogawa, *J. Opt. Soc. Am.* **58**, 129 (1968).
106. M. Zelikoff, K. Watanabe, and E. C. Y. Inn, *J. Chem. Phys.* **21**, 1643 (1953).
107. R. H. Huebner, R. J. Celotta, S. R. Mielczarek, and C. E. Kuyatt, *J. Chem. Phys.* **63**, 4490 (1975).
108. R. E. Rebbert and P. Ausloos, *J. Res. Natl. Bur. Stand., Sect. A* **75**, 481 (1971).
109. K. Watanabe and S. P. Sood, *Sci. Light* (*Tokyo*) **14**, 36 (1965).
109a. W. C. Walker and G. L. Weissler, *J. Chem. Phys.* **23**, 1540 (1955).
109b. H. Sun and G. L. Weissler, *J. Chem. Phys.* **23**, 1160 (1955).
110. K. Watanabe, *J. Chem. Phys.* **22**, 1564 (1954).
111. K. Watanabe and J. R. Mottl, *J. Chem. Phys.* **26**, 1773 (1957).
112. P. H. Metzger and G. R. Cook, *J. Chem. Phys.* **41**, 642 (1964).
113. G. R. Wight and C. E. Brion, *J. Electron Spectrosc. Relat. Phenom.* **4**, 25 (1974).
114. L. de Reilhac and N. Damany, *Spectrochim. Acta, Part A* **26**, 801 (1970).
114a. G. R. Wight, M. J. Van der Wiel, and C. E. Brion, *J. Phys. B* **10**, 1863 (1977).
115. O. P. Rustgi, *J. Opt. Soc. Am.* **54**, 464 (1964).
116. E. E. Koch and M. Skibowski, *Chem. Phys. Lett.* **9**, 429 (1971).
117. W. R. Harshbarger and E. N. Lassettre, *J. Chem. Phys.* **58**, 1505 (1973).
118. R. H. Huebner and R. J. Celotta, Annual Report, ANL-7960, Argonne Natl. Lab., Argonne, Illinois (1971–1972).
119. N. Wainfan, W. C. Walker, and G. L. Weissler, *Phys. Rev.* **99**, 542 (1955).
120. C. Backx, G. R. Wight, R. Tol, and M. J. Van der Wiel, *J. Phys. B* **8**, 3007 (1975).
121. T. Nakayama and K. Watanabe, *J. Chem. Phys.* **40**, 558 (1964).
122. J. C. Person and P. P. Nicole, *J. Chem. Phys.* **53**, 1767 (1970).
123. J. O. Hirschfelder, C. F. Curtiss, and R. B. Bird, "Molecular Theory of Gases and Liquids." Wiley, New York, 1954.
123a. J. A. R. Samson, G.C.A. Tech. Rep. No. 61-15-N. Geophysics Corp. of Amer., Bedford, Massachusetts (1961). "Planetary Atmosphere Studies III: Summary Report; Experimental Results of Vacuum Ultraviolet Spectroscopy."

124. J. C. Person and P. P. Nicole, *J. Chem. Phys.* **49**, 5421 (1968).
125. R. I. Schoen, *J. Chem. Phys.* **37**, 2032 (1962).
126. M. Zelikoff and K. Watanabe, *J. Opt. Soc. Am.* **43**, 756 (1953).
127. J. W. Raymonda and W. T. Simpson, *J. Chem. Phys.* **47**, 430 (1967).
128. B. A. Lombos, P. Sauvageau, and C. Sandorfy, *J. Mol. Spectrosc.* **24**, 253 (1967).
128a. C. E. Klots, *J. Chem. Phys.* **56**, 124 (1972).
129. M. Ogawa and G. R. Cook, *J. Chem. Phys.* **28**, 747 (1958).
130. J. C. Person and P. P. Nicole, *J. Chem. Phys.* **55**, 3390 (1971).
131. J. C. Person and P. P. Nicole, Annual Report, ANL-75-3, Part I. Radiol. Environ. Res. Div., Argonne Natl. Lab., Argonne, Illinois (1973–1974).
132. L. Sanson and N. Damany-Astoin, *C. R. Acad. Sci.* **250**, 4138 (1960).
133. D. R. Salahub and C. Sandorfy, *Chem. Phys. Lett.* **8**, 71 (1971).
134. S. M. Bunch, G. R. Cook, M. Ogawa, and A. W. Ehler, *J. Chem. Phys.* **28**, 740 (1958).
135. J. C. Person, *J. Chem. Phys.* **43**, 2553 (1965).
136. J. A. R. Samson, personal communication (data obtained in 1961).
137. M. Yoshino, J. Takeuchi, and H. Suzuki, *J. Phys. Soc. Jpn.* **34**, 1039 (1973).
138. E. E. Koch and A. Otto, *Chem. Phys. Lett.* **12**, 476 (1972); *Int. J. Radiat. Phys. Chem.* **8**, 113 (1976).
138a. M. Sasanuma, E. Ishiguro, H. Masuko, Y. Morioka, and M. Nakamura, *J. Phys. B* **11**, 3655 (1978).
139. K. Codling, *J. Chem. Phys.* **44**, 4401 (1966).
139a. T. M. Zimkina and V. A. Fomichev, *Sov. Phys.—Dokl.* **11**, 726 (1967).
140. D. Blechschmidt, R. Haensel, E. E. Koch, U. Nielsen, and T. Sagawa, *Chem. Phys. Lett.* **14**, 33 (1972).
141. L. C. Lee, E. Phillips, and D. L. Judge, *J. Chem. Phys.* **67**, 1237 (1977).
142. A. S. Vinogradov and T. M. Zimkina, *Opt. Spectrosc.* (*USSR*) **32**, 17 (1972).
143. R. E. Lavilla and R. D. Deslattes, *J. Chem. Phys.* **44**, 4399 (1966).
143a. A. P. Hitchcock and C. E. Brion, *Chem. Phys.* **33**, 55 (1978).
144. J. A. Simpson, C. E. Kuyatt, and S. R. Mielczarek, *J. Chem. Phys.* **44**, 4403 (1966).
145. R. D. Nelson, Jr. and R. H. Cole, *J. Chem. Phys.* **54**, 4033 (1971).

Chapter VI
PARTIAL CROSS SECTIONS

A. General Considerations

1. Auger, Fluorescence, and Nonradiative Processes

In this chapter we shall attempt to apportion the total photoionization cross sections for the molecular systems considered in the previous chapter† in two ways:

(1) We shall deduce the relative cross sections for formation of the various initial states of the ion, primarily with the aid of photoelectron spectroscopic data. Ideally, such data are required at a large number of incident wavelengths, and measurements should be performed at the "magic angle" of 54° 44′, or angular distribution measurements should be available (see Chapter VII). In practice, this detailed information is available only for a few of the systems considered here. Nevertheless, we have included such information as exists, however sparse, where the detailed data are unavailable.

† The examples are primarily chosen for continuity with the previous chapter, which encompasses those molecules for which the most complete set of absorption coefficients is available. The polyatomics are also representative of (1) alkynes, (2) alkenes, (3) alkanes, (4) alcohols, (5) aromatics, and (6) highly symmetrical inorganic species. Photoelectron spectra have been reported for a vast range of molecules. (See Siegbahn *et al.*[1] and Brundle and Baker[1a] for compilations.) Photoionization mass spectra reported for species not encompassed in this monograph have been tabulated in the first appendix to this chapter, together with the wavelength range and resolution achieved by the authors.

Where possible, we have tried to include correlation, shake-up, or two-electron excitations, and double ionization, as well as the dominant one-electron processes. These relative cross sections have been converted to absolute partial cross sections, using the total ionization cross section results from Chapter V.

(2) The other type of partial cross section considered in this chapter is a measure of the "final" state of the products resulting from ionization. This latter information is primarily deduced from photoionization mass spectrometry or photoelectron–photoion coincidence studies. Photoelectron spectroscopy is sensitive to time scales of $\sim 10^{-14}$–10^{-15} sec, whereas photoionization mass spectrometry samples times of $\sim 10^{-5}$–10^{-6} sec. It is in this sense that we refer to the "final" states.

For atoms, ionization from the valence shell results in both unipositive ions and (due to correlation effects) some double ionization. When a vacancy is created in an inner shell, it may be filled by either a radiative transition or an Auger process. In the Auger case, the energy released is transferred to one or more electrons in one of the outer shells, causing ejection of one or more additional electrons. The final state is then a multiply charged ion. The Auger process is often more probable than radiative deexcitation by emission of x rays, especially for systems with low atomic number. Figure 1 shows the

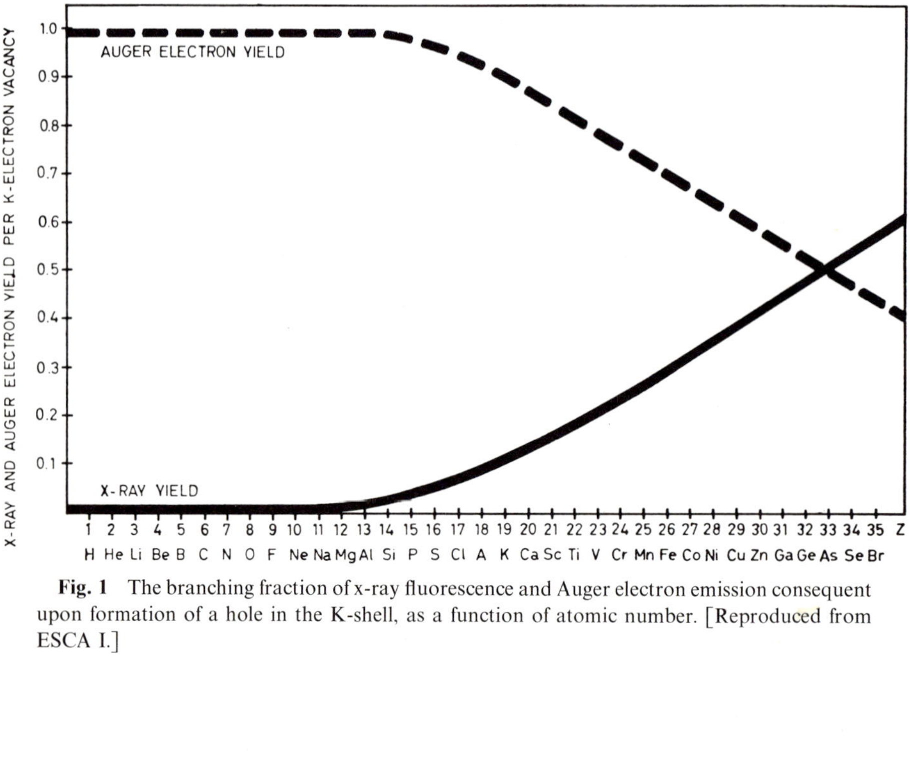

Fig. 1 The branching fraction of x-ray fluorescence and Auger electron emission consequent upon formation of a hole in the K-shell, as a function of atomic number. [Reproduced from ESCA I.]

relative yield of the Auger process and x-ray deexcitation consequent upon the creation of a K-shell hole, for elements up to $Z = 35$. When an atom has several shells, formation of a vacancy in a deep inner shell can result in a vacancy cascade from successive outer shells until all vacancies reach the outermost occupied shell. This can give rise to a highly charged ion. In some cases of not-too-light elements, a vacancy in an inner shell may be filled by a transition from a high-lying subshell of the same primary shell. Auger processes of this type are called Coster–Kronig transitions.

For diatomic, and in general, polyatomic systems, the same considerations apply. However, Auger transitions that result in multiply charged molecular ions are not often detected as such, since the coulombic repulsion between the atomic entities fragments the molecular ion during the time scale of observation.

Ionization from valence shells of molecular ions can also result in dissociation when the excited state lies above the first dissociation limit. Such processes for diatomic (and to some extent, triatomic)[1b] ions usually involve the presence of a state which predissociates the initially formed molecular ion state.

It has been the conventional wisdom in the annals of mass spectrometry that the fragmentation patterns of such small molecules can be interpreted in terms of potential curves, Franck–Condon factors, and asymptotic limits. This established view places large molecules in a different class. The initial ionization act may instantaneously create a molecular ion in one of several electronic states, but a rapid equilibration takes places such that only the electronic ground state, together with its complement of vibrational energies, need be further considered in determining the fragmentation pattern. This approach then proceeds by formulating a "transition state" of the vibrationally excited but electronic ground state of the ion, which has been prepared in some configuration that is about to dissociate along some path, e.g., an elongated bond. One then computes the number of complexions having this configuration, compared to the number of complexions having the same total energy but different configurations to determine the probability of dissociation at that energy. For a recent review, see Rosenstock.[2]

The two approaches (the one considering detailed electronic potential curves, the other completely disregarding them) are seen to be drastically different. In this framework, it is difficult to construct a theory for the intermediate molecule[3] case.

Let us try to test the basic tenets of the statistical theory, staying as close as possible to experimental observations and making use of recent developments in the theory of radiationless transitions. Radiationless transitions play a role even in the cases of diatomic molecules. We refer to them in that context as predissociations. They must compete with other modes of decay, particularly radiative emission, where this is permissible. For those ionization events in which a valence electron is ejected, radiative emission will have characteristic lifetimes of 10^{-6}–10^{-8} sec. Radiationless transitions will have to be at least as fast, and in any event the characteristic measurement times in mass spectrometers are $\sim 10^{-6}$ sec; radiationless transitions must certainly be faster than this to play a significant role. As we consider electrons being ejected from successively deeper orbitals, the resulting fluorescence will become more rapid. For ejection from the K-shell of carbon, for example, the fluorescence rate is $\sim 10^{12}$ sec^{-1}. More important for light elements is the Auger process, which is $\sim 10^3$ times faster for C and O K-shells than fluorescence. The radiationless transitions, which are the mechanisms for equilibrating the excited ionic state into a vibrationally excited but electronic ground state, would have to be at least this fast to compete with Auger and fluorescence processes.

In fact, fluorescence and Auger processes are observed for large molecules ($CH_4, C_2H_6, C_2H_4, C_6H_6$),[4†] and hence they must be successfully competing with radiationless transitions. Fluorescence is also prominent in valence ionic transitions.

Observation of fluorescence in molecular ions is the most direct evidence for the stability of excited states of these ions. When fluorescence can be detected, radiationless transitions and (if thermochemically possible) dissociative processes must be at least weakly forbidden.

The emission spectra of about 50 diatomic ions have been reported.[5,6] Of these, the dissociation energies are fairly well established in about 16 cases, in 6 of which the emission initiates above the dissociation limit. Among triatomic ions, fluorescence has been reported for 10 (isotopic species excluded): CO_2^+, COS^+, CS_2^+, N_2O^+[7]; $ClCN^+$, $BrCN^+$, ICN^+[8,8a]; H_2O^+[9,9a]; H_2S^+[10] and SO_2^+.[10a] Of these, only three (N_2O^+, COS^+, SO_2^+) have emissions emanating from states above a dissociation limit. Quite recently, a number of emission spectra of polyatomic ions (larger than triatomic) have been reported. These include the monohalo- and dihaloacetylene ions,[11] the di-, tri-, and tetraacetylene ions,[12] halides of the diacetylenes,[12a] hexa-, penta-, tetra-, tri-, and diflurobenzenes,[13,13a] three chlorobenzenes,[13b] the cis and trans forms of 1,3,5-hexatriene[14] and 2,4- and 1,3-hexadiyne.[14a] Of these, only the hexatriene and hexadiyne systems have emitting states lying above a dissociation limit.[12a,14a]

The statistical information, while sparse, clearly indicates that nonradiative transitions do not dominate radiative transitions in all cases. Even when dissociative channels are open, fluorescence is sometimes observed. Since it

† See also recent work from the Uppsala group[4a] on NH_3 and NO_2 and numerous references therein to studies of other, larger molecules.

it is difficult to generate molecular ions in large concentration (compared to neutral species), it is likely that those radiative transitions reported for molecular ions must be the dominant decay modes in most instances. For neutral species, it is possible to observe radiative decay when competition from nonradiative processes is strong.[15]

The data compiled above also suggest that in the competition between radiative and nonradiative processes, the latter are favored as the size of the molecular ion increases, as implied by the theory for radiationless transitions (see below).

In summary, the limited experimental evidence would seem to circumscribe the realm of applicability of a statistical theory to molecules containing four or more atoms, in states of excitation implying ejection of an electron not too far below the valence region.

Recent developments in the theory of radiationless transitions, while predominantly oriented to the study of neutral systems, shed additional light on this problem. In order to translate these investigations into the framework of the current study, consider the diagrams in Fig. 2, which are closely related to Fig. 4 of a recent review by Freed.[16]

Ionization of a molecule from its ground state ϕ_0 is assumed to form an ion in the excited state ϕ_{is}. Without being too precise, we shall refer to this state ϕ_{is} as a Born–Oppenheimer state since it is not an eigenstate of the system. (If it were an eigenstate, there would be no such thing as a radiationless transition.) Again, without being too precise, we accept as a deduction of

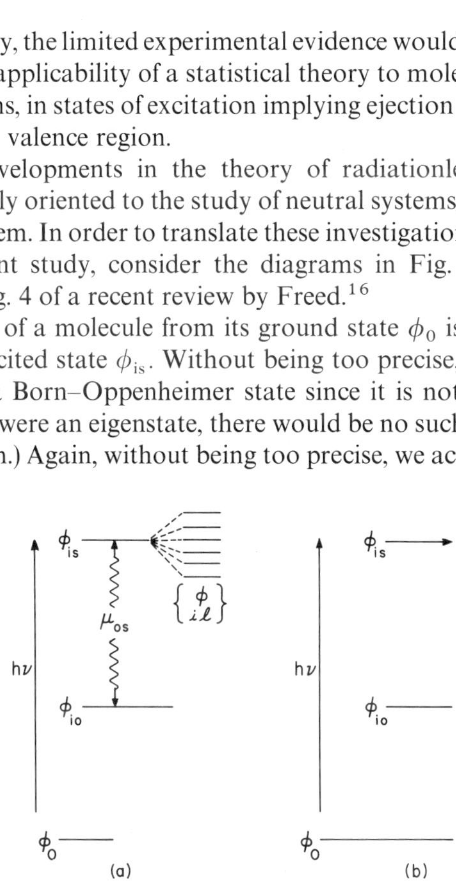

Fig. 2 The molecular energy level model used to discuss radiationless decay processes in polyatomic molecules. (a) Depicts the small molecule limit, (b) the large molecule, or statistical limit. In the present context, ϕ_o is the ground state of the neutral molecule, ϕ_{is} is an excited state of the molecular ion and ϕ_{io} is the ground state of the molecular ion. The set of states $\{\phi_{il}\}$ describe a vibrational manifold of another electronic state of the ion. On (a), competition between fluorescence (μ_{os}) to ϕ_{io} and radiationless transition to the loose manifold $\{\phi_{il}\}$ is illustrated. On (b), the dense manifold $\{\phi_{il}\}$ favors decay by radiationless transition.

the theory of radiationless transitions that the transition rate from the electronically excited but vibrationally unexcited state ϕ_{is} to the electronically less excited but vibrationally excited set of states ϕ_{il} is given by

$$k_{nr} = \frac{2\pi}{\hbar} \sum_{l} |v_{sl}|^2 \, \delta(E_s - E_l), \tag{1}$$

where k_{nr} is the nonradiative decay rate, v_{sl} is a particular matrix element coupling the state s to a state l, and $\delta(E_s - E_l)$ is a Dirac delta function. If, for the moment, we make the cruder approximation that all v_{sl} are the same, we can write

$$k_{nr} = \frac{2\pi}{\hbar} v^2 \rho_l, \tag{2}$$

where ρ_l is the density of states corresponding to a final electronic state ϕ_{il}. This type of equation is sometimes referred to as Fermi's Golden Rule.

In the small molecule limit (see Fig. 2a), except in the case of accidental degeneracies, the energy level spacings are relatively large, and hence ρ_l is small (Leach[7] gives 0.02 states per cm^{-1} for CO_2^+ in the valence region, for example) and the coupling matrix element v is typically not large enough to compensate for the small ρ_l. Hence, as Leach has shown in more detail for CO_2^+, COS^+, CS_2^+, and N_2O^+, fluorescence can be observed to compete effectively with radiationless transitions.

If we return to the more accurate Eq. (1) and allow for the possibility that some of the terms in the summation are large, then we may view part of the transition strength to ϕ_{is} as "crossing over" to some ϕ_{il}, which may either radiate (if allowed) or cross back, or be lost by some other mechanism, such as dissociation.

This is a rationalization of the observation by Douglas[17] that absorption and emission strengths can be drastically different for some small triatomic (in his case, neutral) molecules. The question of recurrence, or irreversibility, arises at this point. If a crossing back from ϕ_{il} to ϕ_{is} takes place, we refer to it as recurrence, and if it does not, we describe the reaction as "irreversible." This situation is related to the familiar paradox of statistical mechanics. Consider a box divided in two by a partition. One side of the partition contains molecules at atmospheric pressure, the other is pumped to a high vacuum. At $t = 0$, we open a plug between the two sections. The pressure rapidly equilibrates. There is a calculable probability for the initial condition to recur, but if the probability for this event is considered, the corresponding time for "recurrence" is astronomically high. In the molecular case being considered here (the microcanonical ensemble) the cut-off time between recurrence and irreversibility is measured in terms of the time scale of the experiment, which is typically $<10^{-5}$ sec. If recurrence is possible, one may expect some interesting interference effects.

The large molecule limit (also called the statistical limit) corresponds to a very large value of ρ_l in Eq. (2), so that even a modest value of v ensures that the radiationless transition occurs rapidly. This is the situation depicted in Fig. 2b.

The theory of nonradiative transitions has received considerable attention in the past several years. One of the useful generalizations resulting from this effort, and from experimental studies of neutral systems, is the so-called "energy-gap law." In Appendix 2 in this chapter, we present a simplified derivation of this law, borrowed from Freed.[18]

It can be expressed as

$$k_{\mathrm{nr}} \propto \exp(-\gamma \cdot \Delta E/h\omega_{\mathrm{M}}). \tag{3}$$

When related to the diagram of Fig. 2, this expression states that the nonradiative rate falls off exponentially with the "energy gap" ΔE between the s and l surfaces, but is favored by high frequency vibrational modes ω_{M}.

The energy gap law is likely to have a profound impact on some inner shell ionization processes, as the following argument will illustrate.

Consider a photoionization process in which a K-shell electron is ejected from N_2. The valence levels of the resulting $(N_2{}^+)^*$ have been shown[19] to behave much like NO^+. In other words, there is a small set of energy levels, spaced by a few electron volts, which have a local "ground state" of $(N_2{}^+)^*$ akin to the ground state of NO^+. However, this set of levels of $(N_2{}^+)^*$ is very highly excited relative to ground state $N_2{}^+$. There will be several other sets of such levels, each set lying (in energy) close to the mean (or vertical) ionization potential corresponding to ejection of an electron from a particular molecular orbital. (We shall see that one-electron processes dominate photoelectron spectra.) Some of these orbital levels are separated from others by hundreds of volts, others by tens of volts. The conversion of electronic to vibrational energy in such magnitudes becomes meaningless, and the energy gap law, even for the smaller gaps of several volts, would imply a very small conversion rate.

The foregoing discussion has implied that the fragmentation behavior of large polyatomic molecules upon ionization can be viewed as a two-stage process.

(1) The "randomization" of the electronic and vibrational energy initially imparted to the molecular ion by the ionization act into all regions of phase space, i.e., all conformations of the molecular ion having that same total energy are accessible during the time of the experiment.

(2) Some of these conformations are favorably disposed toward bond cleavage, and they act as a "sink" for the irreversible dissociation process.

The assumptions of (1) have been called into question by the preceding discussion.

If process (1) is satisfied, the theoretical formulation for process (2) becomes pertinent. We present below a pedagogic development of the statistical theory as applied to molecular ion decomposition, which is often called the quasi-equilibrium theory (QET). We proceed in stages.

2. Quasi-Equilibrium Theory

a. Simplified Kassel[20] Picture. Consider j identical quanta distributed among s identical oscillators. What are the number of distinguishable ways of distributing the quanta among the oscillators? This is the familiar problem of dividing j identical balls among s boxes. There are $s - 1$ partitions. The total number of configurations (balls and partitions) is $(j + s - 1)!$ if all objects are different. However, the j balls are alike, as are the $s - 1$ partitions. Hence we must divide by $j!$ and $(s - 1)!$ to remove duplications. The number of *different* configurations is

$$\frac{(j + s - 1)!}{j!(s - 1)!}. \tag{4}$$

Now consider the case where, of the j quanta, m are concentrated in one selected oscillator. That leaves only $j - m$ quanta to distribute randomly. The result follows directly from the previous considerations:

$$\frac{(j - m + s - 1)!}{(j - m)!(s - 1)!}. \tag{5}$$

If we identify the latter situation as corresponding to an activated complex, and the former as the total number of states having the same energy (quanta), the ratio

$$\frac{\text{No. of states of activated complex}}{\text{Total no. of states}} = \frac{j!(j - m + s - 1)!}{(j - m)!(j + s - 1)!}. \tag{6}$$

The rate constant for unimolecular decomposition, in this simplified but exact formulation, is then given by a constant (having the dimensions of frequency) multiplied by the statistical factor. Thus,

$$k_j = \frac{Aj!(j - m + s - 1)!}{(j - m)!(j + s - 1)!}. \tag{7}$$

In Kassel's development, the constant A is not further specified.

For pedagogic purposes, we now make the assumption that j (number of quanta) $\gg s$ (number of oscillators). Then,

$$\frac{(j-m+s-1)!}{(j-m)!} \rightarrow (j-m)^{s-1}$$

and (8)

$$\frac{j!}{(j+s-1)!} = \frac{1}{j^{s-1}}$$

$$\lim_{j\rightarrow\infty} \frac{j!(j-m+s-1)!}{(j-m)!(j+s-1)!} = \left(\frac{j-m}{j}\right)^{s-1} \tag{9}$$

Substituting $j = E/h\nu$, $m = E_0/h\nu$ = critical energy, we obtain

$$k_j = A\left(\frac{E-E_0}{E}\right)^{s-1} \tag{10}$$

for the situation where the no. of quanta $\gg$ no. of oscillators.

As we shall see below, the approximation $j \gg s$ is equivalent to the Dirichlet integral approximation (Appendix 3) in the more general development of QET.

Let us examine the validity of this approximation. Consider the case of ethane where $s = 18$. Following Rosenstock[21] we take as representative values $m = 20$, $j = 25$.

Exact calculation:

$$k_j = \frac{A(25)!(25-20+18-1)!}{(25-20)!(25+18-1)!}$$

$$k_j = A(1.034 \times 10^{-7}).$$

Approximate calculation:

$$k_j = A\left(\frac{25-20}{25}\right)^{17} = A\left(\frac{1}{5}\right)^{17}$$

$$k_j = A(1.311 \times 10^{-12}).$$

Hence,

$$\frac{k_j(\text{exact})}{k_j(\text{approximate})} = 7.89 \times 10^4.$$

b. Quasi-Equilibrium Theory. Following Rosenstock *et al.*,[22] we consider an isolated system which can have many distinct chemical conformations consistent with a prescribed initial chemical structure and an energy in the range E to $E + \delta E$. The assumption is made that the rate of transition among the various states of the system in this energy range is rapid compared to the rate of unimolecular decomposition.

The approach is similar to that used in the absolute rate theory for a Boltzmann distribution of molecules. One selects a "reaction coordinate" and an "activated complex" defined in terms of a saddle point on a potential surface, such that the change in the reaction coordinate describing the motion of the system through the saddle can be taken to be a translational motion.

The rate of unimolecular decay will be a ratio of states in the activated complex to the total number of states in this energy range (as in the Kassel expression for s identical oscillators) multiplied by the frequency of crossing the barrier en route to decomposition in the translational mode. This latter frequency appeared as an ad hoc empirical factor in the Kassel expression.

The activated complex has a total energy between E and $E + \delta E$, potential energy ε_0, and kinetic energy in the reaction coordinate between ε_t and $\varepsilon_t + \delta\varepsilon_t$. The number of states in the activated complex is given by $\rho^{\ddagger}(E, \varepsilon_0, \varepsilon_t)\rho_t(\varepsilon_t)\,\delta E\,\delta\varepsilon_t$.

From the familiar quantum mechanical expression for the number of states n of a particle with reduced mass μ in a one-dimensional box of dimension l with translational energy ε_t, i.e.,

$$\varepsilon_t = \frac{h^2 n^2}{8\mu l^2}, \tag{11}$$

we can readily calculate

$$\frac{dn}{d\varepsilon_t} = \frac{4\mu l^2}{h^2}\frac{1}{n} = \frac{l}{h}\left(\frac{2\mu}{\varepsilon_t}\right)^{1/2}, \tag{12}$$

which is just $\rho_t(\varepsilon_t)$.

The velocity of the system along the reaction coordinate is simply

$$v = \left(\frac{2\varepsilon_t}{\mu}\right)^{1/2}, \tag{13}$$

and hence the frequency of transversal of the barrier is

$$\frac{v}{l} = \frac{1}{l}\left(\frac{2\varepsilon_t}{\mu}\right)^{1/2}. \tag{14}$$

It is usually assumed that there is equal probability for crossing the barrier in the direction of decomposition and in the opposite direction, and hence half the above frequency is taken.

This frequency is the analog of the ad hoc frequency A in Kassel's theory. It is still not explicit since it contains the translational dimension l, but it is readily seen that this factor will cancel a factor l appearing in the density of translational states. For a narrow range of states $\delta E\,\delta\varepsilon_t$, we have as a partial rate constant

$$k_{\text{partial}} = \frac{\text{“}v\text{”}^{\ddagger}(E, \varepsilon_0, \varepsilon_t)\rho_t(\varepsilon_t)\,\delta E\,\delta\varepsilon_t}{\rho(E)\,\delta E}, \tag{15}$$

where

$$\text{“}v\text{”} = \frac{1}{2}\frac{v}{l} = \frac{1}{2l}\left(\frac{2\varepsilon_t}{\mu}\right)^{1/2},$$

$$k_{\text{partial}} = \frac{\rho^{\ddagger}(E, \varepsilon_0, \varepsilon_t)\dfrac{1}{2l}\left(\dfrac{2\varepsilon_t}{\mu}\right)^{1/2}\dfrac{l}{h}\left(\dfrac{2\mu}{\varepsilon_t}\right)^{1/2}\delta E\,\delta\varepsilon_t}{\rho(E)\,\delta E} \tag{16}$$

$$k_{\text{partial}} = \frac{\rho^{\ddagger}(E, \varepsilon_0, \varepsilon_t)}{\rho(E)\,\delta E}\frac{1}{h}\delta E\,\delta\varepsilon_t. \tag{17}$$

The total unimolecular rate is obtained by integrating over all possible translational energies from zero to the maximum possible, which is $E - \varepsilon_0$. The quantity ε_0 is the energy that must stay in the products after decomposition, i.e., it is the minimum energy for decomposition,

$$k_{\text{total}} = \int_0^{E-\varepsilon_0} \frac{\rho^{\ddagger}(E, \varepsilon_0, \varepsilon_t)}{\rho(E)}\frac{1}{h}\delta\varepsilon_t. \tag{18}$$

In early applications of this expression, the density of states ($\rho^{\ddagger}$ or ρ) was approximated by an expression obtained by use of the Dirichlet integral (see Appendix 3). For the denominator,

$$\rho(E) = \frac{dG(E)}{dE} = \prod_{n=1}^{N}\left(\frac{1}{h\nu_n}\right)\frac{E^{N-1}}{(N-1)!}. \tag{19}$$

Marcus and Rice already noted in 1952[23] that the corresponding expression is less valid for the activated complex since the energy available for vibrational motion ($E - \varepsilon_0 - \varepsilon_t$), which they call nonfixed internal energy, is small and quantum restrictions become important. Nonetheless, proceeding in the spirit of this classical approximation, the numerator becomes

$$\rho^{\ddagger}(E, \varepsilon_0, \varepsilon_t) = \prod_{m=1}^{N-1}\left(\frac{1}{h\nu_m}\right)\frac{(E - \varepsilon_0 - \varepsilon_t)^{N-2}}{(N-2)!}. \tag{20}$$

Note that one degree of freedom has been used as translational motion in the activated complex, which necessitates a corresponding reduction in the vibrational degrees of freedom of the activated complex. We now have

$$k_{\text{total}} = \frac{1}{h}\int_0^{E-\varepsilon_0} \frac{\prod_{m=1}^{N-1}(1/h\nu_m)[(E-\varepsilon_0-\varepsilon_t)^{N-2}/(N-2)!]\,d\varepsilon_t}{\prod_{n=1}^{N}(1/h\nu_n)[E^{N-1}/(N-1)!]}$$

$$= \int_0^{E-\varepsilon_0} \frac{\prod_{n=1}^{N}\nu_n}{\prod_{m=1}^{N-1}\nu_m}\frac{(N-1)}{E^{N-1}}(E-\varepsilon_0-\varepsilon_t)^{N-2}\,d\varepsilon_t$$

$$= \frac{\prod_{n=1}^{N}\nu_n}{\prod_{m=1}^{N-1}\nu_m}\frac{(N-1)}{E^{N-1}}\left[-\frac{(E-\varepsilon_0-\varepsilon_t)^{N-1}}{N-1}\right]_0^{E-\varepsilon_0}$$

$$= \frac{\prod_{n=1}^{N}\nu_n}{\prod_{m=1}^{N-1}\nu_m}\frac{(N-1)}{E^{N-1}}\left(\frac{(E-\varepsilon_0)^{N-1}}{N-1}\right),$$

$$k_{\text{total}} = \frac{\prod_{n=1}^{N}\nu_n}{\prod_{m=1}^{N-1}\nu_m}\left(\frac{E-\varepsilon_0}{E}\right)^{N-1}. \tag{21}$$

This final expression is seen to be very similar to the Kassel expression for a system of s identical oscillators *when the number of quanta is large compared to the number of oscillators.* The ad hoc frequency factor A in Kassel's expression has been given a calculable interpretation as the ratio of products of frequencies of all states in the relevant energy range and frequencies of activated complex states. Since the numerator has one more frequency than the denominator, this preexponential term will have the dimensions of frequency.

The fact that $[(E-\varepsilon_0)/E]^{N-1}$ appears in the above expression as well as in Kassel's approximate expression when j and $j-m$ are large compared to s, together with the previous observation that Kassel's approximate expression can be orders of magnitude off from Kassel's exact factorial expression should immediately make us suspicious of using the above expression [Eq. (21)]. Early attempts[24] to use Eq. (21) arrived at the conclusion that agreement with experiment could only be obtained if N (the total number of oscillators) was reduced by about a factor of two, and hence the expression "effective number of oscillators" came to be used. In the last decade, it has come to be realized that a major part of this discrepancy is the method of evaluating $\rho^{\ddagger}$ and ρ, which is equivalent to Kassel's transition to the classical assumption that the number of quanta is much larger than the number of oscillators.

A number of techniques have since been developed for determining $\rho^{\ddagger}$ (and its integral) and ρ more accurately. They are based on the observation that the conventional thermodynamic partition function has the functional form of a Laplace transform. Hence, operating on this function by an inverse Laplace transform can generate ρ. Several elegant mathematical devices for performing this operation are discussed in a pedagogic review article by Forst.[25] Some provide results within 10% of that achieved by direct counting of states. A numerical inversion of the Laplace transform, using Fourier transform methodology, has been proposed by Hoare and Pal.[25a] Their sample calculations indicate still higher accuracy, down to rather low energies. We shall not discuss these calculations further here, since they require computers and become rather specialized, have little pedagogic value for the understanding of the basic theory, and the interested reader can obtain the relevant references with brief descriptions from the sources given here. It is sufficient to note that, given a structure and the relevant vibrational frequencies, the mathematical task of evaluating ρ and $\rho^{\ddagger}$ can be solved to the required degree of accuracy.

It is of some historical interest to note that about two decades ago a theory of unimolecular decay was advanced by Slater[26] which was based on the philosophy that an activated molecule could still be described by its normal modes of vibration, but that the amplitudes and frequencies of motion along particular bonds (or bond angles) would vary with time if one described these bond motions as linear combinations of the normal modes. Suppose one had reason to identify a particular bond stretch as the "reaction coordinate." When the bond length exceeded some critical value, unimolecular breakup would occur. Then the superposition of normal modes could be used to calculate the frequency with which this bond extension could take place. The mathematical development is rather complicated, involving integration over Bessel functions, and arrives at the rate constant

$$k_E = \bar{\nu}\left(\frac{E - \varepsilon_0}{E}\right)^{N-1}, \tag{22}$$

where

$$\bar{\nu} = \frac{(\bar{\lambda})^{1/2}}{2\pi},$$

N is the number of oscillators as before,

$$\begin{aligned} \bar{\lambda} &= \frac{\Sigma \alpha_{ri}^2 \lambda_i}{\alpha^2} \\ \lambda_i &= 4\pi^2 \nu_i^2 \end{aligned} \tag{23}$$

and the α_{ri} are coefficients relating the important internal coordinate (e.g., bond stretching coordinate) q_r to the normal modes Q_i, i.e. $q_r = \sum_{i=1}^{N} \alpha_{ri} Q_i$. Since the Slater expression has the same energy dependence as the approximate expression derived previously, and found not to agree with experiment, this approach has been abandoned in the last decade.

Most real situations are now known to involve not too many quanta relative to the number of oscillators, and hence classical (and even semi-classical) expressions are bad. However, it can be shown that a set of loosely coupled harmonic oscillators do not achieve ergodicity[27] and even anharmonicity does not assure ergodicity, although it gets better as anharmonicity increases. Few quanta relative to no. of oscillators implies small anharmonicity (low down in the well), and hence the assumption of randomization of vibrational energy is on shaky theoretical grounds. However, a recent assessment[27a] of experimental data on neutral molecules, using laser excitation of vibrational levels, presents a more optimistic view.

Before one can embark upon the technology of counting states in the rate expression, one needs a model for the transition state. This introduces a degree of arbitrariness, which is also encountered in other applications of absolute rate theory, e.g., for bimolecular reactions of neutral Boltzmann populations. Practioners of this art invoke loose and tight complexes, active and inactive vibrations, and reverse activation barriers. (See Hase[28] and Klots[29] for alternative approaches to the handling of this problem.)

It has been fashionable to test the consequences of statistical calculations based on assumptions (1) and (2), the sequence of events described at the end of Section A-1, above, by three types of experiments.

(a) The determination of fragmentation patterns as a function of an ion's internal energy, using threshold laws (see Chapter IV, Section A) or charge-transfer experiments (so-called breakdown graphs). The experimental breakdown diagrams can be compared with corresponding ones deduced from QET calculations.

(b) The observation of mass spectrometric "metastable ions." In a magnetic mass spectrometer, one accelerates all singly charged ions in an electrostatic field to an energy $E = eV$ (V is the accelerating voltage), then deflects them in a magnetic field. It can readily be shown (see Appendix 4) that if a molecular ion is accelerated as a parent ion, then decomposes before magnetic deflection occurs, it is observed in the mass spectrum at an apparent mass

$$m^* = (m_f)^2/m_p,$$

where m_f and m_p refer to the mass of fragment and parent ions, respectively. The observation of metastable ions can be related to a rate constant for unimolecular decay.

(c) Photoion–photoelectron coincidence experiments, which enable one to deduce the fragmentation pattern resulting from a particular level of excitation of the molecular ion.

We shall have occasion to refer to each of these methods in the examples below which involve the decomposition of molecular ions.

B. Atoms

1. Helium

Very few processes need to be considered here. The threshold[30] for ionization occurs at 24.587 eV. Ionization resulting in He^+ ($n = 2$) has an onset at 65.400 eV. A complex *ns–np* Rydberg series converging to this limit has been observed by Madden and Codling.[31] Beyond 65.4 eV, the ratio of He^+ ($n = 2$) to He^+ ($n = 1$) has been investigated experimentally[32] and theoretically[33] with good concordance. However, the calculation predicts a bulge in this ratio between ~70 and 100 eV which had not been observed in the most recent experiments.[33a] The results are plotted in Fig. 3. The double ionization process has a threshold at 79.005 eV. From the data of Schmidt *et al.*,[34] we can follow this cross section from a vanishing value at threshold to a peak in the partial cross section at ~100 eV (Fig. 3) or to a plateau in the He^{2+}/He^+ ratio at ~150 eV. Very recently, this ratio has been remeasured by Holland *et al.*[34a] between 80 and 200 eV. Their results are in good agreement with those of Schmidt *et al.*[34]

In this instance, the correlation between the two electrons is seen to result in an excited singly charged ion with greater probability than double ionization. In neon (see Section B.2) the reverse is true.

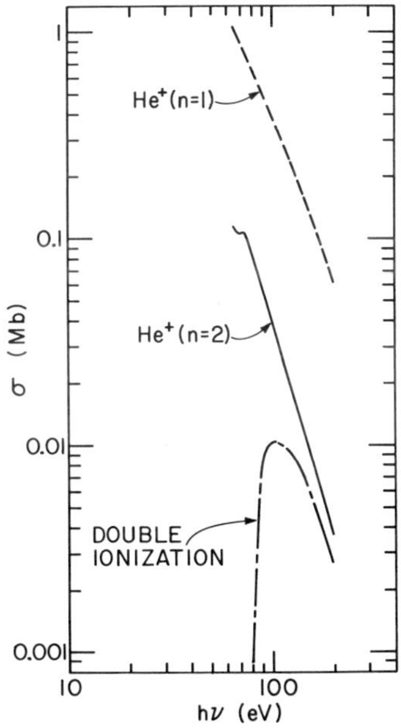

Fig. 3 Partial cross sections for formation of ionized states of helium by photoabsorption.

2. Neon

Apart from the two-electron systems (He and H_2), the investigation of partial cross sections is probably most nearly complete in the case of neon. The electron configuration of neon in its ground state is, of course,

$$(1s)^2(2s)^2(2p)^6.$$

Ejection of an electron from the 2p shell is the only significant process contributing to the absorption cross section from the ionization threshold (21.565 eV) to 48.476 eV, which is the threshold for ejection from the 2s shell. However, the p^5 configuration of Ne^+ is split into two states, $^2P_{3/2}$ and $^2P_{1/2}$, by spin–orbit coupling. The photoabsorption or photoionization behavior in the region between these limits has until recently been the subject of just one investigation[35] using photographic techniques. This is primarily due to the fact that the ionization threshold of neon lies outside the range of the helium continuum (see Chapter VIII). We[36] have now studied this region by photoionization mass spectrometry, employing a 3-m normal incidence monochromator at the Stoughton, Wisconsin, storage ring. In Fig. 4 we can see partially resolved components of two Rydberg series, *n*s and *n*d, converging to the (upper) $^2P_{1/2}$ limit. The characteristic peak shapes of these two series (narrow "*s*" sequence, broad "*d*" sequence) is not readily apparent here because of limited resolution. A detailed description of the matrix elements determining these widths in Ar, Kr, and Xe, as well as Ne, is presented in the aforementioned work.[36] The structure

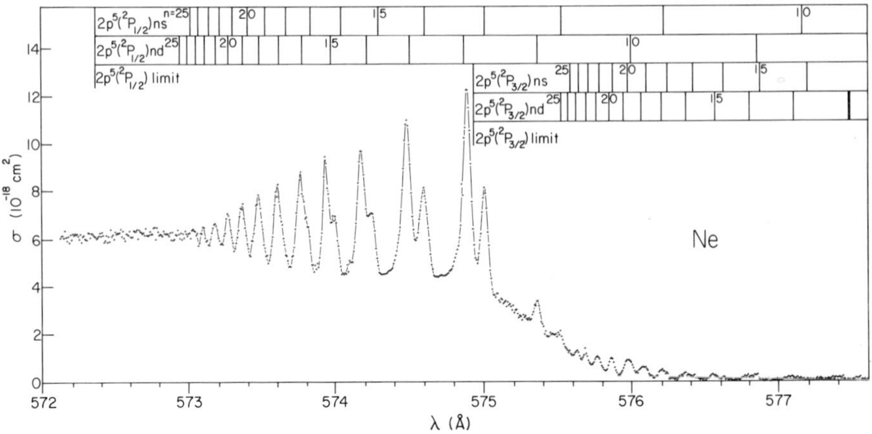

Fig. 4 Photoionization mass spectrum of neon, displaying autoionization between $^2P_{3/2}$ (ground) and $^2P_{1/2}$ (excited) ionic states, and collisional ionization of high Rydberg states below the ionization threshold. [From K. Radler and J. Berkowitz, *J. Chem. Phys.* **70**, 216 (1979).]

between 574.94 and 578 Å is due to the collisional ionization of high Rydberg states, i.e.,

$$\mathrm{Ne^{**} + Ne \rightarrow Ne^{+} + Ne + e^{-}}.$$

Photoelectron spectroscopy of neon in the vacuum ultraviolet has also not been very extensive since the ionization threshold lies beyond the 584 Å resonance line of helium. Measurement of this ratio is also more difficult than for the heavier noble gases because of the small spin–orbit splitting in the 2p shell of neon (0.097 eV). There had been some earlier studies[37] of the $^2P_{3/2}:{}^2P_{1/2}$ intensity ratio, and recently Samson *et al.*[38] have presented data indicating that this ratio is 2.18 from the $^2P_{1/2}$ threshold to ~19 eV beyond threshold. In this energy range, the absorption cross section is fairly constant.

Ejection of a 2s electron becomes energetically possible[30] at 48.476 eV. Rydberg series converging to this limit have been examined by Codling *et al.*[35] In order to apportion the total ionization cross section between the $(2s)^{-1}$ and $(1s)^{-1}$ thresholds into its components, we must take into account the various possible processes and the manner in which the data are reported. In addition to the single-electron processes $(2p)^{-1}$ and $(2s)^{-1}$, there exist two-electron processes, often referred to as shake-up and shake-off. In the shake-up process, the excitation accompanies ionization, whereas shake-off involves the removal of two (or more) electrons.

Shake-up can be observed as peaks in the photoelectron spectrum, whereas shake-off gives rise to a broad smear since the two departing electrons share the excess kinetic energy. This latter process can be observed directly by measuring the double ionization probability with a mass spectrometer. Such a study has been performed by Schmidt *et al.*[34] at the Orsay storage ring. The direct experimental result (see Fig. 5) is a measurement of

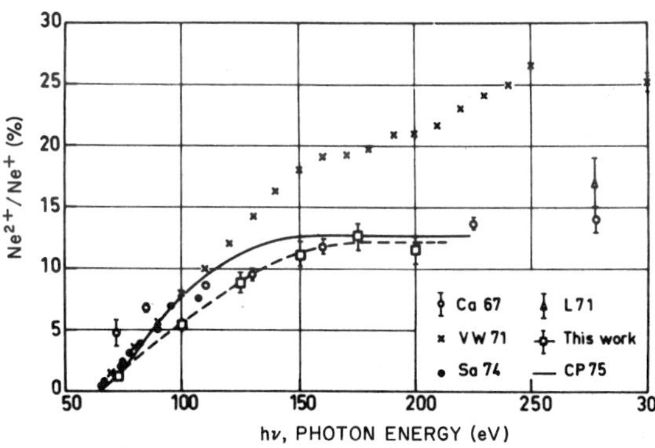

Fig. 5 The probability of double ionization (expressed as percent of single ionization) in the photoionization of neon. [From Schmidt *et al.*, *Phys. Rev. A* **13**, 1748 (1976).]

Ne^{2+}/Ne^+ as a function of photon energy. With this information, together with the total ionization cross section, we can extract that portion which results in *singly* ionized neon. Wuilleumier and Krause[39] have measured the ratio of shake-up involving $(2p)^{-1}$ to the single-electron event, and the corresponding ratio for $(2s)^{-1}$ as a function of photon energy. They also report the ratio of $(2p)^{-1}$ to $(2s)^{-1}$. With this information, and the cross section for forming singly ionized neon, it is possible to relate each of these processes to one of them, and hence to apportion the single ionization cross section. The resulting values are plotted in Fig. 6. Wuilleumier and Krause[39] and Wuilleumier[40] had already performed such an apportionment, but the present evaluation employs more accurate total absorption cross sections for this region, and also has the benefit of the double ionization measurements of Schmidt *et al.*[34] The Ne^{2+}/Ne^+ ratio has also been re-examined recently by Holland *et al.*[34a] They observe lower ratios than reported by Schmidt *et al.*[34] and earlier workers. They also searched for Ne^{3+} between 126.6 eV (Ne^{3+} threshold) and 280 eV without success, concluding that Ne^{3+} cannot contribute more than 1% from L-shell ionization.

Wuilleumier and Krause noted that the shake-up process beyond the threshold region remains a more-or-less constant fraction of the single ionization event. (Also, the various components of a particular shake-up series remain in constant proportion to one another with increasing photon energy.) Hence, for clarity, only the total shake-up for each series has been plotted in Fig. 6.

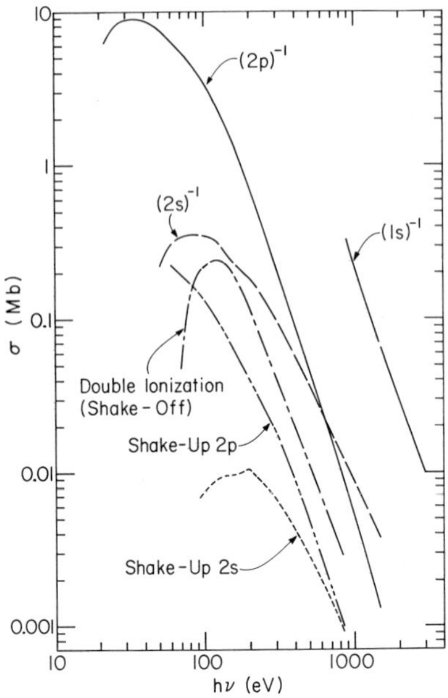

Fig. 6 Partial cross sections for formation of various ionic states of neon by photoabsorption.

This behavior apparently persists when the $(1s)^{-1}$ threshold is traversed. The $(1s)^{-1}$ partial cross section dominates all other processes beyond this edge, at 870.2 eV.[40] Indeed, the shake-up and shake-off processes associated with $(1s)^{-1}$ ionization, amounting to ~15% of the single ionization probability, are more intense than $(2s)^{-1}$ and $(2p)^{-1}$ for $h\nu \gg 870.2$ eV.

In Fig. 6 we have plotted the cross section for $(1s)^{-1}$, but not for the shake-up and shake-off processes that accompany K-vacancy formation. Carlson *et al.*[41] and Krause[42] do present intensities of shake-up processes relative to either $(1s)^{-1}$ or $(2p)^{-1}$, but the intensities of K- and L-shell ionizations do not appear to be directly compared. As an estimate of the behavior of $(2s)^{-1}$ and $(2p)^{-1}$ beyond the K-edge, we have taken the relative intensities $(1s)^{-1}$, $(2s)^{-1}$, and $(2p)^{-1}$ at 1486.7 eV given by Siegbahn *et al.*[43]; corrected these for angular distribution effects; and scaled $(2s)^{-1}$ and $(2p)^{-1}$ to $(1s)^{-1}$. The interpolation of $(2s)^{-1}$ and $(2p)^{-1}$ would appear to traverse the K-edge without discontinuity. Uncertainty in the intensity of the two-electron processes involving the K- and L-shell ionizations is not likely to incur substantial error in the $(1s)^{-1}$ cross section. Wuilleumier[40] also indicates that the L-shell cross sections ($(2s)^{-1}$ and $(2p)^{-1}$ and the corresponding shake-ups and shake-off) do not experience discontinuity at the $(1s)^{-1}$ threshold.

Several observations from Fig. 6 are worthy of note.

(1) The $(2s)^{-1}$ cross section declines less rapidly with energy than $(2p)^{-1}$. The cross over occurs at ~650 eV.

(2) The shake-up processes represent ~10% of the corresponding single ionization events.

(3) Shake-off appears to be somewhat more abundant than shake-up, representing 12–15% of the single ionization events.

Apart from shake-off processes, formation of a vacancy in the $n = 2$ shell of neon can only result in Ne^+. The Coster–Kronig transition (2s ← 2p) does not provide sufficient energy to eject a second electron. The K-shell vacancy primarily results in Ne^{2+} as a consequence of the KLL Auger process,[44] but can yield Ne^{3+} from a double Auger process with a 7.5% relative probability.[45]

3. Argon

The electronic ground state of argon has the configuration

$$(1s)^2(2s)^2(2p)^6(3s)^2(3p)^6.$$

Ionization begins[30] at 15.7597 eV with the ejection of a 3p electron and formation of the $^2P_{3/2}$ ground state of Ar^+. The $^2P_{1/2}$ spin–orbit component has a threshold[30] at 15.9373 eV. The Rydberg series $\cdots 3p^5(^2P_{1/2})ns\,^1P_1$ and $\cdots 3p^5(^2P_{1/2})nd\,^1P_1$ converging to this limit can autoionize when their energies exceed 15.7597 eV, giving rise to the structure seen in Fig. 3 in Chapter 3. The peak shapes are seen to differ from the corresponding ones in neon in that a sharp (*s*) component and broad (*d*) component clearly manifest themselves at the lower *n* values.

Beyond the $^2P_{1/2}$ onset, the ratio $\sigma_{3/2}:\sigma_{1/2}$ has been reported[38] to be 1.93, and to remain rather constant out to 40.8 eV.

Ejection of a 3s electron is possible when $h\nu$ exceeds 29.240 eV. Rydberg states converging to this threshold display window resonances.[46] The partial cross sections of $(3s)^{-1}$ and $(3p)^{-1}$ above 29.240 eV have been apportioned using measured ratios of Samson and Gardner[47] and Houlgate *et al.*[48] Somewhat more precise data[48a] became available to this author after the ensuing analysis was performed. Above 43.386 eV (threshold for Ar^{2+}) the total absorption cross sections have been corrected for double ionization, using the direct mass spectrometric measurements of Schmidt *et al.*[34] Both $(3s)^{-1}$ and $(3p)^{-1}$ cross sections plunge to a Cooper minimum between 40 and 50 eV, beyond which they recover. Data points are available to ~90 eV. The double ionization cross section increases from zero at threshold to a maximum at ~80 eV, corresponding to a plateau in the Ar^{+2}/Ar^{+} ratio (see Fig. 7). The recent data of Holland *et al.*[34a] confirm these observations.

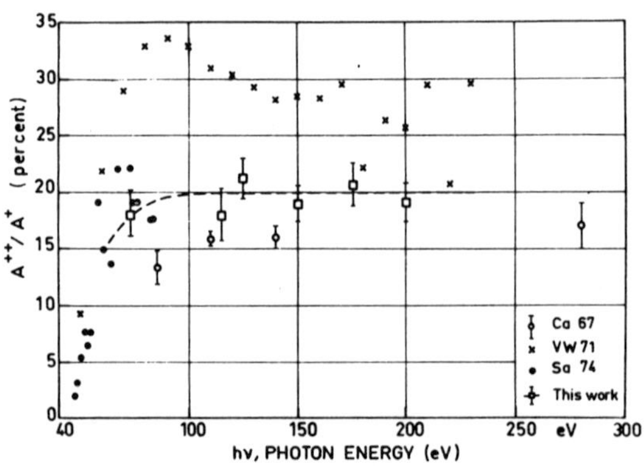

Fig. 7 The probability of double ionization (expressed as percent of single ionization) in the photoionization of argon [From Schmidt *et al.*, *Phys. Rev.* **A 13**, 1748 (1976).]

Isolated energies where photoelectron spectra have been measured[49] are plotted in Fig. 8 at 277 and 1486 eV. Unlike the detailed measurements below 90 eV, these points do not include corrections for angular distribution distortions, but have been corrected for shake-up and shake-off.

An approximation for the $(1s)^{-1}$ partial cross section has been arrived at by subtracting the shake-up and shake-off processes at 4510.8 eV (the Ti K_α line) reported by Krause *et al.*[50] from the total cross section at this energy[51]

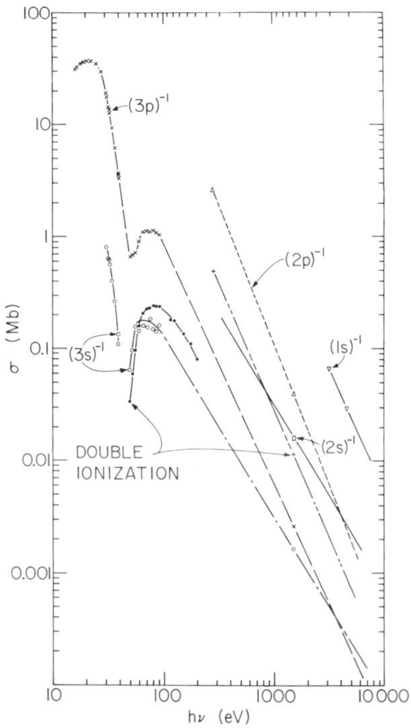

Fig. 8 Partial cross sections for formation of various ionic states of argon by photoabsorption.

and then further subtracting the extrapolated partial cross sections of $(3p)^{-1}$, $(3s)^{-1}$, $(2p)^{-1}$, and $(2s)^{-1}$.

We are now in a position to examine the consequences of vacancy formation in each of the shells. If ejection of an electron occurs from the 3p or 3s subshells, argon remains singly charged since $\Delta E[(3s - 3p)]$ is less than the second ionization potential of argon, and hence Auger processes are energetically forbidden. Formation of $(2p)^{-1}$ results in one Auger process, giving rise to Ar^{2+}. Radiative deexcitation is much less probable in this instance. Creation of a 2s vacancy leads in most cases[52] to a Coster–Kronig transition $(2s) \leftarrow (2p)$ which leaves a 2p hole that is filled by a second Auger process. The net result is formation of Ar^{3+}. Holland *et al.*[34a] report a weak Ar^{3+} intensity (0.01–0.22 Mb) until the L threshold is reached, at which energy this partial cross section increases by a factor of ~50. The Ar^{2+} cross section exhibits a similar jump at 248 eV, in agreement with the aforementioned inferences.

A K-hole in argon can be filled in a variety of ways. As a consequence, a spectrum of charge distributions is obtained (Fig. 9). Calculations[52] that encompass the various possibilities are in rather good agreement with the experimental results. Some contribution to the charge spectrum comes from

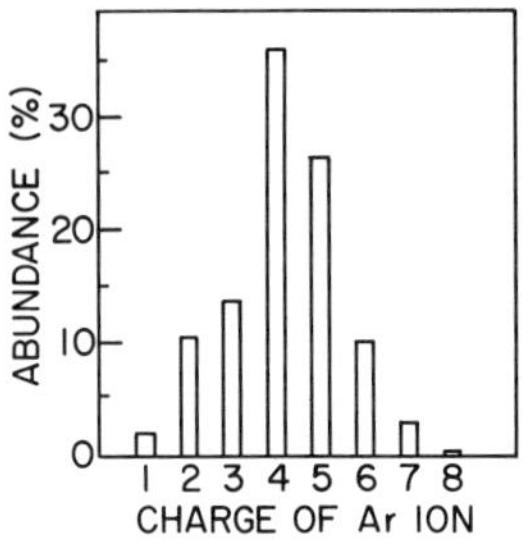

Fig. 9 Spectrum of charged ions obtained after creation of a hole in the K-shell of argon. [Data are from T. A. Carlson and M. O. Krause, *Phys. Rev.* **137**, A1655 (1965).]

higher unfilled shells, but at the photon energy of the experiment (4510.8 eV) it is estimated that 90% of the initial ionization results in $(1s)^{-1}$.

4. Krypton

The electron configuration of krypton in its ground state is

$$(1s)^2(2s)^2(2p)^6(3s)^2(3p)^6(3d)^{10}(4s)^2(4p)^6.$$

Ionization is first detected at 14.00 eV with the ejection of a 4p electron and the formation of the ground $^2P_{3/2}$ state of Kr^+.

In the interval between $^2P_{3/2}$ and its spin–orbit partner $^2P_{1/2}$ (14.00–14.66 eV), typical autoionization features are exhibited by Rydberg states converging to $^2P_{1/2}$ (see Fig. 10). One notes sharp peaks identified as $p \rightarrow ns$ transitions, and broad ones corresponding to $p \rightarrow nd$ transitions, as in the case of argon. Beyond the $^2P_{1/2}$ threshold, the ratio of cross sections $\sigma_{3/2}$: $\sigma_{1/2}$ is less than statistical (2:1) as predicted by the Walker *et al.*[53] theory for a declining partial cross section, but recent experiments by Samson *et al.*[38] have shown a surprisingly constant ratio of 1.77 up to 18 eV above this threshold. The Walker theory would imply a declining value for this ratio as the Cooper minimum is approached.

The onset for ejection of an electron from the 4s shell occurs[30] at 27.514 eV. The Rydberg series converging to this limit has characteristic window resonances.[54] Samson and Gardner[47] have measured the $(4s)^{-1}$:$(4p)^{-1}$ ratio up to ~40.8 eV. In order to place these partial cross sections on an absolute scale, it is necessary to subtract the contribution of Kr^{2+}, which has a threshold[30] at 38.571 eV. Such a correction, using the data of Samson and Haddad,[55] was made before plotting the $(4s)^{-1}$ and $(4p)^{-1}$ partial cross sections in Fig. 11. We note a steep decline in both $(4p)^{-1}$ and $(4s)^{-1}$ as the Cooper minimum is approached, indicating a strong correlation between 4s and 4p. Similar behavior was noted in the case of Ar 3s and 3p. The recent data of Holland *et al.*[34a] indicate that the Kr^+ partial cross section

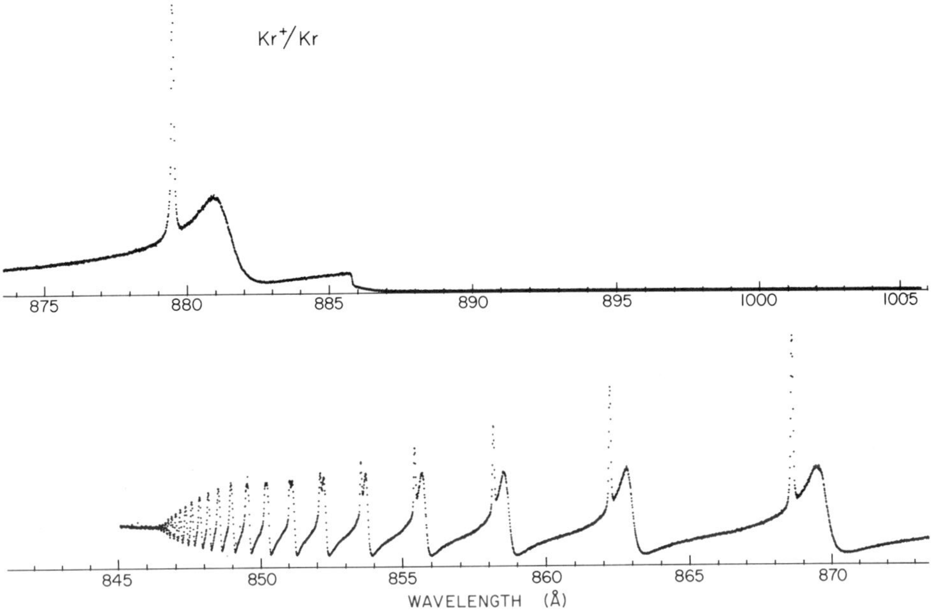

Fig. 10 Photoionization mass spectrum of krypton between the $^2P_{3/2}$ (ground) and $^2P_{1/2}$ (excited) ionic states, displaying sharp "s"-like and broad "d"-like resonances. [Unpublished data obtained by the author, with a photon bandwidth of 0.07 Å.]

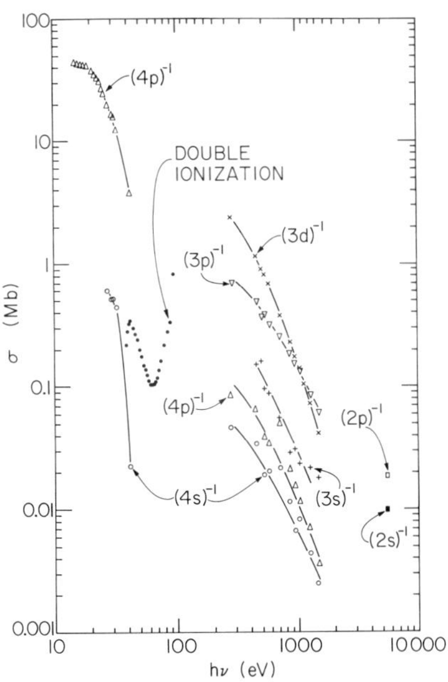

Fig. 11 Partial cross sections for formation of various ionic states of krypton by photoabsorption.

remains small above 80 eV, perhaps displaying small oscillations in the vicinity of the $(3d)^{-1}$ threshold.

The cross section for double ionization[55] increases more-or-less linearly from threshold, as expected. The ratio $\sigma(Kr^{2+})/\sigma(Kr^{+})$ has a plateau between 50 eV $< h\nu <$ 60 eV of $\sim$0.17, in line with the directly measured ratios in neon and argon. However, a dramatic increase in Kr^{2+} begins at $\sim h\nu =$ 70 eV, well below the threshold for $(3d)^{-1}$ at 93.83 eV.[56]

This increase was already noted in the pseudophoton experiments of El-Sherbini and Van der Wiel.[57] It has been attributed to "... interference between competing processes, one or more of which involve the d^{10} shell" by Samson and Haddad.[55] The double ionization cross section up to this energy must mechanistically involve electron–electron correlation since shake-off and Auger processes are energetically forbidden.

There is a gap in our knowledge of partial cross sections until 277 eV, where the x-ray photoelectron spectroscopic studies of Krause[58] are available. Krause reports subshell ratios for $(4s)^{-1}$, $(4p)^{-1}$, $(3d)^{-1}$, $(3p)^{-1}$, and $(3s)^{-1}$ at discrete x-ray lines between 277 and 1487 eV, measured at 90° with respect to the photon propagation direction. These ratios have been corrected to the "magic angle" of 54° 44′, using Krause's angular distribution data supplemented by the calculated values of Kennedy and Manson.[59] Multiple electron transitions have been taken into account, using the information in Krause's paper,[58] but have not been plotted to simplify the figure. The single-electron transitions in this region are dominated by $(3d)^{-1}$ and $(3p)^{-1}$, the partial cross sections for these processes crossing over at $\sim$1020 eV. There is an intriguing tendency for the double ionization cross section discussed previously to join up with the $(3d)^{-1}$ cross section, but more information is required in the 90–277 eV region. The partial cross sections given by Krause are in fair agreement with those given by Siegbahn *et al.*[60] at 1254 and 1487 eV. The spin–orbit components of $(3d)^{-1}$ and $(3p)^{-1}$ have been summed; the latter ratio is nearly statistical, according to Siegbahn *et al.* We have plotted one additional piece of information beyond the $n = 2$ edges, based on Fig. 4.9 in ESCA II, an extrapolation of the previously considered cross sections and an assumed 20% contribution to the shake-up and shake-off processes accompanying ejection from the $n = 2$ shell.

We now consider the charge states of krypton that result from the various subshell ionizations. Ejection from (4s) and (4p) energetically leads to Kr^{+} only, as noted earlier.

The $(3d)^{-1}$ process results primarily in Kr^{2+} as a result of an Auger process, but two-electron ejection[61] occurs in about one-third of the Auger events, resulting in Kr^{3+}. Holland *et al.*[34a] report a slightly higher proportion of $Kr.^{3+}$

The $(3p)^{-1}$ process leads mainly to Kr^{3+} via a Coster–Kronig transition followed by an Auger transition. Some Kr^{4+} is also formed by double-electron ejection in either of the positionization transitions. The 3s vacancy also results in Kr^{3+} and Kr^{4+}.[57] The partial cross section for formation of Kr^{4+} has been directly measured by Holland *et al.*[34a] It appears to increase linearly from ~150 eV to ~220 eV (approximately the $(3p)^{-1}$ threshold) and remains more-or-less constant to 280 eV.

The consequences of creating a vacancy in the $n = 2$ and $n = 1$ shells in krypton have been investigated by Krause and Carlson[62] experimentally and rationalized with a model that incorporates shake-off and various alternative Auger and radiative decays. For the L-shell study, a titanium anode was used, producing primarily the 4510.84 eV Ti K_1 line. The initial vacancy distribution was calculated to be 2s = 0.30; 2p = 0.563; 3s = 0.05; 3p = 0.07; 3d = 0.015; and 4s = 0.003, hence predominantly in the $n = 2$ shell. The experimentally observed charge distribution (Fig. 12) is seen to have major components at Kr^{4+}, Kr^{5+}, Kr^{6+}, with a mean charge of 5.22. Their calculated charge distribution is in good agreement with experiment, but the dominant contributors to the Auger cascade are not mentioned.

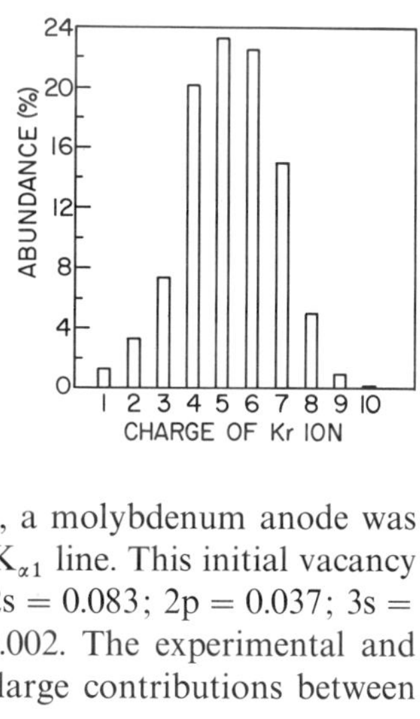

Fig. 12 Charge distribution of krypton ions following creation of a vacancy in the $n = 2$ shell. Minor contributions from vacancies in the $n = 3$ and $n = 4$ shells are included. [Data are from M. O. Krause and T. A. Carlson, *Phys. Rev.* **158**, 18 (1967).]

To induce a primary vacancy in the K-shell, a molybdenum anode was used, primarily generating the 17.479 keV Mo $K_{\alpha 1}$ line. This initial vacancy distribution was calculated to be 1s = 0.865; 2s = 0.083; 2p = 0.037; 3s = 0.007; 3p = 0.004; 3d = 0.002; and 4s, 4p = 0.002. The experimental and calculated charge distributions (Fig. 13) have large contributions between Kr^{5+} and Kr^{9+}, with a mean charge of 6.0.

5. Xenon

The electron configuration of xenon in its ground state is

$$(1s)^2(2s)^2(2p)^6(3s)^2(3p)^6(3d)^{10}(4s)^2(4p)^6(4d)^{10}(5s)^2(5p)^6.$$

Ionization is initially observed at $h\nu = 12.130$ eV with the ejection of a 5p electron and the formation of the (ground) $^2P_{3/2}$ state of Xe^+. In the

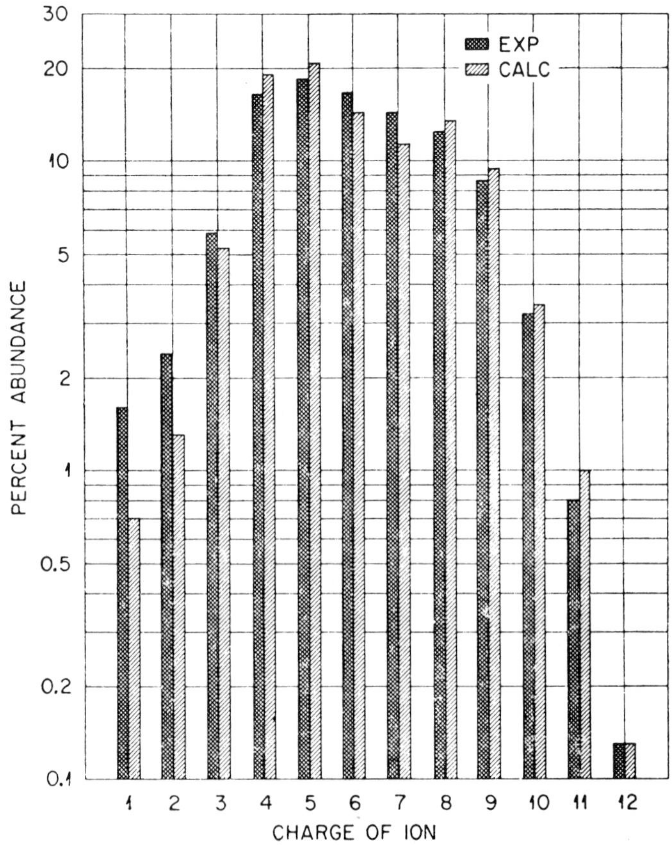

Fig. 13 Charge distribution of krypton ions following creation of a vacancy primarily in the K-shell. [From M. O. Krause and T. A. Carlson, *Phys. Rev.* **158**, 18 (1967).]

interval between $^2P_{3/2}$ and $^2P_{1/2}$ (12.13–13.436 eV), autoionization structure is prominent (see Fig. 14) due to Rydberg states converging to $^2P_{1/2}$ (see corresponding comments regarding krypton).

Beyond the $^2P_{1/2}$ threshold, the ratio of partial cross sections $\sigma_{3/2}:\sigma_{1/2}$ has been reported[38] to be constant at 1.54 out to 40.8 eV. The ratio is less than statistical, as predicted by Walker and Waber,[63] but recent experiments[64] indicate that it has a minimum of ~1.35 at ~30 eV; it then increases to a weak maximum or plateau at ~60 eV, the ratio having approximately the statistical value (2.0) at the higher energy.

The threshold[30] for $(5s)^{-1}$ is at 23.397 eV. The Rydberg series converging to this limit has window resonances akin to those previously noted for argon and krypton. Kemeny *et al.*[64a] have examined the branching ratio $\sigma(^2P_{3/2})$:

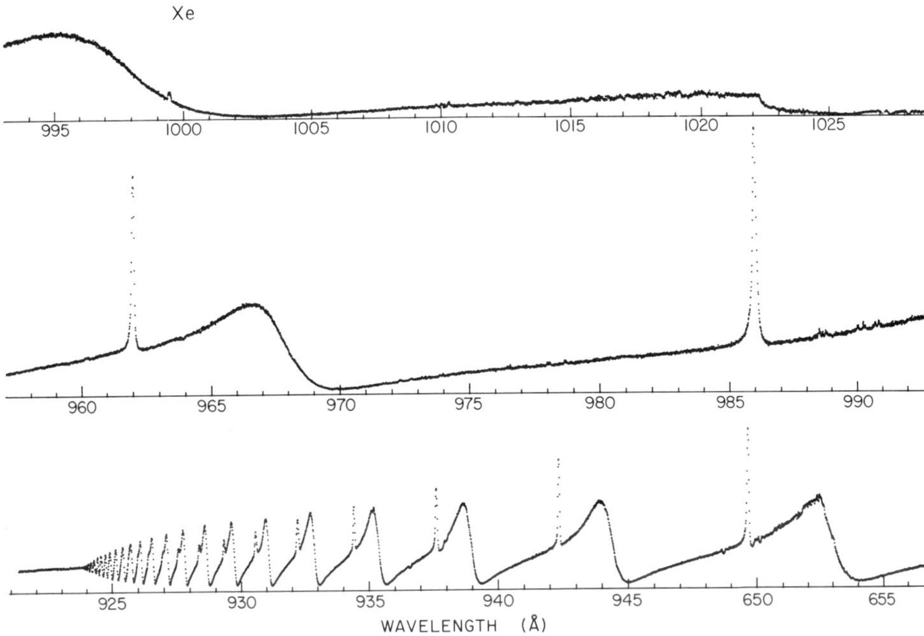

Fig. 14 Photoionization mass spectrum of xenon between the $^2P_{3/2}$ (ground) and $^2P_{1/2}$ (excited) ionic states, displaying sharp "s"-like and broad "d"-like resonances. [Unpublished data obtained by J. H. D. Eland, Argonne National Laboratory, with a photon bandwidth of 0.07 Å.]

$\sigma(^2P_{1/2})$ across one of these resonances, the $5s5p^66p$ ($^1P_1$) at 20.95 eV. The experimental results reveal a rapidly increasing value of this ratio in the wings of the resonance. A theoretical fit which includes auxiliary information on the Fano parameters q, Γ, and ρ^2 (see also Section D.3.b) leads to the conclusion that this ratio is 8.8 ± 0.5 at the center of the resonance, whereas the nearby continuum (as noted above) has a ratio of ~ 1.55. The $(5s)^{-1}$:$(5p)^{-1}$ ratio between 23.55 and 40.8 eV has been measured by Samson and Gardner.[47] Subsequent to the analysis presented here, somewhat more precise and extensive data[64b] became available to the author. The general shape of the partial cross section curve for $(5s)^{-1}$ remains the same, but the Cooper minimum (vide infra) may have even a smaller cross section than indicated earlier, i.e., <0.1 Mb. The partial cross sections have been put on an absolute scale, after correcting for double ionization.[65]

In Fig. 15, both $(5p)^{-1}$ and $(5s)^{-1}$ can be seen to decline sharply with energy as the Cooper minimum is approached. The Xe^{2+} cross section rises more-or-less linearly from its threshold[30] (33.340 eV) and passes through a maximum at about 40.8 eV. After a brief decline, the double ionization cross section increases sharply from about 50 eV, well *before* the threshold for the 4d ejection at 67.55 eV.[66]

This behavior was also noted in krypton and similar behavior has recently been noted in Tl, In, and Ga by Connerade.[66a] The partial cross section pattern can be followed between 60 and 135 eV, with the aid of recent results by West *et al.*[67] The $(5s)^{-1}$ and $(5p)^{-1}$ cross sections are seen (Fig. 15) to recover from the Cooper minimum, whereas the $(4d)^{-1}$ partial cross section rises to a maximum about 30 eV above its threshold. Once again, the double ionization cross section appears to be merging with the $(4d)^{-1}$ cross section, as in the case of krypton.

Recent measurements by Holland *et al.*[34a] indicate that at ~100 eV photon energy (the peak of the $(4d)^{-1}$ absorption), the partial cross sections for formation of Xe^+, Xe^{2+}, and Xe^{3+} are 1.111 ± 0.112, 18.40 ± 1.60, and 7.49 ± 0.67 Mb, respectively. Hence, one can plausibly conclude that the partial cross section for $(4d)^{-1}$ is dominant at this energy and that it primarily results in the formation of Xe^{2+}, as inferred earlier. Although the partial cross section of Xe^+ is relatively small at this energy, it is of some theoretical significance that it exhibits a weak peak at ~100 eV, which has been interpreted[67a] as intershell coupling between 4d and the outer shells 5s and 5p.

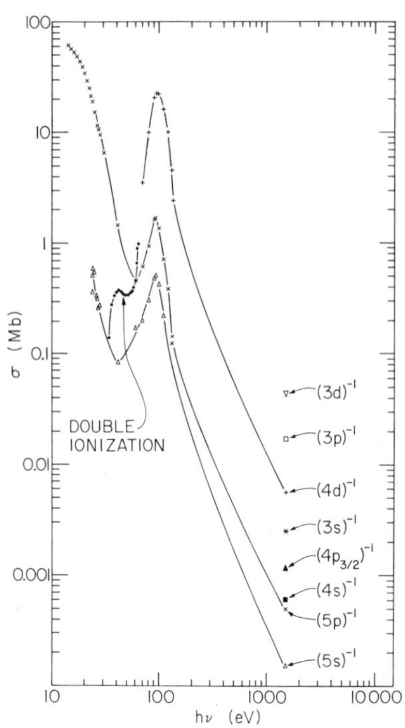

Fig. 15 Partial cross sections for formation of various ionic states of xenon by photoabsorption.

Detailed information is lacking above 140 eV. An isolated point at 1486.7 eV, derived from relative intensities given by Siegbahn *et al.*[68] is indicated in Fig. 15, and provides some hint of the decline in $(5p)^{-1}$, $(5s)^{-1}$, and $(4d)^{-1}$ in this interval. A 20% contribution from shake-up and shake-off has been assumed for this point. Some relative intensities are also given by Siegbahn *et al.*, at 1253.6 and 8047.8 eV, but they are incomplete.

The consequences of inner shell vacancies in xenon are as follows:

$(4d)^{-1}$: Auger decay leads primarily to Xe^{2+}, with about 21% Xe^{3+} due to a double Auger process.[57] The aforementioned data of Holland *et al.*[34a] imply an attribution of ~28% to the process producing Xe^{3+}.

$(4p)^{-1}$: Auger cascade proceeds from $4p^5$ to $4d^9 5p^5 + e^-$ and then to $5p^3 + 2e$ and $5p^2 + 3e^-$. Alternatively, $4p^5$ leads to $4d^9 5p^4 + 2e^-$ and then to $5p^2 + 3e^-$. The net result is Xe^{3+} and Xe^{4+} in approximately equal abundance.[57] The data of Holland *et al.*[34a] are consistent with this view.

$(4s)^{-1}$: Approximately the same as $(4p)^{-1}$.[69]

Product ions from initial vacancies in the $n = 3, 2, 1$ shells have been investigated by Carlson *et al.*,[69] and histograms are shown in Fig. 16. The $n = 1$ results are based on formation of a K-shell vacancy by the nuclear internal conversion process. The lower average charge resulting from this vacancy, compared to the 2s vacancy, is attributed to enhanced radiative transition probability, providing an alternative path for disposal of this large energy.

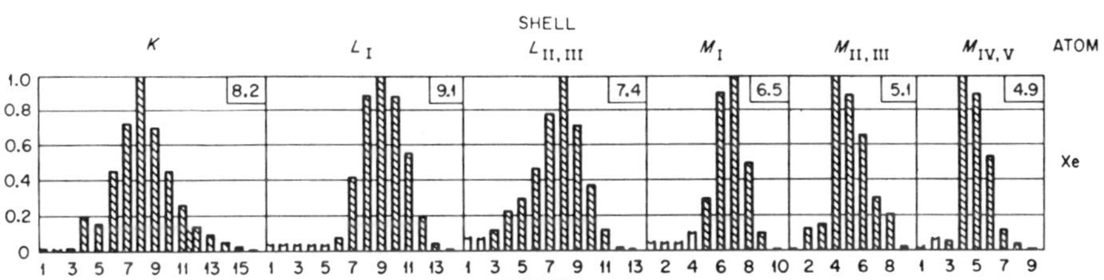

Fig. 16 Charge distribution of xenon ions following creation of vacancies in the K-, L-, and M-shells. [From T. A. Carlson *et al.*, *Phys. Rev.* **151**, 41 (1966).]

6. Mercury

The electronic configuration of mercury in its ground state is

$$(1s)^2(2s)^2(2p)^6(3s)^2(3p)^6(3d)^{10}(4s)^2(4p)^6(4d)^{10}(4f)^{14}(5s)^2(5p)^6(5d)^{10}(6s)^2.$$

The threshold for emission of a 6s electron occurs[30] at 10.437_6 eV. However, autoionization from Rydberg states converging to $(5d)^{-1}$ masks the direct photoionization process corresponding to $(6s)^{-1}$. The onset for 5d ionization, corresponding to the formation of Hg^{+} $^{2}D_{5/2}$, is at 14.840_8 eV. The partial cross section for ionization from the 5d shell increases from threshold to a maximum at $h\nu \approx 40$ eV. Absolute partial cross sections are rather unreliable at this time, but the ratio $\sigma(6s)^{-1}:\sigma(5d)^{-1}$ has been measured at several energies, as follows:

$h\nu$, eV	$\sigma(6s)^{-1}/\sigma(5d)^{-1}$	Ref.
21.2	0.0165	(70)
40.8	0.0151	(70)
132.3	≤ 0.0091	(71)
1486.7	0.0248	(72)

There is a suggestion of a minimum for $\sigma(6s)^{-1}$ at ~ 100 eV, although other considerations[70,70a] indicate a minimum around 18–20 eV. With incident radiation $h\nu \geq 16.705_3$ eV, ejection of an electron from the 5d shell can result in $^{2}D_{3/2}$, as well as $^{2}D_{5/2}$. Walker *et al.*[53,63] have predicted the variation of $\sigma_{5/2}:\sigma_{3/2}$ with excess photon energy on the basis of relativistic Dirac–Slater calculations of the partial cross sections. The spin–orbit subshells have bound state wave functions that are slightly different, the $d_{3/2}$ component being somewhat more attractive to the nucleus. As a consequence, overlap with the continuum wave function near threshold is larger for $d_{5/2}$ than for $d_{3/2}$. In addition, if the wavelength region corresponds to a rising $(d)^{-1}$ cross section, the $d_{5/2}$ will have risen more than the $d_{3/2}$ since its threshold is lower. Hence, two effects, one involving slightly different dipole matrix elements, the other often referred to as a kinetic energy effect, combine to make $\sigma(d_{5/2})^{-1}:\sigma(d_{3/2})^{-1}$ larger than statistical (3:2) when the d-shell partial cross section is rising. When this cross section has passed its maximum (in the case of the Hg 5d, about 40 eV) similar arguments lead to the reverse conclusion, i.e., $\sigma(d_{5/2})^{-1}:\sigma(d_{3/2})^{-1}$ will be less than statistical as the d-shell partial cross section is falling. A summary of experimental and theoretical evaluations of this ratio is depicted in Fig. 17.[71] The agreement is seen to be good, providing strong support for the theoretical arguments.

Svensson *et al.*[72] have measured the photoelectron spectrum of Hg vapor at 1486.7 eV, and identified peaks corresponding to single-electron ejection from all sublevels with principal quantum numbers 4–6. The ionization energies observed, together with crude relative intensities scaled from their figure, are given in Table 1. Note that the relative intensities are not corrected for apparatus transmission, nor have the angular distribution effects been taken into account.

The Hg^{n+} charge distributions resulting from photoionization in each of the principal shells of mercury have been measured by Carlson *et al.*[69]

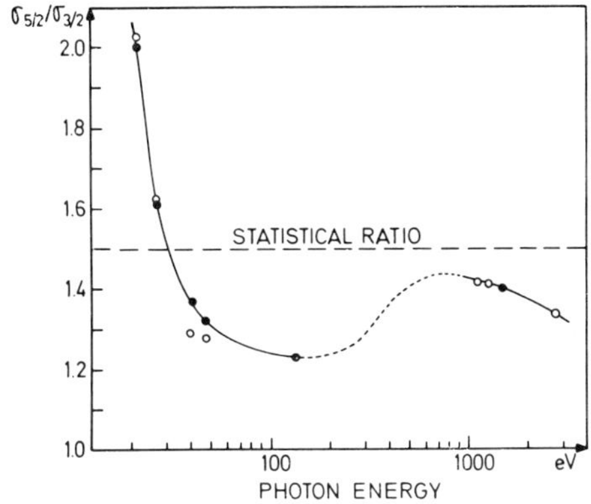

Fig. 17 The ratio of spin–orbit states $^2D_{5/2}$ and $^2D_{3/2}$ of Hg^+ observed upon photoejection of an electron from the 5d shell of mercury. Hg5d: ●, experimental and ○, theoretical. [From R. Nilsson *et al.*, *J. Elctr. Spectr.* **9**, 337 (1976).]

However, the photon energies used by these authors, and the assigned principal shells from which electron ejection took place, do not seem to correspond with the data in Table 1. For example, with $h\nu = 220$ eV, the primary ionization process is identified as $(5d)^{-1}$ by Carlson *et al.*, whereas the data of Table 1 indicate a large contribution by $(4f)^{-1}$, which has its ionization

TABLE 1 *Orbital Ionization Energies of the n = 4, 5, and 6 Levels of Hg and Crude Relative Cross Sections at hν = 1486.7 eV*[72]

Subshell	Ionization energy (eV)	Relative intensity
6s	10.43	1
$5d_{5/2}$	14.91	26.5
$5d_{3/2}$	16.73	18.5
$5p_{3/2}$	71.7	3
$5p_{1/2}$	90.3	2
$5s_{1/2}$	134	3
$4f_{7/2}$	107.06	212.3
$4f_{5/2}$	111.13	160.9
$4d_{5/2}$	366.0	86
$4d_{3/2}$	385.4	54
$4p_{3/2}$	583.8	22
$4p_{1/2}$	687.4	5.3
4s	809.4	1.3

onset at ~110 eV, at the measurement energy of 1486.7 eV. A critical reassessment of these charge distributions must await more detailed knowledge of the partial subshell cross sections as a function of incident photon energy.

The direct double ionization cross section has been measured by Cairns *et al.*[73] from threshold to ~72 eV. The onset for $(6s)^{-2}$ occurs[30] at 29.186 eV; hence, a 5d vacancy alone is energetically insufficient to cause double ionization, and no deeper subshells are within the range of their experiments. The ratio of $Hg^{2+}:Hg^{+}$ (Fig. 18) does not exhibit a distinct plauteau (as do the smaller noble gases) but appears to increase dramatically before the onset of the next (5p) subshell. The absolute partial cross section also increases[73] with photon energy.

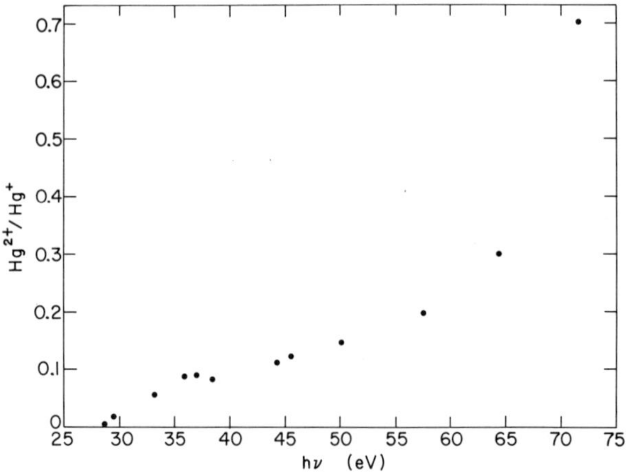

Fig. 18 The ratio of double to single ionization in the photoionization of mercury. [Data are from R. B. Cairns *et al.*, *J. Chem. Phys.* **53**, 96 (1970).]

C. Diatomic Molecules

1. H_2

a. Photoelectron Spectroscopy. In several ways, H_2 is atypical. In its ground state, it has only one occupied molecular orbital $(1s\sigma_g)$.[2] The total photoionization and partial photoelectron cross sections are almost identical, the sole difference being the small but finite probability of two-electron processes. As we have seen, the oscillator strength distribution function has a relatively large contribution from the discrete (subionization) region. Somewhat more than $\frac{1}{2}$ of the $S(-2)$ sum (and hence the contribution to static electric dipole polarizability) is attributable to the subionization region, primarily the Lyman and Werner bands.

The VUV photoelectron spectrum of H_2, using 584 Å incident radiation, has been reported by several investigators.[74–78] It corresponds to the process

$$H_2(X^1\Sigma_g^+, v'' = 0) + h\nu \rightarrow H_2^+(X^2\Sigma_g^+, v' = 0, 1, 2, \ldots) + \bar{e},$$

i.e., a bonding electron has been removed, increasing r_e and giving rise to a broad electronic band. Within the framework of the Born–Oppenheimer approximation, the relative intensities of the various v_n peaks should be governed by Franck–Condon factors involving the Hermite functions of the $X^1\Sigma_g^+$ and $X^2\Sigma_g^+$ states. Since the potential curves of these states are now known to high accuracy,[79] it is possible to calculate correspondingly accurate Frank–Condon factors.[80] These have been compared with the experimental results of several workers in Fig. 19. The general trend of the experiments points to a monotonic increase of vibrational intensity above the Franck–Condon expectations between $v' = 0$ and $v' = 6$ with an apparent decline above $v' = 6$. The experimental data chosen for this plot involved apparatuses which selected the photoelectrons at different angles. Frost *et al.*[75] used a hemispherical retarding field, and hence were observing all photoelectrons in a particular energy, irrespective of angle of ejection. Turner's experiment[74] is most sensitive to electrons ejected at an angle of 90° with respect to the photon propagation direction, i.e., along the electric vector. The data of Berkowitz and Spohr,[78] taken with a cylindrical mirror analyzer, were most sensitive to an angle of ~60° with respect to the photon propagation direction. In the range $v' = 0$ to $v' = 6$, the results are comparable. Angular distribution studies by Carlson and Jonas,[81] and Morgenstern *et al.*[82] provide

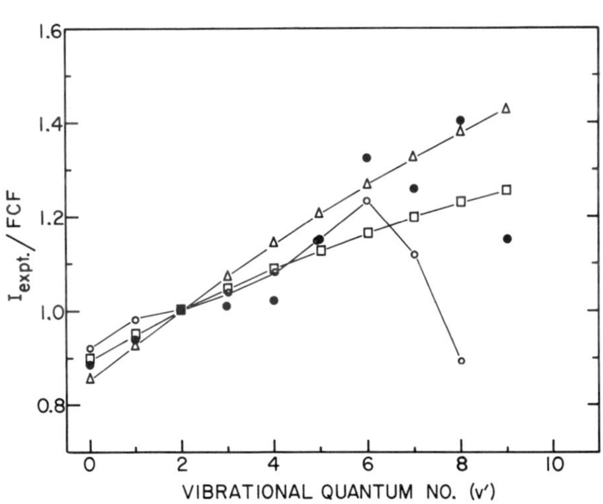

Fig. 19 The relative intensity of vibrational states and calculated Franck-Condon factors in the photoionization of H_2. $H_2 + h\nu \rightarrow H_2^+(v') + e$. ○, Frost *et al.*[75]; △, □, Itikawa[83]; Berkowitz and Spohr. [From J. Berkowitz and R. Spohr, *J. Electr. Spectr.*●, 143 (1973).]

evidence that the angular distribution of photoelectrons from H_2 is slightly dependent on the vibrational state of H_2^+. Hence, the tentative conclusion is that precise Franck–Condon factors do not accurately predict the relative vibrational intensity. Itikawa[83] has examined this problem theoretically, taking into account explicity the dependence of the electronic transition moment on the internuclear distance. He has succeeded in obtaining rather good agreement with experiment in the $v' = 0$ to $v' = 6$ region (see Fig. 19) but his curve continues to increase monotonically between $v' = 6$ and $v' = 10$, whereas the presently available experimental evidence shows a decline. It is conceivable that the transmission of lower energy electrons, corresponding to $v' > 6$, begins to diminish in the electron spectrometers used for these experiments. However, the state $v' = 7$ corresponds to an electron energy of 4.2 eV, and other evidence from at least the cylindrical analyzer experiment leads to the conclusion that serious transmission troubles were not encountered until the electron kinetic energy was less than 1 eV. The low experimental point of Frost *et al.*, for $v' = 8$ is readily attributable to measurement error since it involves a very small step superimposed upon many other steps.

We are thus left with a residual discrepancy for the vibrational intensities above $v' = 6$. At this time, it is not clear whether the discrepancy is to be attributed to experiment or theory.[84]

One of the contributions to the width of vibrational lines in the H_2^+ spectrum is attributable to the thermal motion of the H_2 molecules. The magnitude of this Doppler broadening is proportional to the square root of the kinetic energy of the ejected electron. (In the intensity measurements discussed above, this has been considered by measuring the area of each peak and dividing by resolution width.[85]) Hence, it is possible to significantly narrow the line width by using less energetic incident radiation. Åsbrink[76] succeeded in doing this using NeI radiation ($\lambda = 743.7195$ and 735.8962 Å) and his results are seen in Fig. 20. Note that the vibrational intensities are probably not correctly given here, e.g., $v' = 5$ is shown as the largest line, whereas all other experimental and theoretical results yield $v' = 2$ as the most probable

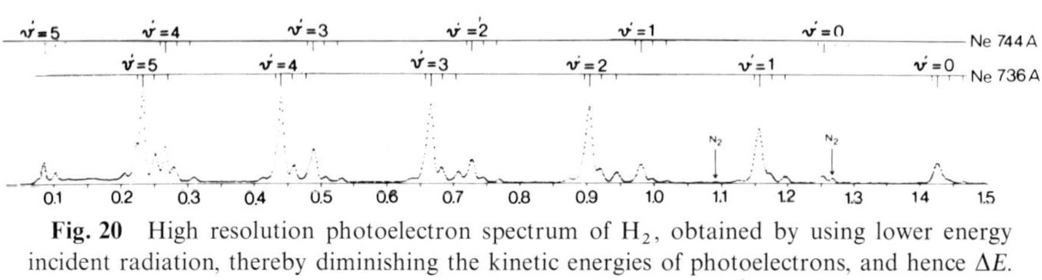

Fig. 20 High resolution photoelectron spectrum of H_2, obtained by using lower energy incident radiation, thereby diminishing the kinetic energies of photoelectrons, and hence ΔE. At this resolution, rotational structure is partly resolved. [From L. Åsbrink, *Chem. Phys. Lett.* 7, 549 (1970).]

transition.† However, there are distinctly resolved shoulders on the vibrational lines which can be identified with alternate *rotational* transitions.

It may be recalled that the even rotational levels of H_2 ($K'' = 0, 2, \ldots$) are coupled to a particular pairing of nuclear spins in H_2, which must be antisymmetric in order to make the total wave function antisymmetric. This subset of states is called parahydrogen. The rotational states $K'' = 1, 3, 5$, which themselves have antisymmetric wave functions, are coupled to nuclear spin functions that are symmetric, this subset being called orthohydrogen. Hence, unless nuclear spin states can be altered in the photionization act, transitions from even K'' states to odd K'' states, or odd K'' states to even K'' states are forbidden. The probability of nuclear spin flip, and hence of ortho–para conversion by the photionization mechanism is extremely small.

However, there does exist a reasonable probability for $\Delta K = \pm 2$, ± 4 transitions, and this has been considered by Sichel[86] and somewhat more specifically by Brehm and Frey.[77] An expanded section of Asbrink's results, in the $v' = 5$ region, are shown in Fig. 21. Here he has attempted to deconvolute the various rotational lines. Although there exist residual overlappings, partly due to the melange of rotational states and partly due to the presence of two NeI lines in the incident radiation, it appears as if the $\Delta K = \pm 2$ (O and S form) transitions are about 10% as intense as the $\Delta K = 0$ transitions (Q forms) with no indication of $\Delta K = \pm 1$ transitions (forbidden by the ortho–para selection rule) or $\Delta K = \pm 4$ or higher transitions. Niehaus and Ruf[87] have actually measured the photoelectron angular distributions corresponding to $\Delta K = 0$ transitions and $\Delta K = -2$ transitions in the $v' = 1$ region of the $H_2{}^+$ spectrum and observed dramatically different behavior. This behavior may play a role in explaining the discrepancy between theory

† A contradictory view has been reported by Samson.[85a] In Fig. 11 of this article, the 736 Å photoelectron spectrum of H_2 is shown with $H_2{}^+$ vibrational peaks increasing in intensity from $v' = 0$ to $v' = 5$. The background also rises in the same direction. I attribute this effect to acceleration of the photoelectrons before they enter the analyzer, which would favor the lower energy electrons and manifest itself as higher apparent intensities for the higher vibrational levels. Data taken at this laboratory[85b] with a cylindrical mirror analyzer and without pre-acceleration produced an intensity distribution at 736 Å very similar to that at 584 Å. Samson (bottom of p. 318 of his article) states "... when transitions into the ionization continuum proceed via autoionization transitions, the normal Franck–Condon factors do not hold. This is the case for 736 Å radiation." I have carefully re-examined the photoionization spectrum of H_2 in the wavelength region of the neon lines at 735.8962 and 743.7195 Å.[85c] Although there are weak autoionization features throughout the region, they project above the direct ionization continuum only by about 10%. No specific autoionization features fall directly at the positions of the neon lines. Hence, the autoionization explanation for the discrepancy between Samson's observation and Franck–Condon-like behavior is most likely invalid.

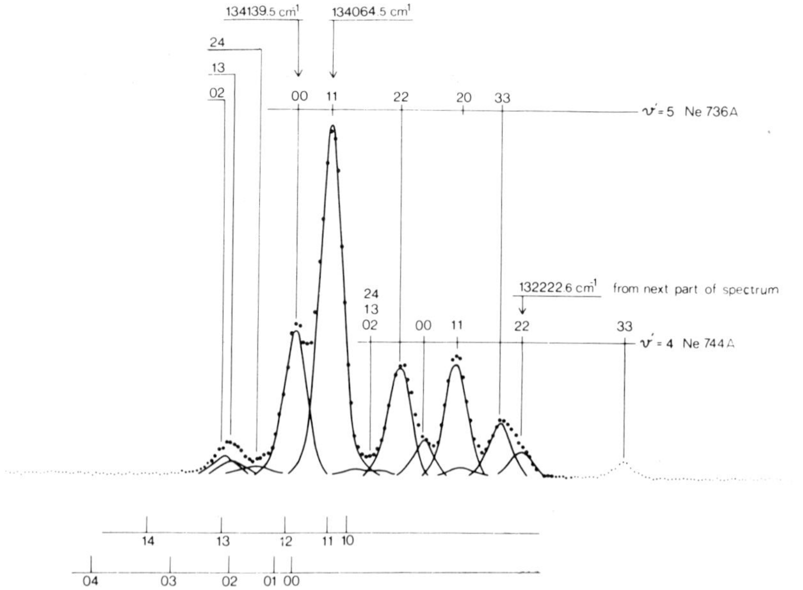

Fig. 21 Expanded scale encompassing the $v' = 5$ vibrational band in the photoelectron spectrum of H_2. In addition to contributions from the two neon lines (incident radiation), the spectrum reveals weaker structure from rotational transitions $\Delta J = \pm 2$, as well as the dominant $\Delta J = 0$ transitions. [From L. Åsbrink, *Chem. Phys. Lett.* **7**, 549 (1970).]

and experiment for the higher vibrational intensities ($v' > 6$) in H_2^+, discussed above.

Samson[88] has studied the photoelectron spectrum of H_2 with 247 Å incident radiation (50.2 eV). He finds evidence for formation not only of $X^2\Sigma_g^+$ of H_2^+, but also some higher repulsive states in the 30–40 eV region, presumably the $^2\Sigma_u^+$ and $^2\Pi_u$ states. He estimates the partial cross sections for the latter processes (which are effectively two-electron excitations in which one electron is ejected and the other excited to a $2p\sigma_u$ or $2p\pi_u$ orbital) to be about 10% of that for formation of $X^2\Sigma_g^+$ at 50.2 eV.

b. Photoionization Mass Spectrometry. The photoionization of H_2 with mass analysis has been obtained by Dibeler *et al.*[89] with an optical resolution of ~1 Å, by Comes and Wellern[90] with a resolution width of ~2 Å, and Chupka and Berkowitz[91] with a resolution width of 0.04 Å (see Chapter VIII for method of acquiring data). One excerpt of this latter study, obtained with parahydrogen at 78 K, is reproduced in Fig. 22. As has been mentioned previously, parahydrogen consists only of even rotational states and at 78 K, 99.3% are in the lowest ($J'' = 0$) level. Hence, even though the initial state is almost entirely in the $v'' = 0$, $J'' = 0$ level, the spectrum is extremely rich.

Most, but not all of the peaks have been identified as vibrationally excited Rydberg series members. Even with 0.04 Å resolution width, the peak widths of most of the lines are determined primarily by instrumental resolution. In addition to the relative photoionization cross section σ_i, this figure also displays a relative absorption cross section σ_a, obtained at about the same time and under similar conditions. Careful perusal of the σ_i curve reveals that there is some underlying direct ionization, which increases with photon energy. The ratio of the integrated ion peak area to the underlying continuum area in this autoionization region (740–803 Å) is 0.500. The validity of such an integration over the peaks depends upon the correctness of the assumption that the target (H_2) is optically thin, i.e., that the expression

$$\sigma_i \propto dI = I_0 n \sigma l \tag{24}$$

is applicable. The ratio σ_i/σ_a for many of the peaks is indicated on the figure, and is based on a similar assumption, and also assumes that the largest such ratio can be set equal to unity.[92] With this information, it is possible to deduce the equivalent peak area for absorption. The ratio of the peak area yielding ionization to the absorption peak area is thus found to be 0.464. Hence, we may roughly conclude that of the total absorption between 804 and 740 Å in H_2, approximately half goes directly to the ionization continuum and half to the absorption peaks. Of the absorption peaks, on the average, half autoionize and the other half predissociate. Therefore, about $\frac{3}{4}$ of the total oscillator strength in this region results in ionization. (This also assumes that there is no significant dissociation continuum in this region.[93]) From our previous analysis of the H_2 absorption spectrum (Chapter V, Section B.1), it was concluded that the oscillator strength to be assigned to the autoionization region (740–803 Å) is about 0.35. The ionized component of this quantity turns out to be 0.255. Taking an average excitation energy of 16.1 eV for this region, the ionization contribution to $S(-1)$ becomes 0.215, which can be added to the continuum (0–740 Å) contribution to $S(-1)$ of 0.563 previously deduced, to obtain 0.799. The experiment of Rieke and Prepejchal,[94] which directly measures this quantity, yields 0.695 ± 0.015. The agreement is not as good as one might hope. If the continuum contribution is correct, Rieke's result implies that the autoionization contribution to $S(-1)$ should be 0.132, and hence that the ionization contribution to "f" in this region should be 0.156 compared to our deduced value, 0.255.

On the other hand, the total $S(-1)$ deduced here for the entire spectrum is 1.518. An accurate calculation of this total $S(-1)$ can be made using the theorem derived in Chapter IV, Appendix 2.

$$S(-1)_{\text{total}} = \int_0^\infty \frac{1}{E}\frac{df}{dE}\,dE = \int_0^\infty \Psi_0^*(\mathbf{r}_1 + \mathbf{r}_2)\Psi_0\,d\tau, \tag{25}$$

where Ψ_0 is the wave function of the ground state of H_2 and yields 1.518–1.55. We are thus left with a situation in which the total $S(-1)$ deduced from photoabsorption studies is approximately correct, but, that portion of $S(-1)$ giving rise to ionization is somewhat higher than the experimental value.

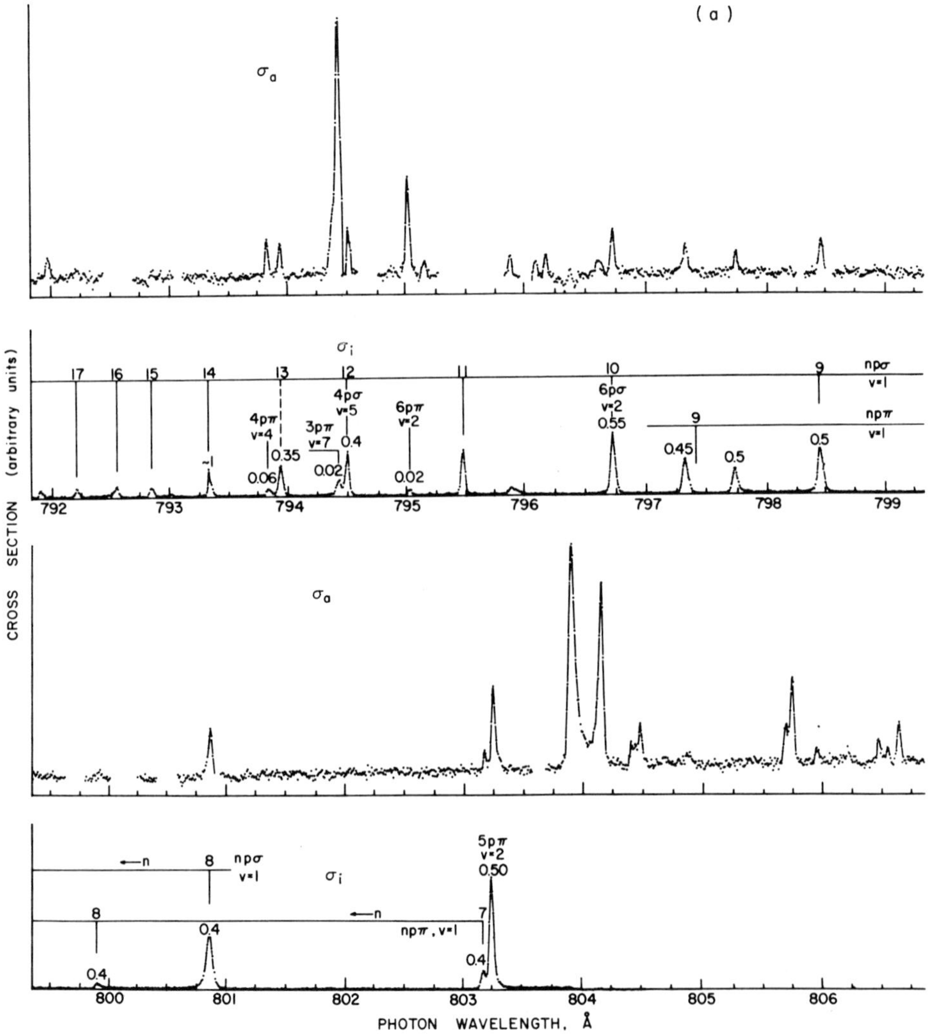

Fig. 22a–c Relative photoionization and photoabsorption cross sections σ_i and σ_a, respectively, for parahydrogen at 78 K, taken with a resolution half-width of 0.04 Å. The numbers above many of the peaks in the photoionization data give the value of η_i, the quantum yield of ionization. [From W. A. Chupka and J. Berkowitz, *J. Chem. Phys.* **51**, 4244 (1969).]

Before leaving the subject of oscillator strength distribution in H_2, it is of some interest to present a more detailed apportioning of this quantity in the autoionizing region, and to compare this with a recent calculation of Shaw and Berry.[95]

The experimental values are based on Fig. 22, normalized to make the entire region (804–740 Å) contribute 0.353 to "f" or $S(0)$. The calculation of Shaw and Berry differs in total cross section by about a factor of two in this

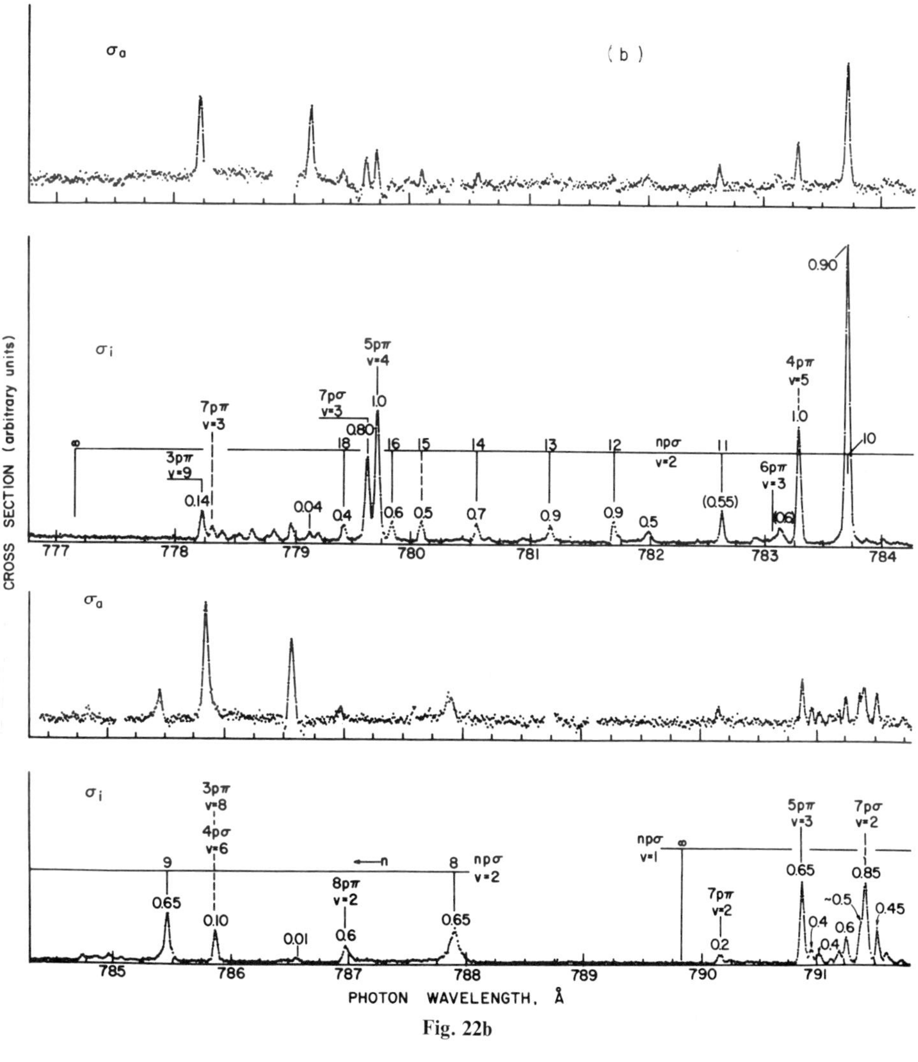

Fig. 22b

region, but their relative apportioning to continuum and quasidiscrete states has also been normalized by making the contribution to $S(0) = 0.353$. The agreement between experiment and calculation shown in Table 2 is reasonably satisfactory, and provides some support for the rough apportionment of oscillator strength deduced above.

There have been a number of attempts made to interpret the spectrum of Fig. 22, in a theoretical framework, with successively increasing accuracy.[95–101] The currently accepted interpretation begins with the understanding that the structure is due to the formation of vibrationally excited Rydberg states which can decay by autoionization. We have already shown that the probability of formation of a Rydberg state from the ground state by photoabsorption goes as $(n^*)^{-3}$ (Appendix in Chapter II). The higher Rydberg levels can be expected to have potential curves approximately

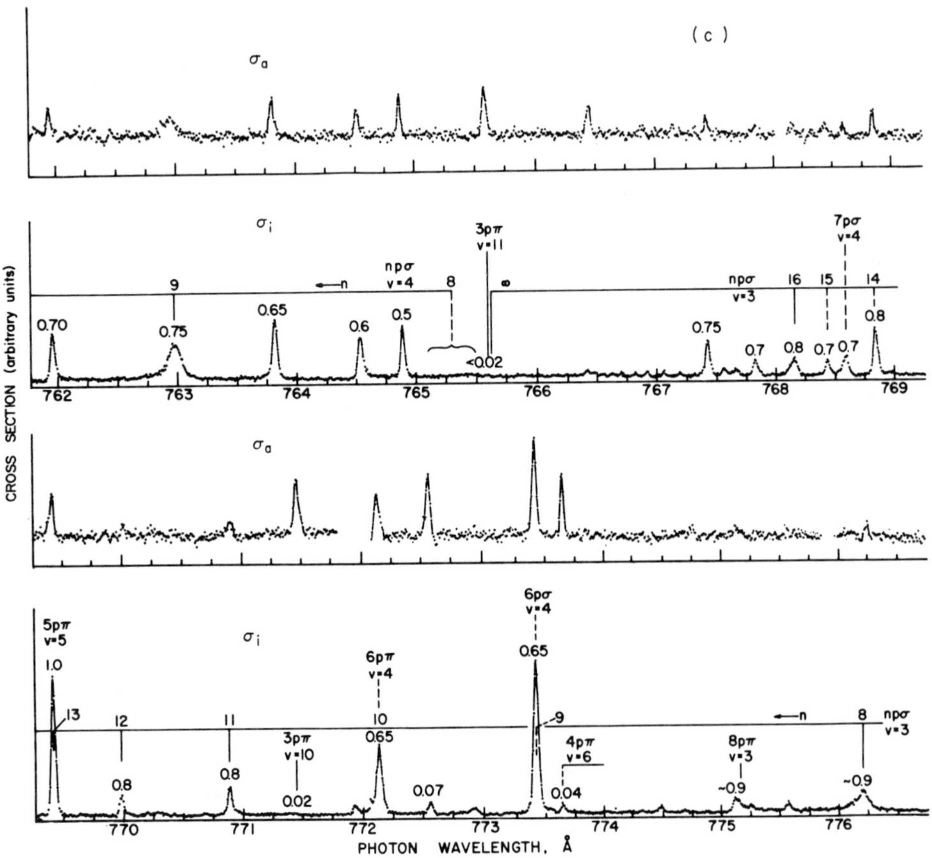

Fig. 22c

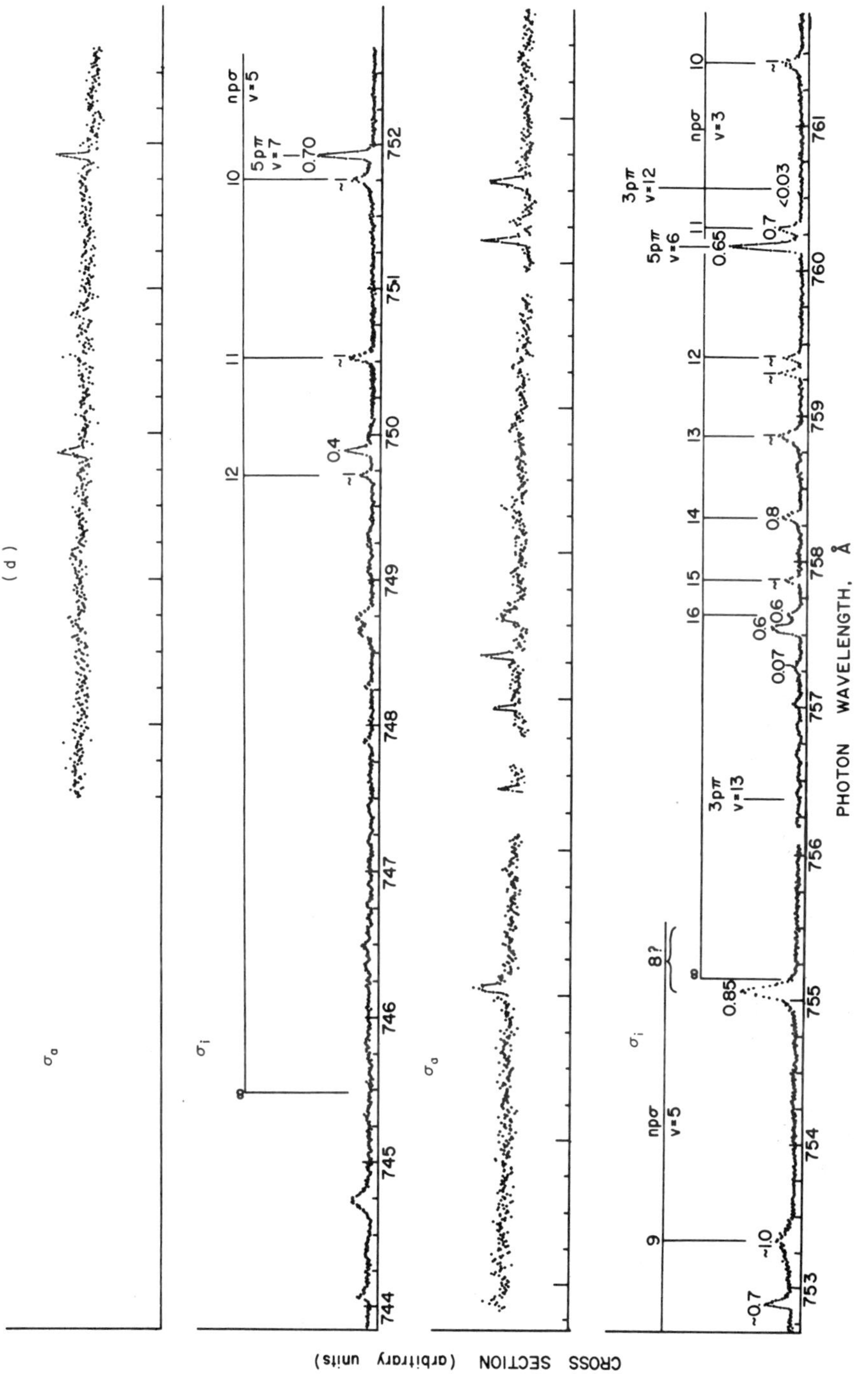

Fig. 22d

TABLE 2 *Partial Oscillator Strength Sums for Absorption to Continuum and Quasi-Discrete States in the Autoionizing Region of the H_2 Spectrum (804–740 Å). The Total f Contribution for this Region is taken as 0.353.*

λ (Å)	Continuum contribution Expt.	Continuum contribution Calculation	Quasi-discrete contribution Expt.	Quasi-discrete contribution Calculation
804	0	0	0	0
800	0.0036	0.0038	0.0135	0.0186
795	0.0083	0.0073	0.0288	0.0291
790	0.0134	0.0106	0.0657	0.0473
785	0.0215	0.0203	0.0912	0.0713
780	0.0328	0.0292	0.1071	0.0784
775	0.0455	0.0454	0.1235	0.1066
770	0.0595	0.0600	0.1420	0.1127
765	0.0750	0.0763	0.1505	0.1293
760	0.0932	0.0952	0.1656	0.1393
755	0.1103	0.1153	0.1753	0.1537
750	0.1290	0.1353	0.1796	0.1593
745	0.1493	0.1561	0.1819	0.1713
740	0.1696	0.1761	0.1834	0.1769

described by the potential curve of $H_2^+(X^2\Sigma_g^+)$. Hence, the relative probability of exciting a particular vibrational level within a particular electronic (Rydberg) transition should be well approximated by the vibrational distribution given in Fig. 23, which shows a maximum for $v' = 2$. (Note that this differs in HD and D_2.) When the sum of vibrational and electronic energy exceeds the ionization potential, the excited molecular state is degenerate with a portion of the ionization continuum but it may be inhibited from ionizing by (a) selection rules, or (b) very weak interaction with this continuum, which may allow competing processes, such as predissociation or reemission of radiation, to dominate.

We shall discuss (a), selection rules, in some detail later. One that has recently come to light, and which we shall just mentioned in passing, is concerned with the observation of fluorescence[102,103] from some states of H_2^* above both the dissociation and ionization limits. Fluorescence lifetimes are of order 10^{-8} sec, and we shall see that autoionization (and predissociation) lifetimes are generally shorter than this. Breton *et al.*[103a] have found that these high-lying fluorescent states are almost entirely of $^1\Pi_u^-$ symmetry, the minus being significant in strongly inhibiting both predissociation (requiring interaction with $^1\Sigma_u^+$ states) and autoionization (requiring a large change in vibrational quantum number). By contrast, very few $^1\Pi_u^+$ or $^1\Sigma_u^+$ states above the ionization potential can be observed to fluoresce. Hence, above the ionization potential, the major competition

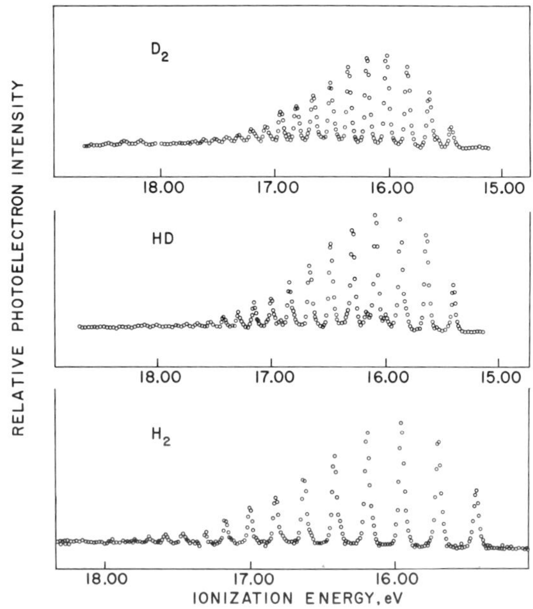

Fig. 23 Photoelectron spectra of H_2, HD, and D_2, obtained with 21.2 eV incident radiation. The relative intensities of the various vibrational states of H_2^+, HD^+, and D_2^+ are rather accurately presented here apart from a $1/E$ correction. [From J. Berkowitz and R. Spohr, *J. Electr. Spectr.* **2**, 143, (1973).]

likely occurs between autoionization and predissociation, especially[103] for n (principal quantum number) greater than four.

Autoionization in this case (H_2) involves conversion of vibrational energy into additional electronic excitation for the "Rydberg electron." This transfer of nuclear kinetic energy into electronic excitation obviously implies a breakdown of the Born–Oppenheimer separation. Starting with the zero order electronic and vibrational wave functions obtained from the Born–Oppenheimer approximation, Berry[96] has shown that the most important operator involved in mixing these states is the nuclear kinetic energy operator, and the most important term is proportional to

$$\langle \psi_f(r, R)\chi_f(R)| \left(\frac{d\chi_i}{dR}\right)\left(\frac{d\psi_i}{dR}\right), \tag{26}$$

where ψ is the electronic wave function and χ the nuclear, or vibrational wave function, R the internuclear distance, and r are electronic coordinates. The operator d/dR acts as a ladder operator on the vibrational wave functions (Hermite polynomials in the harmonic oscillator approximation), raising or lowering them by one unit. This is similar to the behavior of the operator R

on these functions, familiar in infrared spectroscopy. Since vibrational energy must be surrendered in the autoionization process, the vibrational selection rule implied is that χ_i refer to a state that has one more quantum number than χ_f, i.e., $\Delta v = -1$, with lesser probabilities for $\Delta v = -2$, etc., depending upon departure from harmonic oscillator behavior. Berry[96] has called this the vibrational propensity rule. Berry and Nielsen[101] also show from the properties of the Hermite polynomials that higher v states have larger transition probabilities. The electronic term that remains,

$$\left\langle \psi_f(r, R) \middle| \frac{d\psi_i}{dR} \right\rangle, \tag{27}$$

has been recast by Bardsley[97] into the form

$$-\frac{1}{\varepsilon_{if}} \left\langle \psi_f \middle| \frac{dU}{dR} \middle| \psi_i \right\rangle, \tag{28}$$

where ε_{if} is the energy gained by the electron during the autoionization (i.e., Δv), and $U(r, R)$ is the total electrostatic interaction, or effective potential felt by the Rydberg electron. Bardsley argues further that the wave function for a Rydberg state with principal quantum number n in the small region of space where it can interact to cause autoionization is proportional to $(n^*)^{-1.5}$. (This is the same normalization factor obtained in the appendix to Chapter II.) As n increases, the wave function of the Rydberg state spreads out, and a smaller fraction of that wave function is in the small interaction region. Since the transition probability goes as the square of the above matrix element, autoionization rates should be proportional to $(n^*)^{-3}$, which will be recognized as being the same as the photoabsorption probabilities to these states. Hence, the higher the Rydberg state, the smaller the probability of initially forming that state by absorption, and the smaller the probability that this state subsequently autoionizes.

Next, Berry and Nielsen[101] showed that if U is expanded in spherical harmonics, the dominant contribution to the above matrix element comes from the first, or spherically symmetric term, which they call the monopole. The dipole term vanishes for H_2 and the quadrupole term has a small contribution.

In summary, Rydberg states of low n^* [because of the $(n^*)^{-3}$ dependence of the interaction between quasi-discrete state and ionization continuum] and minimum Δv (because of the vibrational propensity rule) are favorable for autoionization. There is also a weaker factor favoring higher v states because of the properties of the Hermite polynomials.

For the competing predissociation process, there is a similar $(n^*)^{-3}$ dependence but in addition this radiationless transition will be affected by a poorer Franck–Condon factor as n^* increases. This is a manifestation of the energy gap law (see Section A). Hence, predissociation is expected to decrease more markedly with increasing n^* than is autoionization.

Using the formalisms described above, Berry and Nielsen[101] have calculated autoionization rates for many of the Rydberg levels corresponding to the peaks in Fig. 22, and have also calculated predissociation rates from these quasi-discrete states into neutral states above their dissociation limits to $H(n = 1) + H(n = 3$ or $n = 4)$. The results are in rough agreement with experiment. The paper by Shaw and Berry[95] uses the above formalism, but with a different (pseudopotential plus orthogonalization) method for determining the relevant electronic wave functions. The experimental requirement is to obtain sufficient resolution to observe natural line widths, and hence absolute measurements of decay rates for the sum of autoionization and predissociation. When this is combined with ionization yield for each peak, the individual autoionization and/or predissociation rate can be inferred. Such resolution has been obtained in absorption spectrographs, but is just nearing that capability in photoionization studies.[104]

Although the average ionization yield in the autoionization region is ~ 0.5, the yield for individual peaks varies from <0.01 to around unity (Fig. 22). If the theoretical framework that has been developed is complete, calculations of the competition between autoionization and predissociation for each peak should provide many tests of the quality of wave functions for these excited levels, and of the validity of a one-center expansion for $U(R, r)$.

A test of the propensity rule has been made by a specific kind of photoelectron spectroscopy.[105] Instead of using a resonance line source as incident radiation, a particular wavelength and bandwidth were selected by a vacuum ultraviolet monochromator to coincide with one of the absorption peaks of H_2 shown in Fig. 22. A retarding field electron energy analyzer then measured the kinetic energy of electrons resulting from ionization of this quasi-discrete state. The electron kinetic energy uniquely determines the vibrational state of H_2^+ which is formed by conservation of energy. Since the vibrational quantum number of the quasi-discrete state initially selected by photoabsorption is also known from spectral analysis, it should be possible to deduce the preferred $\Delta v > 1$ in order to autoionize, on energetic grounds. The primary reason is that the initial absorption is governed by the n^{-3} rule, which strongly favors low-lying Rydberg states. However, in order for these states to have sufficient energy to autoionize, they must have relatively high vibrational quantum numbers. In the region just above the ionization threshold, these states may only have sufficient total energy to form $v' = 0 - 2$ of H_2^+, and hence Δv will tend to be large. Nevertheless, the significant result of the experiment was that in four of five cases studied, autoionization proceeded with the minimum transfer of vibrational energy, and to a single vibrational state of H_2^+. This is consistent with a generalization of the propensity rule. By contrast, a study of N_2 autoionization peaks performed at the same time revealed that several vibrational peaks of N_2^+ were formed during a particular autoionization. The difference in behavior could be readily attributed to the different mechanisms of autoionization in the two cases. The operator d/dR is not the effective one in the N_2 autoionization, and hence the vibrational propensity rule is not applicable (see Section C.2.a).

A few of the autoionization lines of H_2 in the near-threshold region are worthy of special attention. Beutler and Jünger[106] initially arrived at an ionization potential for H_2 by examining the absorption spectrum in the region around the ionization potential. In Fig. 24, the photoabsorption and photoionization spectrum of normal H_2 (room temperature) are shown for this wavelength region. The nomenclature is as follows: *P*, *Q*, and *R* refer

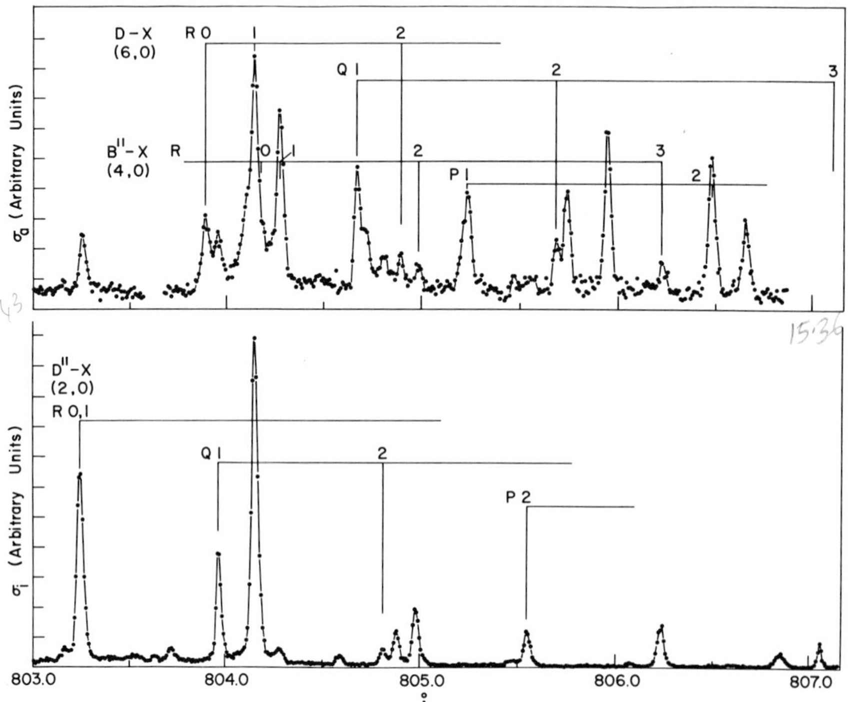

Fig. 24 Relative photoionization (σ_i) and photoabsorption (σ_a) cross sections of ordinary hydrogen at room temperature in the energy region near the ionization threshold. Note that the transition $D - X$ (6, 0), $R(1)$ autoionizes, contrary to the earlier conclusion of Beutler and Junger[106] based on photoabsorption widths. [From W. A. Chupka and J. Berkowitz, *J. Chem. Phys.* **51**, 4244 (1969).]

to absorptions in which $\Delta J = -1$, 0, and $+1$; the parenthetical (2, 0), (4, 0), etc. refer to v'' and v' for photoabsorption; D, D', D'', etc., are notations for Rydberg levels $3p\pi$, $4p\pi$, $5p\pi$, etc. while B'' refers to a Rydberg level $4p\sigma$.

Beutler and Jünger concluded from their absorption spectrum that in the $D - X(6, 0)$ manifold, $R(0)$, $P(2)$, $R(1)$, and $P(3)$ were "sharp," whereas $R(2)$ and $P(4)$ were "unsharp." The apparently critical transition was between $R(1)$ and $R(2)$. If autoionization could not occur for $R(1)$ on energetic grounds and could occur for $R(2)$, one would have bracketed the ionization potential of H_2. There are at least two problems with this type of analysis.

(1) We have already seen that another cause of broadening, predissociation, is prevalent in this region. It would be extremely difficult, even at our current level of understanding, to conclude that a particular broadened line was the result of predissociation or autoionization.

(2) The absorption spectrum σ_a in Fig. 24, which is obtained with higher resolution than that of Beutler and Jünger, does not show more broadening for $R(2)$ than $R(1)$. Tanaka[107] had earlier questioned these observations of Beutler and Jünger.

The photoionization segment of Fig. 24 indicates that $R(1)$ does autoionize, but that $R(0)$ does not. The bracketing of the ionization potential of H_2 will be correspondingly lowered.

A detailed analysis requires consideration not only of energetic criteria, but also of rotational selection rules for autoionization. As in the case of direct ionization, ortho–para conversion is forbidden, and hence the rotational quantum number J must remain even or odd through the autoionization process. In addition, the Rydberg levels accessible by photoabsorption must be ungerade because H_2 in its ground state is gerade. The ground state of H_2^+, which is the product of the autoionization act, is also gerade. Hence, the autoionizing transition occurs between an ungerade Rydberg state and a gerade ionic state. Since the operator d/dR does not affect the g or u symmetry properties, nonvanishing of the matrix element requires that the total final state, $H_2^+ + e^-$, be ungerade, and hence that the wave function of the departing electron be ungerade.

Two other smmetry properties of these states, labeled $(+)(-)$ and (s) symmetric $-(a)$ antisymmetric, must also be considered. A molecular state is called positive or negative according as the total eigenfunction remains unaltered or changes its sign by inversion of all the particles (including the nuclei) at the origin. If the two nuclei are identical, the symmetric–antisymmetric property depends upon the behavior of the wave functions with respect to an exchange of nuclei. For even electronic states (e.g., Σ_g) positive rotational levels are symmetric and negative are antisymmetric; for Σ_u levels the opposite is true. For Π states there is a splitting (Λ-type doubling) caused by coupling of rotational angular momentum with the axial electronic angular momentum. For each J value, one component of the doublet is positive, and other negative, and with each successive J the order of + and − is reversed. The selection rule for the autoionization process is that the overall symmetry be preserved, i.e., $(+a) \rightarrow (+a)$ and $(-s) \rightarrow (-s)$. A detailed discussion of these properties is given in Herzberg.[108]

For our present purposes, the consequences of the selection rules are summarized in Fig. 25, where K is the total angular momentum, which for Σ states is the rotational angular momentum but for Π states is increased by 1.

The energy levels of the rotational components of the $D(v' = 6)$ state are displayed in Fig. 25, together with the rotational components of H_2^+, ${}^2\Sigma_g^+$ $(v'' = 0)$. The relative positions of the scales have been fixed by the observation that $R(1)$ autoionizes, but $R(0)$ does not. These requirements lead to the conclusion that the ionization potential of H_2 lies between 124,416.2 and 124,393.5 cm^{-1}. A similar analysis of the $B'' - X(4, 0)$ band[41] shows that $R(0)$ and $R(1)$ do not autoionize, but $R(2)$ does. This leads

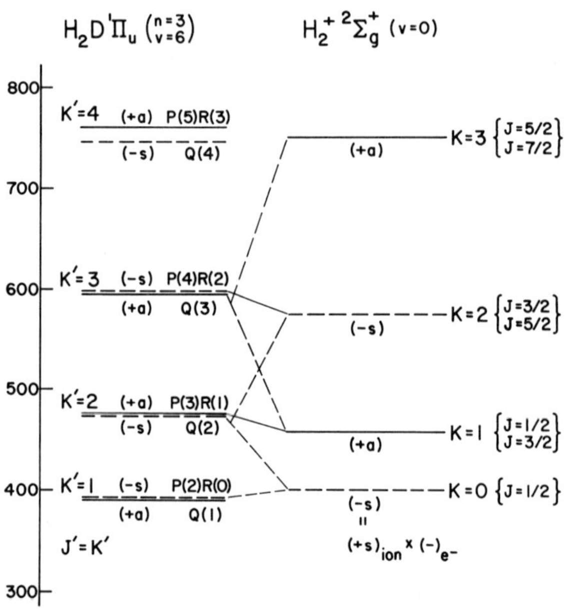

Fig. 25 The energy levels of H_2, $D^1\Pi_u$, $v = 6$, and H_2^+, ${}^2\Sigma_g^+$, $v' = 0$ are juxtaposed. The symmetry designations $+/-$, s/a are indicated (see text) as well as the angular momentum quantum numbers for J and K. Autoionizing transitions that are observed are indicated by solid lines, those not observed by dashed lines. [From W. A. Chupka and J. Berkowitz, *J. Chem. Phys.* **51**, 4244 (1969).]

to the conclusion that IP (H_2) lies between 124,394.6 and 124,407.2 cm^{-1} if the weaker selection rule $\Delta K = \pm 1$ is valid. Other evidence, both experimental[109–111] and theoretical,[112] now points convincingly to a value of 124,417.2 $\pm$ 0.4 cm^{-1} as the ionization potential of H_2. Hence, the conclusion from the study of the $B'' - X(4, 0)$ band is that $\Delta K = -3$ transitions are occurring. This is rather similar to the observation of Åsbrink in the direct photoionization of H_2, which demonstrated in that case that a two quantum rotational change had ~10% as much probability as the zero quantum rotational change.

Two other features of the photoionization spectrum appeared puzzling until recently. (1) As already noted, the primary photoabsorption processes in H_2 promote one electron from a 1sσ_g orbital into an *n*p orbital, σ or π, of ungerade symmetry. We should expect two Rydberg series of σ and π type below the ionization potential and (as we have seen) also above the ionization potential. The photoionization spectrum for moderately high *n* quantum numbers (about 8–12) consists mostly of lines attributable to the *n*pσ series. Only a few weak lines with $n \geq 9$ could be assigned as *R*(0) members of the *n*pπ series. For lower principal quantum numbers on the other hand, *R*(0) lines belonging to the *n*p series are more prominent. (2) In the expanded threshold region shown in Fig. 26, there is a peculiar oscillation of intensity for a fraction of an angstrom around the ionization potential.

Both of these phenomena have been interpreted in Fano's paper on *l*-uncoupling,[113] and have been discussed in great detail by Herzberg and Jungen[111] in connection with their higher resolution absorption studies of H_2 in the subionization and autoionization regions. We shall only sketch the arguments here. The interested reader is urged to read the two original references for a detailed description of the mathematical apparatus and its application to the H_2 spectrum.

We begin by noting that the two Rydberg series are only approximately referred to as *n*pσ and *n*pπ when Hund's case (a) or (b) applies, i.e., the projection of the electronic angular momentum *l* on the internuclear axis is a good quantum number. As *n* increases, an uncoupling of *l* from the internuclear axis takes place, generally described as a transition to Hund's case (d). The electronic angular momentum now couples to the nuclear rotational angular momentum (in this situation, commonly labeled *N*). In our case, for the two Rydberg series, *l* will remain a good quantum number, equal to 1. Hence, there will be three total angular momentum *J* values associated with each *N* value, i.e., $J = N + 1$, N, $N - 1$, except for $N = 0$ when only $J = 1$ occurs. At high *n*, the designation of states as belonging to *n*pσ or *n*pπ series will break down since they will be strongly mixed and ordered solely according to the *N* value of the molecular core. For high *n*, the levels

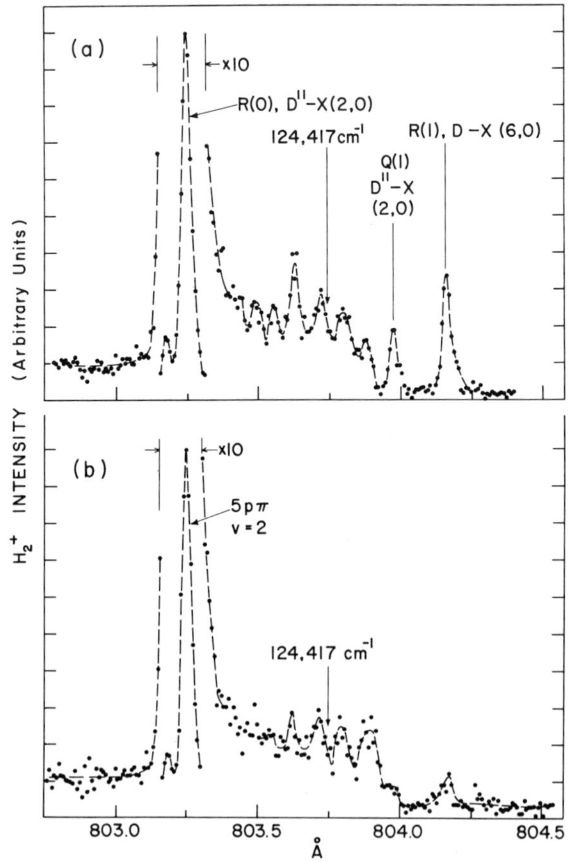

Fig. 26 Two independent scans of the photoionization of parahydrogen at 78 K in the immediate vicinity of the ionization threshold. The small oscillations between ~ 803.5 and 804 Å are manifestations of rotational autoionization (see text). The (a) scan was obtained with a sample containing some orthohydrogen, as indicated by the stronger $R(1)$, $D - X$ (6, 0) peak. [From W. A. Chupka and J. Berkowitz, *J. Chem. Phys.* **51**, 4244 (1969).]

can be termed np0, np1, np2, . . . rather than npσ, npπ. Consider now the case of parahydrogen at 78 K. Since $J'' = 0$ and only $R(0)$ transitions can occur, the upper state must have $J' = 1$. This can occur for $N = 0$ $(+l = 1)$ which we refer to as np0 or $N = 2$ $(-l = 1)$, referred to as np2. The np0 Rydberg series will converge to the ionization potential of H_2, but the np2 series will want to converge to a limit two rotational quanta above the IP, i.e., $N(N + 1)\ B = 6B$. The two Rydberg series np0 and np2 are independent only in a first approximation. In a higher approximation they interact leading to autoionization for levels of the np2 series that lie above the np0 limit, which is the ionization potential. The oscillation seen in the expanded threshold region of Fig. 26 is presumably this autoionization, which can be described as rotational autoionization. Herzberg and Jungen have observed this phenomenon for several higher vibrational onsets as well, and have noted an apparent emission which is identified with a Fano profile parameter, $q = 0$, for this interaction.

Fano has shown how to make the frame transformation from the $n\text{p}\sigma$, $n\text{p}\pi$ basis set to the np0, np2 set. Herzberg and Jungen, applying this transformation matrix to the transition dipole, obtain

$$D_{\text{p}0} = \sqrt{\frac{1}{3}}\,D_{\text{p}\sigma} + \sqrt{\frac{2}{3}}\,D_{\text{p}\pi} \tag{29}$$

$$D_{\text{p}2} = \sqrt{\frac{2}{3}}\,D_{\text{p}\sigma} - \sqrt{\frac{1}{3}}\,D_{\text{p}\pi}, \tag{30}$$

where the absorption intensity $I = 2R|D_{\text{p}n}|^2/(n^*)^3$. For $D_{\text{p}\sigma}$ and $D_{\text{p}\pi}$ they deduce the values $\sqrt{\frac{1}{3}}D_{\text{p}}$ and $\sqrt{\frac{2}{3}}D_{\text{p}}$, respectively. Hence, for very low n, when case (b) is a good approximation, $R(0)$ $n\text{p}\pi$ is predicted to be twice as strong as $R(0)$ $n\text{p}\sigma$. For higher n, we compute instead $D_{\text{p}0} = D_{\text{p}}$ and $D_{\text{p}2} = 0$. Hence, this explains the weakness or absence of what we had been calling higher members of the $n\text{p}\pi$ series in the observed spectrum. This is again a manifestation of the observation, noted in Asbrink's[76] results and in the weak autoionization selection rule, that direct ionization or autoionization prefer (by about an order of magnitude) to retain the same rotational level rather than change by two units.

c. Line Shapes. We have been discussing a very rich structure due to autoionization and predissociation in the photoabsorption of H_2. The interaction of quasidiscrete states with an ionization or dissociation continuum should give rise to (generally) asymmetric line profiles, as shown by Fano.[114] Until recently, very few asymmetric line profiles have actually been reported. The reasons are probably twofold.

(1) The photoionization studies, which in some ways are better suited for profile determinations because intensity is measured directly and problems of saturation of absorption and nonlinearity of peak response are not involved, have in turn been limited in resolution, so that line shapes have been dominated by instrumental functions except in a few cases of naturally broadened lines.

(2) The photoabsorption studies, which have had sufficient resolution, have primarily been photographic, making intensity measurement somewhat uncertain.

Nevertheless, some examples of both autoionization and predissociation line shapes are available. In Fig. 27a, the 8pπ ($v = 1$) and 8pσ ($v = 1$) are

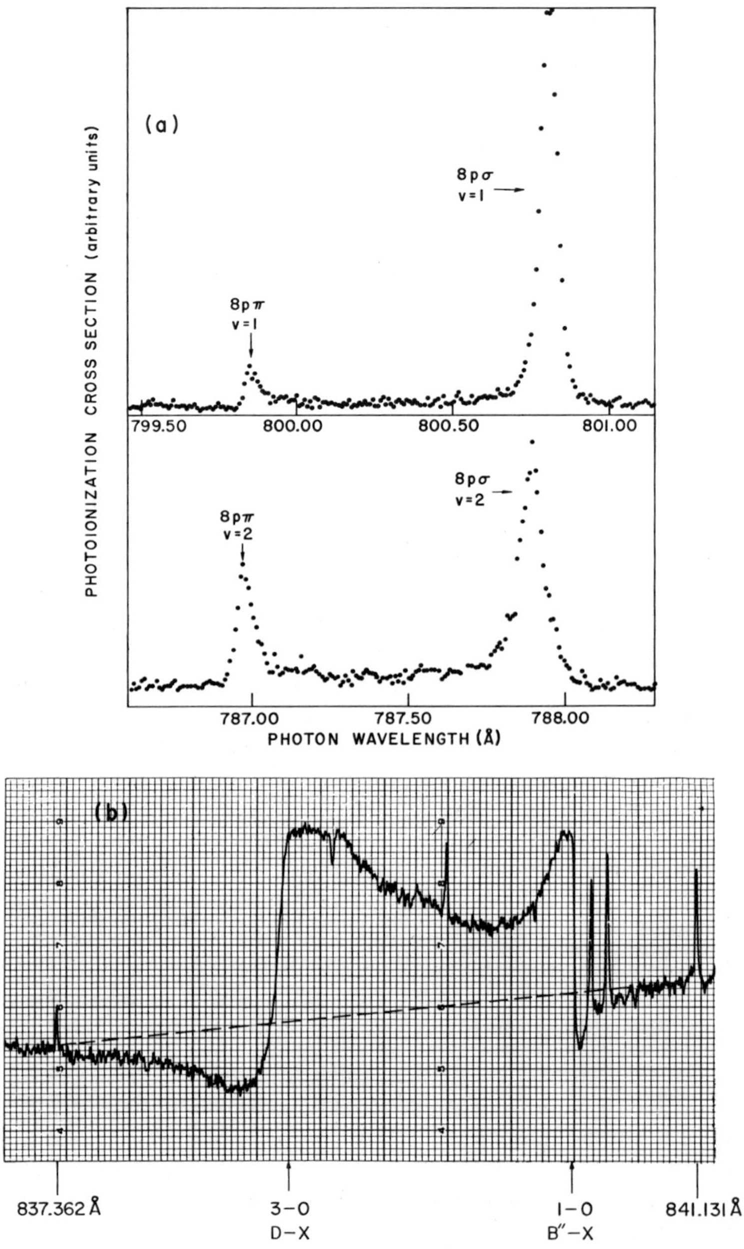

Fig. 27 (a) Asymmetric line shapes of some autoionizing transitions in the photoionization of H_2. The π states are sharper on the high energy (low wavelength) side and suppress the underlying continuum on that side, while the σ states behave in opposite fashion. [From W. A. Chupka and J. Berkowitz, *J. Chem. Phys.* **51**, 4244 (1969).] (b) Asymmetric line shapes of some predissociating transitions in the photoabsorption of H_2. Again, the π state $[D - X\ (3,0)]$ is sharp on the high energy side and dips into the underlying continuum, while the σ state $[B'' - X\ (1,0)]$ behaves in opposite fashion. [From G. Herzberg *in* "Topics in Modern Physics—A Tribute to E. U. Condon," p. 191. Colorado Assoc. Univ. Press, 1971.]

shown juxtaposed, as are the 8pπ ($v = 2$) and 8pσ ($v = 2$), as observed in photoionization.[91] It can be seen that the π states are sharper on the high energy (low wavelength) side, and suppress the underlying continuum on that side, while the σ states behave in opposite fashion. In Fig. 27b, a photoabsorption densitometer trace obtained by Herzberg[115] in the predissociation region is shown. Again, the π state $[D - X(3,0)]$ is sharp on the high energy side and dips into the underlying continuum, while the σ state $[B'' - X(1,0)]$ behaves in opposite fashion. Herzberg reports similar behavior for the same pairs of states at two successively higher vibrational members, one of which $[D - X(5,0)]$ has also been described by Comes *et al.*[116] In Fano's parametrization, the q value is negative in all these cases for the π states, and positive for the σs. Since the q value is essentially the ratio of the transition moment to the discrete state to that of the underlying continuum, it is clear that π and σ states have transition moments of opposite sign. In the *rotational* autoionization regions, Herzberg and Jungen find evidence primarily for apparent emission lines, or "windows" in the absorption curve. This behavior corresponds to $q \approx 0$, and is another indication that the transition moment to $np\pi$ states of very high n is near zero.

d. Effect of Electric Fields and of Collisions on Autoionization and Predissociation. An obvious way of determining the first ionization potential of an atom or molecule is to observe many Rydberg members in its absorption spectrum, and extrapolate to $n^* \to \infty$ by use of the Rydberg formula. It has been known for many years[117] that relatively weak fields can cause the disappearance of very high Rydberg members in the spectrum. For example, the ionization limit of atomic potassium is at 2860 Å (4.33_5 eV). Bakker[118] found that with a field of 200 V/cm the apparent ionization limit was at 2869 Å, with 780 V/cm at 2875.5 Å and with 1600 V/cm at about 2880 Å (4.30_5 eV). For alkali metals adsorbed on some surfaces, where the penetrating electric field is much stronger, the ionization limit is lowered to a much greater degree. In addition to the breaking off of absorption lines, optically forbidden lines now appeared in the absorption spectrum of potassium in the presence of a field. Some selection rules had obviously been broken.

Such phenomena have been observed in the autoionization and predissociation spectra of H_2. In Fig. 28, the photoionization spectra of some peaks straddling the threshold region of H_2 are displayed, at field strengths of 6 V/cm and 22 V/cm present in the ion source region. The $B'' - X(4,0)R(1)$ peak at 804.25 Å is seen to be absent with the lower field, and present at the higher field. The ionization potential of H_2 is now well established at 124,417 cm^{-1}, and this peak lies ~78 cm^{-1} to lower energy. At still higher fields, the $R(0)$ line can also be made to ionize. A few other lines in the H_2 spectrum behave similarly, and introduce a note of caution in determining

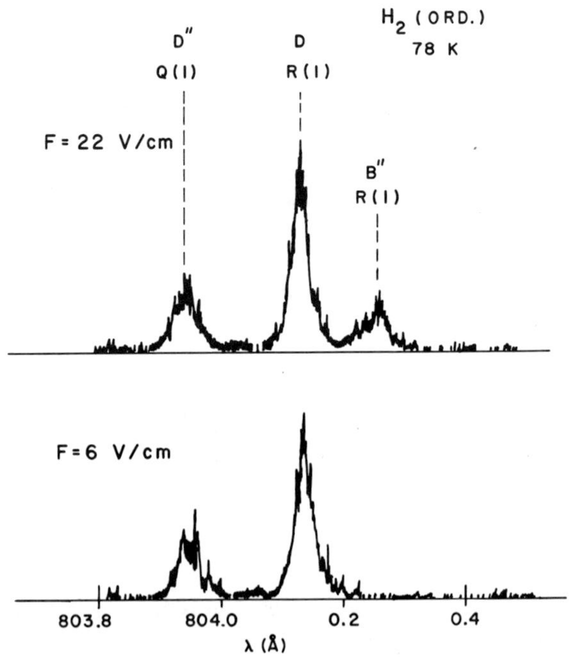

Fig. 28 Two scans of the photoionization of H_2 in the vicinity of the ionization threshold, using different electric field strengths for the ion repeller. The scan at the higher field strength shows ionization of $R(1)$ of $B'' - X$ (4, 0) induced by the field. [From W. A. Chupka and J. Berkowitz, *J. Chem. Phys.* **51**, 4244 (1969).]

precise ionization thresholds by observing the onset of ionization. The $D - X(6,0)R(1)$ peak, which was previously discussed in the bracketing of the IP of H_2, was tested for this behavior and found to ionize even at zero field.

Comes *et al.*[116] have reported enhanced *predissociation* in the region 0.2 eV below the ionization limit by the application of an electric field. They attribute the enhancement to a mixing effect induced by the field. The optically allowed states by absorption from the ground state are of ungerade symmetry. In the presence of the field, they believe gerade states are mixed in, and can then obey the predissociation selection rules.

The enhanced autoionization and predissociation observed with application of electric fields have also been shown to be caused by collisions, by noting the pressure effects on these peaks compared to other peaks.

e. Photo Ion-Pair Formation. One can deduce from either of the combinations $IP(H_2) + D_0(H_2^+)$ or $D_0(H_2) + IP(H)$ that the thermodynamic threshold for formation of H^+ from H_2 should be 18.07663 eV, the electron affinity[119] of H is 0.75421 eV, and hence the threshold for the process

$$H_2 + h\nu \rightarrow H^+ + H^-$$

should be at 17.3224 eV ≡ 715.75 Å.

This process has been observed[120–122] with weak intensity (~0.1% of the H_2^+ intensity in the relevant wavelength range). A portion of this spectrum obtained with normal H_2 and also with para-H_2 is shown in Fig. 29, where the ion being observed is H^-. The threshold occurs at the anticipated place. The spectra from ordinary H_2 and para-H_2 are not identical, indicating a rotational state dependence. The process responsible for the observed spectrum is believed to result from predissociation of Rydberg states converging to H_2^+ $(X^2\Sigma_g^+)$, $v' = 9$, $N = 0$ and 2 in the near-threshold region, and to corresponding states with $v' = 10$ and $v' = 11$ at somewhat higher energies. Since the convergence limit for $v' = 9$ is near the anticipated threshold for ion-pair formation, a piling up of Rydberg levels occurs here, which is as yet unresolved and serves as a quasi-continuum. The detailed nature of the state(s) causing the predissociation is not yet clear, although speculations have been made.[121,122]

In HD, the processes leading to $H^+ + D^-$ and $H^- + D^+$ can be independently observed.[122,123] From energetic considerations, the $H^+ + D^-$ process should have a threshold at 17.3577 eV, while that for $H^- + D^+$ should first occur at 17.3618 eV, the difference being primarily attributable to the difference in ionization potentials of H and D. The two curves have

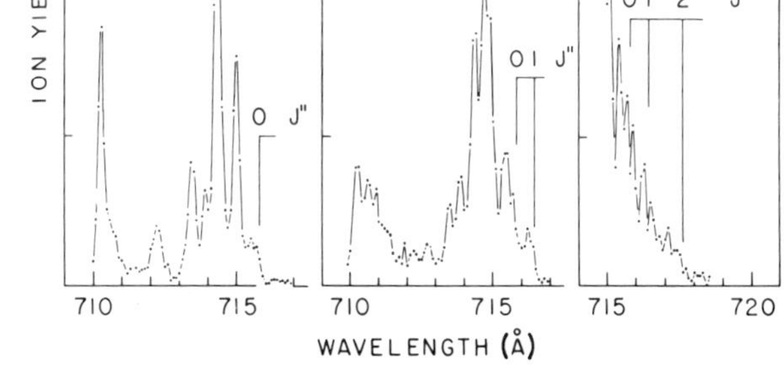

Fig. 29 Ion-pair formation in the photoabsorption of hydrogen, detected by observation of the H^- ion peak. The spectra from ordinary and parahydrogen are not identical, indicating a rotational state dependence. [From K. E. McCulloh and J. A. Walker, *Chem. Phys. Lett.* **25**, 439 (1974).]

virtually identical features, with the H^- channel (the higher energy one) having $\sim$1.4 times the intensity[122,123] of the D^- channel. There is as yet no satisfactory explanation of this observation. See, however the theoretical analysis by Durup.[123a]

Although this ion-pair process is of considerable theoretical interest, it represents a very minor fraction of the oscillator strength in H_2 photo-absorption. The total oscillator strength for the ion-pair process has been estimated[123] to be 0.8×10^{-4}.

f. Dissociative Ionization. The process $H_2 + h\nu \rightarrow H^+ + H + e$ is somewhat difficult to distinguish at the threshold from the residual ion-pair process, but increases beyond threshold to a few percent of $H_2{}^+$, and hence quickly becomes more intense than the ion-pair process.

In the energy region between 18.076 eV (the dissociative ionization threshold) and $\sim$24 eV, the process is very likely direct excitation to $H_2{}^+(X^2\Sigma_g{}^+)$ above its dissociation limit. Above this energy, the repulsive states discussed

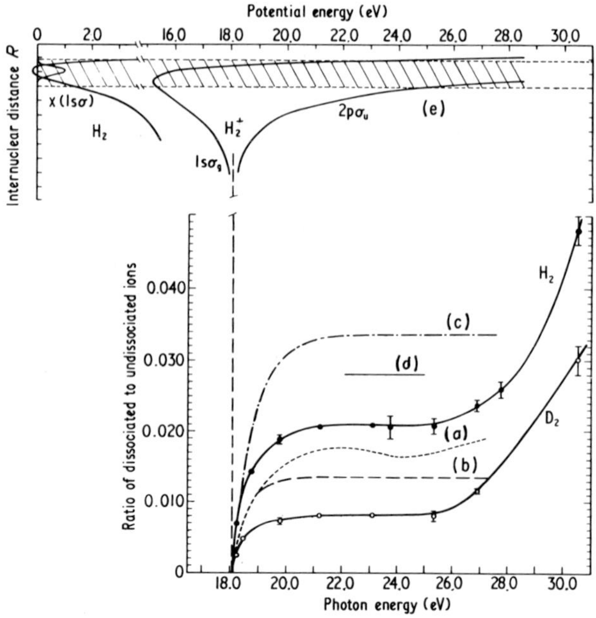

Fig. 30 The ratio $H^+/H_2{}^+$ from photodissociative ionization of H_2 between 18 and 30 eV, and the corresponding ratio from D_2. [●, ○ Experimental data of R. Browning and J. Fryar, *J. Phys. B* **6**, 364 (1973).] The curves labeled (a)–(d) are based on calculations using wave functions of varying sophistication for the ground states of H_2 and $H_2{}^+$. (e) Shows to scale the potential energy curves for H_2, $H_2{}^+$ necessary to interpret the experimental results. The shaded area represents the extended Franck–Condon region associated with the probability distribution shown for H_2. [From R. Browning and J. Fryar, *J. Phys. B* **6**, 364 (1973).]

by Samson[88] ($^2\Sigma_u^+$ and $^2\Pi_u$) become accessible, and the fraction of dissociated ions is seen[124] to correspondingly increase (see Fig. 30). In addition, there is evidence[125] for weak autoionization leading to fragments in the region around 26.9 eV.

g. Isotope Effects. Although absorption spectra of HD and D_2 have been obtained in the relevant spectral regions,[126–129] the focus has been largely on identification of states, rather than measurements of absorption cross sections and effects of isotopic substitution on autoionization and predissociation rates. Clearly, the great range of photoionization and photoabsorption intensity and line width (and shape) studies needs to be extended to the isotopic hydrogen species. The currently best values for the dissociation energies of H_2, HD, and D_2 (in cm^{-1}) are collected in Table 3. The currently best values for the ionization energies of H_2, HD, and D_2 (in cm^{-1}) are summarized in Table 4.

Slight deviations from the expected isotope shifts have been observed in some states and are attributed to electronic isotope effects.[127,128]

2. N_2

a. Photoelectron Spectroscopy. The occupied molecular orbitals of N_2 are (in separated atom terminology)

$$(\sigma_g 1s)^2(\sigma_u 1s)^2(\sigma_g 2s)^2(\sigma_u 2s)^2(\pi_u 2p)^4(\sigma_g 2p)^2$$

or in united atom terminology

$$(1\sigma_g)^2(1\sigma_u)^2(2\sigma_g)^2(2\sigma_u)^2(1\pi_u)^4(3\sigma_g)^2.$$

The two uppermost occupied orbitals are bonding, giving N_2 its triple-bonded structure. Removal of an electron from σ_g 2p increases the internuclear distance from 1.094 to 1.116 Å, and from π_u 2p to 1.177 Å. The 584 Å

TABLE 3 *Dissociation Energies (D_0) of H_2, HD, and D_2*[a]

H_2	HD	D_2
36,117.9[b]	36,405.5[b]	36,748.2[(2)]
<36,118.3[c]	36,406.6[c]	36,748.9 ± 0.4[c]
>36,116.3[c]	36,405.8	

[a] Values are given in centimeters^{-1}.

[b] Theoretical, based on calculation of Kolos and Wolniewicz,[129a] with nonadiabatic correction incorporated.

[c] Experimental, from Herzberg.[129b]

TABLE 4 *Adiabatic Ionization Thresholds for H_2, HD, and D_2*[a]

H_2	HD	D_2
124,417.3[b]	124,568.0[b]	124,745.2[b]
124,417.2 ± 0.4[c]	124,568.7[e]	124,745.3[g]
124,417 ± 1[d]	124,569.0[e]	124,746.0[g]
	124,569.4[e]	124,746.6 ± 0.6[h]
	124,569.5 ± 0.6[f]	124,746.6 ± 0.6[h]
	124,569.6 ± 0.6[f]	

[a] Values are given in centimeters^{-1}.

[b] Theoretical, based on calculation of Jeziorski and Kolos.[112] See Herzberg.[129c]

[c] Herzberg and Jungen.[111]

[d] Takezawa.[109]

[e] Takezawa and Tanaka,[126] using thermochemical cycle involving $D_0{}^0$(HD).

[f] Takezawa and Tanaka,[126] obtained directly from series limits.

[g] Takezawa and Tanaka,[129] using thermochemical cycle involving $D_0{}^0(D_2)$.

[h] Takezawa and Tanaka,[129] obtained directly from series limits.

photoelectron spectrum (Turner *et al.*,[130] p. 46) reflects this change, the first peak being quite narrow, the second encompassing 5–6 vibrational levels.

The most probable ionizing transitions involve single-electron ejection, as expected. In Fig. 31, we have plotted the partial cross sections for $(3\sigma_g)^{-1}$, $(1\pi_u)^{-1}$, $(2\sigma_u)^{-1}$, and $(2\sigma_g)^{-1}$, culled from various sources. Between threshold and 40.8 eV, the branching ratios reported recently by Samson *et al.*[131] have been used. The equally recent synchrotron studies of Plummer *et al.*[132] have been used to fill in the region between 20 and 39 eV.† Data are also now available at 132.3[133] and 151.4 eV.[134] Finally, the x-ray data from ESCA II[135] are available at 1253.6 eV. In each instance, the data plotted in Fig. 31 utilize the given authors' branching ratios, but are normalized to the absolute photoionization cross sections deduced in Chapter V, Section B.2. The data of Samson *et al.*‡ take into account the angular variation of the photoelectron spectra, the others do not. Although this introduces some uncertainty into the resulting graph, thc general features are not likely to be seriously affected. Confirmatory information is available from pseudophotoelectron measure-

† Still another measurement of partial cross sections in N_2 has since appeared (Woodruff and Marr[132a]). Within the combined errors, it does not differ significantly from that of Plummer *et al.*[132]

‡ See also Woodruff and Marr.[132a]

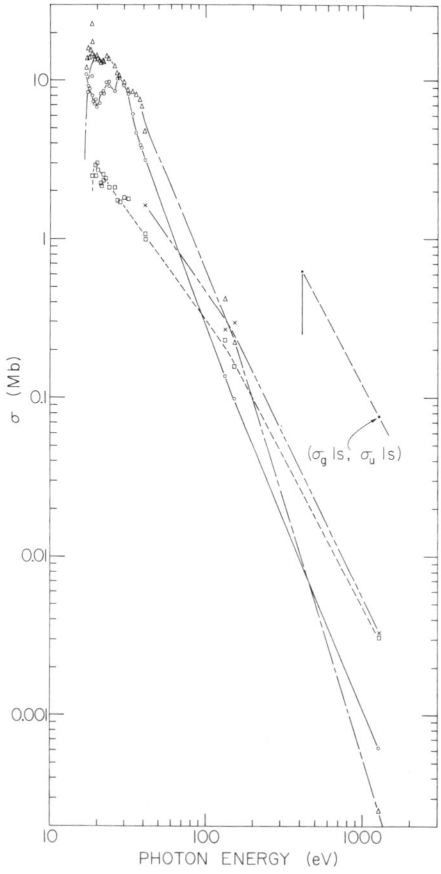

Fig. 31 Partial cross sections for formation of various states of N_2^+ by photoabsorption. ○, $X^2\Sigma_g^+$; △, $A^2\Pi_u$; □, $B^2\Sigma_g^+$, ×, $(2\sigma_g)^{-1}$; ●, $(1\sigma_g)^{-1}$.

ments,[136,137] which we defer to the discussion on photoionization mass spectrometry. The conclusion that may be drawn from the partial cross sections for the single-electron ejection states plotted in Fig. 31 is that, apart from some oscillations in the threshold region, those molecular orbitals that contain strong atomic s character do not fall off in cross section with increasing energy as rapidly as those with strong atomic p character. Price *et al.*[138] have emphasized this point. They interpret this behavior as being due to the greater charge density of s orbitals at smaller radii since it is at these short distances that the major contributions to the transition moment occur for short wavelength photoionization. Several other examples are given in the paper by Price *et al.*[138]

b. Photoionization Mass Spectrometry. The absolute photoabsorption cross section measurements of N_2 in the near threshold range (the autoionization region 740–797 Å) obtained by Carter[139] with 0.04 Å resolution width are displayed in Fig. 32, together with photoionization mass spectrometric measurements[140] obtained with 0.035 and 0.07 Å resolution width. (A somewhat higher resolution photoabsorption spectrum (0.03 Å) has since become available.[140a]) The photoionization measurements, which give relative cross sections, have been converted to absolute cross sections in an earlier work[141] by comparison with data of Samson and Cairns.[142] The ionization yields of the various peaks do not vary greatly from ~ 0.5, unlike the case of H_2, where large variations were noted.

The photoelectron spectra obtained with narrow bandwidth (0.12 Å) incident radiation selected to coincide with some of these peaks has been carried out by Berkowitz and Chupka.[105] They showed that, unlike the case of H_2, several groups of photoelectrons were generally formed from each autoionization peak. A vibrational propensity rule does not seem to be operative, and hence the mechanism of autoionization is almost certainly different. The interpretation that has been given, and that seems reasonably satisfactory, is that the Rydberg states giving rise to autoionization involve excitation of electrons from inner molecular orbitals, e.g., $1\pi_u$ or $2\sigma_u$. The interaction that subsequently causes autoionization may be written in the form

$$\int \phi_R{}^* \chi_R{}^* \frac{e^2}{r_{12}} \phi_i \chi_i \phi_e \, d\tau, \tag{31}$$

where $\phi_R{}^*$, $\chi_R{}^*$ are the electronic and vibrational wave functions of the Rydberg state, ϕ_i, χ_i the electronic and vibrational wave functions of the ionic state finally formed, ϕ_e is the wave function of the departing electron, and e^2/r_{12} is the perturbation operator, which in this case acts only on electronic coordinates, causing configuration interaction.

If this model and separation are appropriate, we are left with an integral over electronic coordinates, whose magnitude determines the rate of autoionization (and hence the autoionization contribution to the peak width) and an integral over vibrational coordinates, a kind of Franck–Condon factor. If sufficient information regarding r_e and ω_e (and possibly $\omega_e x_e$) were available for the Rydberg state and for the resulting ionic state, the vibrational overlap integrals could be carried out and compared to the experimental relative vibrational intensities (observed as relative photoelectron intensities). While such information is available for the ionic states, it is not available for the quasi-discrete Rydberg states, perhaps because autoionization and dissociation broaden these peaks and make rotational analyses very difficult. The calculations have therefore been reversed to determine the internuclear distances (r_e's) of the Rydberg levels, using the experimental relative vibrational intensities and the assumptions discussed above regarding the autoionization mechanism. (The r_e of the Rydberg level is the most sensitive parameter in these calculations.) The resulting values of r_e for the Rydberg levels are plausible, and differ from those of the corresponding convergence limit ionic states in such a direction as to be able to infer that the Rydberg electron acts as a bonding or antibonding entity. For example, the Rydberg electron in Worley's third series (which has an $A^2\Pi_u$ core) is characterized[143] as $ns\sigma$ and correlates with σ_g orbitals of the separated atoms, and is therefore expected to be slightly bonding (Herzberg,[108] pp. 329, 362). The value of r_e deduced for this Rydberg state is 1.164 Å, that of the convergence limit $A^2\Pi_u$ is 1.174 Å, in agreement with the expected behavior.

Bahr *et al.*[144] have obtained similar results, but using much larger photon bandwidths (1.6 Å versus 0.12 Å) and therefore encompassing more of the underlying continuum, which then has to be carefully subtracted.† There exists one glaringly anomalous peak in this spectrum at 781.1 Å. It has not yet been spectroscopically assigned, and analysis of the photoelectron spectrum indicates that the resulting ion is left in the $X^2\Sigma_g^+$ state, with $v' = 0$ being 0.26 as probable as $v' = 1$. The very low value of $v' = 0$ to $v' = 1$ in this case, compared to a value nearly one at 775.84 Å, suggests that a mechanism other than that dominated by Franck–Condon factors may be operative here. However, the vibrational mechanism which was successfully invoked for H_2 encounters difficulty here because vibrationally excited Rydberg levels corresponding to the ionic ground state $X^2\Sigma_g^+$ would have similar vibrational composition to that of the convergence limit, and the $X^2\Sigma_g^+$ state is $\sim 90\%$ $v'' = 0$ and $\sim 10\%$ $v'' = 1$. Hence, this criterion plus the necessity of having a total Rydberg excitation greater than $v'' = 1$ of $X^2\Sigma_g^+$ would require a Rydberg level of high n, which in turn should have a low photoabsorption probability because of the n^{-3} scaling factor. The observation, however, is that the peak is quite intense.

If the alternative interpretation is taken, that configuration interaction is the mechanism for autoionization, it requires that the quasi-discrete state have an internuclear distance considerably larger than that of the convergence limit.

Another anomaly for N_2 in this region, which may be related, concerns the angular distribution of photoelectrons. When the incident radiation has $\lambda = 584$ Å, the photoelectrons corresponding to $v' = 0$ and $v' = 1$ of $X^2\Sigma_g^+$ have different angular distributions.[145] Normally, in treating this problem one makes a Born–Oppenheimer separation and the angular distribution

† Marr and Woodruff[144a] have also performed PES experiments on some autoionizing peaks in N_2, with a photon band width of 2 Å, which again encompasses much of the underlying continuum.

asymmetry parameter (see Chapter VII) depends only on the electronic matrix element, and hence not on the vibrational state. The implication is that there may be some mixing, or dependence of the matrix element on internuclear distance, and this may also play a role in the 781.1 Å peak.

The photoionization mass spectrum is limited to $N_2{}^+$ until 24.294 eV, the thermodynamic threshold for N^+ from N_2. Obviously, the three states observed in the 21.2 eV photoelectron spectrum ($X^2\Sigma_g{}^+$, $A^2\Pi_u$, $B^2\Sigma_u{}^+$) can only contribute to $N_2{}^+$. In the 40.8 eV photoelectron spectrum, $(2\sigma_g)^{-1}$ appears, but prior to that several two-electron excited states are observed.[131,136,137,146] Recent photoionization mass spectrometric work at this laboratory[147] with line sources in the 21.2–40.8 eV range has shown that N^+ appears at 490 Å ≡ 25.3 eV, but not at the next longer line available, 537 Å ≡ 23.1 eV, where it is energetically forbidden. The state of $N_2{}^+$ most likely formed at 25.3 eV is $C^2\Sigma_u{}^+$, a state involving two-electron excitation, and deriving its intensity by configuration interaction with the $B^2\Sigma_u{}^+$ state. The C state is bound, but is evidently predissociated in vibrational levels $v' \geq 3$ (where dissociation into ground state N and N^+ first becomes possible). The nature of this predissociation has been the subject of several theoretical and experimental papers, most recently reviewed by Erman.[148]

The $N^+/N_2{}^+$ ratio rises rapidly from threshold to about 26 eV, at which energy it represents ~5% of total ionization. The next significant increase occurs between 446 Å (27.8 eV) and 405 Å (30.6 eV), which corresponds to passage through the onset of a state of 29 eV designated F by Wight *et al.*[137] At 30.6 eV, the data of ref. 149 imply equal cross sections for the C and F states. At 379 Å (32.7 eV) the $N^+/N_2{}^+$ ratio has increased further, presumably due to a sharply increasing partial cross section for the F state.[137] Finally, at 40.8 eV the $N^+/N_2{}^+$ ratio experiences an additional 37% increase as the $(2\sigma_g)^{-1}$ single-electron state becomes accessible. Browning *et al.*[149] report $N^+/N_2{}^+ = 0.25 \pm 0.03$ at 40.8 eV. The major contributors to N^+ at this energy are probably the $F^2\Sigma$ and $(2\sigma_g)^{-1}$ states.

Fragmentation studies of N_2 in the x-ray region have been performed by Carlson and Krause[150] at 280 and 930 eV. We can test our partial cross section curves against this data, with the following assumptions:

(1) The single-electron excitation states are the dominant contributors.
(2) The states $(2\sigma_g)^{-1}$ and $(1\sigma_g)^{-1}$ predominantly yield N^+.

The latter assumption concerning $(1\sigma_g)^{-1}$ stems from the view that ejection of electrons from such inner orbitals very likely results in an Auger shower, instantaneously creating N_2^{2+} or multiply charged N_2. This entity then disintegrates in a Coulombic explosion, yielding energetic N^+ and possibly some higher charge states.

At 280 eV, we obtain from Fig. 32, the following partial cross sections:

$$\begin{aligned} X^2\Sigma_g{}^+ &= 0.022 \text{ Mb} \\ A^2\Pi_u &= 0.032 \\ B^2\Sigma_u{}^+ &= 0.052 \\ (2\sigma_g)^{-1} &= 0.07. \end{aligned}$$

Only the latter can yield N^+. Hence,

$$\frac{N^+}{N^+ + N_2{}^+} = \frac{0.07}{0.176} = 0.398.$$

The result reported by Carlson and Krause is 0.26. At 930 eV,

$$\begin{aligned} X^2\Sigma_g{}^+ &= 0.00128 \text{ Mb} \\ A^2\Pi_u &= 0.00068 \\ B^2\Sigma_u{}^+ &= 0.0054 \\ (2\sigma_g)^{-1} &= 0.0063 \\ (1\sigma)^{-1} &= 0.135. \end{aligned}$$

The latter two states can result in N^+. We now have

$$\frac{N^+ \text{ (+ other fragments)}}{N^+ + N_2{}^+ \text{ (+ other fragments)}} = \frac{0.1413}{0.14866} = 0.95.$$

Carlson and Krause obtain 0.925 for this ratio. The overall agreement between our partial cross section analysis and photoionization mass spectra, from $\sim 1\%$ fragmentation at 25.3 eV to more than 90% fragmentation at 930 eV, would appear to be fairly good. However, it should be noted that we have counted only a single N^+ from each dissociation whereas a Coulombic explosion can be expected to produce two N^+ entities. The inclusion of this effect would influence the 930 eV calculations (raising the predicted value from 0.95 to 0.97) but would leave the 280 eV predictions unchanged.

3. O_2

a. Photoelectron Spectroscopy. The occupied molecular orbitals of O_2 may be written

$$(\sigma_g 1s)^2(\sigma_u 1s)^2(\sigma_g 2s)^2(\sigma_u 2s)^2(\pi_u 2p)^4(\sigma_g 2p)^2(\pi_g 2p)^2$$

or in united atom terminology

$$(1\sigma_g)^2(1\sigma_u)^2(2\sigma_g)^2(2\sigma_u)^2(1\pi_u)^4(3\sigma_g)^2(1\pi_g)^2.$$

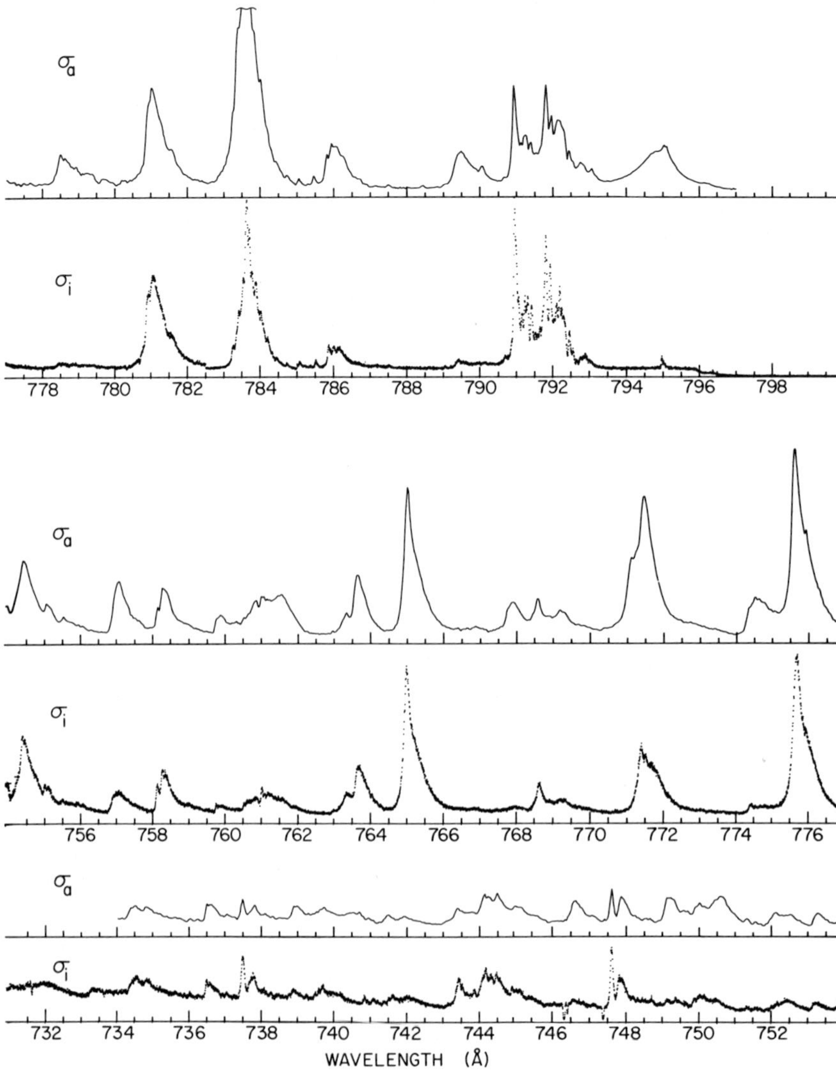

Fig. 32a Relative photoionization (σ_i) and photoabsorption (σ_a) cross sections of N_2 from threshold to 734 Å. The photoionization data (unpublished) were obtained by the author with a wavelength resolution (FWHM) of 0.035 Å. The photoabsorption data are from V. L. Carter, *J. Chem. Phys.* **56**, 4195 (1972), with a reported resolution of 0.04 Å. The author is grateful to Dr. Carter for presenting him with an expanded scan of her original data.

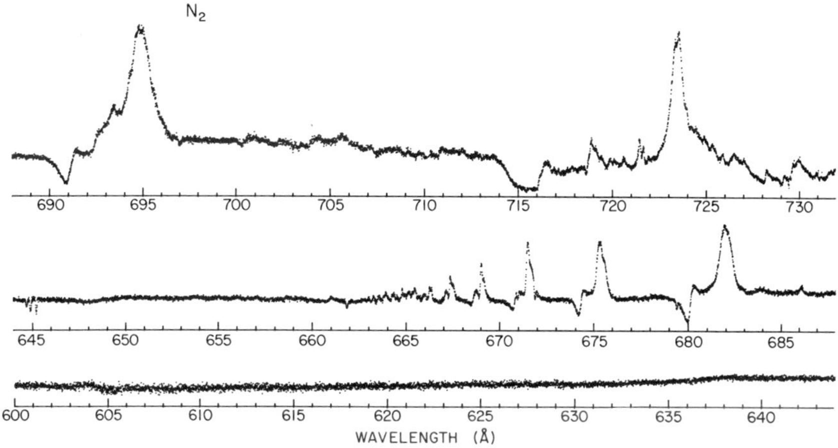

Fig. 32b Relative photoionization cross section of N_2 between 600 and 732 Å, obtained with a wavelength resolution (FWHM) of 0.07 Å. [Unpublished data obtained by the author.]

The 1σ orbitals are essentially atomic in nature, the 2σ orbitals more diffuse. The π_u and $3\sigma_g$ are bonding orbitals, the π_g an antibonding one. This outermost orbital is half-filled, which (from Hund's rule) gives O_2 its characteristic ground state triplet structure. Removal of an electron from this π_g orbital gives rise to the ground state of O_2^+, $X^2\Pi_g$, whereas removal of an electron from any of the other orbitals should make possible a doublet or quartet.

As with NO, the removal of a π_g (or π_g-like) electron decreases the internuclear distance, in this case from 1.2074 Å for O_2 ($X^3\Sigma_g^-$) to 1.1227 Å for O_2^+ ($X^2\Pi_g$). As a consequence, the Franck–Condon envelope for this transition encompasses ~5 significant vibrational peaks (see Turner *et al.*,[130] p. 52). The adiabatic first ionization potential is $12.07_{(1)}$ eV,[151,152] whereas the second, corresponding to the removal of a bonding electron, does not appear until 16.12 eV. This is again similar to the case of NO, and suggests that vibrational autoionization should be observable in the near threshold region.

The remainder of the valence orbital ionization region is a complex intertwining of doublets and quartets that has been sorted out with the aid of a higher resolution 584 Å photoelectron spectrum and a 304 Å spectrum by Edqvist *et al.*[151] For convenience, this spectrum is reproduced in Fig. 33.

Partial cross sections for 8 states between 745 and 304 Å (16.64–40.81 eV) have recently been reported by Samson *et al.*[153] They differ substantially from earlier values of Blake and Carver[154] in the region of overlap. (Note that the unresolved $a^4\Pi_u$ and $A^2\Pi_u$ are treated as a composite.)

Banna and Shirley[133] have measured branching ratios at 132.3 eV, and the 1253.6 eV photoelectron spectrum is available from ESCA II.[155] The spectra

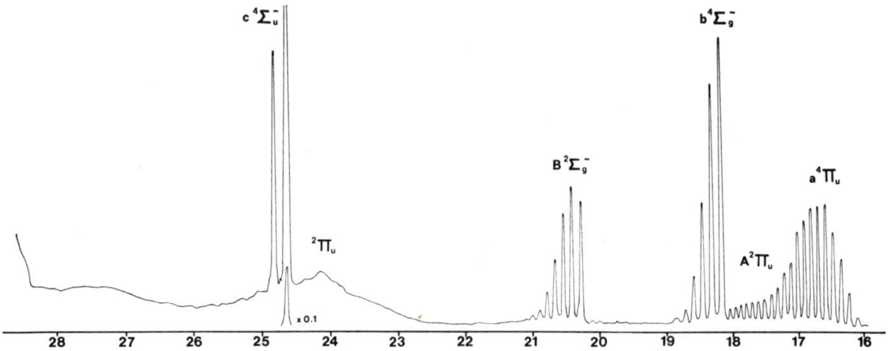

Fig. 33 The photoelectron spectrum of O_2 from 16 to 28 eV, obtained with 304 Å incident radiation. The progression beginning at 12.06 eV has been omitted. [From O. Edqvist *et al.*, *Phys. Scripta* **1**, 25 (1970).]

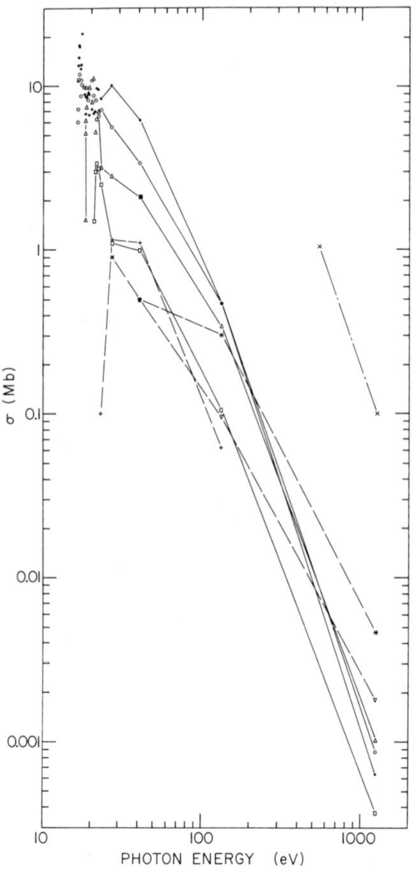

Fig. 34 Partial cross sections for formation of various states of O_2^+ by photoabsorption. ●, $X^2\Pi_g$; ○, $a^4\Pi_u + A^2\Pi_u$; △, $b^4\Sigma_g^-$; □, $B^2\Sigma_g^-$; +, $^2\Pi_u$ (23 eV); ∗, $c^4\Sigma_u^-$; ▽, $^2\Sigma_u^-$; ■, $^{2,4}\Sigma_g^-$; ×, $(\sigma_g/S)^{-1}$ $(\sigma_u/S)^{-1}$.

at 132.3 and 1253.6 eV have been converted to partial cross sections by making use of the total absorption cross sections of Fig. 11 in Chapter V. All of the resulting partial cross sections are plotted in Fig. 34.

b. Photoionization Mass Spectrometry. As we have seen, the envelope of the first electronic band in the VUV photoelectron spectrum of O_2 includes five significant vibrational peaks and covers the energy range from 12.07 to 13.0 eV. The next band does not appear until ~16 eV. We can therefore anticipate some vibrational autoionization in the near-threshold region, i.e., between 1027 and 954 Å. There are complex bands at 965, 973, 983, 993, 1004, and 1023 Å which could be examined for such processes. At shorter wavelengths, electronic autoionization should become prominent, corresponding to excitation of electrons from $3\sigma_g$, $1\pi_u$, and $2\sigma_u$ orbitals to Rydberg states converging upon the complex of states above 16 eV. The influence of predissociation is particularly marked in the 750–850 Å region, as can be seen in Fig. 12 in Chapter V (ionization yield of O_2).

A high resolution photoabsorption spectrum of the threshold region providing cross section information has not yet been published. Probably the best available is that of Matsunaga and Watanabe,[156] taken with 0.3 Å resolution width and used to construct Figs. 35a and b. The photoionization mass spectrum displayed in Fig. 36 was obtained with 0.07 Å resolution (0.14 Å in the near-threshold region). Bahr *et al.*[157] have studied the photoelectron spectra from several of the autoionization peaks between 957.0 and 832.5 Å, and in each instance have found a number of vibrational peaks characteristic of electronic autoionization. This is not surprising in view of our conclusion that vibrational autoionization should have negligible contribution below 954 Å. The photon bandwidth used in their experiments was 1.6 Å, considerably larger than the natural peak width, and hence it was necessary to subtract significant contributions from the underlying continuum. Nevertheless, they have been able to deduce some plausible internuclear distances for the quasi-discrete Rydberg states responsible for the autoionization resonances by a Franck–Condon analysis, as described in the section on N_2.

Tanaka and Tanaka[158] have extended this type of study to longer wavelengths, covering the region 939–1004 Å. They report some success in establishing internuclear distances for the Rydberg series conventionally labeled H, H', and M, whereas Bahr *et al.* focused primarily on the M, J and I progressions.

The dissociation energy of O_2 is 5.1158 eV[159] and the first ionization potential of O is 13.618 eV.[160] Therefore, the onset for dissociative ionization should occur at 18.734 eV. However, the electron affinity of O is 1.47 eV[161] and hence one can anticipate ion-pair formation at ~17.26 eV = 718.3 Å. This process has been observed by Elder *et al.*[162] with 0.07 eV resolution

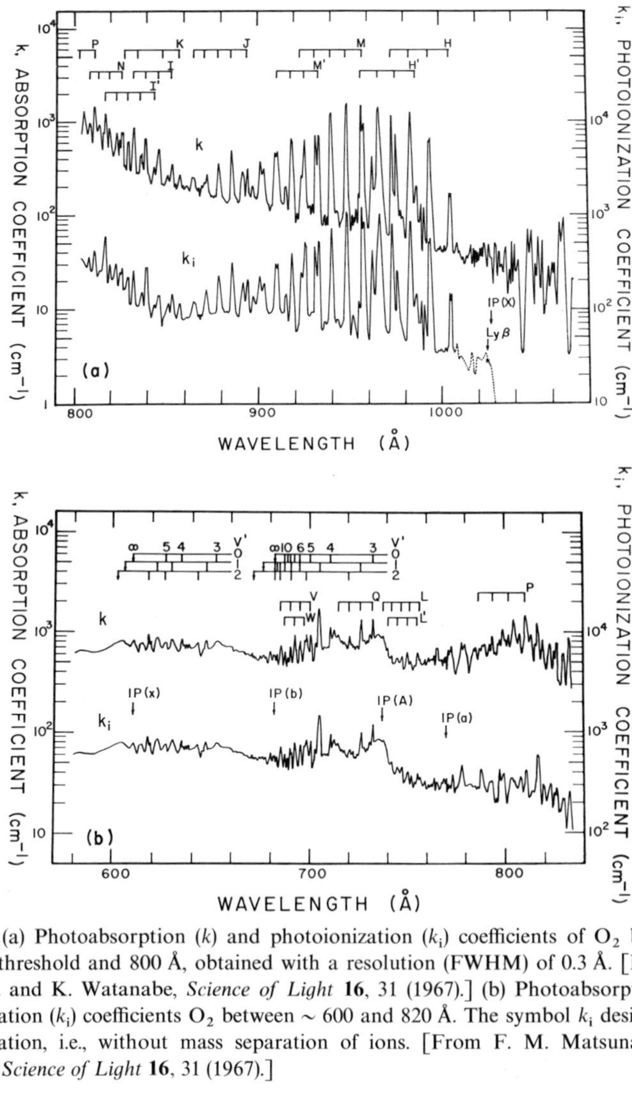

Fig. 35 (a) Photoabsorption (k) and photoionization (k_i) coefficients of O_2 between the ionization threshold and 800 Å, obtained with a resolution (FWHM) of 0.3 Å. [From F. M. Matsunaga and K. Watanabe, *Science of Light* **16**, 31 (1967).] (b) Photoabsorption (k) and photoionization (k_i) coefficients O_2 between ~ 600 and 820 Å. The symbol k_i designates total photoionization, i.e., without mass separation of ions. [From F. M. Matsunaga and K. Watanabe, *Science of Light* **16**, 31 (1967).]

width, and by Dibeler and Walker[163] with about 0.05 eV resolution width. The onset occurs at about the correct energy and the intensity of the process[163] is reported to be 0.03 of the O_2^+ intensity in this region. More recent examination of this region at higher resolution at Argonne[164] has revealed a rather complex structure, indicative of predissociation into ion pairs from Rydberg states whose intensity is not prominent in the O_2^+ spectrum in this wavelength region.

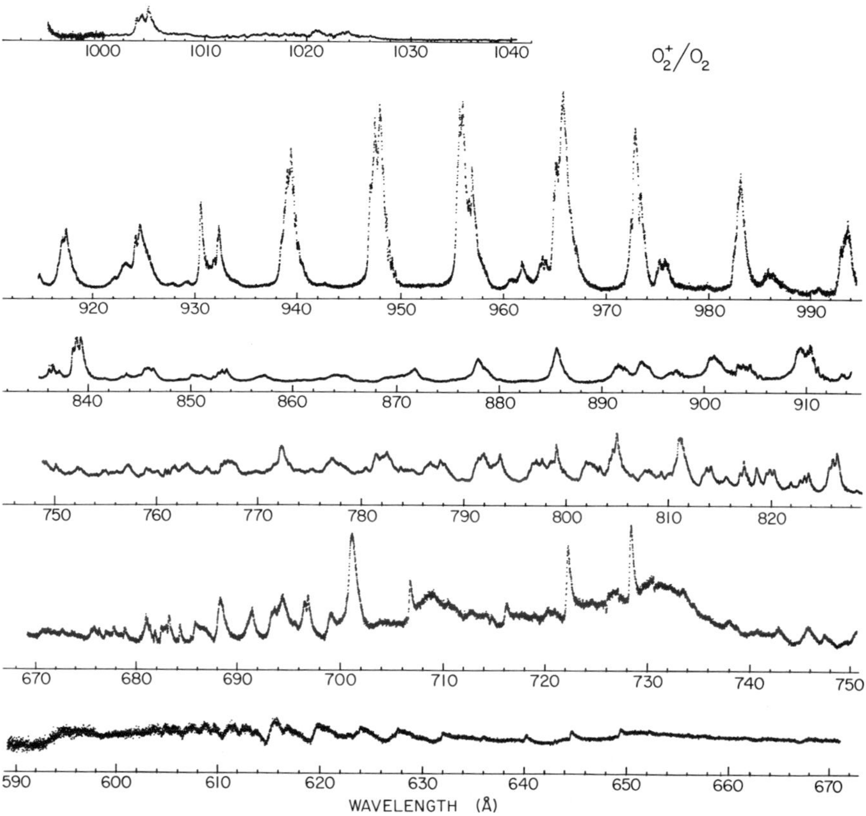

Fig. 36 Photoionization mass spectrum of O_2^+ obtained with a wavelength resolution (FWHM) of 0.07 Å from 590 to 1000 Å and 0.14 Å in the threshold region above 1000 Å. These data (unpublished) were obtained by the author.

The 21.2 eV photoelectron spectrum reveals five states of O_2^+—$X^2\Pi_g$ (IP = 12.071 eV), $a^4\Pi_u$ (IP = 16.101 eV), $A^2\Pi_u$ (IP = 17.15 eV), $b^4\Sigma_g^-$ (IP = 18.171 eV) and $B^2\Sigma_g^-$ (IP = 20.296 eV).

The first O_2^+ state which has sufficient energy to provide a source of O^+ is $b^4\Sigma_g^-$ in $v' \geq 4$. The Franck–Condon factors favor the low vibrational levels so that this source of O^+ will be weak. Even $v' = 4$ can only dissociate in rotational levels $N' \geq 9$ on energetic grounds. Recently, Guyon *et al.*[164a] have studied this region by the photoelectron–photoion coincidence technique and find that

$$\frac{O^+}{O^+ + O_2^+} = 0.4 \pm 0.1, 0.9 \pm 0.1 \text{ and } 0.8 \pm 0.1$$

for $v' = 4$, 5, and 6, respectively. Hence, as soon as energy is available that exceeds the first dissociation limit, virtually total decomposition results. Since the $b^4\Sigma_g^-$ is a bound state, a predissociative mechanism is indicated.

The primary source of O^+ at 584 Å is dissociation of the $B^2\Sigma_g^-$ state. The vibrational envelope of this state covers the range of approximately 20.3–20.8 eV. Therefore, the total kinetic energy release should be 1.6–2.1 eV, divided equally between O^+ and O.

Doolittle *et al.*[165] have measured the kinetic energy of ions produced in the photoionization of O_2 at 572 and 584 Å. In addition to a large thermal energy component (O_2^+), they observe a second group with a most probable energy of 0.95 eV, which can be attributed to O^+ from dissociation of $B^2\Sigma_g^-$. Eland[166] finds a mean energy for O^+ of 0.8 eV, with a 0.34 eV width. Doolittle *et al.* find that approximately 16% of the ions detected have the high energy component. Frost *et al.*,[75] using a spherical retarding field analyzer, observe that 20% of the photoelectron signal in the 21.2 eV photoelectron spectrum corresponds to formation of the $B^2\Sigma_g^-$ state. The recent data of Samson *et al.*[153] utilized in our partial cross section curve provide for 13% in the B state. Browning *et al.* have designed an ion source for photoionization mass spectrometric measurements which employs a very strong extraction field to minimize transmission losses of fragments with large kinetic energies. They report $O^+/O_2^+ = 0.204 \pm 0.018$ at 584 Å. Hence, the relevant measurements quoted above are in substantial agreement with the view that $B^2\Sigma_g^-$ is formed in about $\frac{1}{5}$–$\frac{1}{7}$ of the ionizations at 584 Å and is predissociated more-or-less completely to form O^+ and O in their ground states, with a release of about 0.9 eV kinetic energy to each particle.

Doolittle *et al.* note two additional peaks in the ion kinetic energy distribution curve when the incident photon energy is increased to 25.3 eV, a 3 eV peak with 1% intensity, and a 1.95 eV peak with 5% intensity. Over this energy span, Browning *et al.* report an increase in their O^+/O_2^+ ratio to 29%. From the 304 Å photoelectron spectrum of O_2, it is possible to conclude that this contribution results from ionization to the $c^4\Sigma_u^-$ state (IP = 24.58 eV), which is then predissociated to form $O^+(^4S) + O(^1D)$ having a thermochemical threshold at 20.7 eV. This would account for the kinetic energy component at 1.95 eV.

The recent coincidence measurements of Guyon *et al.*[164a] are a reassuring confirmation of this analysis. Although LeBlanc[167] has identified emission bands arising from the $v' = 0$ level of the $c^4\Sigma_u^-$ state, Guyon *et al.*[164a] find that even this vibrational level is $98 \pm 2\%$ predissociated. They also find sources of O^+ in the 22–24.5 eV region, which appears free of peaks in the 304 Å photoelectron spectrum. Although the contribution to the total O^+ and to the oscillator strength is not significant, this observation may have far-reaching theoretical importance since it suggests sources of fragment ions where states of the parent ion are not readily recognizable.

At 358 Å, Doolittle *et al.* observe an additional component with 4.5–5.0 eV ion kinetic energy. At this wavelength, about one-quarter of all ions are energetic. Of the energetic ions, half are in the 0.95 eV group, one-fifth in the 1.9 eV group, one-tenth in the 3 eV group, and one-fifth in the high energy component.

There is a state of O_2^+ at 27.9 eV $(^2\Sigma_u^-)$[155] which is accessible at 358 Å ≡ 34.63 eV. This is presumably the source of the high energy component.

At 304 Å, Browning *et al.*[149] report $O^+/O_2^+ = 0.42 \pm 0.05$ or $O^+/(O^+ + O_2^+) = 0.296$, which is even higher than the value of Doolittle *et al.* The higher energy states $^4\Sigma_g^-$ and $^2\Sigma_g^-$ accessible at 304 Å likely account for this discrepancy. The recent partial cross sections of Samson *et al.* which include these contributions at 304 Å result in a ratio of cross sections of dissociating states to nondissociating states of 0.44, in good agreement with the directly measured O^+/O_2^+ ratio.

If the partial cross sections of $(\sigma_g 2s)^{-1}$ and $(\sigma_g 1s)^{-1}$, $(\sigma_u 1s)^{-1}$ are apportioned entirely to fragment, as has been done for deep inner shells for the other cases studied, we obtain

	at 280 eV	at 930 eV
$X^2\Pi_g =$	0.051 Mb	0.0015 Mb
$a^4\Pi_u + A^2\Pi_u =$	0.059	0.00202
$b^4\Sigma_g^- =$	0.049	0.00223
$B^2\Sigma_g^- =$	0.016	0.00078
$c^4\Sigma_u^- =$	0.076	0.0082
$^2\Sigma_u^- =$	0.025	0.0030
$(\sigma_g 1s)^{-1}, (\sigma_u 1s)^{-1} =$		0.227

With the assignments previously given, we obtain

	at 280 eV	at 930 eV
$\dfrac{O^+ + \text{high ionizations}}{\Sigma n_i} =$	0.42	0.98

Carlson and Krause[150] obtain 0.39 and 0.96 for this ratio at 280 and 930 eV, respectively. The thermal energy ion measurements of Van Brunt *et al.*[168] give 0.13 and 0.39.

The agreement with Carlson and Krause at 280 eV is quite satisfactory.

At 930 eV, our estimate of the partial cross section for K-shell ionization may be somewhat high. It is based on the jump in total absorption at the K-edge. Only a single O^+ per K-shell hole has been assumed in the calculation.

4. CO

a. Photoelectron Spectroscopy. The occupied molecular orbitals of CO may be written

$$(1\sigma)^2(2\sigma)^2(3\sigma)^2(4\sigma)^2(1\pi)^4(5\sigma)^2.$$

Carbon monoxide is isoelectronic with N_2. The 1σ and 2σ orbitals are essentially the K-shells of oxygen and carbon, respectively, while the 3σ and 4σ have large contributions from O 2s and C 2s, respectively. The 1π and 5σ are bonding valence orbitals, giving CO its characteristic triple bonded structure. As with N_2, removal of an outermost σ electron gives rise to a sharp photoelectron band[169] while removal of an uppermost π electron yields a rather broad band, encompassing ~8–9 vibrational levels. An apparent oddity is to be noted. Although 5σ is a bonding orbital, removal of

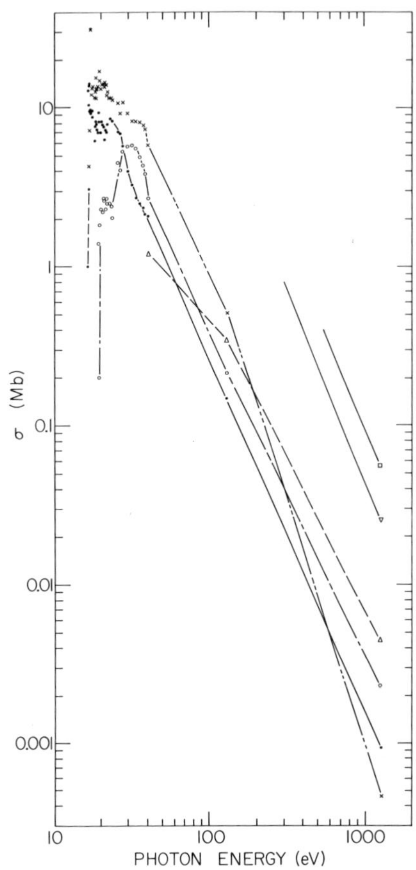

Fig. 37 Partial cross sections for formation of various states of CO^+ by photoabsorption. ●, $X^2\Sigma^+(5\sigma)^{-1}$; ×, $A^2\Pi(1\Pi)^{-1}$; ○, $B^2\Sigma^+(4\sigma)^{-1}$; △, $C^2\Sigma(3\sigma)^{-1}$; ▽, $(2\sigma)^{-1}$; □, $(1\sigma)^{-1}$.

one electron *shortens* the internuclear distance from 1.128 to 1.115 Å.[170] This behavior is even more marked in the next isoelectronic member BF and is connected with the increasing ionicity of the bond.[171]

While some early partial cross section information was available from Bahr *et al.*[172] and Schoen,[173] the most reliable and extensive information has become available quite recently. Samson and Gardner[174] have provided branching ratios at selected lines between 750 and 304 Å (16.53–40.8 eV) while a more detailed picture (albeit with poorer resolution) has been obtained by Plummer *et al.*[132] between 19 and 39 eV using synchrotron radiation. A valuable intermediate point at 132.3 eV has been measured by Banna and Shirley,[133] using the Y $M\zeta$ line, and at 1253.6 eV the x-ray photoelectron spectra from ESCA 2[175] have been utilized. In each instance, the authors' branching ratios have been combined with the total cross sections of Fig. 13 in Chapter V to deduce partial cross sections.

The partial cross sections of the deeper lying orbitals $[(2\sigma)^{-1}$ and $(1\sigma)^{-1}]$ have been estimated from the initial rise at the corresponding x-ray edge and the partial cross sections at 1253.6 eV. The considerations cited above have been utilized to construct Fig. 37, a graph of partial cross sections versus incident photon energy.

b. Photoionization Mass Spectrometry. The photoionization mass spectrum of CO^+ between threshold and 600 Å, obtained with 0.07 Å resolution (FWHM) is displayed in Fig. 38. Absolute photoabsorption cross sections at comparable resolution are not yet available.

The structure observed is largely attributable to Rydberg series converging to the second and third ionization limits of CO^+ (the $A^2\Pi_i$ and $B^2\Sigma^+$ states). An analysis of this spectrum is currently in progress. The relevant potential energy diagrams are displayed in Fig. 39, based on data given by Rosen[5] adapted to Morse curves. There should be very little possibility for vibrational autoionization in the energy interval between the first and second ionic states because Franck–Condon integrals connecting the ground state of CO with the ground state of CO^+ are dominated by the 0–0 transition, and hence Rydberg levels having the character of $CO^+(X^2\Sigma^+)$ should exhibit similar behavior. Bahr *et al.*[144] have obtained photoelectron spectra for a few of these peaks and have interpreted them in terms of an electronic autoionization mechanism (see discussion in N_2 section).

The thresholds for photodissociative ionization of CO are 22.35 eV[30,176] into $C^+ + O + e$ and 24.71 eV into $O^+ + C + e$. [Wight *et al.*[137] report the possibility of a very weak ion-pair process ($O^+ + C^-$) which would have a lower threshold. This process has often been observed in electron impact ionization of CO, most recently by Locht,[176a] who reviews earlier work. Baumgärtel[176b] has recently reported the detection of $O^- + C^+$ between 20.9 and 22.4 eV, and $C^- + O^+$ between 23.5 and 24.2 eV, both processes

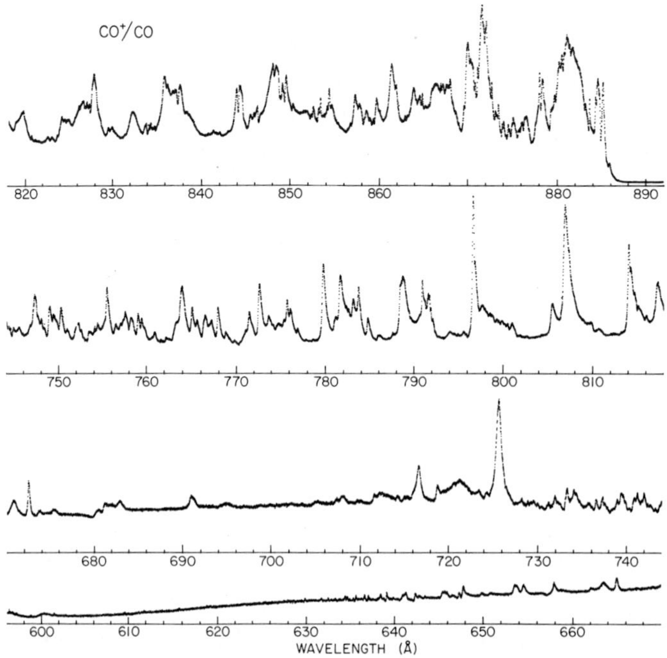

Fig. 38 Photoionization mass spectrum of CO^+ obtained with a wavelength resolution (FWHM) of 0.07 Å from ~ 600 Å to the ionization threshold. These data (unpublished) were obtained by the author.

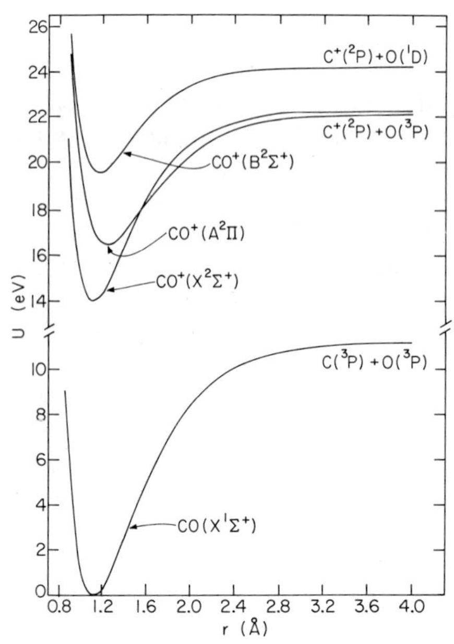

Fig. 39 Morse curves of CO (ground state) and some ionic states prominent in the photoionization of CO. [Based on data given by Rosen.[5]]

extremely weak (10^{-4}–10^{-5} of the absorption cross section).] Hence, the three major one-electron states $X^2\Sigma^+$, $A^2\Pi$, and $B^2\Sigma^+$ appearing in the 21.2 eV photoelectron spectrum cannot contribute to dissociative ionization, and Fig. 38 should scale with total ionization.

Such fragmentation as does occur between 21.2 and 40.8 eV is attributable to two-electron excitation states until one passes through the threshold of $(3\sigma)^{-1}$ at ~38 eV. The weak states appear at 22.7, 23.4, 25.4, 28.1, and 31.8 eV.[146,177] They have been shown to result from electron correlation, which can be described as a breakdown of the single particle picture of photoionization.[177a]

Up to 28.1 eV, the major effect[147] of these weak peaks is to produce $C^+ + O + e$. The 31.8 eV peak divides roughly equally among $C^+ + O$ and $O^+ + C$ and contributes about 7.4% to total ionization at 32.7 eV. From data at 40.8 eV, the effect of the $(3\sigma)^{-1}$ state can be assessed. It contributes ~6.5% to total ionization and favors O^+ to C^+ by about 5:1.

In their ion kinetic energy spectra, Gardner and Samson[178] see only thermal ions at 21.2 and 26.8 eV, but at 40.8 eV a high energy peak is observed, centered at 2.77 eV kinetic energy. Gardner and Samson infer that $(3\sigma)^{-1}$ produces $C^+ + O$, which is at variance with the mass spectrum observed by Kronebusch and Berkowitz.[147] On the other hand, the pseudo photoelectron–ion coincidence studies of Wight *et al.*[137] are in good agreement with the results of Kronebusch and Berkowitz. The partial cross section of $(3\sigma)^{-1}$ at 40.8 eV reported by Samson and Gardner and plotted in Fig. 37 amounts to 10% of total ionization, which is higher than that found by Wight *et al.*[137] (6.5%) but within the error limits.

Mass spectrometric studies on CO in the x-ray range have been reported by Van Brunt *et al.*[168] at 280 eV and by Carlson and Krause[150] at 930 eV. Van Brunt *et al.* primarily collect ions having thermal kinetic energy, and hence they would be discriminating against fragments. They observe

$$\frac{C^+}{\text{total ionization}} = 0.034 \pm 0.010$$

and

$$\frac{O^+}{\text{total ionization}} = 0.020 \pm 0.007.$$

The partial cross section for formation of the one-electron excitation states of CO^+ at 280 eV, obtained from Fig. 37, are as follows:

$$X^2\Sigma^+ = 0.0265, \qquad A^2\Pi = 0.049,$$
$$B^2\Sigma^+ = 0.047, \qquad (3\sigma)^{-1} = 0.08.$$

If we momentarily ignore the two-electron states and assign X, A, and B to parent ion, $\frac{5}{6}$ of $(3\sigma)^{-1}$ to O^+, we deduce

$$\frac{O^+}{\text{total ionization}} = 0.33, \qquad \frac{C^+}{\text{total ionization}} = 0.066.$$

This is already much larger than the fragmentation ratio observed by Van Brunt *et al.*, and furthermore is heavily in the direction of O^+. On the other hand, the two-electron states between 22.7 and 31.8 eV primarily produce C^+, presumably with lower kinetic energy. We infer that the experiments of Van Brunt *et al.* are most sensitive to these states, and quite insensitive to $(3\sigma)^{-1}$.

At 930 eV, Carlson and Krause[150] give

$$\frac{C^+ + C^{2+} + C^{3+}}{\text{total ionization}} = 0.49 \quad \text{and} \quad \frac{O^+ + O^{2+} + O^{3+}}{\text{total ionization}} = 0.45.$$

From Fig. 37, the partial cross sections at 930 eV are

$$\begin{aligned} X^2\Sigma^+ &= 0.00183, & (3\sigma)^{-1} &= 0.0079, \\ A^2\Pi &= 0.0012, & (2\sigma)^{-1} &= 0.0455, \\ B^2\Sigma^+ &= 0.0041, & (1\sigma)^{-1} &= 0.098. \end{aligned}$$

If we assume that Auger cascades ensue after formation of $(2\sigma)^{-1}$ and $(1\sigma)^{-1}$, and that the multiply charged molecular ions formed produce equal amounts of carbon atomic ions and oxygen atomic ions (a kind of statistical assumption for these very high energy processes), we must count $(2\sigma)^{-1}$ and $(1\sigma)^{-1}$ twice to estimate total ionization. Thus,

$$\frac{C^+ + C^{2+} + C^{3+}}{\text{total ionization}} = 0.479, \qquad \frac{O^+ + O^{2+} + O^{3+}}{\text{total ionization}} = 0.497.$$

Without inclusion of the two-electron states, the results favor O^+, which appear contrary to the data of Carlson and Krause. However, it may be noted that O^+ is observed in greater abundance than C^+ by these authors, but C^{2+} makes a more significant contribution than O^{2+}. Carlson and Krause attribute this to the lower ionization potential of C^{2+} compared to O^{2+}.

Recent pseudophotoionization experiments on CO by Van der Wiel *et al.*,[178a,b] using incident elcctrons with 8 keV energy, offer a striking example of the influence of K-shell ionization on the mass spectrum and also provide additional data for comparison with the present analysis. At an absorption energy of 280 eV (comparable to that used by Van Brunt *et al.* and well below the carbon K-edge at 295.9 eV), their ratios of CO^+, O^+, and C^+ to total ionization are 0.46, 0.26, and 0.22, respectively. The values for O^+ and C^+ are much higher than those of Van Brunt *et al.*, as would be expected if the fragment ions had significant kinetic energy. The ratios are in line with our estimates, if allowance is made for the two-electron states between 22.7 and 31.8 eV which primarily produce C^+. At 305 eV, beyond the carbon K-edge, the ratios of CO^+, O^+, and C^+ to total ionization become 0.05, 0.35, and 0.45, respectively. The relative importance of CO^+ has diminished dramatically in traversing the carbon K-edge. The mass spectrum at 930 eV is not very different from that at 305 eV (the relative intensity of O^+ and C^+ ions reverses), indicating that processes resulting from ionization at the oxygen K-edge (542.1 eV) yield similar products to those beyond the carbon K-edge. However, the doubly ionized products (C^{2+} and O^{2+}) have increased in abundance from 11% at 305 eV to 18% at 930 eV.

An interesting sidelight to the measurements of Kay *et al.*[178b] is some suggestive evidence for "post-collision interaction." In a narrow energy region above the carbon K-edge (296–298 eV) they find that the ion yield curves for CO^{2+} and $C^{2+} + O^{2+}$ are proportionately weaker than for $C^+ + O^+$, whereas at higher energy (>298 eV) all three curves appear to have a similar shape. They argue that formation of a carbon K-shell vacancy is followed by an Auger decay that causes ejection of a second valence electron, e.g.,

$$(CO^+)^{**}:\ (1\sigma)^2(2\sigma)^1(3\sigma)^2(4\sigma)^2(1\pi)^4(5\sigma)^2 \rightarrow (CO^{2+}):\ (1\sigma)^2(2\sigma)^2(3\sigma)^2(4\sigma)^2(1\pi)^3(5\sigma)^1 + \bar{e}$$

The initial photoionization, occurring just beyond the K-edge, liberates a rather slow electron. During the brief interval of time when electrons are being ejected (perhaps 10^{-15} sec) a slow electron can be captured into an unoccupied orbital (say 2π) by CO^{2+}, yielding $(CO^+)^*$. In the example cited, $(CO^+)^*$ would be in the $C^2\Sigma$ state. If this occurs, it would detract from the signal in the CO^{2+} and $C^{2+} + O^{2+}$ channels in a narrow energy interval where the kinetic energy of the ejected photoelectron is low, as they purport to observe. The evidence is not entirely convincing, however, since the scatter in their data in the critical region amounts to a substantial fraction of the "missing oscillator strength."

5. NO

a. Photoelectron Spectroscopy. The occupied molecular orbitals of NO are $(1\sigma)^2(2\sigma)^2(3\sigma)^2(4\sigma)^2(1\pi)^4(5\sigma)^2(2\pi)^1$.

The 1σ and 2σ orbitals are essentially the K-shells of oxygen and nitrogen, respectively, while the 3σ and 4σ should have large contributions from O 2s and N 2s, respectively. The 1π and 5σ are nominally the bonding valence orbitals, and the 2π is an antibonding one.

The 584 Å photoelectron spectrum of Turner *et al.*[179] reveals a first ionization threshold, corresponding to formation of $NO^+(X^1\Sigma^+)$ with a threshold at 9.26 eV. The next state does not appear until 15.65 eV. Since the $X^1\Sigma^+$ is formed with five readily distinguishable vibrational peaks, the Rydberg levels converging to this state should have similar vibrational components, making this near-threshold region a likely candidate for vibrational autoionization (see Section B.4.b). The vibrational structure results from the excitation of the antibonding 2π electron, reducing the internuclear distance[180] from 1.1505 Å in $NO(X^2\Pi)$ to 1.06 Å in NO^+ $(X^1\Sigma^+)$.

The remainder of the valence orbital ionization structure (above 15.65 eV) is quite complex and is best described by reference to the relevant potential energy diagrams (Fig. 40) and the analysis of 584 and 304 Å photoelectron spectra of Edqvist *et al.*[181,182] (Figure 1 from Edqvist *et al.*[182] is particularly helpful here.) In general, the removal of an electron from the inner orbitals should give rise to both singlet and triplet states. Edqvist *et al.* interpret the rich spectrum between 15.65 and 20 eV as being due to formation of $a^3\Sigma^+$ ($v' = 0, 1, \ldots, 8$) between 15.65 and 16.8 eV, $b^3\Pi$ ($v' = 0,$

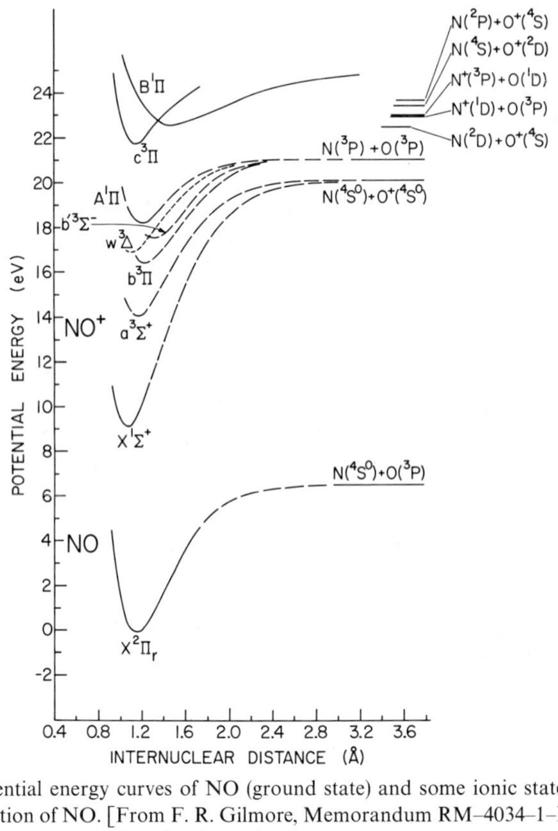

Fig. 40 Potential energy curves of NO (ground state) and some ionic states prominent in the photoionization of NO. [From F. R. Gilmore, Memorandum RM–4034–1–PR, The RAND Corp. (1966), with some additions by the author.]

1, 2) between 16.56 and 16.97 eV, $w^3\Delta$ ($v' = 0, 1, \ldots, 8$) between 16.86 and 18.07 eV, $b'^3\Sigma^-$ ($v' = 0, 1, \ldots, 8$) between 17.59 and 18.75 eV, $A'^1\Sigma^-$ ($v' = 0.1, \ldots, 7$) between 17.82 and 18.84 eV, $W^1\Delta$ ($v' = 0, 1, \ldots, 13$) between 18.07 and 19.72 eV, and $A^1\Pi$ ($v' = 0, 1, 2, 3$) between 19.32 and 18.89 eV. In addition, the 304 Å photoelectron spectrum has been interpreted[181,182] in terms of a $c^3\Pi$ and $B^1\Pi$ at 21.72 eV and $B'^1\Sigma^+$ at ~23 eV (broad peak). Recent calculations[183] have led to the interpretation that the state at 21.72 eV should be assigned to $c^3\Pi$, the broad peak at ~23 eV to $B^1\Pi$, and the very weak peak at 20.41 eV is ignored. There is obviously a disagreement about the assignment of the sharp intense peak at 21.72 eV. We refer to it below as $c^3\Pi$.

Supplementary data on partial cross sections are available from Bahr *et al.*[172] between 584 and 790 Å (15.7–21.2 eV), from Samson[184] at 26.84 eV, and from Siegbahn *et al.*[185] at 1253.6 eV. These partial cross sections, augmented where necessary with total absorption data taken from Fig. 15 in Chapter V, have been combined in Fig. 41.

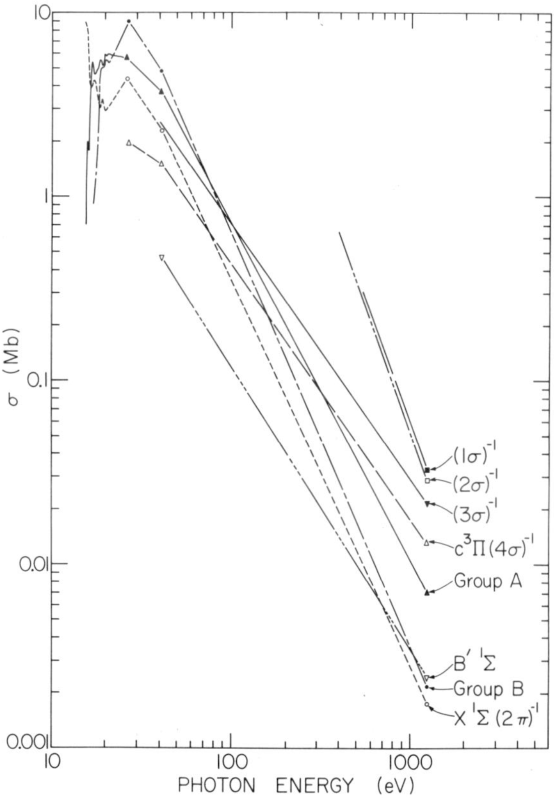

Fig. 41 Partial cross sections for formation of various states of NO^+ by photoabsorption.

Since the high resolution data of Edqvist *et al.*[181] at 40.8 eV cannot be correlated with equally high resolution at other energies, several bands have been combined into groups, as had previously been done by Bahr *et al.*[172] Thus, group A consists of $a^3\Sigma^+ + b^3\Pi$, and group B encompasses $w^3\Delta + b'^3\Sigma^- + A'^1\Sigma^- + W^1\Delta + A^1\Pi$.

b. Photoionization Mass Spectrometry. As previously mentioned, there is a large gap between the first ionization threshold at 9.26 eV and the next one at 15.65 eV. Between 9.26 and ~10.3 eV, one can expect vibrational autoionization. The photoabsorption and photoionization data of Watanabe[186] taken with 0.2 Å bandwidth, are plotted in Fig. 42. The photoabsorption cross section varies by factors of ~2 across the quasi-discrete peaks, while the photoionization cross section displays a more modest modulation. Another striking feature is that the ionization yield appears to depend strongly on the vibrational state of the residual ion $NO^+(X^1\Sigma^+)$, going from ~20% in the $v'' = 0$ region to ~90% in the $v' = 3$ region.

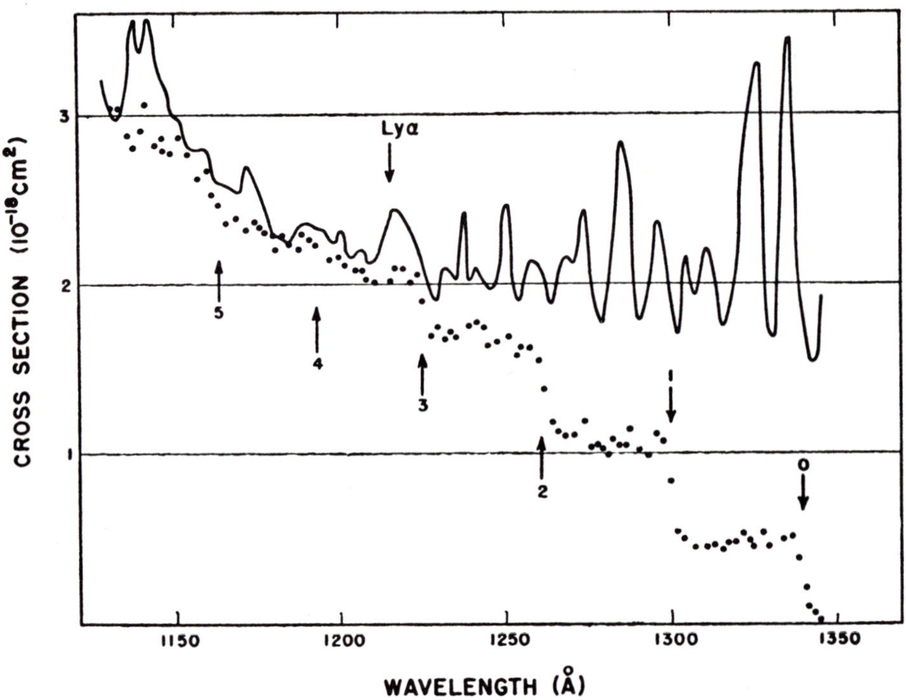

Fig. 42 Photoabsorption (solid line) and photoionization (dots) cross sections of NO in the region of the ionization threshold. Note the contrast between the peaked structure in photoabsorption and apparent steplike structure in photoionization. [From K. Watanabe, *J. Chem. Phys.* **22**, 1564 (1954).]

A photoionization mass spectrum of this region (1220–1345 Å)[187] (Fig. 43) taken with a resolution width of 0.83 Å also shows apparent autoionization structure superposed upon a steplike behavior. This structure is more pronounced when the target NO gas has been cooled.[187a] It had previously been supposed[188,188a] that idealized step-function structure (see Chapter IV, Section A) existed in this region of NO. The evidence now appears to be that predissociation competes very effectively with autoionization in this region (note the quantum yields) and tends to smear out the sharp autoionization features.

As a consequence, vibrational autoionization is apparently strongly suppressed. In the region between the first and second ionization potentials (~1340–900 Å), three groups[144,189,190] have examined the photoelectron spectrum resulting from autoionization at selected lines. The evidence seems to be that above 1130 Å (where autoionization is suppressed) the vibrational distributions agree with Franck–Condon factors connecting the neutral ground state ($X^2\Pi$) and the ionic ground state ($X^1\Sigma^+$). Below 1130 Å, a wider distribution of vibrational levels is observed and varies with the autoionizing state.

In addition, Bahr *et al.*[144] have made such studies on three lines above the threshold of the first excited ($a^3\Sigma^+$) state and find strong preference for autoionization to this state at 783.6 and 771.3 Å. Recent data which confirm this general behavior have been obtained at the wavelengths of the ArI lines (1048 and 1066 Å)[190a] and the NeI lines (736 and 744 Å).[190b]

The mechanism controlling autoionization in all of the above cases is most likely electronic (e^2/r_{ij}) in nature.

Turning now to the photoionization mass spectrum, we note that $D_0(\text{NO}) = 6.50_6 \pm 0.01$ eV[176] IP(O) = 13.618 eV and IP(N) = 14.534 eV.[30]

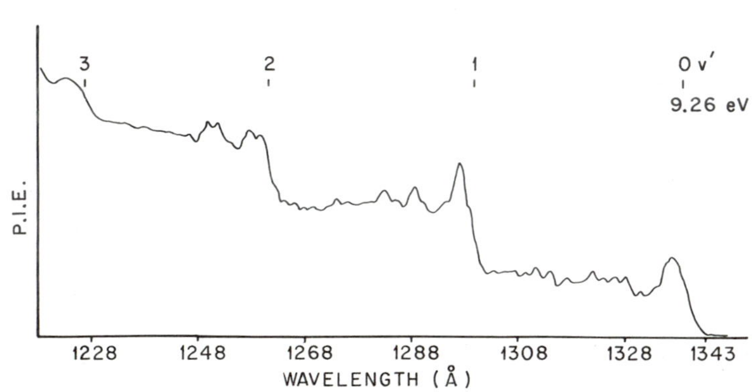

Fig. 43 Photoionization mass spectrum of NO^+ in the threshold region revealing some structure superimposed upon the vibrational steps. [From P. C. Killgoar, Jr. *et al.*, *J. Chem. Phys.* **58**, 803 (1973).]

Therefore, the thermochemical threshold for O^+ from NO should be 20.12 eV and that for N^+, 21.04 eV. (One should also expect $N^+ + O^-$ to appear at 19.55 eV, but this has not yet been observed.)

Thus, most of the states in the 21 eV photoelectron spectrum can contribute only to the parent ion. A recent photoionization mass spectrum of NO^+ from threshold to ~20 eV, obtained with 0.07–0.14 Å resolution, is shown in Fig. 44. In most of this region, it should be proportional to total ionization. At about 612–613 Å, a dip occurs in the NO^+ curve. This wave length corresponds to the threshold for O^+ formation. According to Hertz *et al.*[190c] O^+ does begin to appear here, but the dip in the NO^+ ion yield curve is characteristic of the autoionization process from Rydberg states converging to $c^3\Pi$ or $B^1\Pi$ of NO^+. These authors, using synchrotron radiation, have extended the photoionization mass spectrum of NO to ~550 Å. They find that the aforementioned Rydberg states produce O^+, as well as NO^+. However, at the onset of the $c^3\Pi$ state, O^+ and NO^+ are unaffected, whereas N^+ abruptly rises. Hence, the $c^3\Pi$ state appears to be completely predissociating into $N^+ + O$. Fragmentation has been studied[147] with line sources between 21.2 and 40.8 eV. (Earlier work[191] displays an onset for O^+ below 20.0 eV with an intensity ~2% of NO^+, and hence must be due to impurities, perhaps O_2.) These studies indicate a weak source of O^+ at 21.2 eV. It is possible that the 584 Å line corresponds to some autoionizing structure, since there are a number of states in this region and

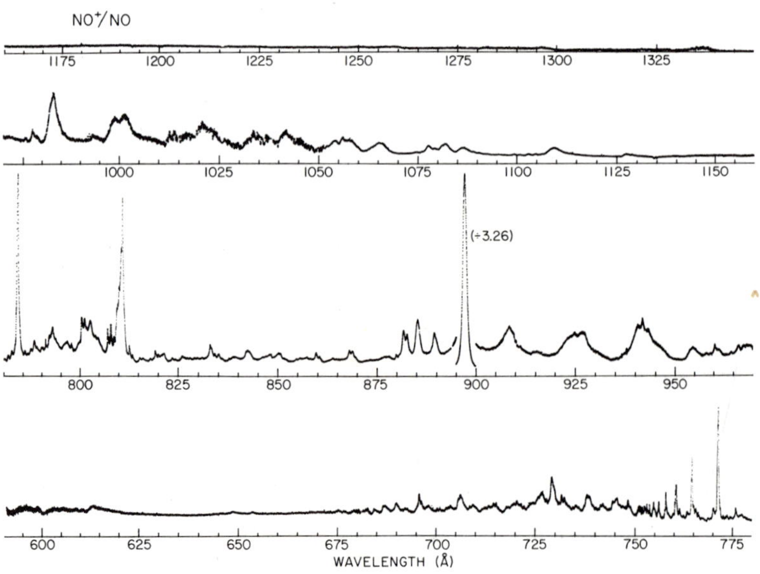

Fig. 44 Photoionization mass spectrum of NO^+ from 600 Å to the ionization threshold, obtained with a wavelength resolution (FWHM) of 0.07 Å. The data (unpublished) were obtained by the author.

the Rydberg manifold may be quite dense. At 537 Å (23.1 eV) there is a dramatic increase in the N^+ channel, strongly suggesting that $c^3\Pi$ predissociates predominantly into $N^+ + O$. There should be little kinetic energy imparted to N^+, since the $c^3\Pi$ peak lies only 0.6 eV above threshold for the decomposition. The influence of $B^1\Pi$ is not detectable in the mass spectrum, which probably is attributable to its small cross section.

The next significant increase occurs between 446 Å (27.8 eV) and 405 Å (30.6 eV). Presumably, this range of energies passes through the onset of one of the 2s-like states with a strong N 2s composition. There is an additional marked increase in the N^+ channel at 304 Å, either due to accessibility of another 2s-like state or an increase in the partial cross section of the first 2s-like state, relative to the sum of the other states.

It is noteworthy that the major fragmentation channels produce $N^+ + O$, although the $O^+ + N$ decomposition into ground state entities has the lower thermochemical threshold.

At 280 eV, there is a significant difference between the results of Carlson and Krause[150] and those of Van Brunt *et al.*[168] not only in the relative yield of fragment to parent ions, but also in the relative yield of N^+ and O^+. The experimental results are as follows:

Carlson and Krause ("all ions")	Van Brunt *et al.* ("thermal" ions)
$\dfrac{N^+ + N^{2+}}{\text{total ionization}} = 0.116$	$\dfrac{N^+ + N^{2+}}{\text{total ionization}} = 0.165$
$\dfrac{O^+ + O^{2+}}{\text{total ionization}} = 0.149$	$\dfrac{O^+ + O^{2+}}{\text{total ionization}} = 0.033$
$\dfrac{NO^{2+}}{\text{total ionization}} = 0.006$	$\dfrac{NO^{2+}}{\text{total ionization}} = 0.105$

Our immediate problem is to rationalize how the N^+ yield, when one is collecting all ions, can be smaller than the N^+ yield when one is collecting only thermal ions. Another is the reversal of importance of the relative intensities of N^+ and O^+ under the two experimental conditions.

The discrepancies are similar at higher energies, as can be seen below.

Carlson and Krause (930 eV)	Van Brunt *et al.* (1250 eV)
$\dfrac{N^+ + N^{2+} + N^{3+}}{\Sigma n_i} = 0.442$	$\dfrac{N^+ + N^{2+}}{\Sigma n_i} = 0.098$
$\dfrac{O^+ + O^{2+} + O^{3+}}{\Sigma n_i} = 0.478$	$\dfrac{O^+ + O^{2+}}{\Sigma n_i} = 0.027$
$\dfrac{NO^{2+}}{\Sigma n_i} = 0.023$	$\dfrac{NO^{2+}}{\Sigma n_i} = 0.687$

The additional and revealing item of information that prominently appears at the higher energy is the relative significance of the NO^{2+} ion. This entity should disintegrate in time, due to Coulomb repulsion to give $N^+ + O^+$. The flight time of ions in the two experiments could be different. If we augment the N^+ and O^+ yields in every instance by $\frac{1}{2}$ the NO^{2+} yield, which then makes these results more directly comparable with the apportionment to either N^+, O^+, or NO^+, we obtain

Carlson and Krause (280 eV)	Van Brunt *et al.* (280 eV)
$\frac{N^+ + N^{2+}}{\Sigma n_i} = 0.119$	0.218
$\frac{O^+ + O^{2+}}{\Sigma n_i} = 0.152$	0.086

and at the higher energy

Carlson and Krause (930 eV)	Van Brunt *et al.* (1250 eV)
$\frac{N^+ + N^{2+} + N^{3+}}{\Sigma n_i} = 0.454$	0.442
$\frac{O^+ + O^{2+} + O^{3+}}{\Sigma n_i} = 0.490$	0.371

The partial cross sections extracted from Fig. 41 are as follows:

	at 280 eV	at 930 eV
$X^1\Sigma^+$	0.041 Mb	0.00323 Mb
Group A	0.109	0.0121
Group B	0.063	0.00425
$c^3\Pi$	0.103	0.0195
$B^1\Pi$	0.024	0.0038
$(3\sigma)^{-1}$	0.167	0.032
$(2\sigma)^{-1}$		0.043
$(1\sigma)^{-1}$		0.0495

We assign the low-lying states composing $X^1\Sigma^+$, group A and group B to NO^+ on energetic grounds, and $c^3\Pi$ to N^+ for reasons discussed earlier. (There is some evidence[192] that high vibrational levels of $w^3\Delta$, $W^1\Delta$ or $A'^1\Sigma^-$ could contribute to N^+, but for our purposes this contribution should be negligible.) The remaining states are equally apportioned to N^+,

O^+, and higher ionization states of these atoms. We thus find

	at 280 eV	at 930 eV
$\frac{N^+ + \text{higher ionization}}{\Sigma n_i}$	= 0.39	0.50
$\frac{O^+ + \text{higher ionization}}{\Sigma n_i}$	= 0.19	0.38

The higher energy results are in rather good agreement with both experimental results, favoring the relative yields of N^+ and O^+ of Van Brunt *et al.* The 280 eV results are also in fair agreement with Van Brunt *et al.* if we recognize that their collection efficiency for higher kinetic energy fragments is small. This does not show up as much in the higher energy results because of the type of normalization, since fewer parent ions contribute to the denominator in this case. The large difference between N^+ yield and O^+ yield at 280 eV is attributable to the $c^3\Pi$ state. Its contribution alone is computed to be 0.20, rather close to the value 0.162 obtained by Van Brunt *et al.* for N^+ yield alone. Hence, a plausible conclusion is that $c^3\Pi$ produces N^+ with relatively low kinetic energy, as we had earlier anticipated.

The major remaining discrepancy is the relatively low N^+ yield of Carlson and Krause at 280 eV, and this does not have any obvious explanation.

D. Triatomic Molecules

1. H_2O

a. Photoelectron Spectroscopy. The water molecule in its electronic ground state has C_{2v} symmetry and the electron configuration is $(1a_1)^2(2a_1)^2(1b_2)^2(3a_1)^2(1b_1)^2$. The $1a_1$ orbital is essentially the K-shell of atomic oxygen, slightly shifted in orbital energy from 543.1 to 539.7 eV,[193] by the change in outer screening due to charge transfer from the hydrogen atoms. Similarly, the $2a_1$ orbital is largely O 2s in nature[194] and has a vertical ionization energy of ~32.5 eV.[134,195,196]

The three remaining occupied orbitals all have O 2p character. These orbitals appear in the 21.2 eV photoelectron spectrum; high-quality spectra have recently been reported by Karlsson *et al.*[197] and by Dixon *et al.*[197a] The spectrum from Karlsson *et al.*[197] is reproduced in Fig. 45. The outermost occupied orbital gives rise to a very narrow band, characteristic of an electron being removed from a nonbonding orbital. It is essentially an

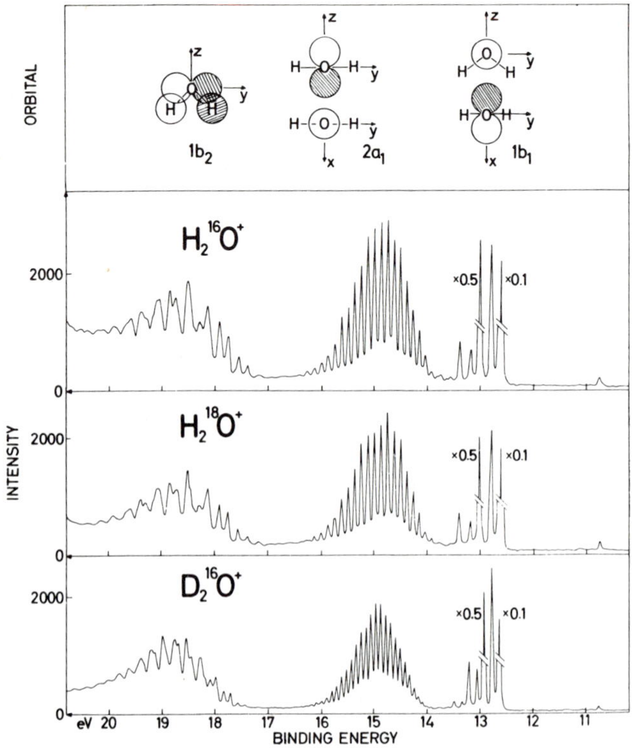

Fig. 45 Photoelectron spectra of $H_2{}^{16}O$, $H_2{}^{18}O$, and $D_2{}^{16}O$ obtained with 21.2 eV incident radiation. [From Karlsson *et al.*, *J. Chem. Phys.* **62**, 4745 (1975).]

oxygen 2p lone-pair orbital, perpendicular to the H_2O plane. The strongest $(0, 0, 0) \leftarrow (0, 0, 0)$ vibrational component of this electronic band has been obtained with sufficiently high resolution to exhibit rotational broadening.[198] While a rotational analysis has not been performed to establish the ionization potential spanning rotationless states, comparison with a Rydberg state of H_2O having a similar band shape[199] indicates that the peak in the asymmetric band shape is very close to the true adiabatic value. This is given as 12.61(6) eV by Brundle and Turner,[200] 12.614 eV by Asbrink and Rabalais[198] and 12.615(1) by Karlsson *et al.*[197]

Ejection of an electron from the second outermost orbital gives rise to a long progression in the bending mode. Karlsson *et al.* conclude that the $^2A_1$ state formed is linear or nearly linear, and hence has the electron removed from an H—H bonding orbital. An emission spectrum corresponding to a transition from this $^2A_1$ state to the ground state of $H_2O^+(^2B_1)$ has recently been observed[9a,96] and analyzed.[201]

Ejection of an electron from the third outermost orbital gives rise to a complex band whose fine structure apparently involves excitation of more than one vibrational progression.

In addition, the onset of dissociation into $OH^+ + H$ occurs in this region, and may partially influence the shape of this band if rapid predissociation occurs. This $1b_2$ orbital is therefore the primary O—H bonding orbital. Table 5 summarizes the experimentally observed energies required to eject electrons from each of these orbitals.

TABLE 5 *Adiabatic and Vertical Ionization Energies for Single Electron Ejection from H_2O*

Orbital	Adiabatic IP (eV)	Vertical IP (eV)	Ref.
$1b_1$	12.615	12.615	(197)
$3a_1$	13.839	14.745	(197)
$1b_2$	17.189(?)	18.51	(197)
$2a_1$	~30	32.2	(43)
		32.4	(195)
$1a_1$		539.7	(43)

In Fig. 46, we have assembled graphically the variation of partial cross section with photon energy for formation of the corresponding H_2O^+ states. The points on this graph have been deduced by taking photoelectron spectroscopic branching ratios from several sources and combining them with our previously deduced photoionization cross sections at the appropriate photon energies. Such cross sections are crude, since the branching ratios have generally been obtained at 90° with respect to the photon direction, and hence the influence of photoelectron angular distributions has been ignored. Also, the transmission of the various electron energy analyzers is rarely well known. Nevertheless, these provisional values do have some semiquantitative value. In the energy region below 21 eV, the branching ratios of Blake and Carver[154] have been used. Between 21 and 60 eV, the branching ratios (but not the absolute cross sections) of Tan *et al.*[202] were employed. Also shown in this region is earlier work by the electron energy loss—pseudophoto-absorption technique.[202a] A factor 2–3 discrepancy in the branching fraction for $(2a)^{-1}$ is exhibited by these two sets of data, which is not clearly explained by Tan *et al.*, although angular distributions are presumed to play a role. At 132.3 eV, the data by Banna and Shirley[195] were incorporated, and at 151.4 eV the values of Allison and Cavell[134] were taken. Finally, at 1253.6 eV the spectrum reported by Siegbahn *et al.*[203] was utilized. Similar, but higher resolution data at 1486.7 eV are reported by Siegbahn.[204]

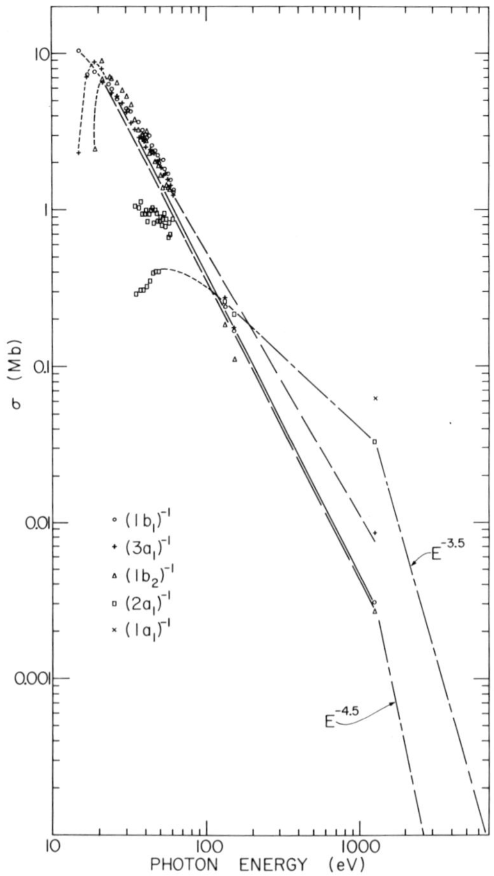

Fig. 46 Partial cross sections for formation of various states of H_2O^+ by photoabsorption.

It is immediately apparent from Fig. 46 that the states formed by electron emission from the s-like orbitals $1a_1$ and $2a_1$ decline less rapidly with photon energy than the states formed from the p-like orbitals $1b_2$, $3a_1$, and $1b_1$. However, the crude slopes inferred at ~1 keV differ significantly from the asymptotic behavior predicted from theory,[205] $E^{-l-7/2}$.

b. Photoionization Mass Spectrometry. From the analysis of the photoelectron spectrum, we can anticipate that the photoionization onset will be sharp, since the $^2B_1(0,0,0) \leftarrow {}^1A_1(0,0,0)$ transition is dominant. The next most abundant vibrational band is $^2B_1(1,0,0)$—excitation of the symmetric stretch v_1—with weaker excitation of v_2 and no evidence for excitation of v_3. Hence, we can expect a weak Rydberg series converging to $^2B_1$ in the $v_1 = 1$

state and a still weaker one to $v_2 = 1$. The higher members of these Rydberg series can vibrationally autoionize, which may give rise to some near-threshold structure. However, the ionization yield in this near-threshold region is about 0.35 (see Fig. 18 of Chapter V). This can suppress the autoionization structure and enhance the direct ionization "step structure."

The gross features of step structure are apparent in the photoionization mass spectrum of H_2O^+ (Fig. 47) between threshold and ~970 Å, but autoionization structure is superposed upon this step in a manner similar to that observed in NO^+.

The long vibrational progression of the second photoelectron band ($^2A_1$ state) should have Rydberg precursors which have equally long vibrational progressions, and these should manifest themselves as autoionization struc-structure in the photoionization mass spectrum converging to ionization potentials in the 14–16 eV region. Such structure is apparent in the photoionization curve of Fig. 47 from ~950 to 800 Å.

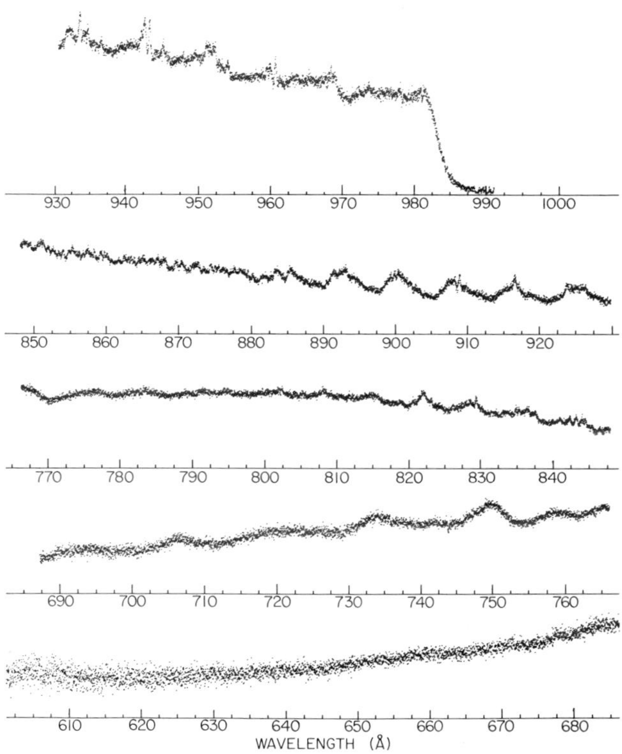

Fig. 47 Mass selected photoion yield curve of H_2O^+ from threshold to 600 Å, obtained with wavelength resolution (FWHM) of 0.14 Å. [From J. H. D. Eland and J. Berkowitz, to be published.]

There is also structure corresponding to Rydberg's converging to the third ionized state $^2B_2$ appearing as broad, irregular features between ~790 and 690 Å. Note that the appearance of the $^2B_2$ band in the photoelectron spectrum is also irregular.

The thresholds for three fragmentation processes fall in the energy range of this third band:

$$H_2O + h\nu \rightarrow OH^+ + H + \bar{e} \qquad 18.11 \text{ eV},$$

$$\rightarrow OH + H^+ + \bar{e} \qquad 18.72 \text{ eV},$$

$$\rightarrow H_2 + O^+ + \bar{e} \qquad 18.65 \text{ eV}.$$

Of these, the OH^+ is quite prominent, the H^+ weaker, and the O^+ barely detectable. There has been some dispute about the relative importance of the latter two processes, in particular, and the results of several groups are summarized in Table 6. There is general agreement that the OH^+/H_2O^+ relative intensity is 0.4–0.5. The H^+/H_2O^+ ratio ranges over a factor four, with the mean value of 0.12–0.13 appearing most convincing at the present time. The high O^+ value reported by Dibeler *et al.*[207] is almost certainly due to an O_2 impurity. The sensitivity of the experiment of Kronebusch and Berkowitz was greater than that of Cairns *et al.*[208] and a correction was made for the residual O_2, so that this weak O^+ intensity is felt to be real.

TABLE 6 *Relative Intensities of Mass Peaks in Photoionization of H_2O at 21.2 eV (Relative to $H_2O^+ = 1.0$)*

Ref.	OH^+	H^+	O^+
(207)	0.4	[a]	0.05
(208)	~0.5	~0.045	<0.001
(147)	0.4	0.07	0.001
(206)	[a]	0.12	[a]
(202)	0.45	0.13	0
(284g)	[a]	0.18	[a]

[a] Not reported.

The experiments of Kronebusch and Berkowitz extend to 40.8 eV. The mass spectral branching ratios obtained with the line sources in this experiment are plotted in Fig. 48, together with a concocted 40.8 eV photoelectron spectrum based upon the partial cross section data of Fig. 46. From this composite graph, we can conclude that the $(1b_2)^{-1}$ state dissociates predominantly to $OH^+ + H$, and to a lesser extent, $H^+ + OH$, with a very weak but measurable $O^+ + H_2$ manifesting itself. This pattern changes only

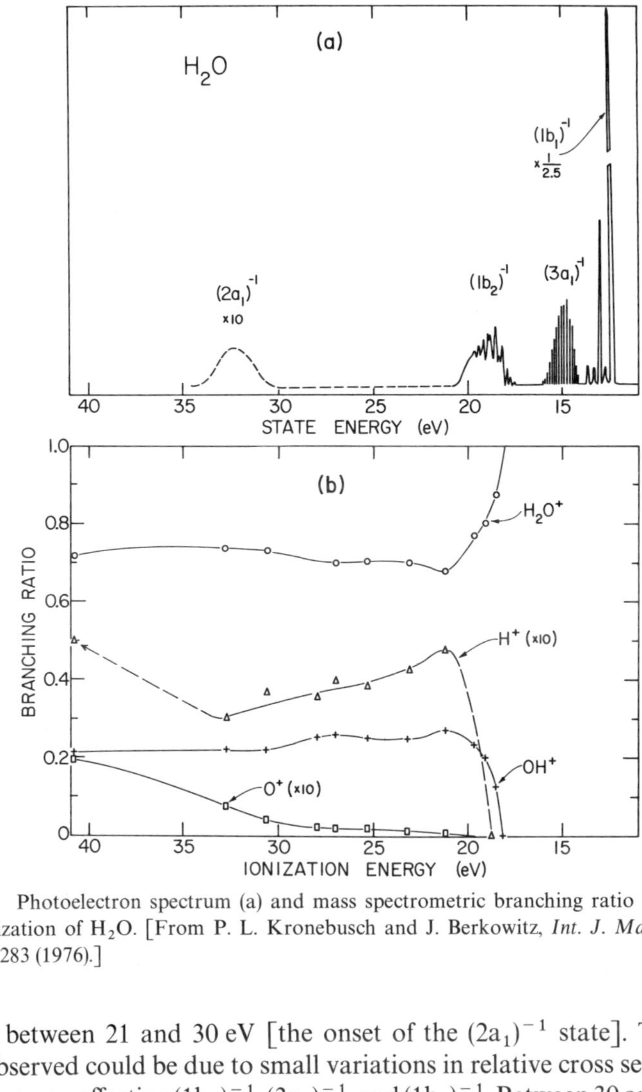

Fig. 48 Photoelectron spectrum (a) and mass spectrometric branching ratio (b) from the photoionization of H_2O. [From P. L. Kronebusch and J. Berkowitz, *Int. J. Mass Spec. Ion Phys.* **22**, 283 (1976).]

slightly between 21 and 30 eV [the onset of the $(2a_1)^{-1}$ state]. The slight trend observed could be due to small variations in relative cross section with photon energy affecting $(1b_1)^{-1}$, $(3a_1)^{-1}$, and $(1b_2)^{-1}$. Between 30 and 40.8 eV, the O^+ intensity increases by an order of magnitude, reaching $\sim 2\%$ of the total intensity at 40.8 eV. The H^+ intensity is also shown increasing, although this measurement is not certain. Tan *et al.*,[202] employing coincidence measurements between electrons of selected energy loss and ions, have recently obtained energy dependent mass spectra similar to the results described above. They find that H_2O^+ produced in the $(2a)^{-1}$ state dissociates completely, producing H^+ and O^+ in the approximate ratio 3:1. Within the uncertainty of all the measurements, we can conclude that $(2a_1)^{-1}$ decomposes to $O^+ + 2H$ and to $H^+ + O + H$. The threshold for these latter processes (they are almost isoenergetic) is 23.13 eV, and it is largely on this basis that the conclusion is drawn that the decomposition from $(2a_1)^{-1}$ occurs into three entities, rather than $O^+ + H_2$ or $H^+ + OH$.

Although no x-ray mass spectrum of H_2O is available, there have been studies of x-ray emission[209] and Auger spectra[210] resulting from the formation of $(1a_1)^{-1}$ in H_2O. The Auger process filling the K-shell of oxygen should occur in $\sim 5 \times 10^{-15}$ sec[211]; the competitive x-ray fluorescence is about three orders of magnitude slower.[211] Hence, the predominant mechanism following K-shell ejection is the Auger process, which forms multiply charged H_2O. The observations of Carlson and co-workers on other molecules[212] leads to the plausible prediction that a Coulombic explosion should ensue, producing $2H^+ + O^+$.

We return briefly to the $(1b_2)^{-1}$ decomposition, in order to examine a recent, detailed photoelectron–photoion coincidence study by Eland.[206] His results can be briefly summarized. The breakdown pattern he obtained is reproduced in Fig. 49. He concludes that both OD^+ and D^+ ions are formed as soon as the energy available exceeds the corresponding threshold. (He used deuterated water for his experiment.) The OD^+ is formed with substantially more kinetic energy than the D^+. The OD^+ kinetic energy always seems to be about 0.3 of the available energy, whereas the D^+ kinetic energy may not be a constant proportion of the total energy.

On this basis, he concludes that the process forming OD^+ acts like a statistical decomposition, whereas the process forming D^+ may be a predissociation.

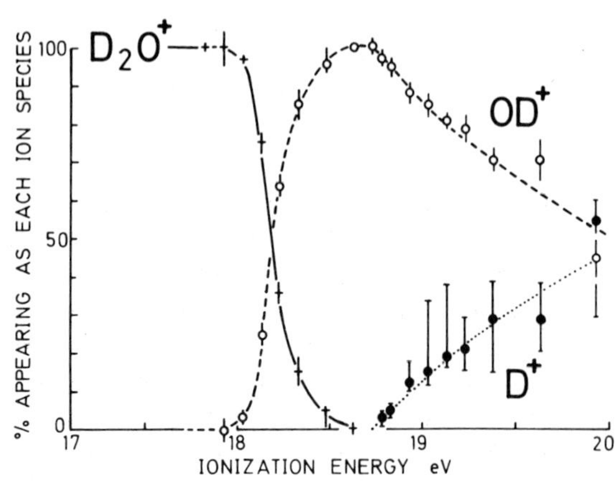

Fig. 49 Breakdown diagram of D_2O^+ in the region of the $\tilde{B}^2B_2$ state obtained by the photoion–photoelectron coincidence method. [From J. H. D. Eland, *Chem. Phys.* **11**, 41 (1975).]

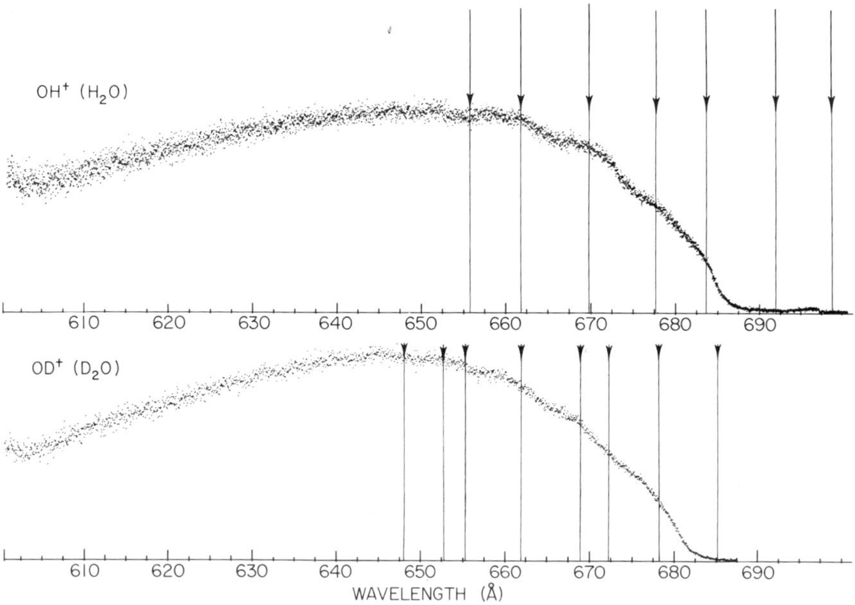

Fig. 50 Photoion yield curves for OH^+ from H_2O and OD^+ from D_2O obtained with a wavelength resolution of 0.14 Å. [From J. H. D. Eland and J. Berkowitz, to be published.]

A theory has recently been suggested by Lorquet and Lorquet[1b] for the decomposition of this third state of H_2O^+, which is based on a predissociation model. This theory predicts a strong isotope effect in the formation of H^+ from H_2O versus D^+ from D_2O, favoring the lighter isotope. No such effect was observed by Konebusch and Berkowitz, since both H^+ and D^+ were found to be about 5% of total ionization. The Lorquets' model predicts a lower yield, which also disagrees with Eland's result.

Figure 50 displays the photoion yield curves of $OH^+(H_2O)$ and $OD^+(D_2O)$. In the corresponding threshold regions, both show broad undulations or rounded steps. These undulations are not noticeable in the parent ions. On each curve, the positions of the peaks of the vibrations observed in the photoelectron spectrum[197] are marked. The peaks in the photoelectron spectrum appear to occur after (i.e., at a higher energy than) the corresponding increases in OH^+ and OD^+. Although the individual vibrations appear to play a role, the mechanism is not yet clear.

2. CO_2

a. Photoelectron Spectroscopy. The occupied molecular orbitals of CO_2 are

$$(1\sigma_g)^2(1\sigma_u)^2(2\sigma_g)^2(3\sigma_g)^2(2\sigma_u)^2(4\sigma_g)^2(3\sigma_u)^2(1\pi_u)^4(1\pi_g)^4.$$

The $1\sigma_g$ and $1\sigma_u$ orbitals are the plus and minus combinations of O 1s orbitals, essentially the K-shell of atomic oxygen. Similarly, $2\sigma_g$ represents the slightly perturbed K-shell of atomic carbon. Ab initio calculations[213,214] indicate that $3\sigma_g$ and $2\sigma_u$ are close in binding energy, the $3\sigma_g$ being slightly more bound. The major atomic orbital eigenvectors for the $3\sigma_g$ orbital are 2s orbitals on the oxygens and carbon, while for $2\sigma_u$ they are oxygen 2s and carbon $2p_z$.[213] The overlap is strong and constructive for both of these orbitals, as is also evidenced by the large increase in orbital binding energy with decrease in internuclear distance.[214] The photoelectron spectrum in the 37 eV region[215,216] shows a broad peak encompassing both $(3\sigma_g)^{-1}$ and $(2\sigma_u)^{-1}$, in accord with the view that electrons have been removed from strongly bonding orbitals. Correlation effects (i.e., a breakdown of the single particle model of photoionization) very likely play an important role here in determining the breadth of these bands.[216a]

The remaining (or uppermost) four orbitals have been studied by a variety of techniques. Rydberg series leading to each of these limits have been observed,[217] providing accurate adiabatic ionization potentials. These values are in good agreement with those obtained from the 584 Å photoelectron spectra.[218] In addition, emission spectra have been reported[216,217,219] for the transitions

$$\tilde{B}[(3\sigma_u)^{-1}] \quad \text{to} \quad \tilde{X}[(1\pi_g)^{-1}] \quad \text{and} \quad \tilde{A}[(1\pi_u)^{-1}] \quad \text{to} \quad \tilde{X} \text{ of } CO_2^+,$$

further verifying the energy separations and symmetry of these states. No emission spectra have been observed to emanate from the $\tilde{C}[(4\sigma_g)^{-1}]$ state; this behavior is strongly linked to the dissociative process involving this state, to be discussed below.

The $4\sigma_g$, $3\sigma_u$, and $1\pi_g$ orbitals are relatively nonbonding. This conclusion follows not only from the rather narrow peaks in the 584 Å photoelectron spectrum[218] corresponding to ejection of electrons from these orbitals, but also from the weak dependence of these orbital energies on internuclear distance.[214] The major bonding orbital in the valence region is $1\pi_u$, which exhibits a strong increase in orbital binding energy with decreasing internuclear distance and a broad photoelectron band, encompassing at least seven vibrational levels. Table 7 summarizes the experimentally observed energies required to eject electrons from each of these nine orbitals.

The variation of the partial cross sections for formation of each of these ionic states with photon energy has been assembled from several sources. Data for the region from ionization threshold (13.77 eV) to 21.2 eV have been taken primarily from Bahr *et al.*[172] At 21.2 eV the analysis of Samson *et al.*[220] is clearly preferable. The arguments presented in that paper in favor of a total absorption cross section at 584 Å of ~33 Mb instead of the much

TABLE 7 *Adiabatic and Vertical Ionization Potentials for Single Electron Ejection in CO_2*

	Adiabatic (eV)	Vertical (eV)	Ref.
$\tilde{X}\ ^2\Pi_g[(1\pi_g)^{-1}]$	13.77	13.78	(130, 219a, 226, 227)
$\tilde{A}\ ^2\Pi_u[(1\pi_u)^{-1}]$	17.31	17.36	(130, 219a, 226)
$\tilde{B}\ ^2\Sigma_u^+[(3\sigma_u)^{-1}]$	18.07_6	18.08	(130, 219a, 226, 227)
$\tilde{C}\ ^2\Sigma_g^+[(4\sigma_g)^{-1}]$	19.39	19.40	(130, 219a, 226, 227)
$(2\sigma_u)^{-1} + (3\sigma_g)^{-1}$		37.6	(215)
$\mathrm{C\ 1s}[(2\sigma_g)^{-1}]$		297.5	(215)
$\mathrm{O\ 1s}[(1\sigma_u)^{-1} + (1\sigma_g)^{-1}]$		541.1	(215)

lower values of Cook *et al.* and Nakata *et al.* are reinforced by the sum rule rule calculations presented in Chapter V. Recently obtained photoelectron spectra at 304 Å,[216] when combined with the total absorption cross section at 304 Å (see Section C.2), provide values of partial cross sections at 40.8 eV. The gap in data between 21.2 and 40.8 eV has been filled, in principle, by recent synchrotron measurements of Gustafsson *et al.*[220a] However, this information cannot be readily utilized. The spectral resolution was insufficient to separate the $\tilde{A}$ and $\tilde{B}$ states. In addition, several higher energy states become populated in this interval, which these authors note but do not incorporate in their branching ratios. The analysis to be presented (vide infra) will be shown to be in better agreement with Gustafsson *et al.*[220a] than with Samson and Gardner[216] regarding the relative abundance of the 37–38 eV feature (presumably $(2\sigma_u)^{-1}$ and $(3\sigma_g)^{-1}$). For the reader specifically interested in the partial cross sections between 21 and 41 eV, the branching ratios of the $\tilde{X}$ and $\tilde{C}$ states given by Gustafsson *et al.* are probably the best currently available, although they have not been used in constructing Fig. 51.

The partial cross sections at the Mg K_α wavelength have been pieced together from the total cross section at this wavelength (Chapter V); the relative cross sections of the six uppermost occupied orbitals are given by Allan *et al.*,[215] while the O(1s) intensity is obtained from a separate figure in that paper. The ratio of $(2\sigma_g)^{-1}$ to the sum of $(1\sigma_g)^{-1}$ and $(1\sigma_u)^{-1}$ can be estimated from figures in the article by Allan and Siegbahn[221] or alternatively from atomic absorption cross sections of Henke[222] and colleagues $[\sigma(\mathrm{C}) = 1/2.88\sigma(\mathrm{O})]$ multiplied by the oxygen abundance. The two estimates are in rough agreement. The overall variation of partial cross sections with energy is plotted in Fig. 51. The slopes of the $(1\sigma_g)^{-1}$, $(1\sigma_u)^{-1}$, and $(2\sigma_g)^{-1}$ partial cross sections are estimated from the rise of the corresponding x-ray edge and the partial cross section at 1253.6 eV.

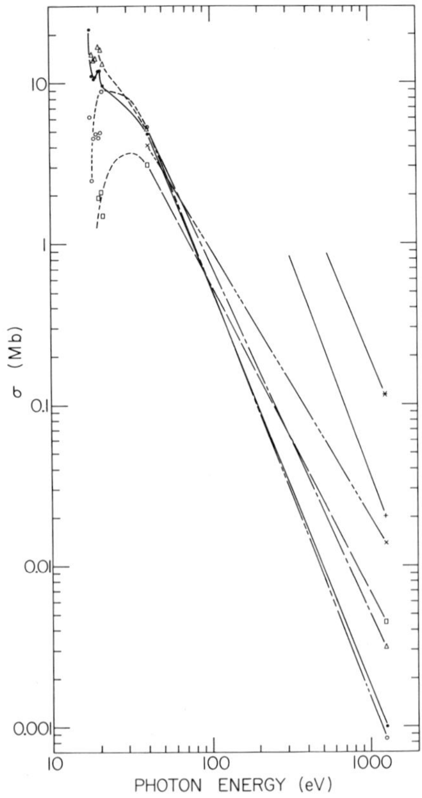

Fig. 51 Partial cross sections for formation of various states of $CO_2{}^+$ by photoabsorption. ● $X^2\Pi_g$, ○ $A^2\Pi_u$, △ $B^2\Sigma_u^-$, □ $C^2\Sigma_g^+$, × $(2\sigma_u)^{-1} + (3\sigma_g)^{-1}$, + $(2\sigma_g)^{-1}$, ∗ $(1\sigma_g)^{-1} + (1\sigma_u)^{-1}$.

However, we forewarn the reader that the branching ratios of Gardner and Samson at 304 Å will be questioned, particularly as regards the importance of $(2\sigma_u)^{-1}$ and $(3\sigma_g)^{-1}$.

b. Photoionization Mass Spectrometry. The photoelectron spectrum[218,223] of the first electronic band displays a strong $(0,0,0) \rightarrow (0,0,0)$ vibrational line and successively weaker lines displaced by 0.16 eV and interpreted as $(0,0,0) \rightarrow (1,0,0)$ and $(0,0,0) \rightarrow (2,0,0)$. Each of the vibrational lines is broader than expected, the broadening being attributed to the $^2\Pi_{g,3/2} - {}^2\Pi_{g,1/2}$ splitting of 19.8 meV deduced from the $CO_2{}^+$ emission spectrum.[224,225] The ionization yield (Fig. 20 in Chapter V) rises rapidly from ~0.26 at threshold to ~0.9 over an energy range of about 0.32 eV, corresponding to two units of vibrational quanta in ν_1. The photoionization mass spectrum (see Fig. 52) displays distinct autoionization features from

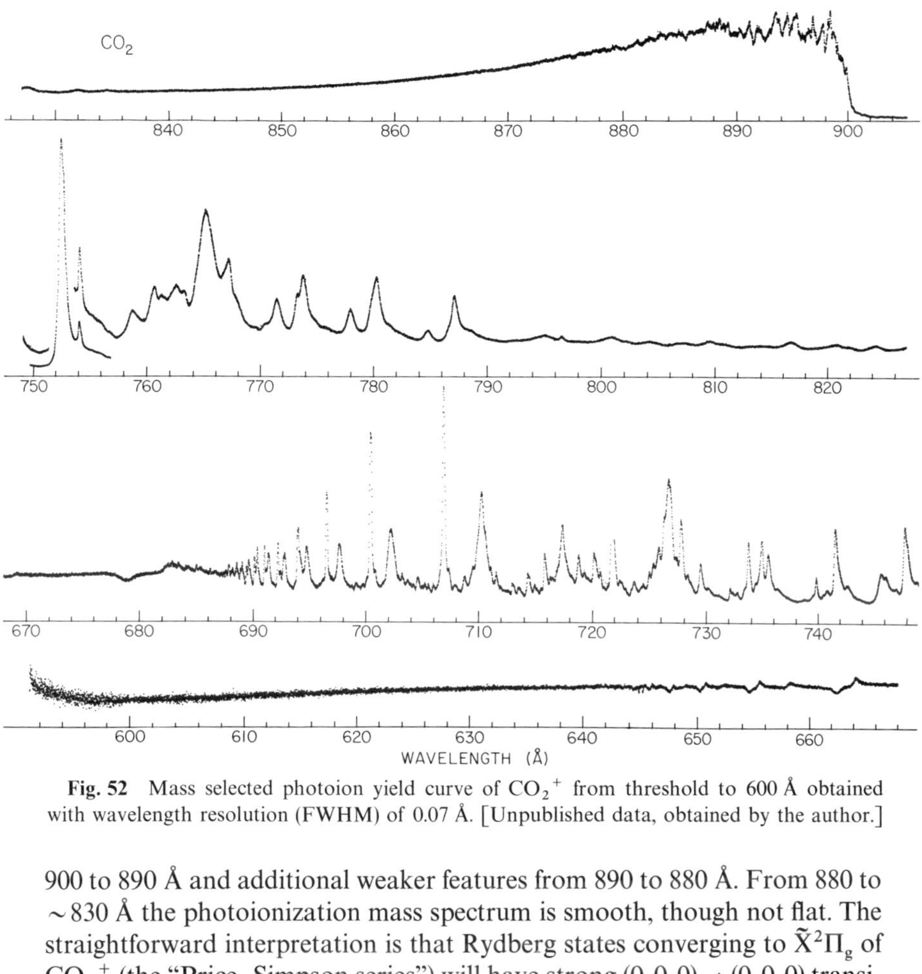

Fig. 52 Mass selected photoion yield curve of CO_2^+ from threshold to 600 Å obtained with wavelength resolution (FWHM) of 0.07 Å. [Unpublished data, obtained by the author.]

900 to 890 Å and additional weaker features from 890 to 880 Å. From 880 to ~830 Å the photoionization mass spectrum is smooth, though not flat. The straightforward interpretation is that Rydberg states converging to $\tilde{X}^2\Pi_g$ of CO_2^+ (the "Price–Simpson series") will have strong $(0,0,0) \rightarrow (0,0,0)$ transitions, the $0 \rightarrow 1$ transition in ν_1 being about 20–30% of the $0 \rightarrow 0$, with the $0 \rightarrow 2$ reduced by another factor of approximately four. The vibrationally excited Rydbergs having a total energy above 13.77 eV can then autoionize by converting vibrational energy into the energy necessary for the Rydberg electron to escape. This process should become negligible, due to Franck–Condon factors in the initial excitation, at an energy corresponding to two vibrational quanta of ν_1 above 13.77 eV or ~14.09 eV ≡ 880 Å, which agrees with the photoionization mass spectrometric observation. Between 830 and 687 Å there exists a complex of autoionization peaks which are most readily attributable to Rydberg series converging to $\tilde{A}^2\Pi_u$ and $\tilde{B}^2\Sigma_u^+$ of CO_2^+. The ionization yield in this region oscillates between ~0.6 and 0.9, evidently

depending upon the relative rates of decay of the quasi-discrete Rydberg states to the ionization continuum or a dissociative continuum. The Rydberg series converging to the $\tilde{B}^2\Sigma_u^+$ state are readily discernible in the photoionization mass spectrum. From the photoelectron spectrum we can expect that the $(0,0,0) \rightarrow (0,0,0)$ transition will have a much larger Franck–Condon factor than other competing vibrational transitions. In fact, the dominant peak in the spectrum is at 752 Å, and corresponds to the $n = 3$ Rydberg state of this "Henning series" (labeled "sharp" by Tanaka and Ogawa[226]). Other prominent members of this sharp series can be seen at 717.4, 706.9, 700.4, 696.5 Å, and still shorter wavelengths approaching the series limit at 685.87 Å = 18.076 eV. These members all correspond to $v' = 0$, and hence the autoionization mechanism must be electronic rather than vibrational. There is also a "diffuse series" converging to this series limit, for which observable members can be seen at 764.7, 726.3, 710.2, 702.2, 697.6 Å, and then merging with the sharp series to the convergence limit at 685.87 Å.

The "new series" of Tanaka and Ogawa, which converges to the $^2\Pi_u$ limits of the $\tilde{A}$ state, can be expected to fill in most of the remaining autoionization peaks in the spectrum. The corresponding photoelectron spectrum of the $^2\Pi_u$ band encompasses at least six vibrational members of significant intensity. Each Rydberg member converging to the $^2\Pi_u$ limit would then have its own complement of vibrational members, with roughly the same Franck–Condon pattern. One can readily assign the remaining prominent members of the spectrum to this series. For example, the peaks at 787, 780.7, and 773.5 Å have been assigned by Tanaka and Ogawa to $n = 3$, $v' = 0, 1, 2$ of the series converging to $^2\Pi_{u,1/2}$; those at 754, 747.5, and 741.3 Å to $n = 4$, $v' = 0, 1, 2$ of the same series. Here again, the presence of $v' = 0$ in the photoionization mass spectrum at a relative intensity commensurate with its Franck–Condon factor implies that the major autoionization mechanism is configuration interaction rather than a mixing of nuclear and electronic terms (Born–Oppenheimer breakdown).

Up to this time, we have not discussed the possible fragments that can result upon photoionization of CO_2. In Table 8, the adiabatic ionization potentials of CO_2 that are prominent in the various photoelectron spectra (i.e., the one-electron transitions) are shown together with various dissociation limits derived from thermochemical data. The first possible photodissociation producing ions is the ion-pair ($CO^+ + O^-$) with a threshold at 18.00_1 eV. McCulloh[227] has looked for this process and has not observed it, which implies an intensity for this channel 0.1% of that producing CO_2^+ at this photon energy. (The products $O^+ + CO^-$ have also not been observed.)

The first fragmentation process seen is that corresponding to $O^+ + CO$. Its onset at 19.07 eV implies that both fragments are in their ground states. The intensity of the O^+ peak is weak ($\sim$0.2% of the major CO_2^+ channel)

TABLE 8 *Threshold Energies for Fragmentation and Fragmentation Ratios for CO_2*

λ (Å)	$h\nu$ (eV)	CO_2^+	CO^+	O^+	C^+	Remarks
740	16.76	1	—	—	—	IP = 13.788 eV
						$O^+ + CO$ (19.067 eV)
						$CO^+ + O$ (19.466 eV)
584	21.10	1	0.025	0.045	—	
			(0.018)[a]	(0.039)[a]		$C^+ + O_2$ (22.685 eV)
537	23.10	1	0.039	0.055	—	
490	25.30	1	0.075	0.088	0.001	
460	26.95	1	0.105	0.104	0.001	
446	27.80	1	0.117	0.112	0.0015	
						$C^+ + 2O$ (27.80 eV)
405	30.60	1	0.123	0.146	0.009	
379	32.70	1	0.09	0.130	0.027	
304	40.80	1	0.08	0.140	0.056	
				(0.14)[a]	(0.22)[a]	

[a] Fryar and Browning.[230]

and structured, indicating an autoionization process (see Fig. 53). At 19.39 eV, which corresponds precisely to the threshold for formation of the $\tilde{C}^2\Sigma_g^+$ state (Table 7), the O^+ intensity dramatically increases by an order of magnitude, while the CO_2^+ intensity shows a very slight decrease. An obvious interpretation is that the formation of CO_2^+ in the $\tilde{C}^2\Sigma_g^+$ state leads predominantly to the products $O^+(^4S) + CO(^1\Sigma)$. Adiabatic correlation rules would require these products to form a quartet potential surface of CO_2^+. Hence, we are drawn to the conclusion that the $\tilde{C}^2\Sigma_g^+$ state is crossed by a quartet state, and predissociation results.

Eland[228] has studied this process in more detail. Upon irradiation of CO_2 with 584 Å photons, the photoelectrons produced were energy analyzed and the ions were mass and energy analyzed. The photoelectrons and photoions were then investigated with a coincidence circuit, in order to deduce which ions resulted from particular states of CO_2^+. He confirmed that $\tilde{X}^2\Pi_g$, $\tilde{A}^2\Pi_u$, and $\tilde{B}^2\Sigma_u^+$ states lead only to parent ions.

The detailed apportionment of the $\tilde{C}$ state to product ions is somewhat complicated. The threshold for formation of $CO^+ + O$ lies ~0.08 eV above (0, 0, 0) of the $\tilde{C}$ state, which means that the (1, 0, 0) and (0, 0, 2) levels (also formed in the photoionization process) provide sufficient energy for this dissociative channel. The relative probability of (1, 0, 0) and (0, 0, 2) to (0, 0, 0) is less than 0.1, but other vibrational excitations recently found[229] are able to rationalize the observed $CO^+:O^+$ ratio of about 0.3. The spectrum of Fig. 53 also suggests that the vibrationally excited levels of the $\tilde{C}$ state lead to $CO^+ + O$. Presumably, this channel competes favorably with the $O^+ + CO$

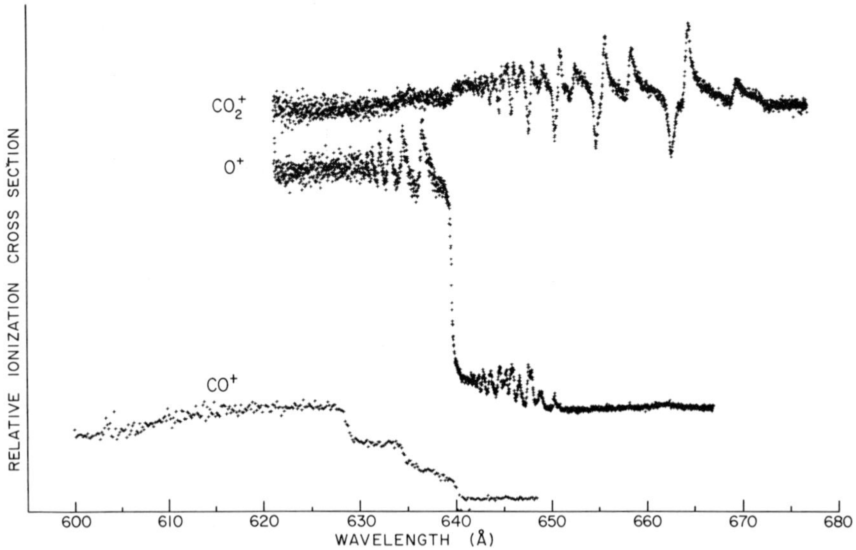

Fig. 53 Relative photoion yield curves of $CO_2{}^+$, O^+, and CO^+ in the vicinity of the dissociative ionization thresholds, which is also in the region of formation of the $\tilde{C}\ {}^2\Sigma_g{}^+$ state of $CO_2{}^+$. Each curve is presented in arbitrary units and the relative intensities of the respective ions are not meaningful. [From J. H. D. Eland and J. Berkowitz, *J. Chem. Phys.* **67**, 2782 (1977).]

channel since $CO^+(^2\Sigma^+) + O(^3P)$ can give rise to a doublet state of $CO_2{}^+$, which could predissociate the $\tilde{C}$ state with higher probability than a quartet state.

For the (0, 0, 0) level of the $\tilde{C}$ state, this competition is forbidden on energetic grounds. Eland has found that the O^+ formed by this predissociation has a kinetic energy distribution implying two groupings corresponding to $O^+ + CO$ ($v = 0$) and $O^+ + CO$ ($v = 1$), the latter comprising 85% of this predissociation mode.

In summary, the $\tilde{C}$ state at (0, 0, 0) leads to $O^+ + CO$, with higher vibrational levels of the $\tilde{C}$ state yielding $CO^+ + O$ about a factor of three lower in intensity. Samson *et al.*[220] find that the $\tilde{C}$ state is produced in 4.5% of the ionizations at 584 Å; Eland has measured the $O^+:CO_2{}^+$ ratio at 584 Å to be 0.033. McCulloh finds 0.018 (at 625 Å) and Fryar and Browning[230] obtain 0.039, while Kronebusch and Berkowitz[147] observe 0.045. Taking into account the weakness of this partial cross section and kinetic energy effects in the transmission of O^+, the agreement is satisfactory, except for McCulloh's value. The earlier and somewhat cruder measurements of Weissler *et al.*[191] are in harmony with these conclusions.

The fact that no metastable ion is observed corresponding to the process

$$(CO_2{}^+)^* \rightarrow O^+ + CO$$

implies that the predissociation of the $\tilde{C}$ state occurs in less than 10^{-6} sec. The transitions $\tilde{C} \rightarrow \tilde{B}$ and $\tilde{C} \rightarrow \tilde{A}$ are optically allowed, but have not been observed. Also, the photoexcitation of the $\tilde{B} \rightarrow \tilde{X}$ fluorescence does not show an enhancement at the photon energy corresponding to excitation of the $\tilde{C}$ state, indicating that the cascade process $\tilde{C} \rightarrow \tilde{B} \rightarrow \tilde{X}$ is a negligible contributor to the $\tilde{B} \rightarrow \tilde{X}$ fluorescence.[231]

Since the radiative lifetime of the $\tilde{C} \rightarrow \tilde{B}$ and $\tilde{C} \rightarrow \tilde{A}$ transitions can be expected to be $\sim 10^{-7}$ sec, we can conclude that predissociation of the $\tilde{C}$ state occurs in less than 10^{-7} sec. On the other hand, the 584 Å photoelectron spectrum[218] does not indicate any measurable broadening of the $\tilde{C}(0, 0, 0)$ peak relative to the $\tilde{B}$ or $\tilde{A}$ peaks. Since a 10 meV broadening would likely be detectable, we can infer a lower limit to the predissociative lifetime of $\sim 10^{-13}$ sec. A possible absorption spectrum corresponding to the transitions $\tilde{A} \rightarrow \tilde{C}$ or $\tilde{B} \rightarrow \tilde{C}$ could be revealing in establishing a narrower bracketing of the rate constant for the predissociation. As an aside, we note that at 584 Å the $\tilde{A}$ and $\tilde{B}$ states of CO_2^+ are formed[216] in the ratio 0.65 ± 0.05. At this same incident wavelength, the $\tilde{A} \rightarrow \tilde{X}$ and $\tilde{B} \rightarrow \tilde{X}$ emissions are observed with an intensity ratio of 2.7:1[219] or 3:1.[216] Perturbations have been observed in the emission spectra, most recently by Gauyacq *et al.*,[232] and the indication is that a radiationless transition is occurring from $\tilde{B} \rightarrow \tilde{A}$ between formation of the ions ($\sim 10^{-15}$ sec) and the observation of fluorescence ($\sim 10^{-7}$ sec). This may be a chance example in a relatively small molecular ion of the radiationless transitions assumed in theories of unimolecular decay of more complex species. Leach *et al.*[232a] present a recent summary of this problem, with additional data.

Proceeding higher in ionization energy, we note that the next prominent ionization process corresponds to ejection of electrons from the $3\sigma_g$ and $2\sigma_u$ orbitals at 37–39 eV. It has already been noted that the ab initio calculations describe these orbitals as strongly bonding the oxygens to carbon, and the 304 Å and MgK_α photoelectron spectra are broad, characteristic of dissociating states. The energies of these states are well above the thresholds for $C^+ + O_2$, $C + O_2^+$, $C^+ + O + O$, and $C + O^+ + O$.

We[147] have recently completed experiments on the photoionization mass spectrum of CO_2 between 21.2 and 40.8 eV, using a hollow cathode lamp which can generate eight lines in this photon energy range. The results of these experiments, summarized in Table 8, clearly indicate that the major decomposition process subsequent to the ejection of electrons from $3\sigma_g$ and $2\sigma_u$ is $(CO_2^+)^* \rightarrow C^+ + O + O$. Some fluorescence studies[216] suggest that

$$(CO_2^+)^* \rightarrow CO^+ + \begin{Bmatrix} O^*(^3S) \\ O(^3P) \end{Bmatrix}$$

may be a significant contributory channel. The $O^+:CO_2{}^+$ intensity is not very different from that at 584 Å, suggesting that $(3\sigma_g)^{-1}$ and $(2\sigma_u)^{-1}$ do not have O^+ as a major channel in their decomposition modes.

Fryar and Browning[230] have measured $CO^+/CO_2{}^+ = 0.14$ and $O^+/CO_2{}^+ = 0.22$ at 304 Å. The corresponding ratios of Kronebusch and Berkowitz[147] (Table 8) are somewhat smaller, presumably because of lesser efficiency of collection of fragments with higher kinetic energy, but these authors have observed the C^+ fragment alluded to above. Even if their collection efficiency were as low as $\frac{1}{3}$ (see Kronebusch and Berkowitz[147] for this choice of figure) it would imply a fractional ionization to $(3\sigma_g)^{-1}$ and $(2\sigma_u)^{-1}$ no larger than 0.18, in contrast to 0.615 given by Gardner and Samson. Hence, Fig. 51 has been adjusted at 40.8 eV using the 0.18 fraction for the higher energy states, rather than the Gardner–Samson value. We note parenthetically that Stockdale *et al.*[233] have reported C^+ from CO_2 by electron impact, with a threshold at 29.8 eV.

A process that has received relatively little attention heretofore, photoionization to a stable, doubly charged molecular ion, has recently been reported by Samson *et al.*[233a] for CO_2^{2+}. The ratio $CO_2^{2+}/CO_2{}^+$ rises linearly from threshold (36.2 eV) to about 65 eV, where it levels off at a value of ~0.02 to the limit of measurement, ~108 eV.

Carlson and Krause[150] report mass spectra from CO_2 upon x-ray irradiation at 930 eV but not at 280 eV. From Fig. 51, the partial cross sections at 930 eV are

$(1\sigma_g)^{-1} + (1\sigma_u)^{-1}$	: 0.235 Mb
$(2\sigma_g)^{-1}$	: 0.045
$(3\sigma_g)^{-1} + (2\sigma_u)^{-1}$	: 0.0227
$(4\sigma_g)^{-1}$	: 0.0079
$(3\sigma_u)^{-1}$	: 0.0059
$(1\pi_u)^{-1}$	: 0.00185
$(1\pi_g)^{-1}$	: 0.00215.

If we assume, as before, that the K-shell ionizations are followed by Auger processes and then a "Coulomb explosion" leading to $C^+ + O^+ + O^+$, then the sum $(1\sigma_g)^{-1} + (1\sigma_u)^{-1} + (2\sigma_g)^{-1}$ can be assigned to C^+, and twice this sum to O^+. From the hints provided by the data of Table 8, we shall divide the contribution of $(3\sigma_g)^{-1} + (2\sigma_u)^{-1}$ equally between C^+ and O^+. The $\tilde{C}$ state, $(4\sigma_g)^{-1}$, has been thoroughly discussed as a source of O^+ and the lower energy states can only give rise to $CO_2{}^+$. With this bookkeeping, we deduce the abundances given in the accompanying tabulation, and compare them with the observations of Carlson and Krause.

	Calculated	Observations (Carlson and Krause)
C^+	33	25 ± 0.2
O^+	66	52 ± 6
CO^+	0.2	8.9 ± 0.5
$CO_2{}^+$	1.1	5.4 ± 0.6

The preponderance of C^+ and O^+ and the $O^+:C^+$ ratio are predicted quite well by this analysis, which is strong support for the dominance of the $(1\sigma_g)^{-1}$, $(1\sigma_u)^{-1}$, and $(2\sigma_g)^{-1}$, partial cross sections at 930 eV and for the Coulomb explosion hypothesis. The very low cross section for CO^+ is attributable to the neglect of the two-electron processes in this analysis. In Table 8 we note a substantial increase in the $CO^+/CO_2{}^+$ ratio between 21.2 and 27.8 eV, i.e., between the one-electron processes $(4\sigma_g)^{-1}$ and $(3\sigma_g)^{-1} + (2\sigma_u)^{-1}$. These two-electron processes must account for the bulk of CO^+ production, which, may also receive some contribution from the $(3\sigma_g)^{-1} + (2\sigma_u)^{-1}$ single-electron ionization band. The small $CO_2{}^+$ intensity predicted may be a consequence of the uncertain ratio of partial cross sections of the valence shells relative to inner shells. It will be recalled that these had to be pieced together from separate figures, perhaps corresponding to different experimental conditions, in the paper by Allan *et al.*[215] It is possible that C^+ and O^+, which are produced with much higher kinetic energies than the molecular ions, may not have been collected as efficiently, but this is unlikely to be the major source of this large discrepancy.

3. N_2O

a. Photoelectron Spectroscopy. In its electronic ground state, N_2O has the linear but asymmetric N–N–O structure. The sequence of molecular orbitals describing its electronic structure may be written:

$$K(O1s)^2KK(N1s)^2(1\sigma)^2(2\sigma)^2(3\sigma)^2(1\pi)^4(4\sigma)^2(2\pi)^4.$$

It is isoelectronic with CO_2, whose outermost occupied orbital is π_g, and hence antibonding. The analogous behavior in N_2O can be perceived in that the orbital (2π) has dominant contributions from O(2p) and terminal N(2p)[213] and the photoelectron spectrum displays[234] a first band (IP = 12.89 eV) which is largely restricted to the (0, 0, 0) transition. However, the second and third bands are inverted in order (compared to CO_2). In this case of N_2O, the second band (IP = 16.39 eV) is formed by photoejection from 4σ, a molecular orbital with dominant contribution from terminal N(2p, 2s). This also is an antibonding orbital, and the photoelectron band is narrow. The third band corresponds to ejection from the strongly bonding 1π

orbital. Its LCAO composition contains significant 2p coefficients from each of the atoms. The photoelectron spectrum corresponding to $(1\pi)^{-1}$ involves a broad vibrational sequence primarily in v_1', but containing some other vibrational components. Its vertical IP is $\sim$18.23 eV. The 3σ orbital is essentially composed of 2s orbitals, which tend to be antibonding between adjacent atoms. This band in the photoelectron spectrum (IP = 20.11 eV) consists almost entirely of the $(0,0,0) \rightarrow (0,0,0)$ transition. The 2σ and 1σ orbitals also are primarily 2s in nature, but are bonding between N–O, with 1σ providing some N–N bonding. In the x-ray photoelectron spectrum[235] the bands corresponding to $(1\sigma)^{-1}$ and $(2\sigma)^{-1}$ are very broad. The remaining one-electron ionizations are essentially 1s orbitals of N and O, respectively.

Partial cross sections between 600 and 750 Å have been deduced from the branching ratios of Bahr *et al.*[172] and the total photoabsorption cross section of Fig. 21. The ionization yield measurements of Cook *et al.*[236] differ substantially from the values $\eta = 0.88$ at $\sim$740 Å[237] and 0.98 at 584 Å.[238] We have chosen here to accept the latter values, and to linearly interpolate for intermediate wavelengths. At 584 Å, the extrapolated branching ratios of Bahr *et al.* are in rather good agreement with the directly measured branching ratios of Berkowitz and Ehrhardt.[239]

The branching ratios of valence orbitals is provided by Gelius[235] at 1486.6 eV, but not their relative intensities to the K-shell ejection process. Our modus operandi here has been to apportion the valence orbitals just below the K-shell jumps (408.75 and 541.2 eV) according to their branching ratios at 1486.6 eV, and to roughly mark the K-shell cross sections at their respective onsets. With the available data and the aforementioned assumptions, we have constructed the partial cross section graph of Fig. 54.

b. Photoionization Mass Spectrometry. The thermochemical thresholds for various dissociative ionization processes are listed below.

Process	$\Delta H_0{}^0$, eV
$N_2O \rightarrow NO^+ + N$	14.19
$\rightarrow O^+ + N_2$	15.29
$\rightarrow N_2{}^+ + O$	17.25
$\rightarrow N^+ + NO$	19.47

The two lowest dissociative ionization thresholds fall between the electronic ground state of N_2O^+ ($\tilde{X}^2\Pi$, IP = 12.889 eV) and the first excited state observed in the photoelectron spectrum ($\tilde{A}^2\Sigma^+$, IP = 16.39 eV). Hence, in order to observe these processes at threshold one must rely on very weak direct ionization processes or autoionization.

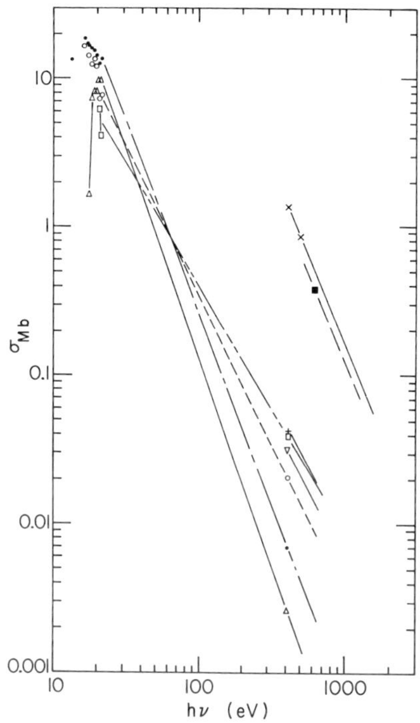

Fig. 54 Partial cross sections for formation of various states of N_2O^+ by photoabsorption. ●, $(2\Pi)^{-1}$; ○, $(4\sigma)^{-1}$; △, $(1\Pi)^{-1}$; □, $(3\sigma)^{-1}$; ▽, $(2\sigma)^{-1}$; +, $(1\sigma)^{-1}$; ×, K(N 1s, both); ■, K(O 1s).

The parent ion N_2O^+ is the only one observed from the ionization threshold (~961.9 Å) to ~840 Å. The photoion yield curve (Fig. 55) displays a barely discernible step above threshold at ~959.1 Å, as well as the primary rise at ~961.9 Å. These correspond to the spin–orbit split $^2\Pi_{3/2}$ and $^2\Pi_{1/2}$

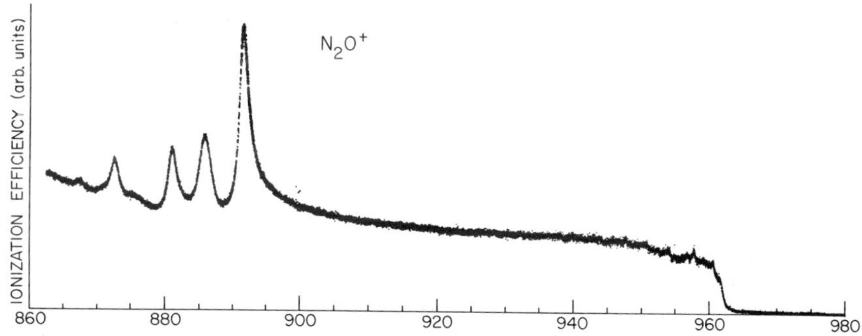

Fig. 55 Mass selected photoion yield curve of N_2O^+ from the ionization threshold to 860 Å. [From J. Berkowitz and J. H. D. Eland, *J. Chem. Phys.* **67**, 2740 (1977).]

components of the ionic ground state. The structure between ~870 and 890 Å represents the earliest members of Rydberg series, to be discussed later.

In the wavelength region between 740 and 840 Å, the ions N_2O^+, NO^+, and O^+ are observed. Their respective photoion yield curves are shown in Fig. 56. This region is seen to be dominated by autoionization structure which is similar, though not identical, for the three ions. This structure has been analyzed[240] into five Rydberg series ($ns\sigma$, $np\sigma$, $np\pi$, $nd\sigma$, $nd\pi$) converging on the $\tilde{A}\,^2\Sigma^+$ state, and hence involving the excitation $4\sigma \rightarrow$ Rydberg. These Rydberg states can autoionize by interacting with either of two continua—$\tilde{X}\,^2\Pi + e$, and $^4\tilde{A}'' + e$. The latter ionic state is not directly observed in photoelectron spectra because of the spin selection rule, but portions of its

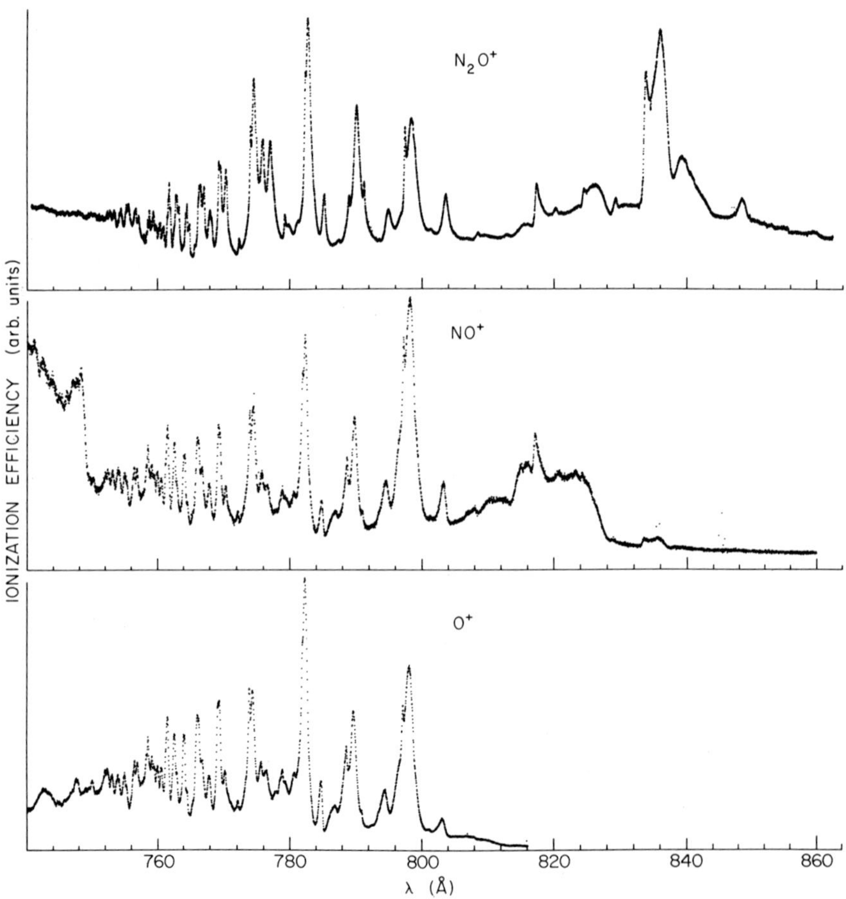

Fig. 56 Photoion yield curves of N_2O^+, NO^+, and O^+ from N_2O between 740 and 860 Å. [From J. Berkowitz and J. H. D. Eland, *J. Chem. Phys.* **67**, 2740 (1977).]

potential surface have been computed.[241] It can correlate either with NO^+ + N or $O^+ + N_2$, whereas $\tilde{X}^2\Pi$ cannot. Hence, the simplest interpretation is that autoionization to $\tilde{X}^2\Pi$ results in N_2O^+, while autoionization to $^4\tilde{A}''$ leads to NO^+, and when energetically possible, to $O^+ + N_2$.

The most stable geometric configuration of $^4\tilde{A}''$ has been computed to be a substantially bent one. We[240] have performed experiments with isotopically labeled $N^{14}N^{15}O$ which indicate that a substantial fraction of the NO^+ produced ($\frac{1}{3}$) in this longer wavelength region is $N^{14}O^+$, i.e., "scrambled." Such a result would be distinctly favored by a bent intermediate state. The $N^{15}O^+$ experiences a marked increase at the onset of the (1, 0, 0) vibrational state of $\tilde{A}^2\Sigma^+$, but not the $N^{14}O^+$ component, nor the O^+.

The $\tilde{A}^2\Sigma^+$ state has a well-known fluorescence[242] not only in its vibrationless state, but also in the (1, 0, 0) state. Eland[243] has shown that the vibrationless state does not dissociate, whereas the (1, 0, 0) vibrational component has a lifetime toward predissociation of 410 ± 90 nsec. This is a sufficiently long lifetime to permit observation of a metastable ion corresponding to this dissociation. We shall return to this aspect shortly.

NO^+ is first observed very weakly at ~ 839 Å $\equiv 14.78$ eV, and more strongly at ~ 828 Å $\equiv 14.97$ eV, both well above the thermochemical threshold of 14.19 eV. This is a relatively rare occurrence, since in most cases fragments seem to appear at their thermochemical thresholds. Metastable ions corresponding to the $N_2O^+ \rightarrow NO^+ + N$ decomposition have been studied by several groups, using electron impact ionization. The threshold energy for formation of these metastables has been reported as 15.1 ± 0.2 eV[244] and 15.7 ± 0.5 eV.[245] Bearing in mind the generalization that electron impact appearance potentials are usually several tenths of an electron volt above photoionization values, it seems safe to conclude that some metastables appear well below the threshold for $\tilde{A}^2\Sigma^+$ (16.39 eV). Therefore, not only the $\tilde{A}$ state predissociation, but also the autoionization mechanism that produces NO^+ at lower energy must proceed by a relatively slow reaction. In this regard, it may be relevant to note that Newton and Sciamanna[245] inferred two distinct lifetimes (~ 90 nsec and ≥ 300 nsec) for this metastable ion in their experiments.

The most satisfactory explanation for the above observations at this time is that autoionization from Rydberg series converging on $\tilde{A}^2\Sigma^+$ populates both $\tilde{X}^2\Pi$ and $^4\tilde{A}''$ states of N_2O^+. Those autoionizations culminating in $\tilde{X}^2\Pi$ are observed in our experiment as N_2O^+. On the other hand, when the autoionization transitions occur to the bent $^4\tilde{A}''$, it decomposes slowly to NO^+ + N and more rapidly to $O^+ + N_2$. Presumable, the Franck–Condon envelope of the autoionization is such that higher vibrational levels of $^4\tilde{A}''$ (those energetically capable of forming $O^+ + N_2$) dissociate readily toward the O^+ product, whereas lower vibrational levels encounter a barrier toward decomposition along the $NO^+ + N$ pathway, even though this process is highly exothermic. The $N^{14}O^+ - N^{15}O^+$ scrambling is apparently a consequence of the bent nature of the $^4\tilde{A}''$ state and the long lifetime toward decomposition, which can involve motion on this potential surface which enables terminal N and O atoms to approach one another. The relative lack of scrambling in the predissociation of the $\tilde{A}^2\Sigma^+$ state is presumably related to the fact that this state is linear.

In the 600–740 Å region, five ion masses (N_2O^+, NO^+, O^+, $N_2{}^+$, and N^+) are observed. The ion yield curves of the first four are displayed in Fig. 57, a–d. Again, this wavelength region is dominated by autoionization, with many features similar in three of the four spectra. The O^+ photoion yield curve displays a series of peaks, separated by roughly 0.11 eV, between 695 and 750 Å, which bear a resemblance to the vibrational structure observed in the photoelectron spectrum in the $\tilde{B}^2\Pi$ region. Hence, we infer that this represents autoionization from the first prominent Rydberg member converging on $\tilde{B}^2\Pi$. The structure is also observed in the $N^{14}O^+$ and $N^{15}O^+$ channels. The autoionization very likely proceeds through the $^4\tilde{A}''$ ionic state to the observed products.

From the threshold of $\tilde{B}^2\Pi$ ($\sim$702.5 Å) to shorter wavelength, there is a smooth increase in the intensity of NO^+ and $N_2{}^+$, but no discernible increase in N_2O^+ or O^+. Hence, we conclude that $\tilde{B}^2\Pi$ decomposes to $NO^+ + N$ and $N_2{}^+ + O$, the branching ratio favoring NO^+ by about 3:1. Orth and Dunbar[246] measured the photodissociation from this state directly by absorption from the ground ($\tilde{X}^2\Pi$) state, and observed NO^+ and $N_2{}^+$ products in the ratio 4:1.

The remainder of the structure in the 600–740 Å region consists of five Rydberg series converging on the $\tilde{C}^2\Sigma^+$ state. Two of these series ($np\sigma$ and $np\pi$) are window resonances, the others being $ns\sigma$, $nd\sigma$, and $nd\pi$. The analyses have been performed[240] and yield IP ($\tilde{C}^2\Sigma^+$) = 20.105 ± 0.001 eV. These series are remarkable in their specificity. For example, the $np\pi$ series appears prominently in photoinduced fluorescence[247] of the $\tilde{A} \rightarrow \tilde{X}$ transition and also in the N_2O^+ photoion yield curve, but only weakly in the NO^+ channel and is not clearly evident at all in the O^+ channel. Hence, the autoionization branching ratio from $np\pi$ strongly prefers formation of $\tilde{A}^2\Sigma^+$, and possibly $\tilde{X}^2\Pi$, but not $^4\tilde{A}''$.

The $np\sigma$ series has the remarkable characteristic of displaying quite different autoionization structure (q values—see Chapter II.C) in different channels. For example, the member at $\sim$657 Å appears as nearly a pure window resonance ($q \approx 0$) in the N_2O^+ ion yield curve, as an asymmetric peak shape in the NO^+ and $N_2{}^+$ curves ($q \sim 1$) and as peak in the O^+ photoion yield curve ($q \geq 4$). The experimental results here are unique in that the individual ion yield curves reflect the branching into corresponding

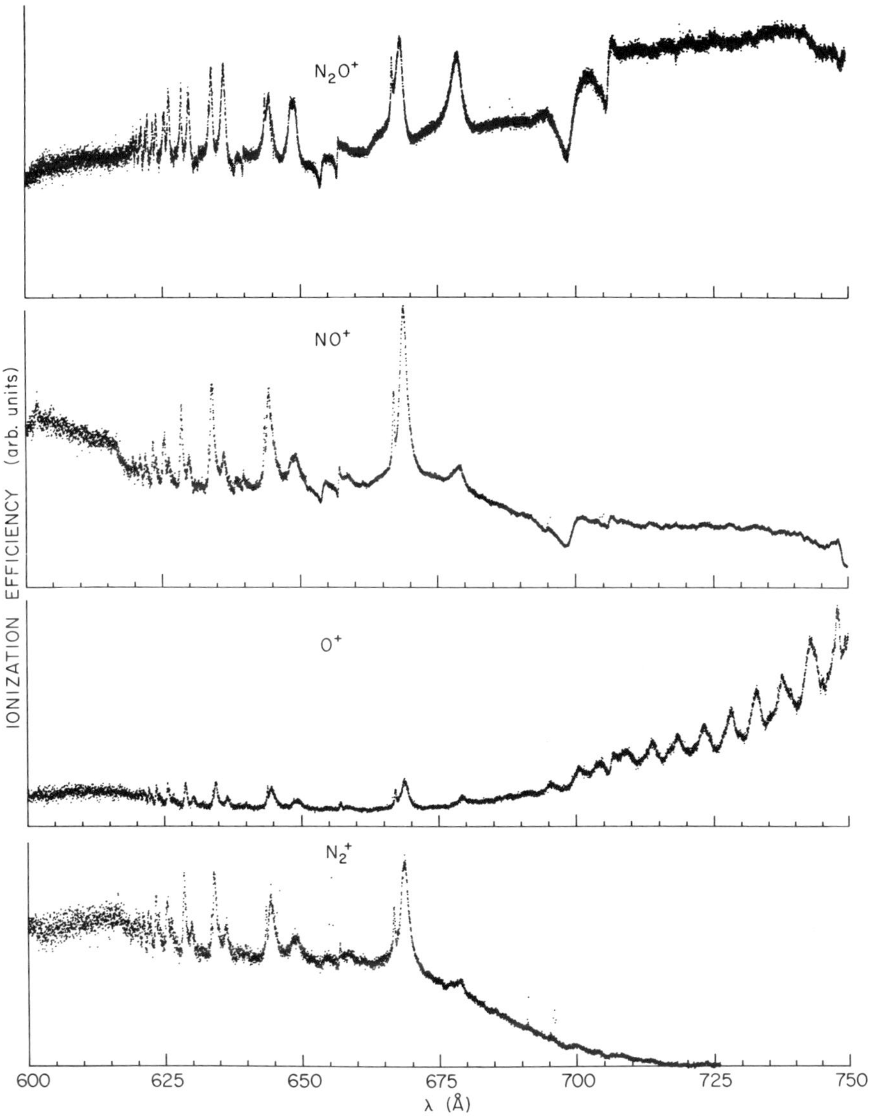

Fig. 57 Photoion yield curves of N_2O^+, NO^+, and O^+ from N_2O between 600 and 750 Å. [From J. Berkowitz and J. H. D. Eland, *J. Chem. Phys.* **67**, 2740 (1977).]

continua, which are not readily observable in most cases. The analysis of the interaction of a quasi-discrete state with more than one continuum has not been explicitly worked out to the extent that it can be compared with the experimental results.

The other series (the peaks) also reveal themselves somewhat differently in the various channels, but they all correspond to relatively high q values and hence the variation in feature is not as dramatic.

The ion N^+ first appears (Fig. 58) at 617.6 Å ≡ 20.08 eV, well above its thermochemical threshold (19.47 eV) but approximately at the onset of the $\tilde{C}^2\Sigma^+$ state. The fact that it first appears slightly below the adiabatic value of the $\tilde{C}$ state deduced from Rydberg series analysis may be significant. The influence of rotational excitation, and perhaps thermal vibrational population of the neutral ground state may be responsible for this discrepancy.

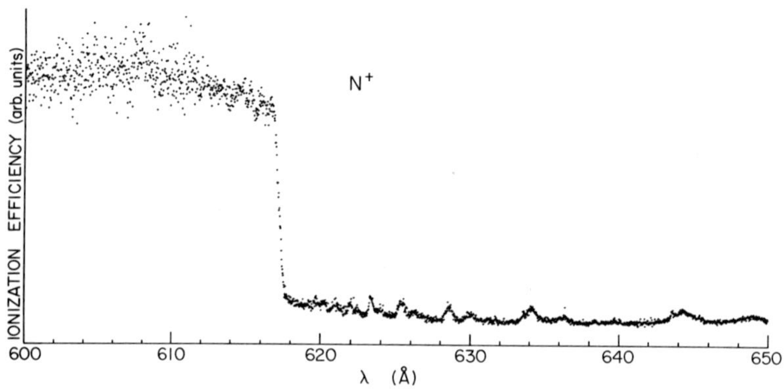

Fig. 58 Photoion yield curve of N^+ from N_2O from its threshold to 600 Å. The weak structure above ~ 618 Å is due to collisional dissociation of N_2O^+. [From J. Berkowitz and J. H. D. Eland, *J. Chem. Phys.* **67**, 2740 (1977).]

Preliminary studies[248] with discrete line sources between 584 and 304 Å indicate that N^+ experiences the largest relative gain in this region, with $N_2{}^+$ and O^+ also gaining slightly. It is perhaps noteworthy that at 584 Å the O^+ yield is only ~1% of the total ion yield. This is a somewhat misleading measure of the decomposition of N_2O^+, since most of the O^+ is produced by autoionization, which is largely bypassed at 584 Å.

The "shake-up" spectrum produced concomitantly with photoejection of oxygen or nitrogen K-shells in N_2O has been reported,[249] as has the x-ray emission spectrum following K-shell excitation.[250] The x-ray photoionization mass spectrum has been studied by Hirsch *et al.*[251] at 0.28, 1.25, and 1.5 keV. Their apparatus is sensitive primarily to ions with thermal kinetic energies, and hence will strongly discriminate against high velocity fragment ions. Recognizing this behavior, we cannot expect good agreement between our predicted ion yields based on partial cross sections and the available experimental data. Nevertheless, we shall pursue this analysis at 0.28 keV, and perhaps provide a basis for comparison with future experimental results.

The partial cross sections obtained from Fig. 54 at 280 eV, and their apportionment to various product ions, are listed below.

Ion state	Partial cross section (Mb)	Apportionment
$\tilde{X}^2\Pi$	0.0184	100% N_2O^+
$\tilde{A}^2\Sigma$	0.043	90% N_2O^+, 10% NO^+
$\tilde{B}^2\Pi$	0.0075	75% NO^+, 25% N_2
$\tilde{C}^2\Sigma$	0.072	50% NO^+, 30% N^+, 18% N_2^+, 2% O^+
$(2\sigma)^{-1}$	0.05	assume 50% N^+, 25% N_2^+, 25% O^+
$(1\sigma)^{01}$	0.072	

With these inferences and assumptions, the fractional abundances are listed below and compared with the results of Hirsch *et al.*

	Predicted	Observed thermal ions[251]
N_2O^+	0.217 (0.401)	0.512
NO^+	0.176 (0.325)	0.228
N_2^+	0.172 (0.104)	0.111
O^+	0.116 (0.01)	0.019
N^+	0.318 (0.159)	0.129

The discrepancy is large, as expected. The atomic ions are relatively very weak in the observed mass spectrum. Our analysis shows that they derive predominantly from $(2\sigma)^{-1}$ and $(1\sigma)^{-1}$. If the contributions of these states are completely ignored, the predicted values shown in parentheses are obtained. Fair agreement is thereby achieved; consequently, the experiments of Hirsch *et al.* are very slightly sensitive to these states, although they account for almost $\frac{1}{2}$ of the total cross section at 280 eV.

E. Polyatomic Molecules

1. NH_3

a. Photoelectron Spectroscopy. The sequence of occupied molecular orbitals of ammonia in its electronic ground state (C_{3v} symmetry) is

$$(1a_1)^2(2a_1)^2(1e)^4(3a_1)^2.$$

The $1a_1$ orbital has a composition (in terms of atomic orbitals) which is almost pure nitrogen 1s.[252] Its binding energy (405.6 eV) is 4.3 eV less than the K-shell binding energy in N_2. Similarly, the $2a_1$ orbital is largely nitrogen 2s in character, although here nitrogen 1s and 2p and hydrogen 1s make some contribution.[252] The corresponding vertical ionization energy has been reported to be 27.0[253] and 27.76_5 eV.[196]

The 1e orbital has about equal contribution from nitrogen 2p and hydrogen 1s, and is the major N—H bonding orbital in this molecule. By contrast, the outermost occupied $3a_1$ orbital is largely a nitrogen 2p lone-pair orbital with very little contribution from hydrogen 1s. Ionization from these latter two orbitals is readily observed in the 21.2 eV photoelectron spectra reported by several investigators.[253–256] The spectrum obtained by Potts and Price is reproduced for illustrative purposes in Fig. 59. The long vibrational progression of the first band is a consequence of a change in geometry. In its neutral ground state, NH_3 is pyramidal, whereas the electronic ground state of NH_3^+ is planar. Hence, the formation of $(3a_1)^{-1}$ is accompanied by excitation of ν_2, the inversion bending mode. The adiabatic ionization potential (10.15 eV) is about six vibrational levels removed from the vertical ionization potential (10.88 eV). This situation is rather similar to that encountered in the second PES band of H_2O.

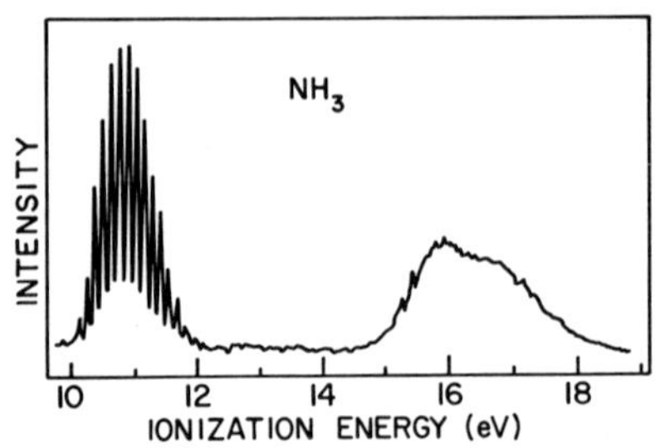

Fig. 59 Photoelectron spectrum of NH_3 obtained with 21.2 eV incident radiation. [From A. W. Potts and W. C. Price, *Proc. Roy. Soc.* (*Lond.*) **A326**, 181 (1972).]

Ejection of an electron from the second outermost occupied orbital (1e) gives rise to a broad band with weak features. Two broadened vibrational peaks appear on the low energy side (assigned as asymmetric stretch excitations) and a tendency toward bimodal structure (interpreted as Jahn–Teller splitting)[253] is apparent. The onset of dissociative ionization into $NH_2^+ + H$ occurs at 15.77 eV, approximately where vibrational structure ceases to be manifest. Table 9 summarizes the experimentally observed energies required to eject electrons from each of these orbitals.

The variation of partial cross section with photon energy, as deduced from a variety of sources, is depicted in Fig. 60. The total photoionization cross section at each energy is that obtained by combining the photoabsorption cross sections and ionization yields deduced in Chapter V. The branching ratios between 20 and 50 eV are taken from the (e, 2e) experiments of Van der Wiel and Brion[257] and Brion *et al.*[257a] The earlier work of this group[257] has lines connecting the points, the latter does not. It can be seen that the later data imply partial cross sections for $(2a)^{-1}$ and $(3a)^{-1}$ which are about a factor of two larger than in the earlier work. The authors attribute some of

TABLE 9 *Adiabatic and Vertical Ionization Energies for Single-Electron Ejection from NH_3*

Orbital	Adiabatic IP (eV)	Vertical IP (eV)	Ref.
$3a_1$	10.15 (10.073)	10.88	(253, 256)
1e	14.98 (14.725)	16.0	(253, 256)
$2a_1$	~26	27.0	(253)
		27.76_5	(196)
$1a_1$		405.6	(43)

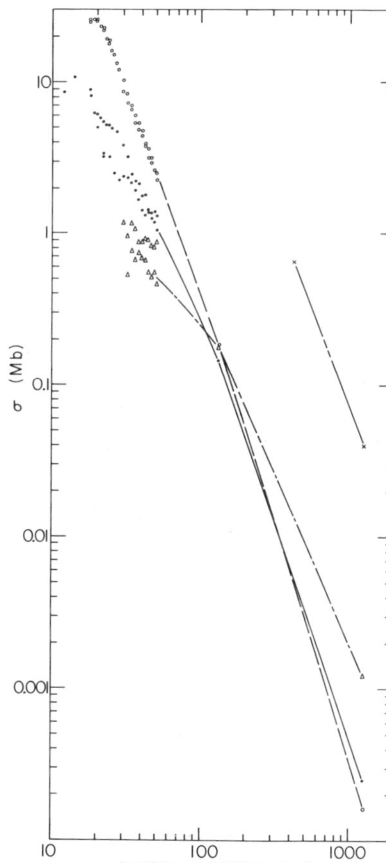

Fig. 60 Partial cross sections for formation of various states of NH_3^+ by photoabsorption. ●, $(3a_1)^{-1}$; ○, $(3e_1)^{-1}$; △, $(2a_1)^{-1}$; ×, $(1a_1)^{-1}$.

the discrepancy to angular distribution effects. Nevertheless, the current situation, depicted in Fig. 60, must be considered to retain systematic and statistical error for the partial cross sections of $(2a)^{-1}$ and $(3a)^{-1}$. Hence, we regard them as uncertain by about a factor two. The ratio of $(1e)^{-1}$:$(3a_1)^{-1}$

of about 5 at 21.2 eV is in rough agreement with the photoelectron data at that energy. No reliable photoelectron spectrum has been published at 40.8 eV; Banna and Shirley quote a branching ratio of about 2:1, whereas the Van der Wiel and Brion value is closer to 3:1. The branching ratios at 132.3 and 1253.6 eV are taken from Banna and Shirley.[196] Once again, we note that photoejection from the 2s-like $2a_1$ orbital declines less rapidly with increasing photon energy than from the 2p-like 1e and $3a_1$ orbitals. Some structure appears in the $(3a_1)^{-1}$ partial cross section curve, which should be tested in a direct photoelectron experiment using synchrotron radiation.

b. Photoionization Mass Spectrometry. The cross section for formation of NH_3^+ increases slowly from threshold. There are indications in the data of Watanabe and Sood[258] and Dibeler *et al.*[207] of steplike behavior, attributed to the formation of successive vibrationally excited states of NH_3^+. The structure in the photoabsorption curve in this wavelength region is apparently unrelated to the photoionization structure.[258] The ionization yield is very low.[258] These observations are consistent with a mechanism in which the excited states characterized by peaks in the photoabsorption spectrum predissociate much more rapidly than they can autoionize. In Fig. 61, a recently obtained photoionization mass spectrum of NH_3^+ obtained at higher resolution and with better statistics is shown. Autoionization superposed upon the "direct" step structure is readily apparent.

Rydberg states converging to the second ionization potential would be expected to autoionize. There is some indication of this behavior in the

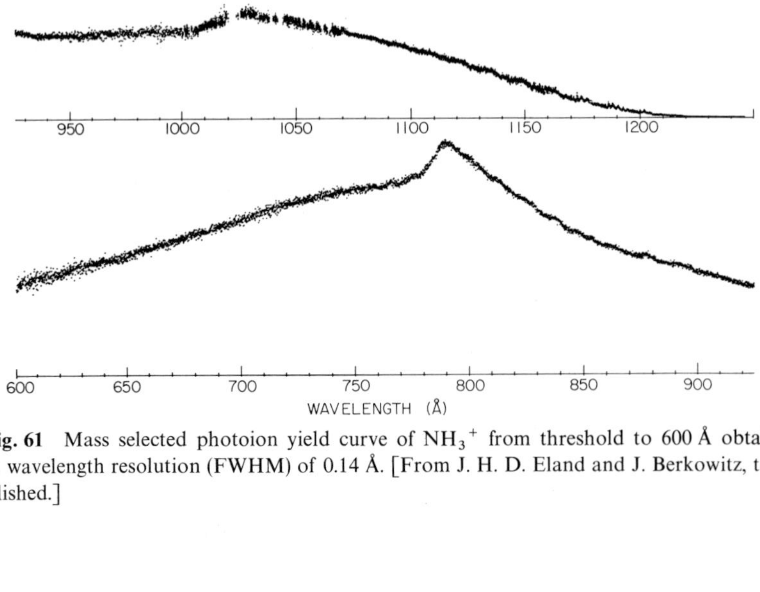

Fig. 61 Mass selected photoion yield curve of NH_3^+ from threshold to 600 Å obtained with wavelength resolution (FWHM) of 0.14 Å. [From J. H. D. Eland and J. Berkowitz, to be published.]

photoionization spectrum of Fig. 61 beginning at about 925 Å. The structure is broad, as would be expected from the breadth of the second photoelectron band, but the variation in intensity with increasing photon energy is unexpected. The first Rydberg member is weak, and higher members seem more intense. The ionization yield is increasing in this region, and apparently modifies the anticipated variation of intensity of successive Rydberg states. Stated another way, predissociation is relatively more competitive in the lower Rydberg members of this series.

The first dissociative ionization channel, corresponding to the process

$$NH_3 + h\nu \rightarrow NH_2{}^+ + H + e \qquad (32)$$

has a threshold at 15.768 ± 0.008 eV, according to McCulloh.[259] This energy occurs approximately one-third of the way across the second band, and appears to correspond to the vanishing of vibrational structure, as previously noted. If one takes the ratio of partial cross sections of $(1e)^{-1}:(3a_1)^{-1}$ to be about 5 at 21.2 eV, and assigns $\sim\frac{2}{3}$ of $(1e)^{-1}$ to $NH_2{}^+$, one can account roughly for the $NH_2{}^+:NH_3{}^+$ ratio observed at 21.2 eV. This is tantamount to assuming that the entire portion of $(1e)^{-1}$ which is energetically capable of dissociating does indeed fragment. Photoion–photoelectron coincidence studies on this system[260] have drawn essentially the same conclusion.

The thresholds for higher fragmentation processes are not accurately known, but can be estimated. The endoergicity of the neutral decomposition process

$$NH_3 \rightarrow NH + H_2 \qquad (33)$$

is 3.92 eV.[176] Hence, the thermochemical threshold for $H_2{}^+(NH_3)$ should be ~19.3 eV, that for $NH^+(NH_3)$ roughly 16 eV (assuming an ionization potential for NH similar to that of NH_2). The ionization potential of NH is likely to be in the range of its isoelectronic analogues CH_2(IP = 10.396 eV)[261] and O(IP = 13.6 eV).[30] Hence, both of these processes are energetically possible at 21.2 eV, although the $(1e)^{-1}$ state does not extend substantially beyond 18 eV.

From Eq. (32), one can deduce that the thermochemical threshold for $H^+(NH_3)$ is about 18.2 eV. The onset energy for $N^+(NH_3)$ corresponds to the reaction

$$NH_3 \rightarrow N^+ + H_2 + H + e \qquad (34)$$

and amounts to slightly more than 22 eV. Hence, of all possible fragments, only N^+ is forbidden to appear in the 21.2 eV photoionization mass spectrum on thermochemical grounds. Kronebusch and Berkowitz[147] have recently examined the photoionization mass spectrum of NH_3 between 21.2

and 40.8 eV. The results are displayed in Fig. 62, as branching ratios, together with a manufactured 40.8 eV photoelectron spectrum based on the considerations in Section E.1.a. Noteworthy features include:

(1) The variation of NH_2^+ and NH_3^+ branching ratios above 21.2 eV. The *decline* of NH_2^+ and corresponding *rise* of NH_3^+ parallel the variation of partial cross sections of $(1e)^{-1}$ and $(3a_1)^{-1}$, respectively. Schweig and Thiel[262] report a similar experimental variation in partial cross sections between 584 and 304 Å.

(2) Although NH^+ has a relatively low threshold, it accounts for $\leq 1\%$ of total ionization at 21.2 eV[207] and slowly rises to about 3.5–4% at 40.8 eV. It does not show any noticeable increase when the $(2a_1)^{-1}$ state becomes accessible. [Note that there is a discrepancy between two recent publications of the pseudophotoabsorption group in this regard. Wight *et al.*[262b] agrees with the conclusions of (2), while Brion *et al.*[257a] does not.]

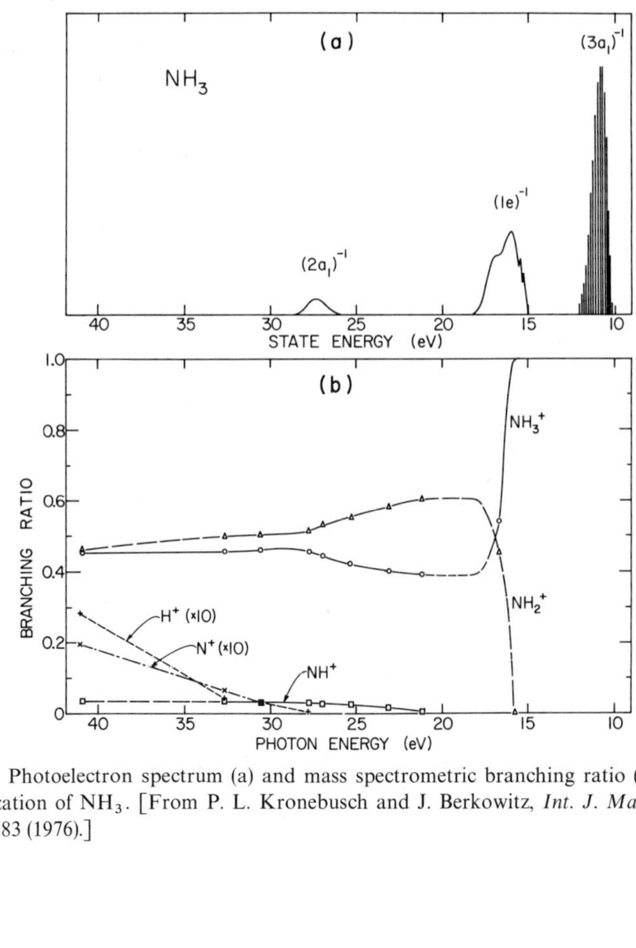

Fig. 62 Photoelectron spectrum (a) and mass spectrometric branching ratio (b) from the photoionization of NH_3. [From P. L. Kronebusch and J. Berkowitz, *Int. J. Mass Spec. Ion Phys.* **22**, 283 (1976).]

(3) Both H^+ and N^+ apparently start to increase when $(2a_1)^{-1}$ becomes accessible, although the correlation between the onset of $(2a_1)^{-1}$, its variation of partial cross section with increasing $h\nu$, and the corresponding increase in H^+ and N^+ is not precise. At the present time, the characterization of the bandwidth of $(2a_1)^{-1}$ is uncertain, and the photoionization data are based on three points in a region where light intensity is weak. The correlation between the sum of N^+ and H^+ intensities relative to total ionization at 304 Å (about 5%) and the ratio of $(2a_1)^{-1}$ to total ionization at 304 Å (about 10%) is highly suggestive of the conclusion that the $(2a_1)^{-1}$ state disintegrates to $H^+ + 2H + N$ and $N^+ + 3H$, the H^+ being favored by its somewhat lower ionization potential. The $H^+ + H_2 + N$ and $N^+ + H_2 + H$ channels must also be considered as possible contributors.

Although x-ray photoionization mass spectrometry has not been reported for this molecule, both x-ray emission[250] and Auger[263] spectra have recently been obtained. The Auger process is expected to dominate upon creation of a K-shell vacancy in this low Z system. The experimental results of Shaw *et al.*[263] and the calculated Auger spectrum[264] are in good agreement, and hence enable one to deduce the energies of various states of NH_3^{2+}. These states are likely formed in a time scale of 10^{-14}–10^{-15} sec. (The experimentally observed line widths are consistent with this assumption.) Most of these states are not likely to survive as such during the time scale ($\sim 10^{-6}$ sec) of typical mass spectrometric observation, and hence the fragment ions likely to result are NH^+, H^+, and N^+, with perhaps some $NH_2{}^+$, with the release of substantial kinetic energy.

2. CH_4

a. Photoelectron Spectroscopy. The electronic ground state of methane has tetrahedral symmetry, with the configuration $(1a_1)^2(2a_1)^2(1t_2)^6$.

Removal of an electron from the $1t_2$ orbital leaves the ion in a triply degenerate $^2T_2$ state which is subject to strong Jahn–Teller distortions. The geometry of the $CH_4{}^+$ resulting from $(1t_2)^{-1}$ ionization has been the subject of several theoretical investigations.[265–271]

The ab initio calculation by Dixon of the coupling of the $^2T_2$ state to the various vibrational modes of $CH_4{}^+$ concludes that a splitting into three states should ensue, each separated from its neighbor by ~ 0.7 eV. The most stable of these states has D_{2d} symmetry, corresponding to the compression of the enveloping cube of the tetrahedral CH_4 across a pair of opposite faces. This structure approaches, but does not quite attain, the square-planar geometry. Grimm and Godoy[270] reach a similar conclusion on the basis of a CNDO calculation. It is the ν_2 bending mode which is responsible for this distortion, and hence one can anticipate that the ν_2 mode will be excited when ionization occurs from the tetrahedral CH_4 to CH_4^+ in the D_{2d} geometry.

Rabalais *et al.*[272] also reach this conclusion from an analysis of their 21.2 eV photoelectron spectrum of CH_4. (Several PES spectra of the first band in CH_4 have been reported in the literature; the most recent ones, displaying the most structure which can be considered significant, are those of Rabalais *et al.*[272] and Potts and Price.[272a] These latter spectra are virtually identical.) They purport to show that excitation of an e vibrational mode (ν_2) in a T_2 electronic state produces vibronic components that fit the experimental pattern, whereas excitation of t_2 vibrational modes do not. A consequence of their analysis is that the $\nu_2 = 0$ level of this ionic state is outside the Franck–Condon region and therefore the adiabatic ionization potential is not observed in photoionization, but can be estimated by extrapolation. We shall return to this question in the subsection on photoionization.

The $(2a_1)^{-1}$ band has been observed with 40.8 eV excitation by Potts and Price, and by Brundle *et al.*[273] It is a fairly broad band, with well-defined vibrational structure corresponding to excitation of the totally symmetric ν_1 vibration. This orbital has a large C 2s contribution, but the breadth of the photoelectron band indicates that it contributes considerable C—H bonding character. The reduction in magnitude of the ν_1 vibrational frequency in the $(2a_1)^{-1}$ state, relative to its value in the neutral ground state of CH_4 provides corroboration for this conclusion.

The relative intensity of $(1t_2)^{-1}$ and $(2a_1)^{-1}$ at 40.8 eV is given as 7:1 by Brundle *et al.* The corresponding ratio at 132.3 eV has been reported by Banna and Shirley[196] to be 2.69_8:1. In addition, Allison and Cavell[134] have obtained a spectrum at 151.4 eV, from which we extract an area ratio of 4:3.

The ejection of an electron from the C 1s-like 1a orbital has been studied with AlK_α radiation by Hamrin *et al.*[274] and by Banna and Shirley. The latter authors report an intensity ratio at 1486.6 eV of $(1a_1)^{-1}:(2a_1)^{-1}:(1t_2)^{-1}$ of 30:1:0.12. A recent high resolution spectrum of this band has been reported by Gelius.[275] (vide infra) The adiabatic and vertical ionization potentials of CH_4 observed in photoelectron spectroscopy are summarized in Table 10.

TABLE 10 *Adiabatic and Vertical Ionization Energies for Single Electron Ejection from CH_4*

Orbital	Adiabatic IP (eV)	Vertical IP (eV)	Ref.
$1t_2$	12.61_5	~14.35	(272, 272a, 279)
$2a_1$	22.39	~23	(272a, 273)
$1a_1$		290.69	(275)

By combining our knowledge of the photoabsorption cross section and ionization yield of CH_4 (see Chapter V Section D.2) with the branching ratios noted in the aforementioned photoelectron spectra, we have constructed the partial cross section curves shown in Fig. 63. Also shown are the deductions of Backx and Van der Wiel[276] in which we have used their reported partial cross sections and our total cross sections. Their results are in fair agreement with the conclusions based directly on photoelectron spectroscopy, although their $(1t_2)^{-1}$:$(2a_1)^{-1}$ ratio appears rather low in the 40–80 region. A more direct (e, 2e) measurement of the partial cross sections of $(1t_2)^{-1}$ and $(2a_1)^{-1}$ between 25 and 50 eV has also been reported by Van der Wiel *et al.*[277] with essentially the same results as those deduced by

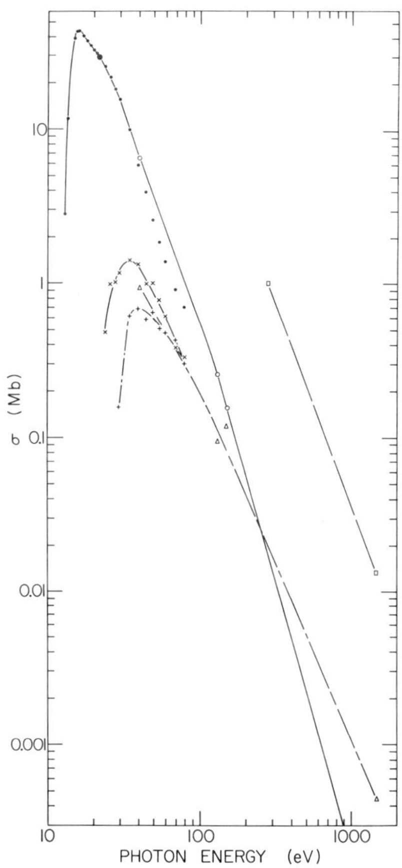

Fig. 63 Partial cross sections for formation of various states of CH_4^+ by photoabsorption. ○, $(1t_2)^{-1}$ PES; △, $(2a_1)^{-1}$ PES; □, $(1a_1)^{-1}$ PES; ●, $(1t_2)^{-1}$ pseudoelectron ion; ×, $(2a_1)^{-1}$ pseudoelectron ion; +, higher states pseudoelectron ion.

Backx and Van der Wiel from relative photoion data. From the photoelectron generated curves, it is again apparent that the s-like $(2a_1)^{-1}$ curve declines less rapidly with increasing photon energy than does the p-like $(1t_2)^{-1}$ curve.

b. Photoionization Mass Spectrometry. In Fig. 29 in Chapter V, the ionization yield of CH_4 can be seen to approach zero at the ionization threshold. The absorption cross section is substantial in the region just above threshold, but most of the absorption apparently leads to neutral dissociation products. Nishikawa and Watanabe[278] have made a study of this region, and conclude that the excitations are to Rydberg states distorted by the Jahn–Teller effect, which converge upon the various components of the $(1t_2)^{-1}$ state. Their calculated ionization yield is in good agreement with the experimental data of Fig. 29 in Chapter V.

The photoionization mass spectrum of CH_4^+ in the near-threshold region is compared with the 21.2 eV photoelectron spectrum in Fig. 64. The photoelectron spectrum at 16.8 eV, which is identical, has been reported by Potts and Price.[195]

In the photoelectron spectrum[272] the first detectable peak is at 12.616 eV, and in the photoionization study[279] the first departure from the background level is at 12.615 ± 0.010 eV. The next substantial peak in the PES is at 12.753 eV, and in photoionization there is a break, or step, at 12.76 eV, again in close correspondence. The third significant increase in the PES is at 12.887 eV, and in the photoionization curve one sees a clearly defined step at 12.88 eV. The agreement between the two very different techniques is excellent. The interpretation of both is more difficult.

A simple and direct interpretation is that the successive breaks, or peaks, correspond to successive excitations of the ν_2 frequency in the ground state of CH_4^+. The mean ν_2 frequency thus deduced (~ 1100 cm^{-1}) is reasonably close to Dixon's[271] calculated value for this frequency, ~ 1300 cm^{-1}. It is in the numbering of the ν_2 levels that a problem arises. In the PES, substructure can be detected within each vibrational band. Rabalais *et al.* argue that this pattern of substructure can be matched by considering the effect of successive ν_2 excitations distorting the tetrahedral $^2T_2$ state, if the first band is associated with $\nu_2 = 1$, and hence that $\nu_2 = 0$ is undetectable because it lies outside the Franck–Condon region. Their extrapolated adiabatic ionization potential for CH_4 is then ~ 12.51 eV. On the other hand, the agreement between detailed portions of the PES and photoionization curves suggests that the adiabatic ionization potential may actually be attained.

The photoionization experiment is capable of detecting not only direct ionization to CH_4^+, but also autoionization from the excited Rydberg states. However, the ionization yield has been seen to be extremely low in the near-threshold region, so that the autoionization contribution may itself be too

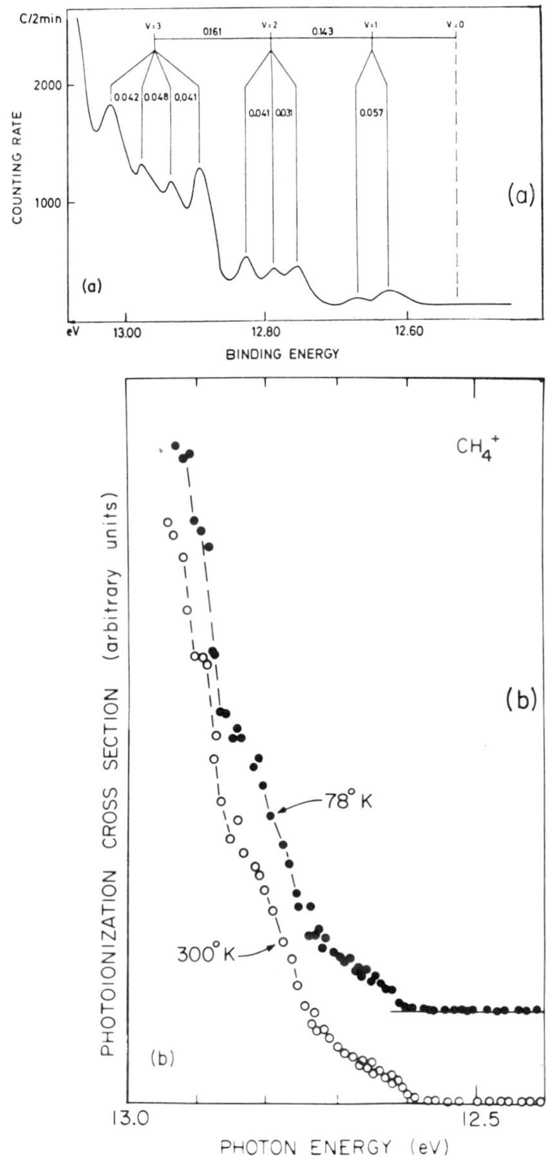

Fig. 64 (a) Photoelectron spectrum of CH_4 in the threshold region obtained with 21.2 eV incident radiation. [From Rabalais *et al.*, *Phys. Scr.* **3**, 13 (1971).] (b) Mass selected photoion yield curve of CH_4^+ in the threshold region with the gas sample at 78 and 300 K. [From W. A. Chupka and J. Berkowitz, *J. Chem. Phys.* **54**, 4256 (1971).] Note: The energy scales of 64a and b have been matched so that corresponding structures and thresholds can be directly compared.

weak to enhance a lower threshold, if it exists. At the present time, the matter remains unresolved.

The intensity of CH_4^+ increases with $h\nu$ from threshold to 14.3 eV, at which energy CH_3^+ begins to appear.[280,281] There appears to be a cusp in the CH_4^+ intensity just before the CH_3^+ onset, which has been attributed to autoionization.[281] Between 15 and 17 eV, the CH_4^+ intensity remains approximately constant, and thereafter slowly declines. The CH_3^+ intensity increases from threshold (14.3 eV) to ~17 eV, and then slowly declines. CH_2^+ begins to appear at ~15.2 eV, increases in intensity to about 16.5 eV, and then declines. This characteristic behavior of the three ions can be understood by reference to the shape of the $(1t_2)^{-1}$ photoelectron band. Additional states (and hence sources for formation of these ions) continue to be made available until the energy breadth of this band is exhausted somewhat beyond 16 eV.[253,272,273] The decline in ion intensity of all three peaks to higher energy is attributable to the decline in partial cross section for $(1t_2)^{-1}$ with increasing $h\nu$ beyond this energy.

Stockbauer[282] has examined the decomposition of the molecular ionic states in the $(1t_2)^{-1}$ band by a photoelectron–photoion coincidence method. His experiments, in principle, determine the number of parent ions produced with known internal energy and the number and mass of the fragment ions resulting from these parents. The number of parent ions with internal energy $h\nu - \mathrm{IP}$, where $h\nu$ is the photon energy and IP is the adiabatic ionization potential, is determined by measuring the number of threshold electrons, i.e., those electrons with near zero initial kinetic energy. The number of fragment ions from these parents is proportional to the number of ions detected in coincidence with the threshold electrons. A breakdown curve is obtained which is the ratio of the number of threshold electron–ion coincidences in each mass channel to the number of threshold electrons, as a function of photon energy. After making corrections for ion transmission as a function of kinetic energy, imperfect mass resolution, accidental coincidences, and stray backround, he reports the breakdown diagram shown in Fig. 65. The dashed lines in this diagram are calculations of Vestal[283] based on quasi-equilibrium theory (QET). The points are the results of charge exchange measurements by von Koch.[284] Stockbauer concludes that QET qualitatively describes the fragmentation of the $(1t_2)^{-1}$ state of the CH_4^+, but believes that quantitative agreement could be improved by using more accurate thermochemical data in the QET calculations.

While CH_4 is a relatively small molecule for application of QET, the Stockbauer experiment is limited to the $(1t_2)^{-1}$ band, with its interwoven Jahn–Teller states which interact vibronically. Hence, since autoionization is largely avoided in his experiment, Stockbauer is primarily testing the ergodicity of the vibrational phase space.

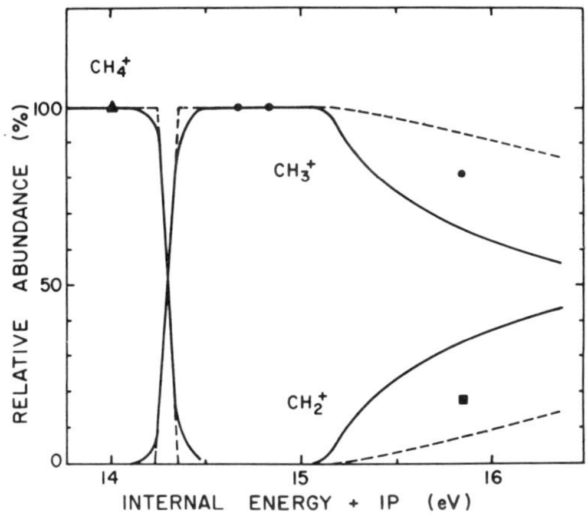

Fig. 65 Breakdown diagram of CH_4^+. ——— Photoion–photoelectron coincidence measurements by R. Stockbauer, *J. Chem. Phys.* **58**, 3800 (1973). ●, ▲, ■ Charge exchange measurements by H. von Koch, *Ark. Fys.* **28**, 529 (1964). --- QET calculations by M. Vestal, *in* "Fundamental Processes in Radiation Chemistry" (P. Ausloos, ed.), pp. 59–118. Wiley, New York, 1968. [From Stockbauer, cited here.]

However, more detailed predictions of QET† run counter to recent experimental results. On the one hand, it has been found[284a] that the rotational thermal energy, $\frac{3}{2}kT$, is fully available for the dissociation process. Other experiments[284b,c] have focussed on a metastable ion observed, corresponding to the process

$$CH_4^+ \rightarrow CH_3^+ + H.$$

The kinetic energy imparted to CH_3^+ has been found to increase with the temperature of the initial CH_4[284b] between 400 and 500 K; at 1500 K, the distribution curve of kinetic energies displays shoulders or discrete structure.[284c] Solka *et al.*[284b] have rationalized their observation with a model involving a rotational (centrifugal) barrier which increases with the average rotational state and hence with temperature. From the viewpoint of the reverse reaction, a larger barrier implies that those traversals (or tunnelings) of the barrier which are successful result in a greater release of kinetic energy. Hence, at least for the metastable ion, increasing rotational energy is largely manifested as an increase in the kinetic energy of the products.

† There are a number of formulations of QET. In a limited form, it involves the fraction of phase space of the transition state, at a particular energy. Hence, no prediction of the internal energies or kinetic energies of *the products* is forthcoming from this description. However, extensions of QET, either via the equipartition principle or a consideration of the reverse reaction, can generate such predictions.

Flamme *et al.*[284c,d] have made model calculations using the centrifugal barrier idea in a quasi-diatomic picture. The CH_3^+–H potential is taken as the sum of ion-induced dipole and centrifugal terms. In their model, the stable vibrational levels situated just below the dissociation asymptote become quasi-bound levels when the rotational energy is taken into account. They calculate the tunneling rates of the various levels and purport to show that the subset of states with lifetimes in the 10^{-5}–10^{-7} sec range (the experimental range) can be related to their observed structure in the kinetic energy distribution of the metastable ion. They conclude that the "observed structure in the metastable peaks must be related to a lack of coupling between the vibro-rotational modes of the CH_4^+ ion. The basic condition of the QET, i.e. energy randomization by fast coupling between the vibrational modes, would not be fulfilled."[284c] However, they also note that "In experiments where no lifetime selection is effective, the dissociation results as the total evolution of all the vibronic levels of the ion: a statistical model could then be more appropriate."[284d]

Although some statistical models[284e] have explicitly included rotational effects, the magnitude of the kinetic energy release calculated is smaller than observed.

Stockbauer[284f] has recently measured the dependence of fragment kinetic energy on the internal energy of CH_4^+ using photoion–photoelectron coincidence. He finds again that the kinetic energy observed is substantially larger than predicted by calculation, and furthermore, that the rate of increase of kinetic energy of the CH_2^+ fragment is twice as large as that of the CH_3^+ fragment, although calculations indicated a similar rate.

Regardless of the mechanism(s) governing the $(1t_2)^{-1}$ decomposition, it is reasonable to assume that the branching ratio from this state will not change appreciably as the photon energy is increased beyond ~17 eV, the high energy limit of the photoelectron band. (Small differences may be expected if the Jahn–Teller states do not have the same cross section dependence with excess photon energy.) In Table 11, the CH_3^+:CH_4^+ and CH_2^+:CH_4^+

TABLE 11 *Fragmentation Ratios CH_3/CH_4^+ and CH_2^+/CH_4^+ from Methane at 16.8 and 21.2 eV*

	CH_3^+/CH_4^+		CH_2^+/CH_4^+	
Ref.	16.8 eV	21.2 eV	16.8 eV	21.2 eV
(147)	1.04	1.07	0.0611	0.0780
(284h)	1.09	1.04	0.0470	0.048
(280)	0.80	—	0.30	—
(276)	0.97	1.02	0.05	0.06

ratios at 16.8 and 21.2 eV obtained by several investigators are compared. The agreement is satisfactory, except for the $CH_2^+:CH_4^+$ ratio of Dibeler *et al.* at 16.8 eV. The fragmentation behavior of the $(2a_1)^{-1}$ state can be studied by extending the photon energy range through this band. The results of such an investigation, performed by Kronebusch and Berkowitz,[147] are displayed in Fig. 66 as branching ratios versus photon energy. Also shown on this diagram is a 40.8 eV photoelectron spectrum, manufactured from previously described photoelectron spectra and partial cross sections.

The thermochemical threshold for the process

$$CH_4 \rightarrow CH_3^+ + H + e \tag{35}$$

at 0 K is 14.320 eV, while the corresponding threshold for

$$CH_4 \rightarrow CH_2^+ + H_2 + e \tag{36}$$

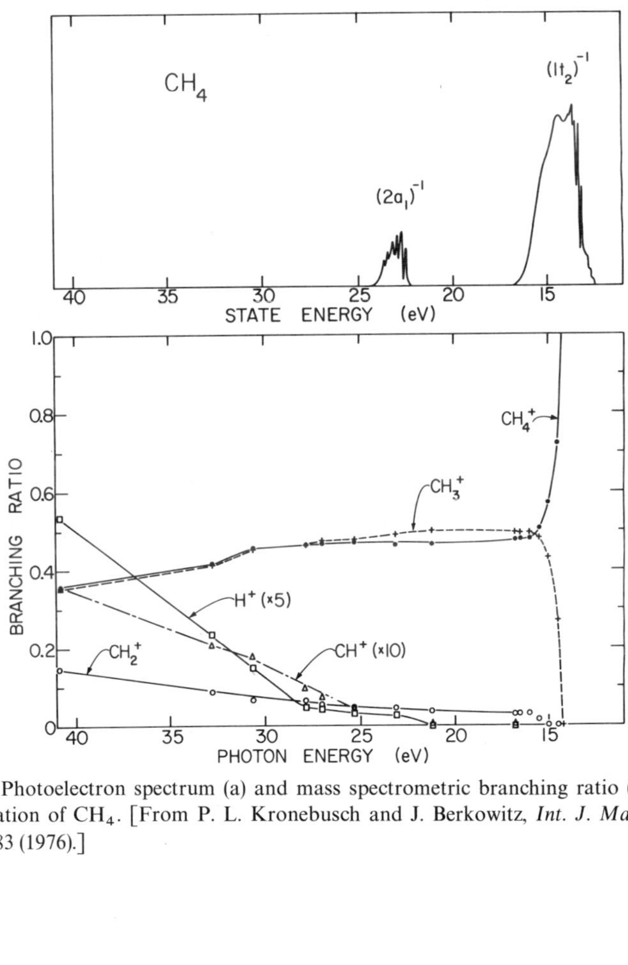

Fig. 66 Photoelectron spectrum (a) and mass spectrometric branching ratio (b) from the photoionization of CH_4. [From P. L. Kronebusch and J. Berkowitz, *Int. J. Mass Spec. Ion Phys.* **22**, 283 (1976).]

is 15.09 eV. These are essentially the values that one can deduce from the photoionization results,[284a] when allowance is made for the internal (rotational) energy of the initial molecule.

The process

$$CH_4 \rightarrow CH_3 + H^+ + e \tag{37}$$

has a threshold at 18.072 eV, which is beyond the breadth of the $(1t_2)^{-1}$ band. Hence, it is not surprising that H^+ does not begin to appear significantly until the $(2a_1)^{-1}$ threshold is attained. Decomposition into 3 particles after ionization, i.e.,

$$CH_4 \rightarrow CH_2 + H^+ + H + e \tag{38}$$

is thermochemically permitted above 22.77 eV, which occurs within the $(2a_1)^{-1}$ band. Hence, without coincidence measurements it is difficult to distinguish between reactions (37) and (38) as the primary sources of H^+, although there does appear to be a change of slope in the H^+ branching ratio suggesting the onset of a new process. [A measurement of the kinetic energy of H^+ from this decomposition has recently been performed by Browning *et al.*,[284g] who conclude that process (37) is dominant here.] Similarly, the additional mode for formation of ${CH_2}^+$, i.e.,

$$CH_4 \rightarrow {CH_2}^+ + 2H + e \tag{39}$$

can occur beyond 19.57 eV, and ${CH_2}^+$ does appear to enhance its relative yield beyond the $(2a_1)^{-1}$ onset, indicating that this must be one of the significant decomposition channels of $(2a_1)^{-1}$.

The first threshold for CH^+,

$$CH_4 \rightarrow CH^+ + H + H_2 + e \tag{40}$$

occurs thermochemically at 19.69 eV. From Fig. 66, we note that the branching ratio of CH^+ increases more-or-less linearly from the onset of the $(2a_1)^{-1}$ state up to 40.8 eV.

In summary, the $(2a_1)^{-1}$ state exhibits processes (38)–(40) as prominent decomposition channels, probably in that order of relative importance. After subtraction of the ${CH_2}^+$ component from process (36), the sum of these fragments amounts to about $\frac{1}{4}$–$\frac{1}{5}$ of the ions, suggesting that the relative intensities of $(1t_2)^{-1}$:$(2a_1)^{-1}$ at 40.8 eV is somewhat smaller than the 7:1 value given by Brundle *et al.*, and more in line with the value deduced by Backx and Van der Wiel. Transmission efficiencies for CH^+ and H^+ smaller than unity could make the $(1t_2)^{-1}$:$(2a_1)^{-1}$ ratio still smaller. We should note here, however, that Backx and Van der Wiel,[276] and Van der Wiel *et al.*[277] report the presence of weak two-electron excited states at ~27 and ~31 eV which could contribute to the higher energy fragmentation processes, and hence confuse the identification of fragment product intensities with partial photoelectron cross sections. One of these states has been assigned to the configuration

$$(1a_1)^2(2a_1)^2(1t_2)^4 3a_1$$

by Appell.[285]

The decomposition reaction

$$CH_4 \rightarrow CH_2 + H_2^+ + e \quad (41)$$

has a thremochemical onset at 20.12 eV, but H_2^+ never becomes significant in the energy range considered.

The ejection of an electron from the $1a_1$ orbital and the processes that ensue have been examined by a variety of techniques. A high resolution x-ray photoelectron spectrum of this region (Fig. 67) shows a distinctly asymmetric line, which has been interpreted as vibrational excitation on the high-energy side, resulting from a Franck–Condon excitation in which $C^*H_4^+$ has *shorter* C—H bonds than in the neutral ground state. Indeed, Meyer[286] has calculated the potential curve of $C^*H_4^+$ with a C 1s hole and found the equilibrium C—H distance to be 0.05 Å shorter than in the neutral ground state. This small displacement is sufficient to account for the two additional excited vibrations (presumably symmetric stretch) seen in the spectrum.

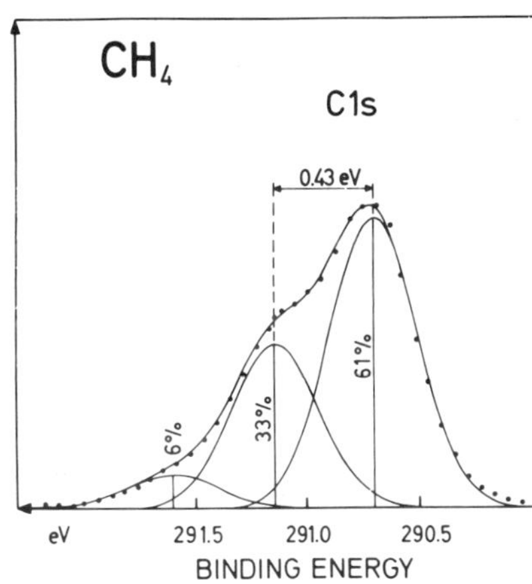

Fig. 67 High resolution x-ray photoelectron spectrum of CH_4 in the energy region corresponding to ejection from the carbon 1s-like orbital. The asymmetric profile has been deconvoluted into three vibrational components, corresponding to Franck–Condon transitions between neutral ground state CH_4 and CH_4^+ in which the excited ion has *smaller* internuclear spacings than the neutral molecule. ●, experimental and ———, least squares fit. [From U. Gelius, *J. Electr. Spectr.* **5**, 985 (1974).]

The decay of $C^*H_4^+$ occurs by fluorescence and the Auger process, the latter being about 300 times faster than the former. Nevertheless, the fluorescence spectrum has been observed[4] (Fig. 68). The main peak occurs at 276 eV, which agrees with the difference between $(1a_1)^{-1}$ (290.7 eV) and $(1t_2)^{-1}$

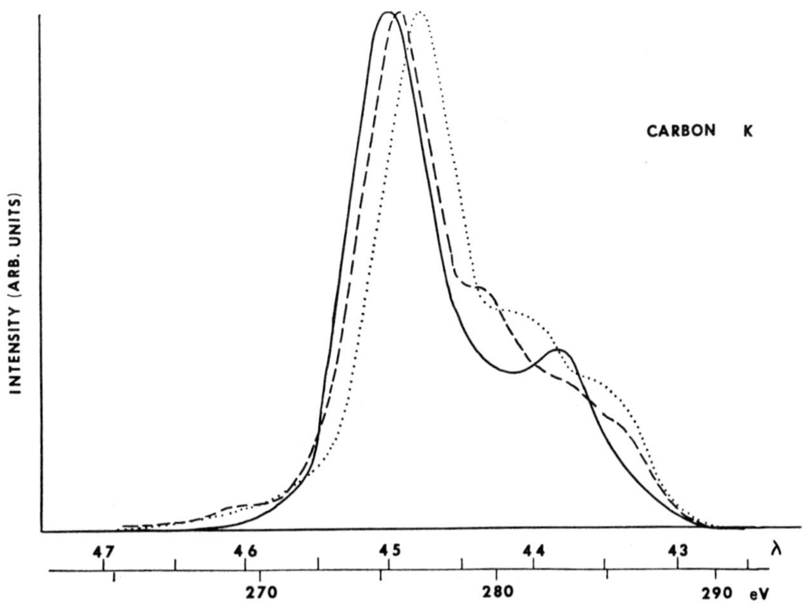

Fig. 68 X-ray fluorescence following creation of a vacancy in the carbon 1s-like orbital of CH_4 (and other molecules). CH_4, ———; CH_3Cl, ---; and CH_2Cl_2, ····. [From R. C. Ehlert and R. A. Mattson, *J. Chem. Phys.* **48**, 5465 (1968).]

($\sim$14 eV vertical IP). There is a satellite at an energy about 6 eV higher, which is possibly due to an initial excitation process leading to $1a_1 2a_1 (1t_2)^6$, which decays to $(1a_1)^2 2a_1 (1t_2)^5$. (The initial excitation in this case was performed by electron impact.)

The Auger process has been reported by Spohr *et al.*[287] (see Fig. 69). The main peak is due to the process

$$1a_1(2a_2)^2(1t_2)^6 \rightarrow (1a_1)^2(2a_1)^2(1t_2)^4 + \bar{e}$$

and liberates an electron with 250 eV kinetic energy. This implies that the process $CH_4{}^+ \rightarrow CH_4^{2+} + \bar{e}$ requires $276 - 250 = 26$ eV energy. Auxiliary peaks to lower kinetic energy correspond to the processes

$$1a_1(2a_1)^2(1t_2)^6 \rightarrow (1a_1)^2 2a_1(1t_2)^5 + \bar{e}$$

(kinetic energy of electron $\cong$237 eV, ionization of $2a_1$ orbital requiring 39 eV) and

$$1a_1(2a_1)^2(1t_2)^6 \rightarrow (1a_1)^2(1t_2)^6 + \bar{e}$$

(kinetic energy of electron $\cong$230 eV, ionization of $2a_1$ orbital from this configuration requiring $\sim$38 eV.)

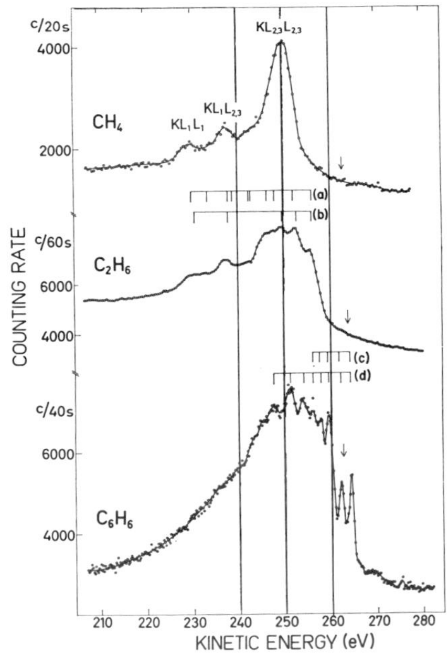

Fig. 69 Auger spectrum (designated KLL) following creation of a vacancy in the carbon 1s-like orbital of CH_4 (and other molecules). [From R. Spohr *et al.*, *Phys. Scripta* **2**, 31 (1970).]

The dominant process thus leaves us with CH_4^{2+}, which is $\sim$40 eV excited relative to neutral CH_4. On the other hand, from process (35) and the ionization potential of atomic hydrogen (13.6 eV) we can estimate the threshold for $CH_4 \rightarrow CH_3^+ + H^+ + \bar{e}$ to be $\sim$28 eV. Hence, the CH_4^{2+} should be unstable with respect to proton ejection. The ion CH_4^{2+} has not been reported in mass spectra, and the breadth of the Auger line in Fig. 69 has in fact been attributed to a short lifetime of the final state of the Auger transition due to this decomposition.

Other decomposition processes should also be possible from CH_4^{2+}, either from the $(1t_2)^{-2}$, $(2a_1)^{-1}(1t_2)^{-1}$ or $(2a_1)^{-2}$ states. Using 10 keV electron excitation, Backx and Van der Wiel report the following relative abundances of CH_4^{2+} decomposition:

$H^+ + CH_3^+$	1.0,	$H^+ + C^+$	0.64,
$H^+ + CH_2^+$	1.42,	$H_2^+ + CH_2^+$	0.32.
$H^+ + CH^+$	0.97,		

Finally, we take note of recent photoabsorption studies[288] on CH_4, in which weak structure is observed in the wavelength region between 550 and 640 Å. It has been interpreted in straightforward fashion as a Rydberg series converging on $(2a_1)^{-1}$. Fluorescence has been seen[288] which correlates with this structure. We[289] have subsequently measured the photoionization mass spectra of CH_4^+, CH_3^+, and CH_2^+ very carefully in this wavelength region, and find no sign of structure. Hence, these Rydberg states apparently predissociate with high probability, producing excited fragments which fluoresce.

3. C_2H_2

The structure of this linear molecule ($D_{\infty h}$ symmetry) in its electronic ground state has the molecular orbital sequence

$$(1\sigma_g)^2(1\sigma_u)^2(2\sigma_g)^2(2\sigma_u)^2(3\sigma_g)^2(1\pi_u)^4.$$

The adiabatic ionization potential $[(1\pi_u)^{-1}]$ has been reported[290] to occur at 11.398 ± 0.005 eV. In early experiments[291,292] it appeared that "pure" step structure corresponding to excitation of a stretching vibration of $C_2H_2^+$ characterized the threshold region. More recently, Dibeler and Walker[290] have observed autoionization features superposed upon the steps (see Fig. 70). The autoionizing Rydberg series apparently converge to the vibrationally excited states of the electronic ground state of $C_2H_2^+$, $\tilde{X}^2\Pi_u$. The gross step-like features (which correlate reasonably with Franck–Condon factors)[290] imply that predissociation is a significant competitor of autoionization for the decay of the Rydberg states in this interval. However, the ionization yield

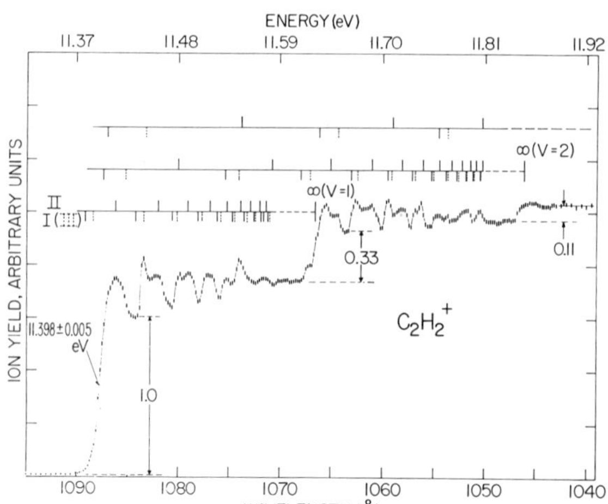

Fig. 70 Mass selected photoion yield curve of $C_2H_2^+$ in the vicinity of the ionization threshold, revealing autoionization structure superimposed upon the steplike vibrational structure. The autoionizing features appear to correspond to states converging upon vibrationally excited $C_2H_2^+$ in its ground ($\tilde{X}\ ^2\Pi_u$) state, implying that the vibrational autoionization mechanism is involved. [From V. H. Dibeler and J. A. Walker, *Int. J. Mass Spec. Ion Phys.* **11**, 49 (1973).]

is 0.92[237] at 740 Å and is unity at 584 Å.[293] The onset for $(3\sigma_g)^{-1}$ is at 16.36 eV.[294] A broader photoelectron band is associated with electron ejection from this molecular orbital, implying that some bonding character has been disturbed. Incompletely resolved vibrational structure is observed[294] and has been assigned to excitation of the C≡C and C—H stretching modes.

The onset of the first fragmentation process

$$C_2H_2 + h\nu \rightarrow C_2H^+ + H + \bar{e}$$

occurs at 17.44 ± 0.01 eV[295] which is in the energy range of this band. Figure 71 displays the photoionization yield of $C_2H_2^+$ and C_2H^+ from 1000 to 600 Å, taken at higher resolution than that of Dibeler *et al.* The incomplete resolution of vibrational structure cannot be attributed to rapid dissociation, since it manifests itself below the apparent threshold for fragmentation.

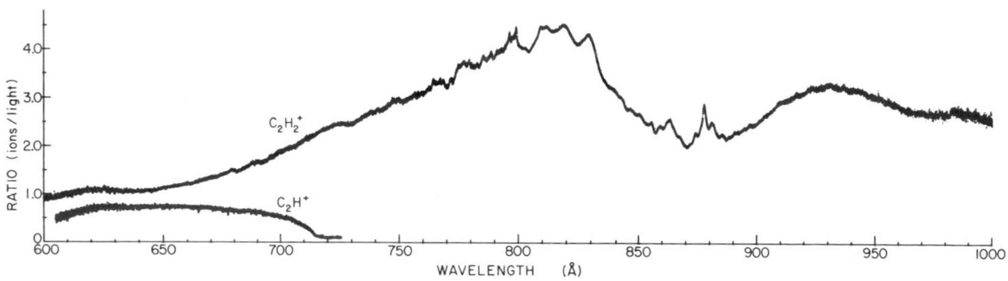

Fig. 71 Photoion yield curves of C_2H^+ and $C_2H_2^+$ from C_2H_2 between 600 and 1000 Å, at a wavelength resolution (FWHM) of 0.14 Å. [Unpublished data, obtained by the author.]

The $(2\sigma_u)^{-1}$ ionization has a threshold at 18.38 eV.[294] Although broad, and well above the energy of the first fragmentation, it also exhibits vibrational structure assigned to the same stretching modes as in the second band.

The $(2\sigma_g)^{-1}$ ionization has been observed to occur at 23.5 eV with both 40.8[296] and 151.4 eV[297] photon sources.

The branching ratios can only be crudely derived from the existing data, and corrections for angular distribution variations are unavailable. At 21.2 eV, Thiel and Schweig[298] have obtained for $(1\pi_u)^{-1}$:$(3\sigma_g)^{-1}$:$(2\sigma_u)^{-1}$ the proportions 100 : 78 : 14. The branching ratios at 40.8 eV have been estimated from the published spectra of Streets and Potts[296] with the additional assumption that $(1\pi_u)^{-1}$ (not reported by these authors) bears the same relationship to $(3\sigma_g)^{-1}$ at 40.8 eV as at 21.2 eV; at 151.4 eV values are derived from the data of Cavell and Allison.[297] Spectra appropriate for such estimates in the x-ray region are presently unavailable, although the binding energy of the C 1s-like $1\sigma_g$ and $1\sigma_u$ have been measured[299] to be 291.14 eV.

The existing branching ratios, combined with the total absorption cross sections (Fig. 30 in Chapter V), yield the rather limited partial cross section

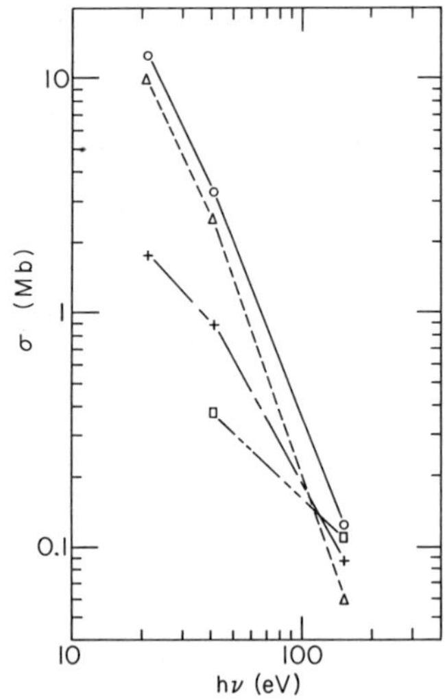

Fig. 72 Partial cross sections for formation of various states of $C_2H_2{}^+$ by photoabsorption. ○ $(1\pi_u)^{-1}$, △$(3\sigma_g)^{-1}$, $+(2\sigma_u)^{-1}$, and □$(2\sigma_g)^{-1}$.

graph of Fig. 72. Even in this limited domain, the $(2\sigma_u)^{-1}$ and $(2\sigma_g)^{-1}$ partial cross sections (primarily composed of C 2s orbitals) are seen to decline less rapidly than $(1\pi_u)^{-1}$ and $(3\sigma_g)^{-1}$. Cavell and Allison note that at 1253.6 eV the latter two cross sections are "at least one and perhaps two orders of magnitude" less than $(2\sigma_u)^{-1}$ and $(2\sigma_g)^{-1}$.

The treatment of the decomposition of $C_2H_2{}^+$ is ambiguous since it occupies an intermediate region between the small and large polyatomic molecular ions. One might expect fluorescence from the first excited $(^2\Sigma_g)$ ionic state to the ground $(^2\Pi_u)$ state, since the selection rules are favorable and similar transitions have been observed in the mono- and dihaloacetylenes.[11] Also, a large portion of the excited state band lies below the first dissociation limit. No observation of such emissions has yet been reported, perhaps due in part to the more difficult region (~2400 Å) where this transition can be expected. The larger energy gap tends to disfavor radiationless transitions, which also require vibronic mixing for the appropriate selection rules to apply. Hence, the validity of this first assumption for the applicability of QET remains to be tested.

Haarhoff[300] has made such statistical calculations and utilized the results to test H—D isotopic effects in decomposition, comparing his calculations with electron impact mass spectra. His calculated "breakdown diagram" (fractional products versus internal energy of molecular ion) is shown in Fig. 73. He concludes that the QET "provides a satisfactory description of the main processes which determine the decomposition pattern" but "there are also other indications that electronic excitation energy of ions is not completely randomly distributed."

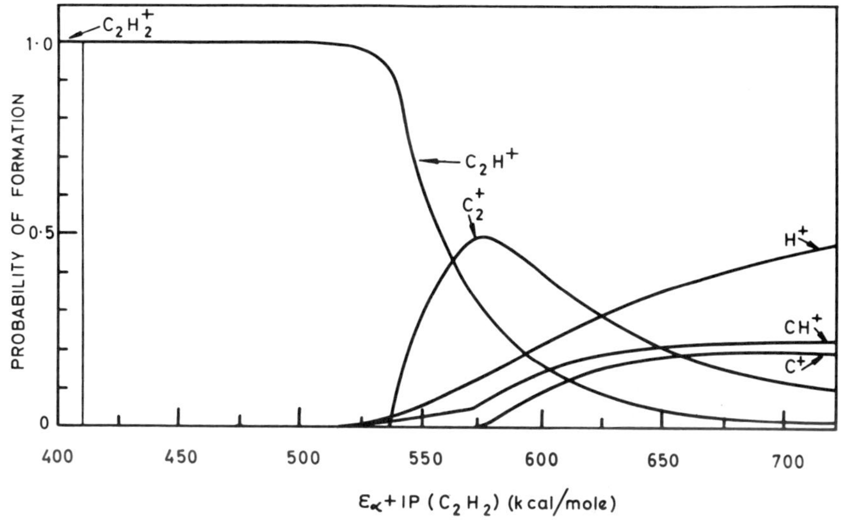

Fig. 73 Calculated breakdown diagram of $C_2H_2^+$, employing QET. [From P. C. Haarhoff, *Mol. Phys.* **8**, 49 (1964).]

Vestal[301] has performed QET calculations which "show quite unequivocably that if the reaction ($C_2H_2^+ \rightarrow C_2H^+ + H$) occurs from the ground electronic states (i.e., with complete randomization) the metastable peak corresponding to the reaction should be observed (calculated as 0.15% of total ionization). On the other hand, if the reaction occurs from the excited state, without randomization to the ground state, the calculation predicts that the metastable should not be observed."

Cooks *et al.*[302] report such a metastable ion having both unimolecular and collision induced contributions (see Fig. 74). It appears to this observer that the presence of the unimolecular component is not firmly established by this measurement. Von Puttkamer[303] has performed photoion–photoelectron coincidence studies on this system and observed a rise in parent ion yield above the fragment threshold (which would violate QET) but his results are suspect (because of improper correction of false coincidences).

The thermochemistry of this decomposition is in doubt, and introduces an ambiguity into the interpretation of the kinetic behavior. The uncertainty involves $\Delta H^\circ_{f0}(C_2H)$ and $IP(C_2H)$.

A direct measurement of $\Delta H_f(C_2H)$ based on equilibrium studies with a Knudsen cell[304] has yielded $\Delta H^\circ_{f298}(C_2H) = 130 \pm 3$ kcal/mole, and an electron impact ionization potential of 11.6 ± 0.5 eV. Okabe and Dibeler[305] have studied the photodissociation threshold

$$C_2HCN + h\nu \rightarrow C_2H + CN^*$$

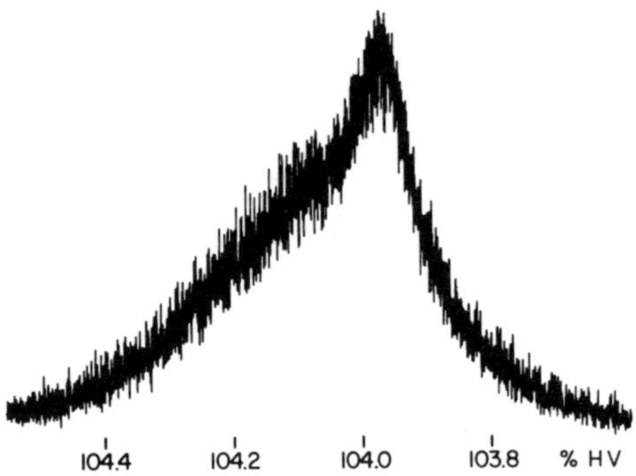

Fig. 74 A scan of the metastable ion region corresponding to the decomposition $C_2H_2^+ \rightarrow C_2H^+ + H$. [From R. G. Cooks *et al.*, "Metastable Ions," p. 82. Elsevier, Amsterdam, 1973. The authors contend that this scan represents a superposition of the unimolecular decomposition and a collision-induced decomposition.]

for CN emission at 3880 Å, and obtain 9.41 ± 0.04 eV for this onset. Subtracting the excitation energy of CN leaves 6.21 eV for the minimum energy of bond rupture. These authors[305] have also measured the photodissociative ionization threshold for the reaction

$$C_2HCN + h\nu \rightarrow C_2H^+ + CN + \bar{e},$$

and report 18.19 ± 0.04 eV for this onset.

Since $\Delta H_f(C_2HCN)$ is not accurately known, the most direct interpretation of these experiments is that the difference in thresholds (18.19 − 6.21 = 11.98 eV) corresponds to the ionization potential of C_2H. This value falls within the error limit of the directly determined electron impact value, but is ~0.4 eV higher. Usually, measurements of ionization potentials by photoionization fall several tenths of an electron volt below electron impact determinations. It should also be mentioned that Okabe and Dibeler have noted that both photodissociation and photodissociative ionization processes were weak, and the latter (being a higher energy process) may experience a substantial kinetic shift.

An alternative measurement aimed at determining $IP(C_2H)$ has been performed by Miller and Berkowitz.[306] These authors studied two thresholds for photodissociative ionization of C_2HBr, i.e.,

$$C_2HBr + h\nu \rightarrow C_2H^+ + Br + e \tag{42}$$

and

$$C_2HBr + h\nu \rightarrow C_2H + Br^+ + e. \tag{43}$$

Although $\Delta H_f(C_2HBr)$ is not well established, the difference in thresholds should approximate the difference in ionization potentials of C_2H and Br, since both processes are simple bond ruptures. These authors observed ~ 15.56 eV for reaction (42), and ~ 15.90 eV for reaction (43). Since IP(Br) = 11.847 eV,[30] we deduce $IP(C_2H) \cong 11.51$ eV, in surprisingly good agreement with the direct electron impact value, but on the low-energy side, as is more common. On the basis of the available evidence, this would appear to be the best current estimate for $IP(C_2H)$.

The photodissociative ionization of C_2H_2 to yield C_2H^+ discussed earlier[295] (see Fig. 71) was found to have a 0 K threshold at 17.44 eV. Using $IP(C_2H = 11.51$ eV) and well-known heats of formation for C_2H_2[176] and H,[176] we compute $\Delta H^\circ_{f0}(C_2H) = 139.4$ kcal/mole from the photoionization determination.

Maier[307] has measured the dissociative ionization threshold by endothermic ion charge transfer for the reaction

$$Ar^+ + C_2H_2 \rightarrow C_2H^+ + H.$$

With his standard treatment of thresholds for ion impact, he deduces 16.72 ± 0.12 eV for this onset, but adds" . . . that $I(C_2H) + D(C_2H - H) \leq 16.97$ eV in any case." If there are selection rules which prevent the attainment of the true thermochemical threshold in photoionization (vide infra), they could be invalidated in the charge transfer process. With this upper limit and $IP(C_2H) = 11.51$ eV, we obtain $\Delta H^\circ_{f0}(C_2H) \leq 128.6$ kcal/mole.

This value is in very good agreement with $H^\circ_{f0}(C_2H) = 129 \pm 3$ kcal/mole determined by high temperature equilibria in a Knudsen cell. In turn, it implies that photodissociative ionization of C_2H_2 has a vanishingly small probability at threshold, and requires an excess energy of at least 0.47 eV.

Such behavior is rare, but does occasionally occur. (See NO^+ from N_2O, Section C.3.)

In 1967, Fiquet-Fayard[308] arrived at this conclusion from less reliable evidence, and examined various possible causes for such behavior. Fayard examined the possible states of C_2H^+, based on the perturbation of known states of C_2 by a proton. She concluded that the ground state of C_2H^+ is $^2\Sigma_g{}^-$, making the asymptotic form of $C_2H^+ + H$ have either $^4\Sigma_g{}^-$ or $^2\Sigma_g{}^-$ symmetry.

A second element in her analysis, based on correlation rules and energetics, is that predissociation (rather than direct dissociation) is the mechanism for fragmentation near threshold. Examination of the selection rules for spin–orbit interaction or Born–Oppenheimer breakdown as mechanisms for predissociation lead (in first approximation) to the conclusion that only $^2\Sigma_g{}^+$ and $^2\Sigma_u{}^+$ of $C_2H_2{}^+$ are predissociated. By analogy with the cases of H_2O and H_2S, she concluded that $^2\Sigma_u{}^+$ is completely predissociated by a $^4\Pi_g$ state, forming C_2H^+ in the excited $^3\Pi$ state.

The $^2\Sigma_g{}^+$ state can be predissociated by $^4\Sigma_g{}^-$, yielding C_2H^+ in its $^3\Sigma_g{}^-$ ground state. Fayard believed that this may not occur (i.e., the states may not cross) and hence that the onset for C_2H^+ may be forced to occur with C_2H^+ in its excited $^3\Pi$ state. This implies that $^2\Sigma_u{}^+(C_2H_2{}^+)$ be the progenitor of the fragmentation process. However, the adiabatic threshold of $^2\Sigma_u{}^+$ is 18.36 eV[294] about 1 eV above the fragmentation threshold for C_2H^+. This would require that the Franck–Condon domain in the photoelectron spectrum is far from the minimum of $^2\Sigma_u{}^+$. An alternative, and perhaps more attractive mechanism, involves the predissociation of $^2\Sigma_g{}^+$ by $^4\Sigma_g{}^-$, with the curve crossing occurring 0.47 eV or more above the asymptotic $C_2H^+(^3\Sigma_g{}^-) + H(^2S)$.

If subsequent studies verify that C_2H^+ is formed ~0.5 eV above its thermochemical threshold, it would vitiate the application of a fully statistical theory since states of $C_2H_2{}^+$ are formed at the thermochemical threshold for fragmentation.

The next decomposition process anticipated on energetic grounds is

$$C_2H_2 + h\nu \rightarrow C_2{}^+ + H_2 + \bar{e}$$

at a threshold energy of ~18.2 eV.

The $(2\sigma_u)^{-1}$ band could conceivably decay to $C_2{}^+$ on this basis, but the mass spectrum at 21.2 eV[309] has a negligible yield (0.26%) of $C_2{}^+$. The formation of $C_2{}^+$ by the decomposition

$$C_2H_2 + h\nu \rightarrow C_2{}^+ + 2H + \bar{e}$$

has a thermochemical threshold[310] at 22.7 ± 0.6 eV. Unpublished experimental results at Argonne[309] with discrete line sources between 21.2 and 40.8 eV show that $C_2{}^+$ achieves an intensity (relative to $C_2H_2{}^+$) of ~0.04 at 460 Å ≡ 26.95 eV. This is roughly the ratio of partial cross sections of $(2\sigma_g)^{-1}$ to $((1\pi_u)^{-1} + (3\sigma_g)^{-1})$ at this photon energy. Hence, we can conclude that a major (if not predominant) channel for the decomposition of $(2\sigma_g)^{-1}$ is to $C_2{}^+ + 2H$. This does not necessarily imply that the decomposition occurs directly from $(2\sigma_g)^{-1}$ without internal conversion, since Haarhoff's QET calculation (see Fig. 73) also predicts that the $C_2{}^+$ fragment is important at this energy. A photoion–photoelectron coincidence experiment isolating the $(2\sigma_g)^{-1}$ state would be revealing on this point.

The photodissociative ionization of C_2H_2 to form CH^+ and CH has a threshold at 20.53 eV, but the intensity of CH^+ is only 0.1 that of $C_2{}^+$ at $h\nu = 26.95$ eV. Since the primary progenitor of both $C_2{}^+$ and CH^+ is $(2\sigma_g)^{-1}$, centered at ~23.5 eV, the ratio of these fragments at this energy is roughly in line with Haarhoff's calculation (Fig. 73).

No information has come to light, either in the form of x-ray emission, Auger, or mass spectrometry, regarding the processes pursuant to carbon K-shell electron ejection in acetylene.

4. C_2H_4

The sequence of molecular orbitals describing the electronic ground state of this molecule (symmetry D_{2h}) is

$$(1a_g)^2(1b_{3u})^2(2a_g)^2(2b_{3u})^2(1b_{2u})^2(3a_g)^2(1b_{1g})^2(1b_{1u})^2.$$

In earlier photoelectron studies[311] a reversal of the ordering of $3a_g$ and $1b_{1g}$ had been suggested, but the above sequence is strongly supported by ab initio calculations and x-ray PES of the valence shells, combined with Gelius' intensity model.[312] Photoelectron spectra from which partial cross sections may be evaluated are available at 21.2, 40.8, 132.3, and 1253.6 eV. The ionization energies of the various orbitals are summarized in Table 12.

TABLE 12 *Adiabatic and Vertical Ionization Energies of C_2H_4*

	Vertical (eV)	Adiabatic (eV)	Ref.
$1b_{1u}$	(10.51)	10.51	(314)
$1b_{1g}$	12.85	12.45	(314)
$3a_g$	14.66	14.43	(314)
$1b_{2u}$	15.87	15.74	(314)
$2b_{3u}$	19.1	18.8	(314)
$2a_g$	23.68	~22.5	(316)
			(312)
$1b_{3u}$, $1a_g$ } C_{1s}	290.88		(312)

At 584 Å, spectra have been reported by several groups of investigators. (See, e.g., Branton *et al.*,[313] Brundle and Brown,[314] and White *et al.*[315]) For evaluation of intensities, the data of White *et al.*[315] are most useful. The 304 Å data are taken from Lindholm *et al.*[316] while the branching ratios at 132.3 and 1253.6 eV are given by Banna and Shirley.[133] When combined with the previously selected photoabsorption cross sections (Chapter V, Section D.4), we deduce the partial cross sections shown in Fig. 75. We note that the decline in partial cross section with increasing energy is more precipitous with ejection from the uppermost four orbitals than for the next two. This is consistent with the view that orbitals having dominant p character decline more rapidly than those with dominant s character. The uppermost orbital is of π type, and as noted by Turner *et al.*[311] "The average of vibrational peaks associated with this band (i.e., a strong 0–0 component followed by a short series of weaker peaks) is found to be fairly characteristic of π ionization." However, this vibrational fine structure involves some Fermi resonance

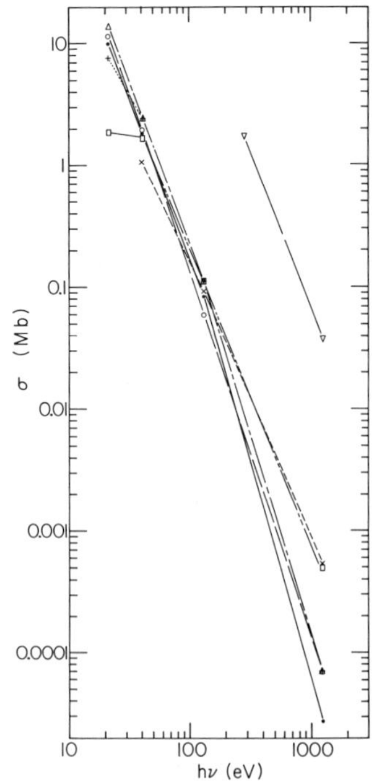

Fig. 75 Partial cross sections for formation of various states of $C_2H_4^+$ by photoabsorption. ● $(1b_{1u})^{-1}$, ○$(1b_{1g})^{-1}$, △$(3a_g)^{-1}$, + $(1b_{2u})^{-1}$, □$(2b_{3u})^{-1}$, × $(2a_g)^{-1}$, and ▽$(1b_{3u})^{-1}$ + $(1a_g)^{-1}$.

interactions, as noted by Brundle and Brown[314] and Stockbauer and Inghram.[317]

A rather large peak (not included in Table 12) encompassing the 25–30 eV range has been attributed[312] to shake-up involving a $2b_{3u}$ hole. The higher energy portion of this band overlaps the domain of $C_2H_4^{2+}$ states observed in double charge transfer studies and corresponding calculations.[318] Hence, it is conceivable that an Auger transition following formation of the shake-up state could produce $C_2H_4^{2+}$. Such an occurrence could have a profound effect on the decompositions ensuing from formation of the initial state. (See Cederbaum *et al.*[216a] for an alternative mechanism of formation for these states.)

The decomposition of $C_2H_4^+$ in its various states between threshold and ~20 eV provides a textbook example of the success of QET. In Fig. 76, photoelectron–photoion coincidence experiments of Stockbauer and Inghram[319] are compared with QET calculations performed by Vestal and Lerner[320] in a conventional breakdown diagram. The processes incorporated

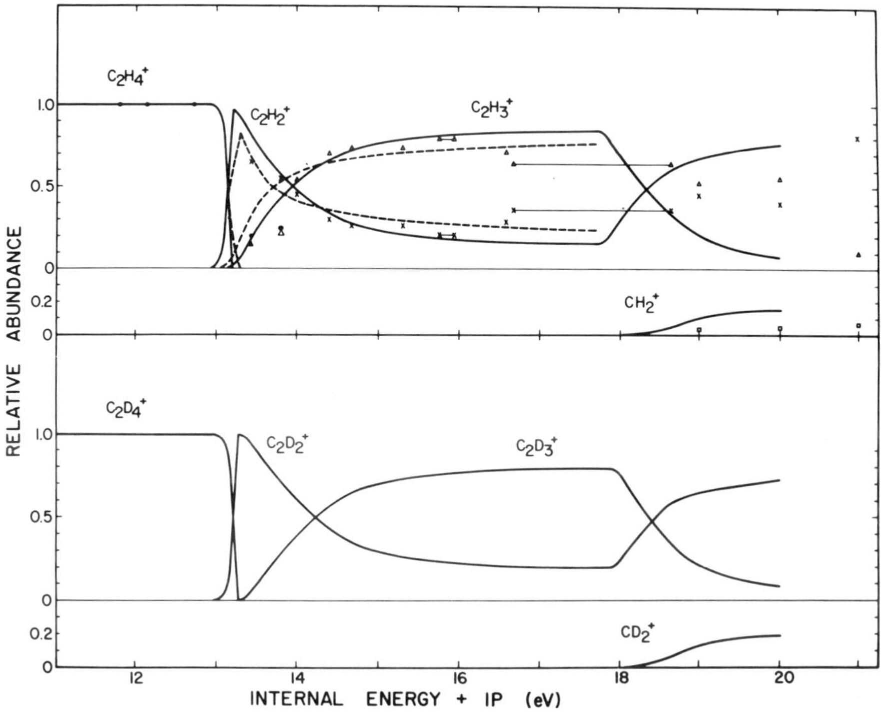

Fig. 76 Breakdown diagrams of $C_2H_4^+$ and $C_2D_4^+$. ——— Photoion–photoelectron coincidence measurements by R. Stockbauer and M. G. Inghram, *J. Chem. Phys.* **62**, 4862 (1975). --- QET calculations by M. Vestal and G. Lerner, Aerospace Research Laboratory Report 67-0114, Jan. 1967. ●, △, ×, □ Charge exchange experiments by I. Szabo, *Ark. Fys.* **31**, 287 (1965).

in this figure are as follows:

Process	Thermochemical threshold, eV
$C_2H_4 \rightarrow C_2H_2^+ + H_2 + \bar{e}$	13.125
$C_2H_4 \rightarrow C_2H_3^+ + H + \bar{e}$	13.22
$C_2H_4 \rightarrow C_2H_3^+ + H + \bar{e}$	
$\quad\quad \llcorner\!\!\longrightarrow C_2H_2^+ + H$	17.601
$C_2H_4 \rightarrow CH_2^+ + CH_2 + \bar{e}$	17.911

The fragmentation patterns appear to be insensitive to the successive appearance of ionic states. Of particular relevance are the $(3a_g)^{-1}$ and $(1b_{2u})^{-1}$ bands. No abrupt change in fragmentation occurs when these states become accessible. This provides strong support for the essential prerequisite of QET, i.e., that internal conversion occur rapidly, and that the vibrational energy and vibrational state density determine the decomposition pattern. Another interesting feature is the initial decline of $C_2H_2^+$, and its subsequent increase at ~18 eV. The initial decline is a consequence of the competition between a simple bond rupture (forming $C_2H_3^+ + H$) and the slower process which permits two hydrogen atoms to combine and be expelled (forming $C_2H_2^+ + H_2$). With sufficient internal energy in the parent ion, however, the $C_2H_3^+$ fragment retains enough energy after the first decomposition to decay further, into $C_2H_2^+ + H$. This accounts for the decrease in $C_2H_3^+$ and increase in $C_2H_2^+$ beyond ~18 eV.

Some minor differences between the experimental and calculated breakdown curves are attributed[319] to slightly incorrect thermochemical values (for $C_2H_3^+$) used in the calculations, and to some revisions required in the activated complex. Further evidence for the statistical nature of the decompositions in $C_2H_4^+$ is provided by the isotopic scrambling studies and metastable ion intensities of Nenner *et al.*[321] For example, the decomposition of HDC=CDH to form acetylene ion and hydrogen molecule produces different metastables if H_2, HD, or D_2 is eliminated. The metastable corresponding to H_2 elimination is 7–8 times more intense than that corresponding to HD elimination, and that corresponding to D_2 elimination is too weak to be observed. (Little difference is observed between cis and trans isomers.) Nenner *et al.*[321] calculate decomposition probabilities by QET in the time range of mass spectrometric flight (10^{-6}–10^{-5} sec) which semiquantitatively match the experimental observations.

At $h\nu = 21.2$ eV the photoionization mass spectrum is reported to be[322]

$$C_2H_4^+:C_2H_3^+:C_2H_3^+ = 100:137:76.$$

The thermochemical thresholds for higher energy processes are shown below.

Process	Energy, eV
$C_2H_4 \rightarrow C_2H^+ + H_2 + H + \bar{e}$	19.144
$\rightarrow C_2H^+ + 3H + \bar{e}$	23.622
$C_2H_4 \rightarrow C_2^+ + 2H_2 + e$	19.963
$\rightarrow C_2^+ + H_2 + 2H + \bar{e}$	24.441
$\rightarrow C_2^+ + 4H + \bar{e}$	28.919
$C_2H_4 \rightarrow CH^+ + CH + H_2 + \bar{e}$	22.251
$\rightarrow CH^+ + CH + 2H + \bar{e}$	26.729
$\rightarrow CH^+ + C + 3H + \bar{e}$	30.199

Preliminary studies[309] with discrete line sources at shorter wavelengths indicate that the contribution of $(2a_g)^{-1}$ at 23.7 eV, and of the shake-up band at ≈27 eV provides enhancement for CH_2^+ and C_2H^+, each of which attains ~6% of the parent ion intensity at 40.8 eV. No clear indication was found for the possible influence of doubly ionized states at ~30 eV, but effects due to such states would presumably be most readily seen in ion kinetic energy measurements.

Information about the processes following carbon K-shell electron emission is currently lacking.

5. C_2H_6

Although some controversy exists regarding the sequence of valence orbitals,[323,324] the most recent studies[325,326,326a] agree on the following ordering in staggered (D_{3d}) symmetry:

$$(1a_{1g})^2(1a_{2u})^2(2a_{1g})^2(2a_{2u})^2(1e_u)^4(3a_{1g})^2(1e_g)^4.$$

Table 13 lists the ionization energies for formation of the corresponding ionic states resulting from single-electron ejection and their atomic character.

The first photoelectron band encompasses the 12–14 eV energy range and includes both $(1e_g)^{-1}$ and $(3a_{1g})^{-1}$ transitions. Incompletely resolved vibrational structure is observed,[325] which serves as the basis for the analyses of Rabalais and Katrib[325] and Richartz *et al.*[326a] that vibronically induced Jahn–Teller interaction (involving a degenerate vibrational mode) splits $(1e_g)^{-1}$ into two substates.

Richartz *et al.*[326a] raise questions about the assignment of the 11.56 eV band to the lowest (i.e. 0–0) vibrational transition. The ambiguity here is

TABLE 13 *Vertical Ionization Energies for Single-Electron Ejection from C_2H_6 and the Atomic Character of the Orbital from Which Electron Emission Occurs*

State	Atomic character	Vertical ionization energy (eV)	Ref.
$(1e_g)^{-1}$	C(2pπ) & H(1s)	12.00 } Jahn–Teller 12.72 } components	(325)
$(3a_{1g})^{-1}$	C(2pσ) & H(1s)	13.5	(325)
$(1e_u)^{-1}$	C(2pπ) + H(1s)	15.0 } Jahn–Teller 15.8 } components	(325)
$(2a_{2u})^{-1}$	C(2s)	20.16	(325)
		(20.3)	(274)
$(2a_{1g})^{-1}$	C(2s)	23.6	(325)
		(23.9)	(274)
		24.1	(327)
$(1a_{2u})^{-1}$ $(1a_{1g})^{-1}$	C(1s)	290.5	(274)

similar to that in methane discussed earlier, where it had been argued by Rabalais *et al.*[272] that the true 0–0 transition lies outside the Franck–Condon region. The variable wavelength photoionization spectrum of ethane[327] displays the same vibrational structure and threshold as does the photoelectron spectrum. Since the photoionization spectrum includes (in principle) autoionization processes which can populate ionic states outside the Franck–Condon region, but despite this possible augmentation still has the same threshold as given by PES, the assignment given above seems plausible.

All other bands except the one between 20.1 and 20.8 eV do not display vibrational fine structure.

This latter band $(2a_{2u})^{-1}$ is similar to the 23 eV band in CH_4 (and in fact with corresponding bands in other methyl methanes[327]) in that vibrational fine structure exists well above the threshold for dissociation, and also because the corresponding orbitals consist of C(2s) combinations. From the half-width of the vibrational structure, one can infer a lifetime $\geq 10^{-14}$ sec. Hence, although these states are not inherently repulsive, the lifetime limit deduced cannot permit one to distinguish between a very long-lived state (surviving for $\sim 10^6$ vibrations) or one predissociated after a single vibration.

The branching ratio at 21.2 eV has been deduced from Turner *et al.*,[328] dividing the respective areas by the electron kinetic energy. At 40.8 eV, an average branching ratio, weighting the spectra of Potts *et al.*[327] and Lindholm *et al.*[316] equally, was selected. The high-energy branching ratios at 1486.6 eV are from Hamrin *et al.*[274] When combined with the total photoabsorption cross sections (Chapter V, Section D.5), the partial cross sections plotted in Fig. 77 are obtained. No corrections for angular distribution effects or ionization yields have been included.

The thermochemical thresholds for the various photodissociative ionization processes are as follows:

	ΔH_0°, eV
$C_2H_6 \rightarrow C_2H_5^+ + H + \bar{e}$	12.45
$\rightarrow C_2H_4^+ + H_2 + \bar{e}$	11.86
$\rightarrow C_2H_4^+ + 2H + \bar{e}$	16.33
$\rightarrow C_2H_3^+ + H_2 + H + \bar{e}$	14.53
$\rightarrow C_2H_3^+ + 3H + \bar{e}$	19.01
$\rightarrow C_2H_2^+ + 2H_2 + \bar{e}$	14.47
$\rightarrow C_2H_2^+ + H_2 + 2H + \bar{e}$	18.95
$\rightarrow C_2H_2^+ + 4H + \bar{e}$	23.43
$\rightarrow CH_3^+ + CH_3 + \bar{e}$	13.65

There is also evidence for a weak[329] ion-pair process, producing $C_2H_5^+ + H^-$, whose threshold should occur at 11.70 eV.

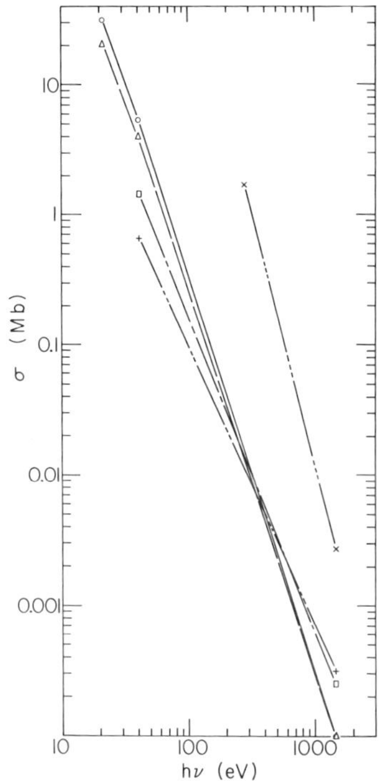

Fig. 77 Partial cross sections for formation of various states of $C_2H_6^+$ by photoabsorption. △ $(1e_u)^{-1}$, □ $(2a_{2u})^{-1}$, + $(2a_{1g})^{-1}$, ○ $(1e_g)^{-1} + (3a_{1g})^{-1}$, and × $(1a_{1g})^{-1} + (1a_{2u})^{-1}$.

The photoionization mass spectrum in the region of the first fragmentation (Fig. 78[329]) reveals that the parent ion continues to increase for about 1 eV beyond the onset for $C_2H_4^+$, which is contrary to the predictions of QET. This behavior does not appear in the photoion–photoelectron coincidence studies[282] which discriminate against autoionization processes. Hence, the discrepancy with QET is most likely due to autoionization populating states of $C_2H_6^+$ below the fragmentation threshold. There is also some structure in the threshold region of the $C_2H_5^+$ ion yield curve attributable to the pair formation process.

The photoionization mass spectra[330] between 12 and 20 eV are shown in Fig. 79. Relative intensities at 750 Å are given in Table 14. Each successive fragment rises from threshold to a plateau, and remains flat thereafter, as predicted by QET. The onsets for $C_2H_2^+$ and $C_2H_3^+$ are close to the adiabatic threshold for $(1e_u)^{-1}$, which could conceivably imply direct dissociation from this electronic state without internal conversion. This appears unlikely in view of the overall agreement with QET (vide infra) although the argument given by Stockbauer[331] does not seem very convincing.

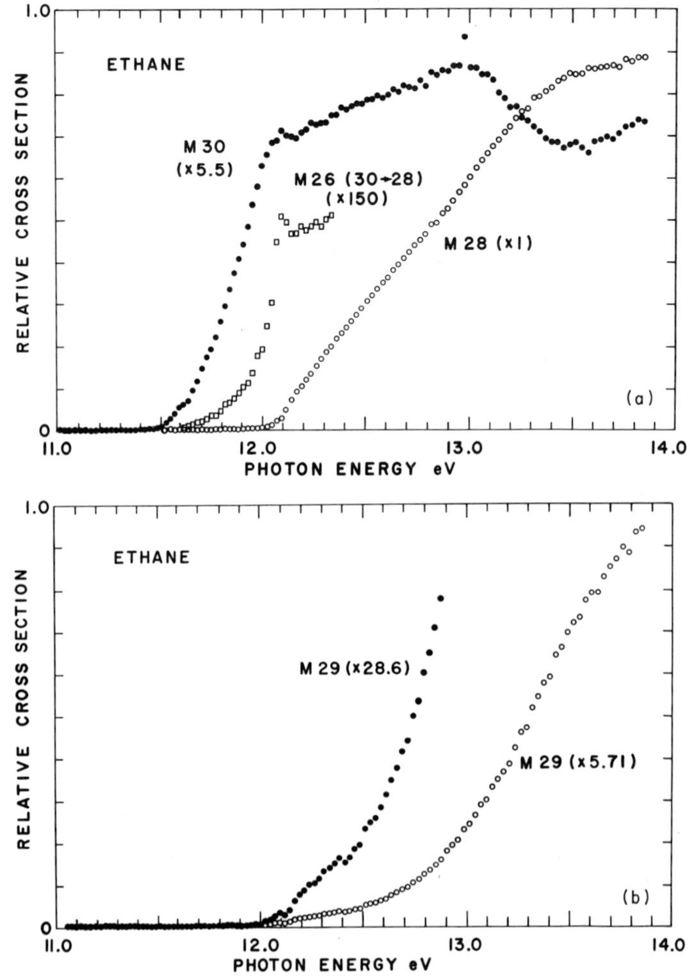

Fig. 78 (a) Photoion yield curves of $C_2H_6^+$ (M 30), $C_2H_4^+$ (M 28), and the metastable ion corresponding to the decomposition $C_2H_6^+ \rightarrow C_2H_4^+ + H_2$ (which should appear at M 26.1) from C_2H_6 between 11.0 and 14.0 eV. (b) Photoion yield curve of $C_2H_5^+$ (M 29) from C_2H_6 between 11.0 and 14.0 eV. [From W. A. Chupka and J. Berkowitz, *J. Chem. Phys.* **47**, 2921 (1967).]

The photoion–photoelectron coincidence spectrum of C_2H_6 is shown in Fig. 80, and in Fig. 81, the smoothed breakdown diagram is displayed, together with QET calculations by Prasil and Forst.[332]

The correspondence is generally satisfactory, although there are shifts (attributable to incorrect thresholds employed in the calculations) and the yield of CH_3^+ appears to be underestimated by the calculations.

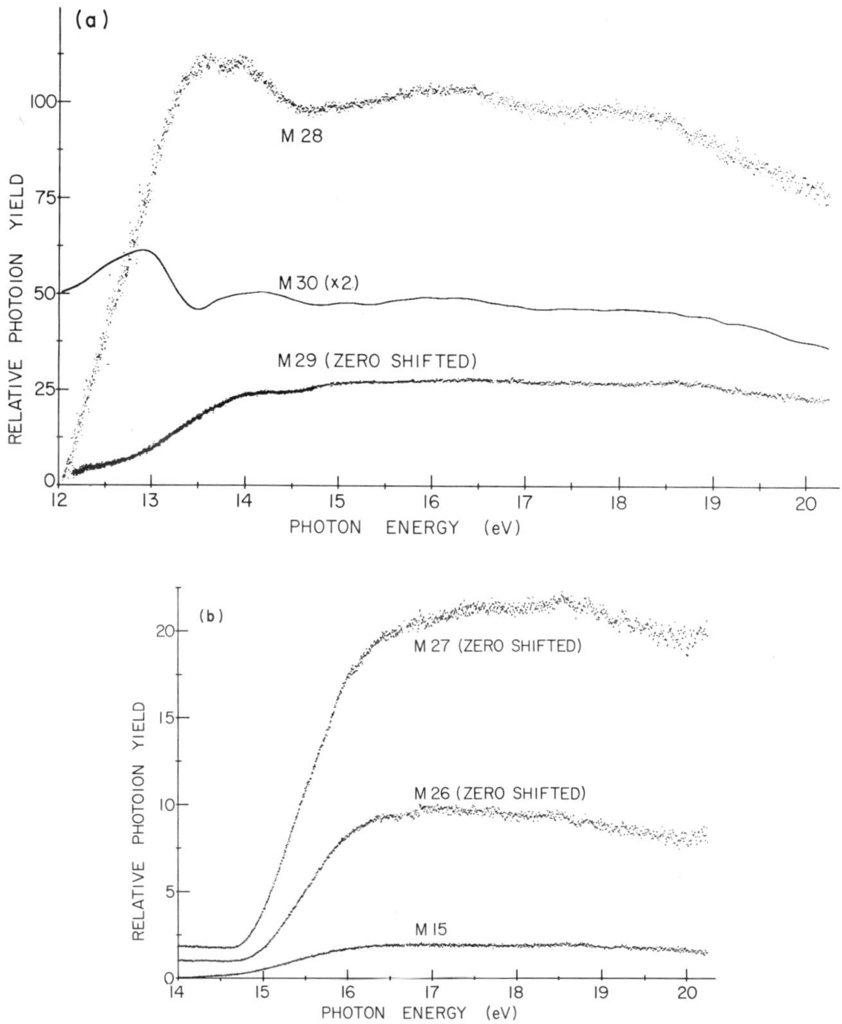

Fig. 79 (a) Photoion yield curves of $C_2H_6^+$ (M 30), $C_2H_5^+$ (M 29), and $C_2H_4^+$ (M 28) from C_2H_6 between 12 and 20 eV. (b) Photoion yield curves of $C_2H_3^+$ (M 27), $C_2H_2^+$ (M 26), and CH_3^+ (M 15) from C_2H_6 between 14 and 20 eV. (Unpublished data, obtained by the author.]

Studies[309] with discrete line sources between 21.2 and 40.8 eV indicate small contributions from C_2H^+, CH_2^+, and C_2^+, not exceeding $\sim 1\%$ of the parent ion, throughout this energy range, with the relative proportions of the major peaks remaining approximately constant. In particular, there does not appear to be any evidence for a departure from QET behavior attributable to $(2a_{2u})^{-1}$ which, it will be recalled, exhibited vibrational fine structure.

TABLE 14 *Relative Intensities of Parent and Fragment Ions in the Photoionization of* C_2H_6[a]

Ion	Relative intensity
$C_2H_6^+$	24.1_4
$C_2H_5^+$	22.7
$C_2H_4^+$	100.0
$C_2H_3^+$	18.0_3
$C_2H_2^+$	8.2_7
CH_3^+	1.9_3

[a] At 750 Å = 16.53 eV.

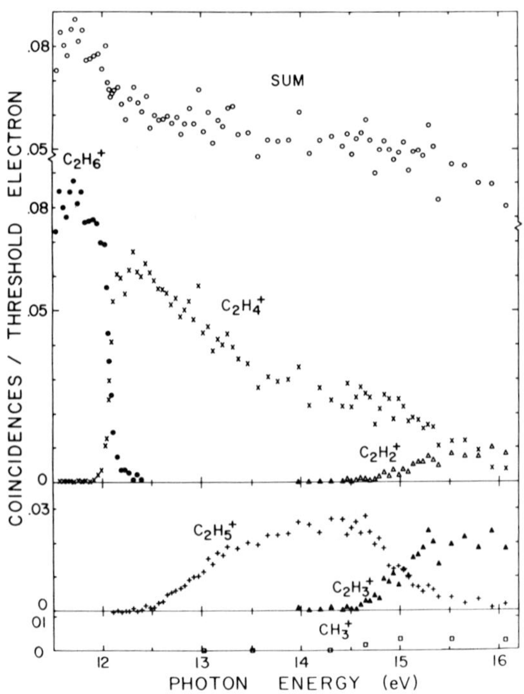

Fig. 80 Experimental breakdown diagram for $C_2H_6^+$, displaying number of ion coincidences per threshold electron in each channel, with appropriate background subtraction. [From R. Stockbauer, *J. Chem. Phys.* **58**, 3800 (1973).]

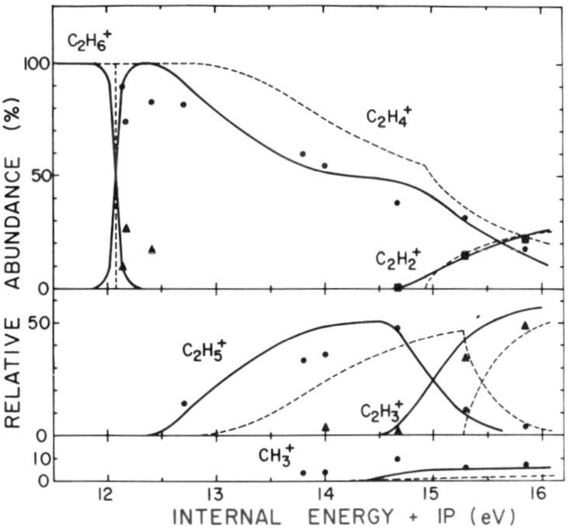

Fig. 81 Breakdown diagram of $C_2H_6^+$. ——— Smoothed photoion–photoelectron coincidence measurements by Stockbauer (cited in Fig. 80). ●, ▲, ■ Charge exchange measurements by H. von Koch, *Ark. Fys.* **28**, 559 (1965). --- QET calculation by Z. Prasil and W. Forst, *J. Phys. Chem.* **71**, 3166 (1967).

By contrast, the fragmentation of $C_2F_6^+$ has been interpreted[333] as proceeding directly from an electronic state without internal conversion, and hence in violation of QET.

The Auger spectrum following carbon K-shell excitation in C_2H_6 has been measured by Spohr *et al.*[287] Their analysis concludes that $C_2H_6^{2+}$ is quite unstable. However, a mass spectrum revealing the nature of the fragments is unavailable.

6. CH_3OH

a. Photoelectron Spectroscopy. Methanol has only C_s symmetry, and hence its molecular orbitals can only be classified as a′ and a″. The molecular orbital sequence and corresponding ionization energies are given in Table 15. The partial cross sections for formation of the corresponding ionic states have been computed from the recorded spectra of Robin and Kuebler[334] at 21.2 eV, Potts *et al.*[335] at 40.8 eV,[336] and Siegbahn *et al.*[337] at 1253.6 eV. The ionization yield at 584 Å has been reported to be 0.98,[238] and was assumed to be unity at higher energy. Branching ratios were measured as (area/E), and the total photoabsorption cross sections taken from Chapter V, Section D.6. The bands at 15.21 and 15.64 eV are strongly overlapped, and were treated together. Figure 82 displays the resulting variation of absolute partial cross sections. As noted in previous cases, $(4a')^{-1}$ and $(3a')^{-1}$ [which have C(2s)

TABLE 15 *Vertical Ionization Energies for Single-Electron Ejection from CH_3OH and the Atomic Character of the Orbital from Which Electron Ejection Occurs*

Molecular orbital	Dominant character	Vertical ionization energy (eV)	Ref.
2a″ }	O(pπ)	10.96	(334)
7a′ }		12.62	
6a′	C(p) + O(p) + H(s)	15.21	(334)
1a″ }	C(pπ)	15.64	(334)
5a′ }		17.62	
4a′	C(2s)	22.65	(334)
3a′	O(2s)	32.1, 32.2	(43, 335)
2a′	C(1s)	292.3	(43)
1a′	O(1s)	538.9	(43)

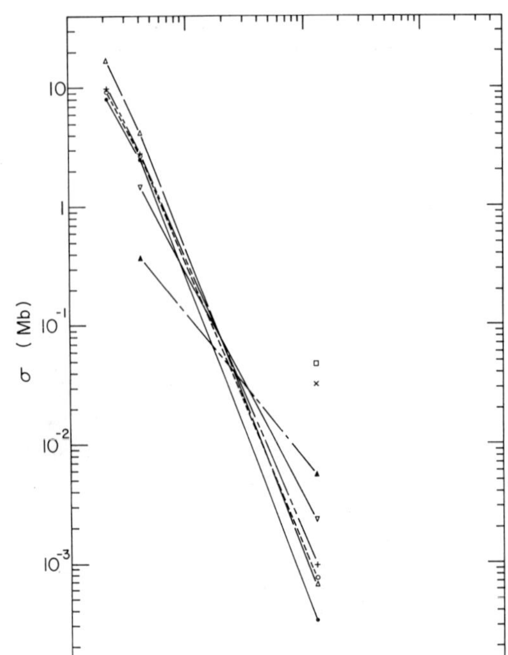

Fig. 82 Partial cross sections for formation of various states of CH_3OH^+ by photoabsorption. ● $(2a'')^{-1}$, ● $(7a')$ (5a′ if O 1s, C 1s not counted), △ $(6a')^{-1} + (1a'')^{-1}$, + $(5a')^{-1}$, ▽ $(4a')^{-1}$, ▲ $(3a')^{-1}$, × $(2a')^{-1}$ (C 1s), □ $(1a')^{-1}$ (O 1s).

and O(2s) character] decline less markedly with increasing photon energy than do those valence orbitals with p character.

b. Photoionization Mass Spectrometry. The photoionization mass spectra of the several ions observed[338,339] in the ionization of CH_3OH between threshold and 20 eV are shown in Fig. 83a and b. The corresponding thermochemical thresholds for the various ions, obtained from standard reference sources, are listed in Table 16. The relative intensities observed at 650 Å = 19.075 eV are shown in Table 17. It can be seen from this latter table that CO^+, H_2O^+, OH^+, $CH_4{}^+$, and/or O^+ occur weakly, and may have contributions from background impurities. Hence, we ignore these weak fragmentation paths in the subsequent analysis and discussion.

At the outset, we must contend with some ambiguities. The first fragment observed (M31) can be written CH_2OH^+ or CH_3O^+. Potapov and Sorokin[340] have studied the photoionization of CD_3OH, CH_3OH, and CH_3OD, and conclude that the first fragment has the structure CH_2OH^+.

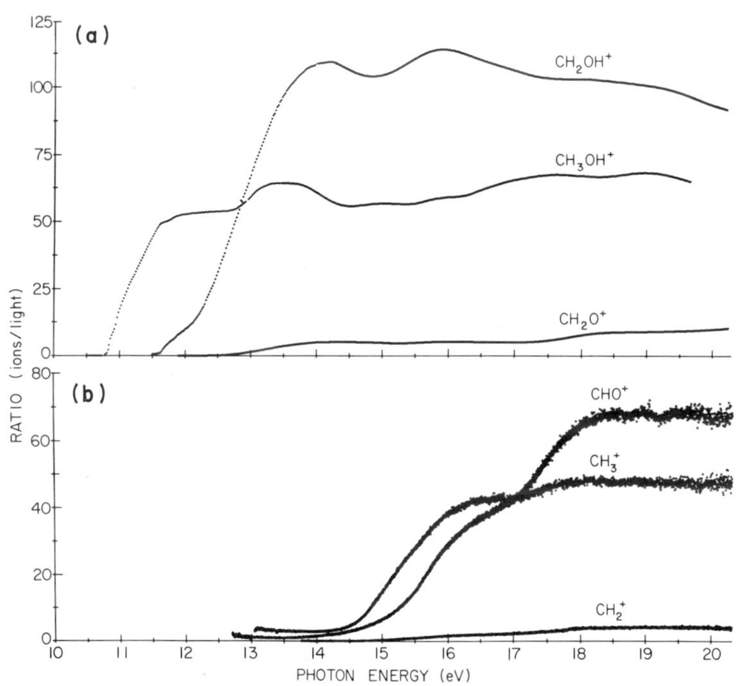

Fig. 83 (a) Photoion yield curves of CH_3OH^+, CH_2OH^+, and CH_2O^+ (or $CHOH^+$) from CH_3OH between the ionization threshold and 20 eV. (b) Photoion yield curves of CHO^+ (or HCO^+), $CH_3{}^+$, and $CH_2{}^+$ from CH_3OH between their respective thresholds and 20 eV. [From J. Berkowitz, *J. Chem. Phys.*, **69**, 3044 (1978).]

TABLE 16 *Thermochemical Thresholds for Formation of Various Ions from CH_3OH*

Products	Energy (eV)	Ref.
$CH_3OH^+ + \bar{e}$	10.85	(333a)
	10.84	(339)
$CH_2OH^+ + H^{\cdot} + \bar{e}$	11.67	(339)
	11.55	(344b)
$CH_2O^+ + H_2 + \bar{e}$	11.67	(176, 333b–d)
$CH_2O^+ + 2H + \bar{e}$	16.16	(176, 333b–d)
$HCO^+ + H_2 + H + \bar{e}$	12.73	(176, 333b–d)
$CO^+ + 2H_2 + \bar{e}$	14.80	(176, 333b,c,e)
$CO^+ + H_2 + 2H + \bar{e}$	19.28	(176, 333b,c,e)
$CH_3{}^+ + OH + \bar{e}$	13.76	(176, 333b,c)
$CH_3 + OH^+ + \bar{e}$	16.91	(176, 333b,c)
$CH_2{}^+ + H_2O + \bar{e}$	13.89	(176, 333b,c)
$CH_2 + H_2O^+ + \bar{e}$	16.11	(176, 333b,c)
$CH_4{}^+ + O + \bar{e}$	16.45	(176, 279, 333b,c)
$CH_4 + O^+ + \bar{e}$	17.45	(176, 333b,c)

TABLE 17 *Relative Ion Intensities in the Photoionization of CH_3OH*[a]

Mass no.	Ion	Intensity
32	CH_3OH^+	67.52
31	CH_2OH^+, CH_3O^+	100.0
30	$CHOH^+$, CH_2O^+	8.64
29	CHO^+	67.99
28	CO^+, $(N_2{}^+)$	very weak
18	H_2O^+	1.77
17	OH^+	0.82
16	$O^+/CH_4{}^+$	very weak
15	$CH_3{}^+$	46.16
14	$CH_2{}^+$	4.19

[a] At 650 Å (19.075 eV).

They note that the formation of CD_3O^+ ions from CD_3OH proceeds at a higher energy (>12.0 eV) with low yield, and that these ions apparently readjust to the CH_2OH^+ structure.

The second fragment (M30) can be written $HCOH^+$ or H_2CO^+, and a still higher energy process as HCO^+ or COH^+. We have performed experiments with isotopic samples in order to distinguish between alternative structures, which we shall discuss subsequently. First, let us turn to the more general question of decay mechanisms.

Photoion–photoelectron coincidence experiments have been performed[341] on this molecule, but these studies were limited to the energy region of the first fragmentation. In this region, more-or-less normal behavior (according to the expectations of quasi-equilibrium theory) was observed. The parent ion remained constant in intensity beyond the threshold for the first fragment, in the *integral* form of the experiment. In the absence of more extensive coincidence studies, we shall resort here to a somewhat old-fashioned and dubious approach. Despite these apprehensions, the method has some pedagogic value for the present purposes.

Let us assume that the partial cross sections for formation of the various ionic states rise abruptly at their corresponding thresholds to some finite values, and remain at these magnitudes with excess photon energy. This is the so-called step-function approximation. (Fig. 82 clearly shows that this assumption is incorrect for large energy changes—the partial cross sections decrease with excess energy. Nevertheless, we hope that a small, smooth decrease over a limited energy range will not affect our conclusions seriously.)

If the partial cross sections behaved as postulated, the first derivative with respect to energy of the total ionization cross section would yield sharp peaks at the corresponding ionization thresholds. Only atoms and selected states of small molecules can be expected to have such sharp onsets. For larger molecules, we can anticipate sloping steps, whose first derivatives would yield broader peaks.

To test this level of the theory, we compare the first derivative of the total ionization cross section with the 21.2 eV photoelectron spectrum in Fig. 84a and b. The 21.2 eV photoelectron spectrum[342] displays a sharp peak with vibrational fine structure at ~11 eV, and broad peaks at ~13, 15.5, and 17.7 eV. The curve of the energy derivative of total ionization displays strikingly similar structure. Even the relative intensities are not greatly disparate.

Some mild doctoring has gone into the construction of Fig. 84b and should be clarified before we proceed. Brehm *et al.*[341] had shown in their paper that the derivative of total ionization displayed fine structure peaks, some of which were due to the effects of autoionization. On the coarser scale of Fig. 84b this is still apparent. Furthermore, if the assumption of quasi-equilibrium theory was valid and autoionization was not prominent, the parent peak should remain constant (or perhaps decline slowly, monotonically) with increasing energy. In Fig. 83, we can see that M32 undergoes some mild undulations, even at higher energy. In the construction of Fig. 84b, we have ignored the contributions of M32 beyond ~13 eV, working on the hypothesis that the undulations are manifestations of autoionization which lead to ground state ions, and hence to CH_3OH^+ only. Consequently, their inclusion would only confuse the ensuing analysis. Figure 84b was constructed by

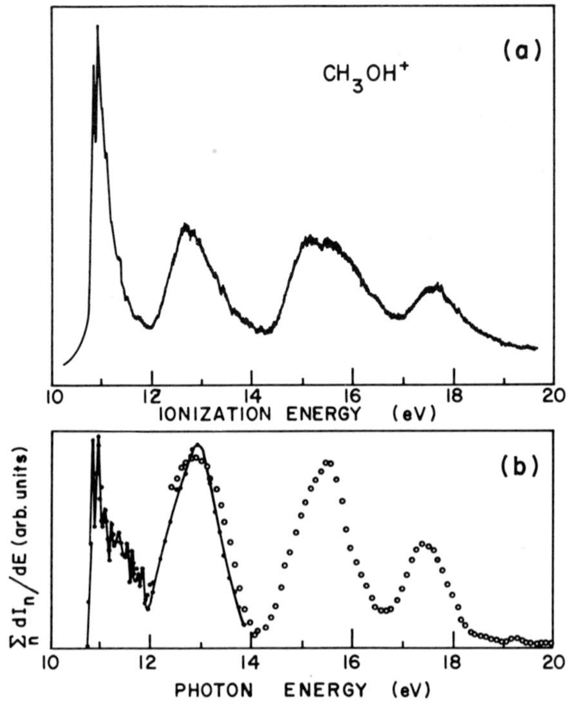

Fig. 84 (a) Photoelectron spectrum of methanol, obtained with 21.2 eV incident radiation. [This spectrum was taken in the laboratory of W. C. Price and was published by J. H. D. Eland *in* "Photoelectron Spectroscopy," p. 22. Butterworths, London 1974.] (b) Energy derivative of total ionization for methanol. [Both figures appear in J. Berkowitz, *J. Chem. Phys.*, **69**, 3044 (1978).] The energy scales are matched so that direct comparison of the features can be made.

smoothing each of the ion yield curves of Fig. 83 using a cubic spline, least-squares procedure, and differentiating the smooth curves. The differentials were then summed at each energy. The differentiated curves dipped negative in some regions. At these energy values, the negative differential was set equal to zero.

The reasonable agreement between Fig. 84a and b gives us some confidence in the procedure. We are now in a position to construct the breakdown diagram. For this purpose, we plot the fractional contribution to each ion at a given internal energy. Within the framework of the above assumptions, this implies plotting the energy derivative of each ion, normalized to the energy derivative of total ionization, as a function of internal energy. Such a plot has been constructed, and is shown in Fig. 85. (The energy scale is not internal energy, but incident photon energy. To convert the energy scale to internal energy, one must substract the first ionization potential of CH_3OH, 10.84 eV.)

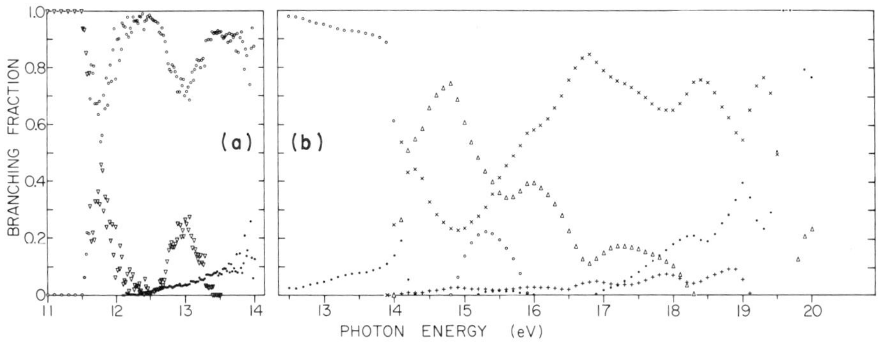

Fig. 85 Experimental breakdown diagram for CH_3OH^+. ▽ CH_3OH^+ (M 32); ○ CH_2OH^+ (M 31); ○ $CHOH^+$, CH_2O^+ (M 30); △ $CH_3{}^+$ (M 15); × HCO^+ (M 29); + $CH_2{}^+$ (M 14). (a) was obtained with a hydrogen many-line light source, (b) with a helium continuum lamp. [From J. Berkowitz, *J. Chem. Phys.*, **69**, 3044 (1978).]

We note a rather abrupt tendency to "cross over" from CH_3OH^+ as the dominant ion to CH_2OH^+ at ~11.6 eV, but a peculiar reversal at ~11.8 eV. This behavior, an apparent violation of quasi-equilibrium theory, is a manifestation of the low-energy autoionization alluded to earlier. This occurs again at ~13.0 eV and is probably due to the same reason.

The coarse grained features reveal CH_2OH^+ as the dominant ion between ~12.0 and 14.0 eV, with M30 slowly growing in this interval. At ~14 eV, CHO^+ abruptly takes over from CH_2OH^+, but $CH_3{}^+$ makes a dominant appearance by ~14.5 eV. At still higher energy, CHO^+ resumes its pre-eminence and remains so till ~19.5 eV.

In Fig. 86, we reproduce a breakdown diagram of CH_3OH^+ calculated by Vestal and Lerner[320] using the assumptions of quasi-equilibrium theory. We note the CH_3OH^+–CH_2OH^+ crossover at ~1.1 eV, corresponding to ~11.9 eV on our experimental breakdown diagram. The calculated crossover occurs at somewhat higher energy, although the experimental results are confused by autoionization. At ~4 eV internal energy (~14.84 eV photon energy) the next major crossover is predicted to occur, with CHO^+ assuming dominance. We observe a crossover at ~14.0 eV, with CHO^+ and $CH_3{}^+$ remaining competitive for ~2.5 eV. The calculated breakdown diagram predicts negligible importance for $CH_3{}^+$ in this energy interval. At higher energies, $CH_2{}^+$ becomes competitive with a weaker $CH_3{}^+$, and a crossover between these two ions occurs at a photon energy of ~18.15 eV. The calculation predicts such a crossover at ~9 eV internal energy or ~19.85 eV photon energy.

In summary, the calculation bears a gross resemblance to the experimental breakdown diagram. CH_3OH^+ is dominant for the first eV (or major

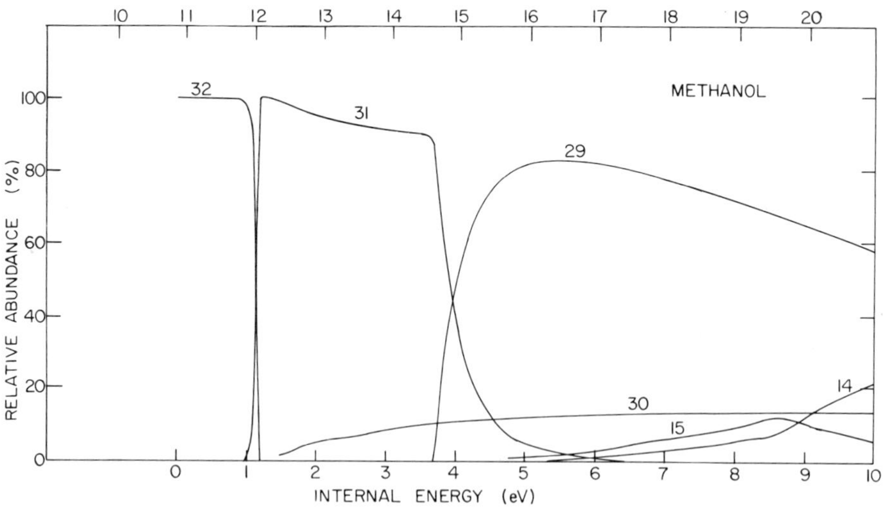

Fig. 86 Calculated breakdown diagram of CH_3OH^+, using QET. [From M. Vestal and G. Lerner, Aerospace Research Laboratories Rept. ARL 67–0014 (1967).]

fraction thereof), CH_2OH^+ then becomes dominant and yields to CHO^+ at about 3 eV higher internal energy.

There are many features not adequately described by the calculation, however. The large fractional contribution of $CH_3{}^+$ at ~14.5 eV is unaccounted for. It may be influenced by autoionization effects, but scattered data from other experiments qualitatively support the observations presented in Fig. 85. Ikuta *et al.*[343] and Jonsson and Lind[344] report breakdown diagrams of methanol obtained by charge exchange. While few points are presented, the combined evidence of these authors seems to be that $CH_3{}^+$ rivals CHO^+ in fractional intensity at ~15.3 and 15.76 eV. The earlier, but more extensive study of Wilmenius and Lindholm[344a] results in a breakdown diagram rather similar to that of Fig. 85. Mass 15 is shown there to compete effectively with mass 29 between 14 and 16 eV. The dominance of $CH_3{}^+$ between 14.2 and 15.4 eV does not appear in as dramatic a form in the charge exchange experiments. The behavior of the mass 30 fragment is revealingly similar, making its appearance between 12 and 15 eV, dropping to zero at 16 eV, then reappearing at ~17 eV. In Fig. 85, mass 30 drops to zero at 14.3 eV and reappears at 16.9 eV. In the Wilmenius–Lindholm diagram, mass 14 begins to rise, and mass 15 to diminish, at ~20 eV, whereas Fig. 85 displays the crossover at 18.2 eV.

The calculated breakdown diagram (Fig. 86) indicates that $CH_3{}^+$ is approximately two orders of magnitude weaker than HCO^+ in the 4–5 eV internal energy range, and that $CH_3{}^+$ always remains much weaker than

CHO^+. Mass 30 does not display the bimodal behavior observed experimentally, but continues to increase with internal energy. The "hole" in the experimental breakdown diagram for mass 30 may be related to the abrupt appearance of $CH_3{}^+$ in just this energy range. These comparisons are not affected by autoionization phenomena, as the charge-exchange measurements should be insensitive to such processes.

Since HCO^+ and CH_3 make their initial appearance at nearly the same energy, quasi-equilibrium theory would predict that their rates of formation should be comparable. The observation of a metastable ion is a kind of rate measurement for a given internal energy. We have observed a metastable peak corresponding to the process

$$CH_2OH^+ \rightarrow HCO^+ + H_2,$$

but none is seen corresponding to the process

$$CH_3OH^+ \rightarrow CH_3{}^+ + OH.$$

A possible explanation is that $CH_3{}^+$ is formed by a rapid, direct dissociation process, before equilibration has occurred among electronic and vibrational modes. As support for this hypothesis, we note that the orbital from which an electron has been removed at ~ 15 eV is a C—O bonding orbital, and that $CH_3{}^+$ begins to contribute to the breakdown diagram almost precisely at the threshold for this particular ionization. Furthermore, Jonsson and Lind[344] quote radiolysis measurements in pure methanol in liquid and gas phase which result in very similar G values for CH_4, but quite different ones for CO. (The G value is the number of species produced per 100 eV of absorbed energy.)

CH_4 is a diagnostic for $CH_3{}^+$ formation, since the reaction

$$CH_3{}^+ + CH_3OH \rightarrow CH_4 + CH_2OH^+$$

is expected to occur at high pressures. Similarly, the reaction

$$CHO^+ + CH_3OH \rightarrow CO + CH_3OH_2{}^+$$

enables CO to be used as a measure of CHO^+ production. Johnsson and Lind contend that the slower process forming CHO^+ is affected by rapid collisional deactivation in the liquid phase, where the mean collision time may be $\sim 3 \times 10^{-13}$ sec, whereas the lack of significant change in G value for CH_4, in going from gas to liquid, implies that dissociation to $CH_3{}^+$ has occurred in less than $\sim 3 \times 10^{-13}$ sec.

By itself, this latter argument may be suspect, since one can question whether the states of CH_3OH^+ produced in ionization of isolated molecules in the gas phase can be regarded as the same states in hydrogen-bonded, liquid phase methanol. Nevertheless, the prevailing evidence from the various arguments presented would seem to favor direct dissociation as a major mechanism for CH_3^+ production at ~14–16 eV.

Isotopic studies[338] with samples of CD_3OH and CH_3OD have confirmed that the first major fragment primarily appears as CD_2OH^+ from CD_3OH_3 but some CD_3O^+ is also observed, and its fraction increases with internal energy. The reappearance of M31 at ~14.9 eV in Fig. 85 has substantial contribution from the CD_3O^+ form, suggesting that a methoxy ion may occur at higher energy, in addition to an isotopic rearrangement of the CD_3O^+ to CD_2OD^+.

Studies with both isotopic samples yield consistent results regarding the mass 30 entity in Fig. 85. It is virtually entirely $HCOH^+$ between threshold and ~15 eV, but above ~16.9 eV the H_2CO^+ structure makes a significant contribution. There are two possible rationalizations for this behavior, which may be related. From Table 16, we note that the thermochemical threshold for the reaction

$$CH_3OH \rightarrow CH_2O^+ + 2H + \bar{e}$$

occurs at 16.16 eV. If the above equation represents the process responsible for the rise in the yield curves of CD_2O^+ from CD_3OH and CH_2O^+ from CH_3OD above ~16.9 eV, then we can identify this increase with the opening of a new thermochemical channel. Although the energy required for this process is substantially greater, the steric constraints (tight complex) associated with H_2 elimination would no longer hold sway.

The approximate onset for H_2CO^+ formation also corresponds rather closely to the threshold for the highest energy band in the 21.1 eV photoelectron spectrum (Fig. 84a). Hence, the H_2CO^+ formation could be interpreted as a kind of direct dissociation of this ionic state, or it may represent a sequential decay from this state, i.e.,

$$(CH_3OH^+)^* \rightarrow CH_2OH^+ \rightarrow CH_2O^+.$$

The observation of metastable ions corresponding to one or the other decay could in principle establish the decomposition path. The formation of $HCOH^+$ at a lower energy than H_2CO^+ is somewhat surprising, since the neutral species (formaldehyde) certainly has the H_2CO structure. Our current observation does not constitute proof that $HCOH^+$ is more stable than H_2CO^+. The appearance potential for $HCOH^+$ from methanol (~12.5 eV) is about 0.8 eV higher than the thermochemical threshold for H_2CO^+. (Warneck[344b] gives this threshold at 12.0 eV, but it corresponds to the end of a thermal tail.) This difference has previously been interpreted as a kinetic shift brought about by the slow unimolecular decay of CH_3OH^+ to $H_2CO^+ + H_2$. From our present vantage point, we can attribute some of this 0.8 eV shift to kinetic effects, but some may be due to an $HCOH^+$ structure *less* stable than H_2CO^+. If a barrier exists to equilibration of $HCOH^+$ with H_2CO^+, the $HCOH^+$ may still be favored kinetically, though not thermochemically.

The structure of methanol[344c] consists of a rather characteristic methyl group joined to hydroxyl through a C—O—H angle of 108.5°. Two of the carbon-bonded hydrogen atoms are separated by 1.776 Å in neutral methanol. In the freely rotating but vibrationless molecule, the minimum distance between a carbon-bonded hydrogen and the hydroxyl hydrogen is 2.078 Å. If this is representative of the lower states of CH_3OH^+, then proximity of the hydrogen atoms (which must come close enough to one another to form H_2) could provide the steric effect which favors $HCOH^+$ over H_2CO^+, perhaps overcoming an energy effect favoring the latter species. In any event, the evidence presented above establishes that $HCOH^+$ has a heat of formation <0.8 eV greater than H_2CO^+. Recent calculations by Dunning[344d] using a double zeta basis set augmented with polarization functions lead to the conclusion that H_2CO^+ is only 0.16 eV more stable than $HCOH^+$. Within the uncertainty of Hartree–Fock calculations, these structures can be considered nearly degenerate, but a barrier to interconversion is likely. The evidence from isotopic experiments regarding the structure of HCO^+ (COH^+) is similar to that regarding CH_2OH^+(CH_3O^+). There seems little question[344e,344f] that HCO^+ is the more stable structure, and the mass spectrum[338,339] of CD_3OH at a photon energy of 16.0 eV reveals that DCO^+ (mass 30) is ten times as abundant as HCO^+ (mass 29). However, the energy derivatives of the ion yield curves are strikingly different for DCO^+ and HCO^+, the former showing a large contribution from the 15 eV peak and a small one from the 17 eV ionic state, whereas HCO^+ has the reverse pattern. Hence, either the rearrangement (scrambling) is strongly enhanced from the 17 eV ionic state, or the new ionic structure COH^+ begins to make its appearance here.

The primary source of DCO^+ appears to be well explained as the product of the successive decomposition $CD_3OH^+ \rightarrow CD_2OH^+ \rightarrow DCO^+$. This means that HD must be eliminated in the second stage, involving some steric hindrance, which could provide some support for a competitive COH^+ path at higher energy.

7. C_6H_6

a. Photoelectron Spectroscopy. The most probable sequence of occupied molecular orbitals describing benzene in its electronic ground state, with D_{6d} symmetry, is as follows:

$$(1a_{1g})^2(1e_{1u})^4(1e_{2g})^4(1b_{1u})^2(2a_{1g})^2(2e_{1u})^4(2e_{2g})^4(3a_{1g})^2$$
$$(2b_{1u})^2(1b_{2u})^2(3e_{1u})^4(1a_{2u})^2(3e_{2g})^4(1e_{1g})^4.$$

The vertical and (where possible) adiabatic ionization energies corresponding to single-electron ejection from these various orbitals are summarized in Table 18, together with the atomic orbital character of these orbitals. Figure 87 from Åsbrink *et al.*[345] is a photoelectron spectrum at 40.8 eV displaying the valence orbitals.

Early molecular orbital calculations of semiempirical and ab initio type[346] predicted different ordering, and probably led to three major controversies in the interpretation of the experimental photoelectron spectra, which are described below.

(1) The first two IPs are at 9.24 and 11.48 eV. The Liege group[347,348] purported to find two ionization states between these two levels, which they argued could not be observed at 584 Å because of low partial cross sections at this photon energy. Samson[349] largely disposed of this argument by examining the benzene photoelectron spectrum at 11.1, 11.83, and 16.85 eV, finding no evidence for those states in each case.

(2) The ordering of orbitals given above is that first proposed by Jonsson and Lindholm.[350] In retrospect, it is based largely on correlating Rydberg excitations in the subionization region with photoelectron spectra, comparing vibrational envelopes and deducing quantum defects. The quantum defects are then related to the types of transitions most probably associated with excitations emanating from specific molecular orbitals.

TABLE 18 *Adiabatic and Vertical Ionization Energies for Single-Electron Ejection from C_6H_6 and the Atomic Character of the Orbital from Which Electron Emission Occurs*

Orbital	Atomic character[350,353,354]	Vertical IP (eV)	Adiabatic IP (eV)	Ref.
$(1e_{1g})^{-1}$	C(pπ), weak C—C bond		9.24	(344g)
$(3e_{2g})^{-1}$	C(p) planar, weak C—C and C—H bonding	11.8	11.48	(130)
$(1a_{2u})^{-1}$	C(pπ), strong C—C bond	12.3		(357)
$(3e_{1u})^{-1}$	C(p) + H(1s), strong C—H bond, C—C nonbond	14.0		(357)
$(1b_{2u})^{-1}$	C(p) planar, strong C—C bond	14.8		(357)
$(2b_{1u})^{-1}$	C(2s) + H(1s), strong C—H bond	15.4		(357)
$(3a_{1g})^{-1}$	C(p) planar + H(1s), strong C—H bond, weak C—C bond	16.9	16.84	(130)
$(2e_{2g})^{-1}$	C(2s), weak C—H bond, weak C—C antibond	19.0	18.4	(345)
$(2e_{1u})^{-1}$	C(2s)	22.7	~22	(345)
$(2a_{1g})^{-1}$	C(2s)	26.0		(352)
$(1a_{1g})^{-1}$, $(1e_{1u})^{-1}$ $(1e_{2g})^{-1}$, $(1b_{1u})^{-1}$	C(1s) combinations	290.2		(352)

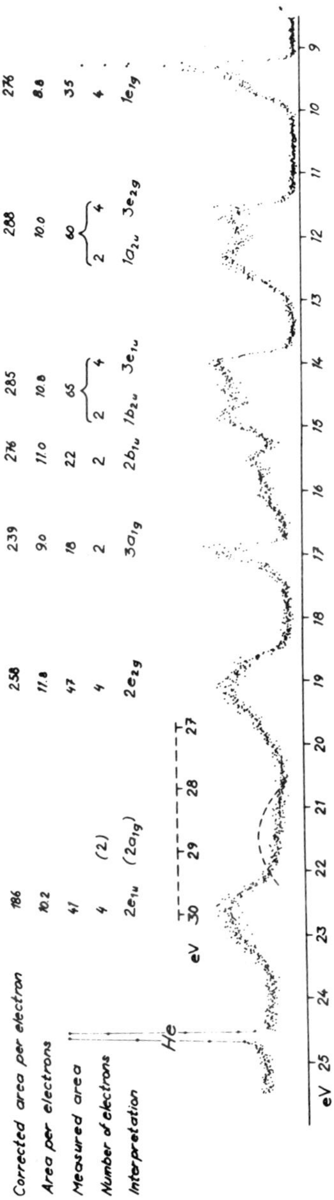

Fig. 87 Photoelectron spectrum of benzene, obtained with 40.8 eV incident radiation. [From Åsbrink *et al.*, *Chem. Phys. Lett.* **5**, 192 (1970).]

At the Faraday Discussions[351] Potts *et al.* proposed reversing the order of $(3e_{2g})^{-1}$ and $(1a_{2u})^{-1}$, and also $(2b_{1u})^{-1}$ and $(3a_{1g})^{-1}$. Their arguments were largely based on vibrational fine structure changes in C_6D_6 compared to C_6H_6. The $(2b_{1u})^{-1}$–$(3a_{1g})^{-1}$ order could be subjected to a simple, plausible test since $(2b_{1u})^{-1}$ has C(2s) character, $(3a_{1g})$ has C(2p) character, and with high incident photon energy the orbital having C(2s) character should dominate. The high resolution 1486.6 eV PES[352] showed that $2b_{1u}$ had the lower ionization energy.

(3) The $(3e_{2g})^{-1}$–$(1a_{2u})^{-1}$ ordering is more difficult to establish unambiguously. Several calculations[353–356] have been cited by Jonsson and Lindholm in support of their ordering, and Gelius[352] has found this sequence gives good concordance of his intensity model with the 1486.6 eV PES intensities, but in this case the sensitivity of this method is weak. The β (angular distribution) parameters of these states are significantly different at 21.2 eV, but β parameters for molecular systems are not readily calculable (see Chapter VII).

The branching ratios at 21.2 eV have been taken from Carlson and Anderson.[357] The 90° spectra were used (i.e., the β effect was not taken into account, although given) since similar data are not available at 40.8 or 1486.6 eV. The 40.8 eV data are given by Åsbrink *et al.*[345] and the 1486.6 eV spectra are taken from Gelius,[352] but the relative intensity of C(1s) levels and valence levels is not given. The corresponding values given in Fig. 88 are

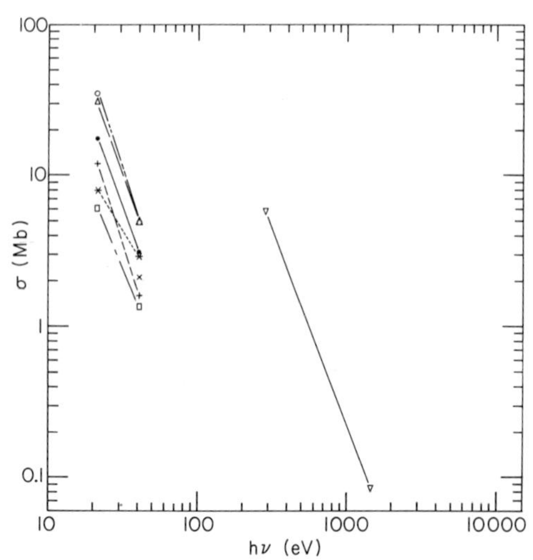

Fig. 88 Partial cross sections for formation of various states of $C_6H_6^+$ by photoabsorption. ● $(1e_{1g})^{-1}$, ○ $(3e_{2g})^{-1} + (1a_{2u})^{-1}$, △ $(3e_{1u})^{-1} + (1b_{2u})^{-1}$, + $(2b^1{}_u)^{-1}$, □ $(3a_{1g})^{-1}$, ∗ $(2e_{2g})^{-1}$, × $(2e_{1u})^{-1}$, ▽ $(1a_{1g})^{-1} + (1e_{1u})^{-1} + (1e_{2g})^{-1} + (1b_{1u})^{-1}$.

based on the carbon K-edge jump at 290 eV, and the absorption cross section at 1486.6 eV, both taken from the C_6H_6 photoabsorption curve (Fig. 39 in Chapter V). This jump is ~ factor 20 greater than the sum of valence levels.

b. Photoionization Mass Spectrometry. The detailed understanding of the decomposition of benzene ions is even less certain than the assignment of the ionic states. The first requirement is a firm knowledge of the decomposition energies. Table 19 represents the current best estimate,[358] but the values in parentheses are particularly poorly known. Rosenstock and Krauss[359] initially questioned the validity of QET for this system, on the grounds that the range of appearance potentials for the reactions listed in Table 19 (and also $C_6H_6 \rightarrow C_5H_3^+ + CH_3 + \bar{e}$) was too large to be consistent with the observation of metastable ions corresponding to all five reactions. Vestal[360] calculated the rates for these reactions using QET and showed that the very large density of states for this molecule, together with its stability, causes the rate of unimolecular decomposition to increase very slowly with energy. This circumstance requires a considerable excess of energy to be deposited in the molecular ion before the rate of decomposition is rapid enough for a metastable ion (10^{+5}–10^{+6} sec^{-1}) or a fragment ion ($>10^6$ sec^{-1}) to be observable in a conventional mass spectrometer. This excess of energy above the thermochemical threshold is often referred to as a kinetic shift. His estimate of metastable intensities was crudely in agreement with experiment.[361,362] Ottinger[361,362] developed an apparatus for observing metastable ions in a range of lifetimes by forming the ions in an extremely narrow region of well-defined field and extracting them with a strong and variable field. By employing charge transfer as the mechanism for depositing energy, Andlauer and Ottinger[363,364] were able to

TABLE 19 *Thermochemical Enthalpies for Photodissociative Ionization of C_6H_6*

Reaction	ΔH_0° (eV)
$C_6H_6 + h\nu \rightarrow C_6H_5^+ + H + \bar{e}$	12.91
$\rightarrow C_6H_4^+ + H_2 + \bar{e}$	(12.94)[a]
$\rightarrow C_4H_4^+ + C_2H_2 + \bar{e}$	(13.53)[b]
$\rightarrow C_3H_3^+ + C_3H_3 + \bar{e}$	13.79

[a] From Rosenstock *et al.*[358]
[b] From Eland *et al.*[367]
Note: The $C_3H_3^+$ ion is in the cyclopropenyl form and $\Delta H_f(C_4H_4^+)$ also strongly suggests a cyclic structure.

deduce the rate of $C_4H_4{}^+$ formation at two well-defined internal energies, and the rate of $C_6H_5{}^+$ formation at three separate energies. Their results are shown in Fig. 89. With a fixed internal energy, Fig. 89 shows that the rates of decomposition for forming $C_6H_5{}^+$ and $C_4H_4{}^+$ are very different. (For example, at ~5.7 eV excitation energy the rates differ by almost an order of magnitude.) An equilibrated system may have two or more unimolecular decay channels, with branching ratios proportional to their individual rate constants k_1, k_2, etc., but each will exhibit a rate characteristic of the total decomposition rate, $\sum_i k_i$.

Andlauer and Ottinger concluded that processes involving breakage of a C—C bond and those involving hydrogen loss appear to originate from different types of excited states, between which energy and structural randomization is not possible within the time (~10^{-6} sec) available for decay.

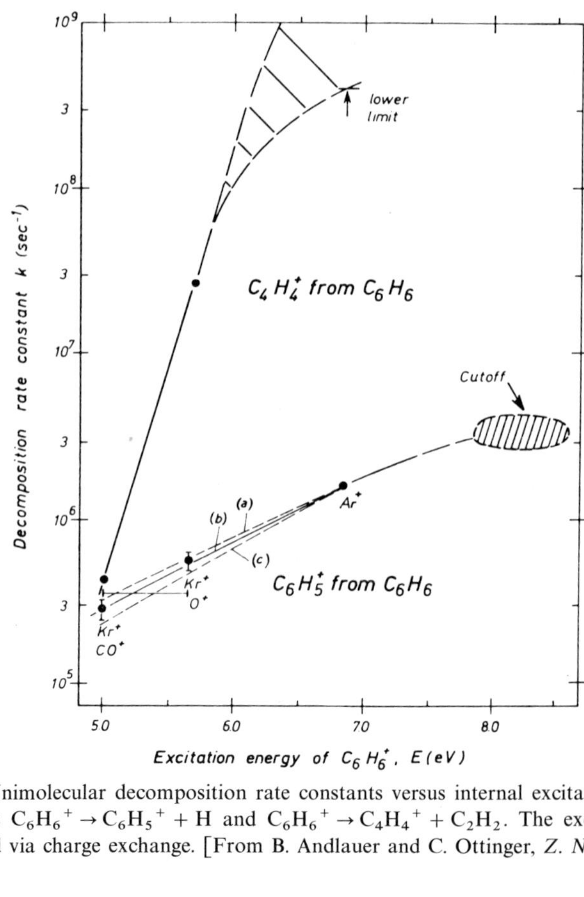

Fig. 89 Unimolecular decomposition rate constants versus internal excitation energy for the reactions $C_6H_6{}^+ \rightarrow C_6H_5{}^+ + H$ and $C_6H_6{}^+ \rightarrow C_4H_4{}^+ + C_2H_2$. The excitation energy was provided via charge exchange. [From B. Andlauer and C. Ottinger, *Z. Naturf.* **27a**, 293 (1972).]

Eland and Schulte[365] have subsequently confirmed the rate of $C_4H_4^+$ formation as a function of energy (they also infer that 0.28 eV more is deposited in the molecular ion by charge transfer, presumably due to additional momentum transfer). They also showed that the intensity of the metastable ion peak corresponding to $C_6H_4^+$ formation varies much less rapidly with internal energy than would be predicted if this process were competitive with the $C_4H_4^+$ decomposition.

The Andlauer–Ottinger observation led Rosenstock et al.[358] to undertake a series of more-or-less conventional QET calculations encompassing a wide range of transition state parameters to see how QET related to the observed rate constants. They were able to show (see Fig. 90) that the variation of rate constant with internal energy for $C_3H_3^+$ formation (and presumably also $C_4H_4^+$ formation) could be rationalized if this decomposition was occurring from an *excited* state, rather than the ground state. The $C_6H_5^+$ (and $C_6H_4^+$) rates could not be fitted with any reasonable QET model.†

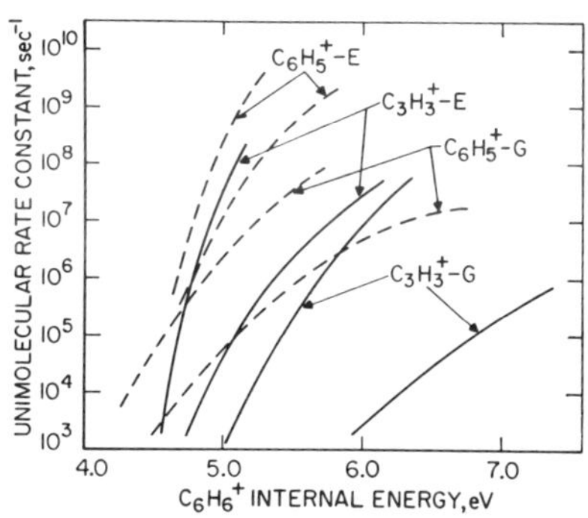

Fig. 90 Calculated unimolecular decomposition rate constants versus internal excitation energy for the reactions $C_6H_6^+ \rightarrow C_6H_5^+ + H$ (– – –) and $C_6H_6^+ \rightarrow C_3H_3^+ + C_3H_3$ (———), where the decomposition path is either assumed from the ground (G) or first excited (E) state of $C_6H_6^+$. [From Rosenstock *et al.*, *Int. J. Mass Spec. Ion Phys.* **11**, 309 (1973).]

Rosenstock *et al.*[358] were further able to show that the photoionization yield curves for $C_3H_3^+$ and $C_4H_4^+$ near threshold (2–3 eV above) were in good agreement with a modified QET model in which these decompositions

† A puzzling feature to one who does not perform QET calculations routinely is a calculation by Klots[366] of the rate constant for $C_4H_4^+$ formation which is in reasonable agreement with the experimental results, although Klots assumes that this decomposition is occurring from the $C_6H_6^+$ *ground* state in his calculation. Presumably he chose a "very loose" transition state, and Rosenstock *et al.* a "tight" one. In any event, it demonstrates the flexibility of the model. Klots also computed the decomposition rate to $C_6H_5^+$. His results, as those of Rosenstock *et al*,. showed a much more rapid increase of this rate with internal energy than was exhibited by the experimental results.

occur from an excited state. The agreement was much poorer if they occurred from the ground state. These two fragments ($C_3H_3^+$ and $C_4H_4^+$) appear to have very nearly the same activation energies for their formation[358,367] and their relative intensities correspond to the statistical degeneracies of the products.

By contrast, the photoionization yield curve for $C_6H_5^+$ was in much better agreement with calculations assuming that the reaction was occurring from the ground state, but there was an abrupt departure at ~855 Å (see Fig. 91).

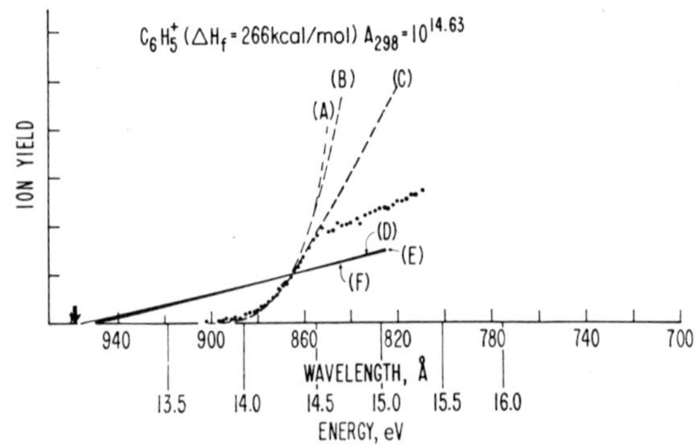

Fig. 91 Calculated and experimental photoion yield curves for $C_6H_5^+$ from C_6H_6. ●●● Experimental data, --- three model QET calculations (A), (B), (C) in which the decomposition occurs from the ground state of $C_6H_6^+$, ——— three model QET calculations (D), (E), (F) where decomposition is assumed to occur from the first excited state of $C_6H_6^+$. All curves are normalized at 865 Å. [From H. M. Rosenstock *et al.*, *Int. J. Mass Spec. Ion Phys.* **11**, 309 (1973).]

In the following paragraphs, we offer a possible rationalization of these observations which, while humble and tedious, sticks as closely as possible to experimental results.

From a comparison of Tables 18 and 19, we note that the first three states of $C_6H_6^+$ populated by photoionization, and a substantial portion of the fourth state, can only give rise to $C_6H_6^+$, the parent ion. Chizhov *et al.*[368] have measured branching ratios for these states from 1300 to 600 Å. We have previously deduced (Chapter V, Section D.7) best values for the total photoabsorption cross sections and ion yields for C_6H_6 in this region. Upon combining these experimental results, we obtain the partial cross section for formation of the first three states (or alternatively, first four states) shown in Fig. 92, in absolute units. (The point at 736 Å represents an average of the branching ratios of Samson[349] and that deduced from integration of Fig. 2 of Chizhov *et al.* Values to shorter wave-

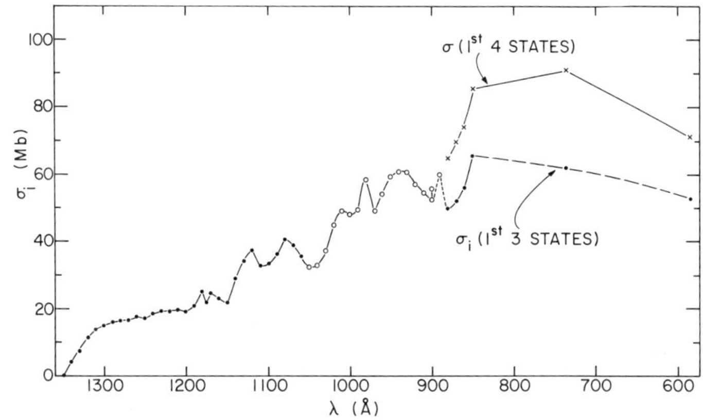

Fig. 92 Total photoionization cross section for formation of the sum of the first three states (or alternatively, first four states) of $C_6H_6^+$.

length are not extractable from the graph of Chizhov *et al.*, since the higher excited states are not included in their Fig. 1. The data point at 584 Å was previously discussed.)

The absolute partial cross section for $C_6H_6^+$ formation (Fig. 93) was obtained by normalizing the photoion yield curve of $C_6H_6^+$ obtained by Dibeler and Liston (see Rosenstock *et al.*[358]) between 930 and 760 Å at

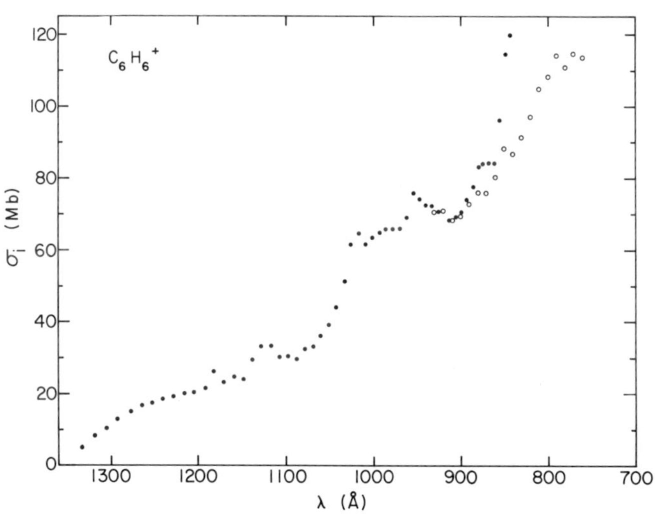

Fig. 93 Photoion yield curve of $C_6H_6^+$ from C_6H_6, obtained by normalizing the photoion yield curve of Dibeler and Liston[358] at 880 Å to the absolute photoionization cross section. The data for $\lambda > 930$ Å are from Brehm[359] normalized to the Dibeler–Liston data at 920–930 Å. 920–930 Å.

880 Å to the absolute photoionization cross section. The longer wavelength region, extracted from Brehm's graph[369] was normalized to the Dibeler–Liston data at 920–930 Å. The agreement is fair. Differences can be attributed to poorly known ionization yields in some regions. Also, the authors performing the photoion yield experiments may not have been paying close attention to certain factors (such as constancy of pressure, accurate measurement of the photodetector response with wavelength) which were not of primary interest at that time.

One inference from Fig. 93 is the possible existence of a kink at 855 Å. (Points are lacking between 850 and 740 Å, but the indications from Fig. 1 of Chizhov *et al.* are that the inference is reasonable.) The kink corresponds to the fact that, although the total photoionization cross section continues to increase at a fairly rapid rate below 855 Å, that portion representing the first four states does not. We had previously noted a kink at that wavelength (Fig. 91) in the $C_6H_5^+$ photoion yield curve.

Photoion–photoelectron coincidence measurements (see Fig. 94) show that $C_6H_6^+$ has a significant contribution from the band at 14 eV $(3e_{1u})^{-1}$, but that the higher energy portion of this band contributes to $C_6H_5^+$ and $C_6H_4^+$. The angular distribution measurements (Fig. 95) correlate well with the

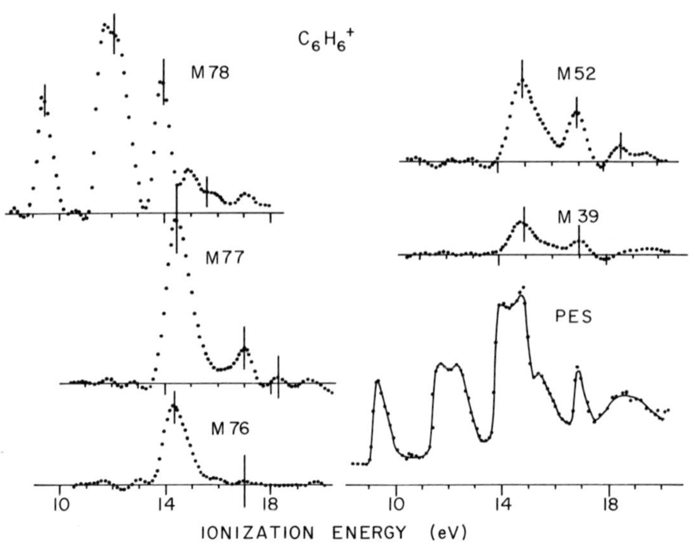

Fig. 94 Photoion–photoelectron coincidence measurements on benzene. $C_6H_6^+$ (M 78) is the sole ion from the 10 and 12 eV bands, and the first portion of the 14 eV band. The higher energy portion of the 14 eV band contributes to $C_6H_5^+$ (M 77) and $C_6H_4^+$ (M 76). The 14.8 eV band appears to contribute primarily to $C_4H_4^+$ (M 52) and $C_3H_3^+$ (M 39). [From a Diplomarbeit thesis of H. Schulte, Physikalishes Institute der Universitat Freiburg (1975).]

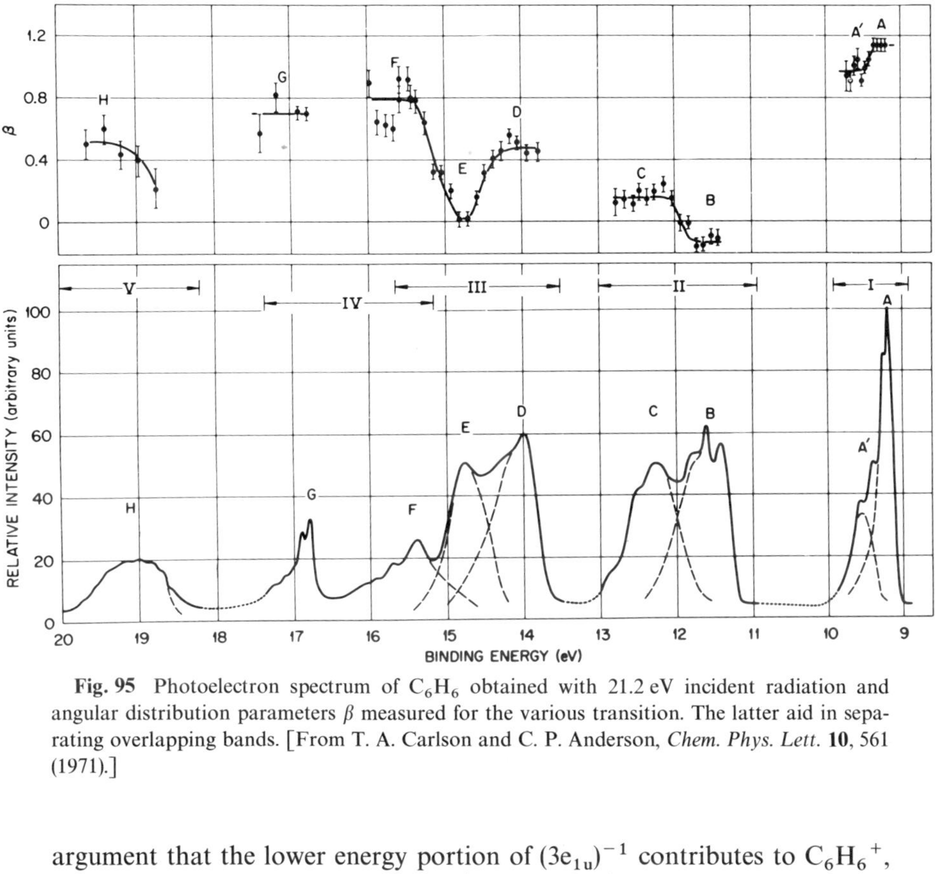

Fig. 95 Photoelectron spectrum of C_6H_6 obtained with 21.2 eV incident radiation and angular distribution parameters β measured for the various transition. The latter aid in separating overlapping bands. [From T. A. Carlson and C. P. Anderson, *Chem. Phys. Lett.* **10**, 561 (1971).]

argument that the lower energy portion of $(3e_{1u})^{-1}$ contributes to $C_6H_6{}^+$, the higher energy portion to $C_6H_5{}^+$ and $C_6H_4{}^+$. However, the band at 14.8 eV $(1b_{2u})^{-1}$ (see Fig. 95) appears to contribute primarily, if not exclusively, to $C_4H_4{}^+$ and $C_3H_3{}^+$. Our inferences are as follows:

(1) The $(1b_{2u})^{-1}$ and $(3e_{1u})^{-1}$ states, although very close in energy, do not interconvert.

(2) The $(3e_{1u})^{-1}$ state has a very slow rate of internal conversion to the ground state, on the time scale of mass spectrometric measurements. Hence, the rate constant for unimolecular decay is the product of an internal conversion rate and the conventionally calculated QET rate. This can explain the slow increase of rate constant for formation of $C_6H_5{}^+$ with increasing internal energy, and the persistence of $C_6H_6{}^+$ at internal energies where it should be absent.[365] While we do not purport to understand why the internal conversion rate should be low, we note that the energy gap to lower states is fairly high and the symmetries of lower states are different from $(3e_{1u})^{-1}$.

(3) In order to rationalize the deductions of Rosenstock *et al.*,[358] the $(1b_{2u})^{-1}$ state should internally convert to the $(3e_{2g})^{-1}$ state. Again, we do not claim to understand the specificity of this radiationless transition.

(4) The 16.8 eV state $(3a_{1g})^{-1}$ appears to contribute to $C_4H_4{}^+$, $C_3H_3{}^+$, and (less so) to $C_6H_5{}^+$. It may internally convert via several channels.

(5) The 19 eV state $(2e_{2g})^{-1}$ probably contributes mostly to the new channel $C_4H_3{}^+$, which now becomes accessible. This is strictly a thermochemical deduction, which may well be explicable by QET.

(6) The photoionization mass spectrum at 584 Å (see three measurements in Table 20), while not constituting strong proof, is in harmony with the above arguments regarding partial cross sections for formation of various states and the resultant mass spectrum.

(7) The $C_6H_6{}^+$ photoabsorption investigation of Freiser and Beauchamp[370] to the extent that it is relevant, is in agreement with the above inferences both as to orbital ordering and the decomposition mechanism. The 304 Å mass spectrum[309] reveals some additional peaks corresponding to $C_6H_3{}^+$, and $C_4H_2{}^+$, at about the 10% level compared to $C_6H_6{}^+$.

The multiply charged ions $C_6H_6^{2+}$ and $C_6H_6^{3+}$ have been observed by electron impact mass spectrometry,[371] the former with a relatively high abundance. The threshold observed for formation of the doubly charged ion (26.1 ± 0.2 eV) suggests that it may arise by formation of $(2a_{1g})^{-1}$, followed by an Auger transition, although the quadratic "threshold law" behavior implies a direct double ionization process.

Both the Auger[287] and the x-ray emission spectra[372] resulting from electron ejection from one of the several molecular orbitals which together comprise the carbon K-edge have been reported. However, mass spectro-

TABLE 20 *Photoionization Mass Spectra of C_6H_6*[a]

Ion	Larson[b]	Schulte[c]	Rindfleisch[d]
$C_6H_6{}^+$	1.0	1.0	1.0
$C_6H_5{}^+$	0.475	0.372	0.334
$C_6H_4{}^+$	0.111	0.121	0.117
$C_4H_4{}^+$	0.331	0.270	0.281
$C_3H_3{}^+$	0.177	0.104	0.107
$C_5H_3{}^+$	0.024	0.026	0.027
$C_4H_3{}^+$	0.134	0.113	0.114

[a] At 584 Å = 21.2 eV.
[b] Larson.[309]
[c] Schulte.[367a]
[d] Rindfleisch.[367b]

metric measurements of the ionic fragments ensuing from this process are lacking.

8. SF_6

a. Photoelectron Spectroscopy. This molecule has octahedral symmetry (O_h). Several different proposals[373–377] have been made for the ordering of the molecular orbitals. The one we adopt here is based on an $X\alpha$ calculation,[376] the most accurate ab initio Hartree–Fock calculation attempted,[377] a many-body approach,[377a] and the intensity arguments of Gelius.[387] A recent experimental study of the variation in partial cross sections with wavelength reaches somewhat different conclusions.[378a]

The sequence of molecular orbitals, and where clear, the atomic character is as follows:

$$\underbrace{(1a_{1g})^2}_{S(1s)}\underbrace{(2a_{1g})^2(1e_g)^4(1t_u)^6}_{F(1s)}\underbrace{(3a_{1g})^2}_{S(2s)}\underbrace{(2t_{1u})^6}_{S(2p)}(4a_{1g})^2(3t_{1u})^6(2e_g)^4$$

$$(5a_{1g})^2(4t_{1u})^6(1t_{2g})^6(3e_g)^4\underbrace{(1t_{2u})^6(5t_{1u})^6(1t_{1g})^6}_{F(2p,\ \text{lone pairs})}$$

The evaluation of partial cross sections is confronted by a number of problems. In the first place, the ionization yield in the valence region is not established beyond a doubt. Sasanuma *et al.*[379] report an ionization yield not exceeding 0.30 between 800 and 300 Å, which is most unusual. Furthermore, their ionization yields characteristically decline below 600 Å for several simpler molecules, as well as SF_6, the former contrary to other observations. Also, the ionization yield curve for SF_6 appears to be insensitive to the large variations in photoabsorption cross sections occurring between 800 and 300 Å. Hence, we arbitrarily assume an ionization yield of unity in the computation of the partial cross sections at 584 Å and 304 Å, recognizing that the absolute values may be in error by at least a factor two at these energies, although the relative cross sections should be approximately correct.

A number of groups have published 584 Å spectra, but the derived relative intensities are either difficult to arrive at, or inconsistent. There are 4 distinct bands in this region (see Fig. 96, taken from Turner *et al.*,[130] p. 379), the third outermost displaying some additional splitting. Frost *et al.*[380] produced a spectrum using a spherical retarding field, which prima facie would appear to be the most direct measurement of relative intensitives, free from transmission and angular distribution effects. The ratio of the four-band intensities as measured from this spectrum is approximately 1:3.63:2.06:2.31, the outermost being unity. The overall spectrum of Turner *et al.* (Fig. 96) is not

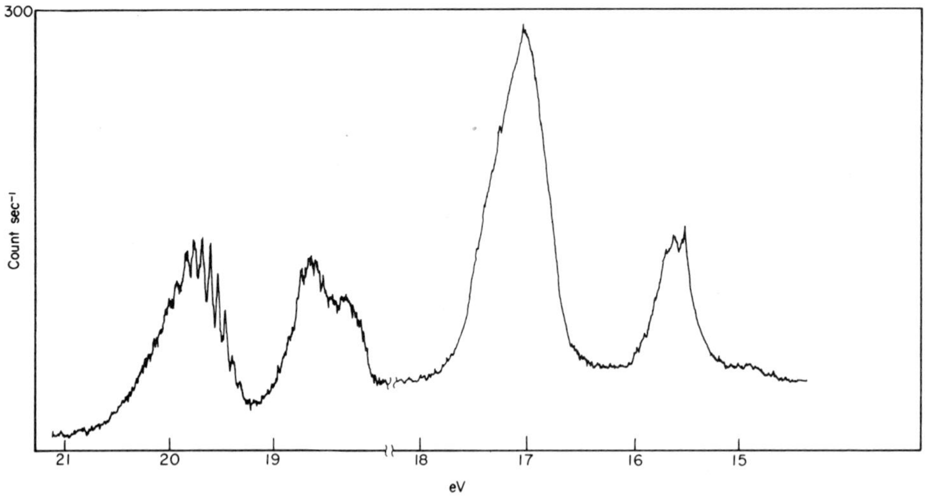

Fig. 96 Photoelectron spectrum of SF_6, obtained with 21.2 eV incident radiation. Note the break between 18 and 19 eV. [From Turner *et al.*, "Molecular Photoelectron Spectroscopy," p. 379. Wiley (Interscience), London, 1970).]

applicable for intensity measurement since the outer two and inner two bands are clearly measured under different sensitivity conditions. An earlier spectrum[381] from Turner's laboratory, using a magnetic deflection instrument, implies relative band intensities of 1:4.21:0.74:0.43 after correcting areas by $1/\sqrt{E}$, the appropriate correction for this type of analyzer. A 584 Å spectrum given by Gelius[378] is highly questionable for intensity purposes, since the fourth band is much stronger than the third, and suggests that the spectrum was recorded with accelerating fields. The corrected (area/E) relative intensities of the first three bands are 1:17.1:1.61.

A spectrum recorded in our laboratory with a cylindrical mirror analyzer, employing the area/E correction, yields 1:6.8:1.9:3.6.

At 22 eV, Gustafsson *et al.*[378a] have obtained relative intensitives of 1:5.7:0.9:1.5. The first two bands are in fair agreement with our cylindrical mirror results, but the latter two are lower, and in fact lower than the corresponding ratios in the earlier works cited. However, we have elected to use the Gustafsson data in constructing partial cross sections, since it provides continuity between ~20 and 54 eV.

In earlier studies,[374] it had been noted that the 304 Å photoelectron spectrum displayed a marked reversal in the relative intensities of the first two bands, compared to the 584 Å spectrum. The recent data of Gustafsson *et al.*[378a] provide a more detailed picture of the dramatic undulations in partial cross sections occurring in SF_6, presumably due to resonant states.

The branching ratio in the x-ray region has been obtained by combining the information contained in two measurements:

(1) A relatively low resolution spectrum[373] (obtained with $h\nu = 1253.6$ eV) which relates the intensity of valence shell ionization to inner shell ionization, but does not resolve the valence shell.

(2) A high resolution spectrum[382] (Fig. 97) (obtained with $h\nu = 1486.6$ eV) which adequately resolves the valence shell ionization states.

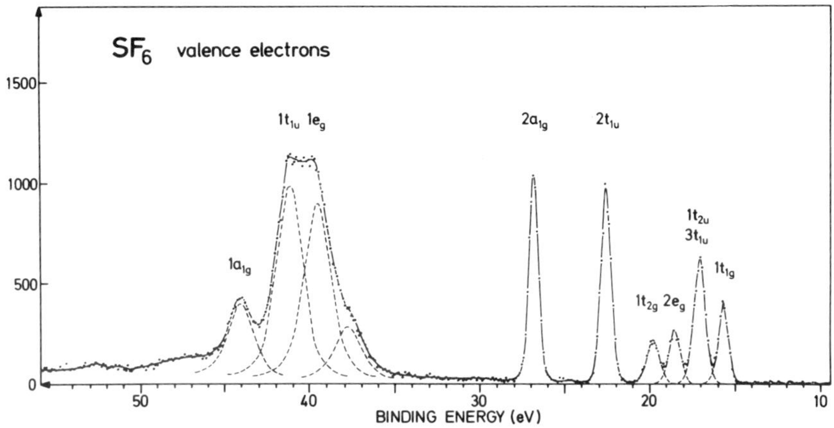

Fig. 97 The complete valence region photoelectron spectrum of SF_6 excited by monochromatized Al K_α radiation. [From U. Gelius, *J. Electr. Spectr.* **5**, 985 (1974).] The principal quantum numbers for orbital designations differ here from the text because of a different convention for the core orbitals.

The partial cross sections deduced from the total photoabsorption cross sections (Fig. 41 in Chapter V), the branching ratios discussed above and an assumed ionization yield of unity are plotted in Fig. 98.

b. Photoionization Mass Spectrometry. The photoionization mass spectrum (Fig. 99) reveals three prominent ions in the region from threshold to 21.1 eV—SF_5^+, SF_4^+, and SF_3^+. The SF_6^+ ion is observed with extremely weak intensity (3×10^{-4} relative to SF_5^+)[383] at 21.2 eV. The apparently weak stability of SF_6^+ is similar to the behavior of XeF_6^+,[384] CF_4^+,[385] and other heavily halogenated (particularly fluorinated) systems. The case of CF_4, where the relevant thermochemical data are available, provides some insight into the pattern of instability for these ions. The bond energy CF_3—F is 5.56 eV[176] and the ionization potential of CF_3 is ~9.2 eV[385] so that the thermochemical threshold for the reaction

$$CF_4 \rightarrow CF_3^+ + F + e$$

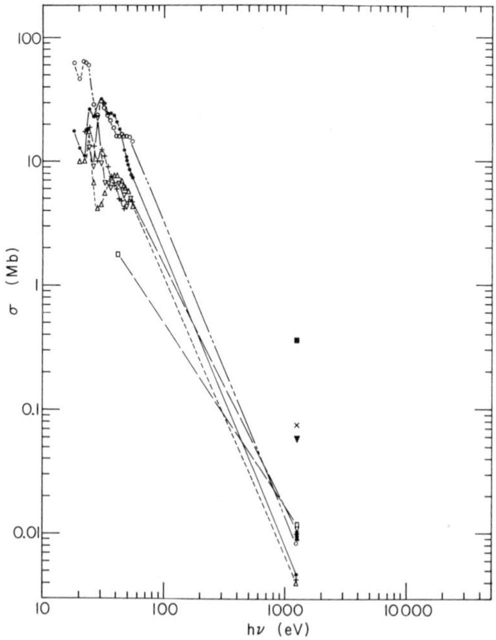

Fig. 98 Partial cross sections for formation of various states of SF_6^+ by photoabsorption. ● $(1t_{1g})^{-1}$ (15.8 eV band), ○ $(5t_{1u})^{-1} + (1t_{2u})^{-1}$ (16.9 eV band), △ $(3e_g)^{-1}$ (18.6 eV band), + $(1t_{2g})^{-1}$ (19.9 eV band), ▽ $(4t_{1u})^{-1}$ (22.5 eV band), □ $(5a_{1g})^{-1}$ (26.8 eV band), ∗ $(2e_g)^{-1}$ (39.3 eV), ▼ $(3t_{1u})^{-1}$ (41.2 eV band), ▲ $(4a_{1g})^{-1}$ (44.2 eV band), × $(2t_{1u})^{-1}$ (180.4–181.7 eV), ■ $(1t_{1u})^{-1} + (1e_g)^{-1} + (2a_{1g})^{-1}$ (694.6 eV).

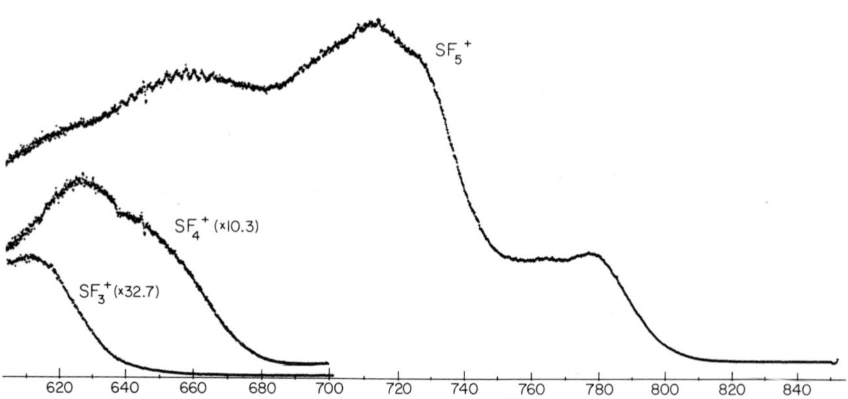

Fig. 99 Photoion yield curves of SF_5^+, SF_4^+, and SF_3^+ for SF_6 between threshold and 600 Å. [Unpublished data, obtained by the author.]

is 14.76 eV. On the other hand, the Franck–Condon maximum for formation of CF_4^+ in its ground state occurs[386] at an energy slightly greater than 16 eV, with an extremely weak, if detectable, signal below 15 eV. It is very unlikely that the configuration of maximum stability is very different for the molecular ion than the neutral molecule, since relatively nonbonding (covalent sense) fluorine orbitals are involved. The apparent instability of CF_4^+ is related to the large difference in ionization potential between CF_4 and CF_3. By contrast, the ionization potentials of CH_4 and CH_3 are 12.61 and 9.84 eV, much closer to one another, and both CH_4^+ and CH_3^+ are prominent in the photoionization of methane.

Viewed from a different perspective, the large electronegativity difference in the C—F or S—F bonds is an indication of significant ionic bonding, in contrast to the C—H bond. We have already noted that the uppermost occupied orbitals are localized on the fluorine atoms. In the limit of pure ionic bonding, an S^+F^- or C^+F^- bond would be drastically weakened if electron removal from the fluorine occurred, whereas the largely covalent C—H bond would be very slightly affected. In this context, it is noteworthy that the highly ionic alkali[387] and alkaline earth halides[388] tend also to have weak parent ions, and to preferentially lose a halogen.

SF_4^+ is initially observable at ~18.1 eV (Fig. 99). Hence, the first two bands in the photoelectron spectrum result in the formation of SF_5^+. The contribution of each of these bands is clearly seen in the photoionization yield of SF_5^+ (Fig. 99). A rather large kinetic energy release [(up to 1.10 eV from the ground state, 2.15 eV from the first excited state(s)] accompanies this fragmentation.[389] Simm *et al.*[389] have offered the following explanation for this behavior:

Stretching one S—F bond from initially octahedral symmetry leads to C_{4v} symmetry and ultimately to an SF_5^+ ion in square pyramidal configuration. Other evidence[389] points to the trigonal bipyramid configuration as the most stable for SF_5^+. The substantial geometric changes that must ensue could lead to the large kinetic energies observed. Simm *et al.* conclude that at the onset of SF_5^+ formation, more than 0.76 eV is released as kinetic energy.

The onset for SF_4^+ formation is rather close to the adiabatic threshold for the third photoelectron band of SF_6 at ~18 eV. However, a series of bands in the photoabsorption curve of SF_6 between 620 and 680 Å (18.23–20.0 eV) is also observed in the SF_5^+ channel (Fig. 99) suggesting that autoionization from this quasidiscrete series leads to formation of those ionic states characteristic of the lower two bands.

The correlation of the onset of SF_4^+ with that of $(3e_g)^{-1}$, together with the lack of influence of this state on SF_5^+, and the leveling off of SF_4^+ intensity at higher photon energy strongly suggest that the $(3e_g)^{-1}$ state predominantly decays to SF_4^+, probably with the concomitant formation of two fluorine atoms.

The ion SF_3^+ rises in exponential fashion from a threshold roughly correlating with the adiabatic ionization potential corresponding to $(1t_{2g})^{-1}$. At 584 Å, Dibeler and Walker[383] report $SF_5^+:SF_4^+:SF_3^+ = 1.0:0.11:0.14$. (Our own measurements yield 1.0:0.12:0.13.) The arguments attributing SF_5^+ exclusively to the first two bands, SF_4^+ to the third and SF_3^+ to the fourth thus imply that the partial cross sections of these four bands are roughly in line with those measured at 22 eV by Gustafsson *et al.*[378a] However, possibly poor collection efficiencies for SF_3^+ and SF_4^+ (if produced with higher kinetic energy) could influence these conclusions. Mass spectra obtained with several discrete line sources between 584 and 304 Å at this laboratory[309] indicate that SF_3^+ continues to increase its relative intensity through $(4t_{1u})^{-1}$ and possibly $(5a_{1g})^{-1}$, whereas SF_4^+ does not. SF_2^+ is first detectable above background at 26.95 eV, suggesting that this is one of the channels for decay of $(5a_{1g})^{-1}$. The first significant appearance of SF^+ occurs at 32.7 eV in the tail of the broad and complex peak $[(2e_g)^{-1}, (3t_{1u})^{-1}]$ seen (Fig. 97) in the x-ray photoelectron spectrum at ~40 eV. At 40.8 eV, SF^+ surpasses SF_2^+ in intensity, amounting to ~4.5% of SF_5^+, whereas SF_2^+ achieves ~2.8% of SF_5^+.

Pullen and Stockdale[390] have studied the formation of F^+ from SF_6 by electron impact. They observe an appearance potential for F^+ of 37.5 ± 1.0 eV, which again correlates with the broad, complex photoelectron peak centered at ~40 eV. A very broad kinetic energy distribution of F^+ is observed, extending to 8 eV with incident electron energy of 45 eV. It appears as if the states associated with this broad peak disintegrate in several channels, prominently including SF^+ and F^+. The exceptionally large kinetic energy release observed in the SF^+ channel suggests that a doubly ionized moiety is formed in the decomposition sequence.

Although the thermochemical data for the fragments of SF_6 are not sufficiently accurate for applying the quasi-equilibrium theory, the large kinetic energies measured for SF_5^+ [389] and F^+ [390] would be very difficult to rationalize by most statistical theories, which predict that relatively little energy should be released into translation in the dissociation of a polyatomic ion, except in the presence of a reverse activation energy.[391]

Hitchcock *et al.*[391a] have recently provided us with a rare glimpse of the fragmentation processes subsequent to inner shell vacancy production in a polyatomic molecule. Using 8 keV incident electrons and coincidence measurements between selected electron energy losses and positive ions formed, they have examined the region between 165 and 225 eV energy loss, encompassing the S(2p) threshold at ~181 eV. After subtracting the underlying valence-shell processes, they find that the fragmentation ratio (fraction of each ion observed) is constant beyond the S(2p) edge, even across the resonances beyond threshold. By contrast, a different fragmentation pattern was observed for the pre-threshold resonant state at 172 eV. The dominant ions observed were F^+, S^+, SF^+, and SF_2^{2+}, with much weaker fractions at $SF_3{}^+$, $SF_2{}^+$, and SF^+. No $SF_5{}^+$, $SF_4{}^+$, or SF_4^{2+} could be attributed to decays consequent upon formation of an S(2p) hole.

For the entire post S(2p) threshold region, including resonances, the authors conclude that a rapid Auger process ensues formation of the S(2p) hole, followed by dissociation, primarily into two singly charged moieties, i.e.

$$SF_6^{2+} \rightarrow A^+ + B^+ \qquad \text{or} \qquad A^{2+} + B^+$$

Doubly charged ions account for $<15\%$ of S(2p) ionization events. Thus $\sim 85\%$ of S(2p) ionization results in two singly charged ions.

X-ray emission spectra following electron ejection from molecular orbitals of predominantly S(1s), S(2s), and F(1s) type have been observed and analyzed,[392] but corresponding Auger spectra and mass spectrometric analysis of ionic fragments have not been reported.

Appendix 1

TABLE A.1 *Wavelength Dependent Photoion Yield Curves for Molecular Species Not Encompassed in this Monograph*

Molecule	λ Range, Å	Resolution, Å	Ref.
Atoms			
O	650–920	0.42	(393)
N	600–870	0.83	(394)
Zn	247–1242	2–4[a]	(395)
	191–453		(396)
Cd	649–1379	2.5	(397)
	150–1378	2–4[a]	(395)
Fe	1150–1600	1–2	(398)
Cu	1240–1435	1.5	(399)
Ag	1280–1580	2.0	(399)
Au	1205–1340	2.3	(399)
Eu	1350–2189	3	(400)
Tm	1350–2006	3	(400)
Yb	1350–2000	3	(401)
$Yb \rightarrow Yb^{++}$	584–689	5	(402)
Diatomics			
HF	610–790	0.83	(403)
	610–790	1.0	(404)
DF	610–790	0.83	(403)
HCl	600–990	1.0	(405)

TABLE A.1 (*continued*)

Molecule	λ Range, Å	Resolution, Å	Ref.
Diatomics			
HI	600–1205	0.14	(229)
	1117–1205	~5.0	(406)
DI	600–1205	0.14	(229)
F_2	600–815	1.0	(407)
			(404)
	620–800	0.4	(403)
Cl_2	850–1090	0.7	(408)
Br_2	950–1210	0.7	(408)
I_2	1175–1355	0.7	(408)
	1117–1410	~5.0	(406)
	920–1370	2.5–7.0	(409)
ClF	600–1030	0.7	(410)
ICl	1000–1240	0.7	(408)
IBr	1050–1280	0.7	(408)
NaI	1000–1680	~3.5	(411)
RbCl	860–1590	2.5–5	(412)
RbBr	860–1600	2.5–5	(412)
RbI	860–1760	2.5–5	(412)
CsF	880–1610	2.5–5	(413)
CsCl	865–1570	2.5–5	(413)
CsBr	875–1675	2.5–5	(413)
CsI	865–1735	2.5–5	(413)
TlF	865–1745	2.5–5	(414)
TlCl	865–2000	2.5–5	(414)
TlBr	865–2150	2.5–5	(414)
TlI	1020–2220	~3.5	(411)
S_2	1180–1336	2.5–3.0	(415)
	700–1330	0.8–2.5	(416)
Se_2	625–1440	2.5	(416)
Te_2	920–1610	~3.5	(416)
Triatomics			
H_2S	600–1200	1.0(?)	(417)
HOF	600–1000	0.83	(418)
HCN	600–830	0.42–0.83	(419)
	600–925	~1.0	(420)
COS	600–1120	1–2	(163)
CS_2	600–1235	1–2	(163)
CSe_2	600–1370	0.15	(421)
O_3	600–1000	0.44	(422)
	600–1000	?	(423)
SO_2	600–1020	1.0(?)	(417)
S_2O	625–1190	0.44	(424)
S_3	945–1290	2.5–3	(415)

TABLE A.1 *(continued)*

Molecule	λ Range, Å	Resolution, Å	Ref.
NO_2	1220–1295	0.83	(425)
	600–1300	1.0(?)	(426)
	885–1265	6	(427)
	500–1100	8[a]	(191)
F_2O	600–954	0.83	(428)
KrF_2	600–1100	2.5	(429)
XeF_2	600–1100	0.4	(384)
	918–1018	?	(430)
FCN	600–940	1.0	(431)
ClCN	600–1025	1.0	(431)
BrCN	700–1070	1.0	(431)
ICN	700–1165	1.0	(431)
MgI_2	1020–1320	1.7	(411)
Free radicals			
CH_3	750–1300	1.6	(432)
	1130–1278	6	(433)
C_2H_5	1130–1515	6	(433)
n-C_3H_7	1170–1625	6	(433)
iso-C_3H_7	1230–1680	6	(433)
C_6H_5	1180–1570	6	(434)
C_7H_7	1580–1700	8	(435)
	1600–2000	2.5	(436)
CF_3	860–1420	1.6	(437)
piperidone radical	1100–1900	?	(438)
Alkanes			
C_3H_8	1050–1145	6.7	(439)
	890–1160	1.66	(329)
C_3H_8	500–1180	8[a]	(440)
n-C_4H_{10}	1065–1195	6.7	(439)
	890–1215	1.66	(329)
iso-C_4H_{10}	1065–1195	6.7	(439)
n-C_5H_{12}	1070–1220	6.7	(439)
iso-C_5H_{12}	1070–1220	6.7	(439)
neo-C_5H_{12}	1080–1230	6.7	(439)
n-hexane	1080–1235	6.7	(439)
2-methyl pentane	1080–1250	6.7	(439)
3-methyl pentane	1080–1260	6.7	(439)
2,3-dimethyl butane	1080–1260	6.7	(439)
2,2-dimethyl butane	1080–1280	6.7	(439)
n-heptane	1080–1250	6.7	(439)
	860–1255	1.0	(369)
n-octane	1080–1270	6.7	(439)
3,4-dimethyl hexane	1080–1290	6.7	(439)

TABLE A.1 (*continued*)

Molecule	λ Range, Å	Resolution, Å	Ref.
Alkanes			
cyclopropane	475–1290	≥0.7	(441)
	550–1180	8[a]	(440)
cyclohexane	920–1265	2	(442)
Alkyl halides			
CH_3F	600–1000	1.0	(405)
CH_3Cl	1030–1065	2.0	(443)
CH_3Cl (CH_2DCl,CHD_2Cl,CD_3Cl)	650–1250	1.0	(405)
CH_3Br	750–1200	1.0	(405)
CH_3I	1115–1320	~5	(406)
CH_2Cl_2	970–1110	2.0	(443)
$CHCl_3$	1025–1115	2	(443)
CF_4	790–815	0.25	(385)
CCl_4	1035–1105	2	(443)
CBr_4	1140–1220	2	(443)
C_2H_5Cl	1080–1320	10–15	(444)
	930–1160	2	(445)
C_2H_5Br	920–1220	2	(445)
C_2H_5I	930–1350	2	(445)
	950–1350	10–15	(444)
$(CH_3)_2CHF$ (2-fluoropropane)	960–1160	1	(446)
$(CH_3)_2CF_2$ (2,2-difluoropropane)	990–1117	1	(446)
Cyclo-$C_6H_{11}Cl$	950–1470	10–15	(444)
Alkenes and their halides			
1-propene	1070–1305	6.7	(439)
	475–1330	≥0.7	(441)
1-butene	1060–1315	6.7	(439)
1-pentene	1060–1320	6.7	(439)
1-hexene	1080–1320	6.7	(439)
1,3-butadiene	950–1380	3	(447)
	855–1370	1	(369)
1,2 butadiene	920–1380	3	(447)
allene (C_3H_4)	920–1290	3	(447)
C_2H_3F (vinyl fluoride)	550–1215	≥0.7	(448)
C_2H_3Cl (vinyl chloride)	565–1240	≥0.7	(448)
C_2H_3Br (vinyl bromide)	540–1280	≥0.7	(449)
$C_2H_2F_2$	565–1240	≥0.7	(448)
C_2F_4	600–1305	0.25	(385)

TABLE A.1 *(continued)*

Molecule	λ Range, Å	Resolution, Å	Ref.
Alkynes			
propyne (C_3H_4)	930–1250	3	(447)
	1000–1210	1.3	(450)
Amines and other nitrogen containing compounds			
$CH_3CH_2NH_2$ (ethylamine)	1130–1420	8.3	(451)
	1140–1420	6–12	(188)
$CH_3CH_2CH_2NH_2$ (propylamine)	1140–1425	6–12	(188)
n-butyl amine	1020–1450	?	(452)
methylamine, ethylamine, *n*-propylamine, *tert*-butylamine	No spectra given	?	(452)
diethylamine	990–1550	?	(453)
triethylamine	No spectra given	?	
n-hexylamine	(950–1550)	?	(454)
n-decylamine	No spectra given		
dimethyl aniline	1060–1730	~10	(455)
benzylamine	1080–1425	~10	(455)
trimethylamine $(CH_3)_3N$	885–1590	?	(456)
tetramethyl diamino methane	950–1600	?	(456)
methyl ester of *N,N'*-dimethyl aminoacetic acid	950–1600	?	(456)
N-dimethyl-*p*-methyl benzylamine	1030–1650	?	(457)
dimethylallylamine	885–1550	?	(457)
dimethylpropargylamine	980–1520	?	(457)
nitrotrimethylamine	900–1425	?	(457)
$CH_3N{=}NCH_3$ (transazomethane)	960–1480	1	(458)
CH_3ONO_2 (methyl nitrate)	1000–1550	?	(459)
CH_3NO_2 (nitromethane)	No spectrum given	?	(459)
C_2N_2	600–950	1.0	(431)
NF_3	600–975	?	(460)
ONF_3	600–960	?	(460)
N_2H_4	1025–1460	10	(461)
$CH_3N_2H_3$	1050–1570	10	(461)
$(CH_3)_2N_2H_2$	1060–1630	10	(461)
$(C_2H_5)_2N_2H_2$	1060–1630	10	(461)
$C_4H_9CH_3N_2H_2$	1060–1630	10	(461)
C_5H_5N	970–1350	?	(462)

TABLE A.1 *(continued)*

Molecule	λ Range, Å	Resolution, Å	Ref.
Amines and other nitrogen containing compounds			
$C(NO_2)_4$	885-Threshold	(No spectra shown)	(463)
$C(NO_2)_3F$	885-Threshold		(463)
$C(NO_2)_3Cl$	885-Threshold		(463)
$C(NO_2)_3Br$	885-Threshold		(463)
pyrrole 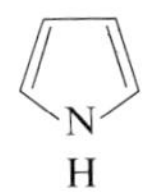	1030–1520	?	(464)
	1050–1520	?	(465)
N-methyl pyrrole	1000–1540	?	(465)
2-methyl pyrrole	1050–1610	?	(465)
2,2,6,6-tetramethyl-*N*-oxyl piperidine	1100–1850	?	(438)
Aldehydes			
$CH_2O(CHDO,CD_2O)$	650–1150	0.83	(333d)
CH_3CHO	1090–1230	6–12	(188)
	Only short region around threshold given in *differentiated* form		(340)
	475–1250	≥0.7	(441)
	930–1225	2–3	(466)
Alcohols			
C_2H_5OH	845–1195	1.66	(339)
n-C_3H_7OH	855–1250	1.66	(339)
	1080–1235	6–12	(188)
i-C_3H_7OH	855–1250	1.66	(339)
Ketones			
$(CH_3)_2CO$	950–1290	2–3	(466)
	1090–1290	6–12	(188)
	1090–1340	6.7	(467)
	1090–1340	6.7	(468)
	1148–1285	1.5	(468a)
$(CD_3)_2CO$	1148–1285	1.5	(468a)
2-butanone	1100–1350	6.7	(467)
	1100–1350	6.7	(468)
	1100–1325	6–12	(188)
2-pentanone	1100–1380	6.7	(467)
3-methyl-2-butanone	1100–1380	6.7	(467)
2-hexanone	1100–1380	6.7	(467)
4-methyl-2-pentanone	1100–1400	6.7	(467)
2,3-pentanedione ($CH_3COCOC_2H_5$)	1100–1420	6.7	(468)
Bi-acetyl ($(CH_3CO)_2$)	1100–1405	6.7	(468)

TABLE A.1 (*continued*)

Molecule	λ Range, Å	Resolution, Å	Ref.
$(CH_3)_2CO$	Only short region around threshold given in *differentiated* form		
$C_2H_5COCH_3$			
n-$C_3H_7COCH_3$			(340)
Sulfides			
S_5	920–1500	2.5–3	(415)
S_6	920–1400	2.5–3	(415)
S_7	920–1460	2.5–3	(415)
S_8	920–1410	2.5–3	(415)
CH_3SH	950–1350	2–3	(469)
$(CH_3)_2S$	900–1460	2–3	(469)
$(C_2H_5)_2S$	940–1500	2–3	(469)
$(C_2H_5)S(C_3H_7)$			(470)
$(C_3H_7)_2S$	No spectra given		(470)
$(CH_3)S(C_5H_{11})$			(470)
propyl butyl sulfide			(471)
propyl isobutyl sulfide	No spectra given		(471)
dibutyl sulfide			(471)
butyl isobutyl sulfide			(471)
ethyl amyl sulfide			(472)
ethyl tert-amyl sulfide	No spectra given		(472)
ethyl isobutyl sulfide			(472)
thiophene 	1010–1415	?	(464)
Aromatic hydrocarbons and aromatic halides			
toluene	590–950	3.5	(473)
toluene	980–1365	2–2.5	(474)
toluene	900–1410	6	(475)
	1060–1430	~10	(455)
o-xylene	1040–1480	2–2.5	(474)
m-xylene	900–1450	6	(475)
p-xylene	980–1475	2–2.5	(474)
p-xylene	900–1480	6	(475)
ethyl benzene	900–1420	6	(475)
C_6H_5Cl	950–1400	2	(476)
C_6H_5Br	950–1400	2	(476)
C_6H_5I	950–1450	2	(476)
benzyl chloride	940–1400	2	(435)
	900–1365	6	(475)
Organic acids and esters			
HCOOH	920–1130	3	(477)
CH_3COOH	940–1180	3	(478)
$HCOOCH_3$	940–1165	3	(477)
$HCOOC_2H_5$	950–1180	3	(477)

TABLE A.1 *(continued)*

Molecule	λ Range, Å	Resolution, Å	Ref.
Organic acids and esters			
$HCOOC_3H_7$-*n*	950–1210	3	(477)
CH_3COOCH_3	950–1230	3	(478)
$CH_3COOC_2H_5$	950–1250	3	(478)
$CH_3COOC_3H_7$-*n*	950–1260	3	(478)
$CH_3COOC_4H_9$-*n*	950–1250	3	(478)
Boron compounds			
BF_3	600–810	?	(479)
BCl_3	600–1080	1	(480)
B_2F_4	600–1050	?	(479)
B_2Cl_4	600–1230	1	(480)
Miscellaneous			
XeF_4	625–1000	0.4	(384)
XeF_6	600–1045	0.4	(384)
$Zn(CH_3)_2$	600–1390	1	(481)
$Cd(CH_3)_2$	600–1470	1	(481)
$Hg(CH_3)_2$	600–1395	1	(481)
$Si(CH_3)_4$	600–1300	1	(482)
furan 	1025–1410	?	(464)
p-benzoquinone	950–1380	1.5–4	(483)
methyl-*p*-benzoquinone	950–1380	1.5–4	(483)
duroquinone	950–1380	1.5–4	(483)
$(CH_2)_2O$ (ethylene oxide)	475–1215	≥0.7	(441)
$Fe(CO)_5$	600–1560	1	(484)
$Ni(CO)_4$	600–1500	1	(484)
van der Waals clusters			
Ar_2	750–855	1.2	(485)
Kr_2	850–965	1.2	(485)
Xe_2	965–1118	1.2	(486)
ArKr	790–930	1.3	(487)
ArXe	790–1065	1.3	(487)
KrXe	790–1065	1.3	(487)
$(HF)_2$	600–840	2.5	(488)
$(NO)_2$	1210–1460	2.5	(489)
$(H_2O)_2$	950–1100	2.5	(498)
$(CO_2)_2$	500–950	1.90	(490)
$[(CH_3)_2CO]_{2,3,4}$, $[(CD_3)_2CO]_2$	1200–1370	1.5	(468a)

[a] Line sources used.

Appendix 2. Derivation of the "Energy Gap Law" Simplified

Consider the states described in Fig. 2b, particularly the initial electronically excited state $\phi_{is}(=\phi_{si})$ and the manifold to which it decays $\{\phi_{il}\}$, which are described below by the wave functions ϕ_{ij}. We make the Born–Oppenheimer approximation for both sets of states.

$$\phi_{si}(q,Q^s) = \phi_s(q,Q^s)\chi_{si}(Q^s) \tag{1}$$

$$\phi_{lj}(q,Q^l) = \phi_l(q,Q^l)\chi_{lj}(Q^l). \tag{2}$$

Here, q denotes electronic coordinates, Q the nuclear coordinates.

Assume the electronic part of the matrix element is slowly varying.

$$v_{si,lj} = \int \phi_s{}^*\chi_{si}^* v \phi_l \chi_{lj}\, dq\, dQ \approx \beta_e \int \chi_{si}^* M(Q)\chi_{lj}\, dQ. \tag{3}$$

Here, β_e is the effective electronic interaction; $M(Q)$ is a Q-dependent operator, which, for the ensuing discussion, can be taken as unity.

The nonradiative rate can be written

$$k_{nr} = \frac{2\pi}{\hbar}\beta_e{}^2 \sum_{lj} |\langle\chi_s|\chi_{lj}\rangle|^2 \rho_{lj}(E_{si}). \tag{4}$$

For simplicity, consider that only one vibrational level of the initial electronically excited system (the s system), the $v = 0$ level, is populated. We then use a harmonic oscillator wave function for this χ_s. On the other hand, the χ_l are highly excited vibrational states, and for tractability the harmonic oscillator functions are still used. The rationale appears to be that anharmonic oscillators can be described as linear combinations of harmonic oscillators. Thus, a given harmonic oscillator wave function can be thought of as a linear combination of true vibrational states (taking the inverse linear combination) over a range of energies of order δV about E_l, where δV is a measure of the anharmonic coupling in this energy region.

If the normal modes remain identifiable and unchanged in the $s \rightarrow l$ transition, but change in frequency ω and equilibrium position, we can more explicitly write the nonradiative rate as

$$k_{nr} = \left(\frac{2\pi}{\hbar}\right)\beta_e{}^2 \sum_{n_j}\prod_j |\langle\chi_0|\chi_{n_j}\rangle|^2 \times \delta\left[\Delta E - \sum_j \hbar\omega^l\left(n_j + \frac{1}{2}\right) + \sum_j \frac{\hbar\omega_j{}^s}{2}\right], \tag{5}$$

where χ_{n_j} is the vibrational wave function of one of the many vibrational states in the l manifold whose frequency is ω_j^l. The last term represents the zero-point vibrational energy in the initial s state.

The density of states ρ_l has now become a somewhat fuzzy Dirac delta function.

Now another simplification is introduced—the frequencies in the s and l manifolds are assumed equal

$$(\omega_j^s = \omega_j^l = \omega_j),$$

while the mode displacements are described by the dimensionless parameter

$$X_j = \frac{1}{2}\left(\frac{\mu_j\omega_j}{\hbar}\right)(Q_{j,0}^l - Q_{j,0}^s)^2. \tag{6}$$

Here, μ_j is the reduced mass for mode j, $Q_{j,0}^l$ is the equilibrium position for mode j in the l manifold (and correspondingly for $Q_{j,0}^s$). A standard solution is available for the displaced harmonic oscillator Franck–Condon factors and yields

$$k_{\text{nr}} = \left(\frac{2\pi}{\hbar}\right)\beta_{\text{e}}{}^2 e^{-\Sigma_j X_j}\left[\prod_j \sum_{n_j=0}^{\infty} X_j^{n_j}\Big/(n_j)!\right]\delta\left(\Delta E - \sum_j n_j\hbar\omega_j\right). \tag{7}$$

Freed[18] has noted the analogy of this expression to that encountered in treatments of Boltzmann statistics in elementary texts on statistical thermodynamics. Defining

$$F\{n_j\} = \prod_{j=1}^{N} \frac{X_j^{n_j}}{n_j!} \tag{8}$$

subject to the constraint $\Delta E = \sum_j n_j\hbar\omega_j$, we seek the maximum of $\ln F\{n_j\}$.

From Stirling's approximation,

$$\ln F\{n_j\} \cong \sum_{j=1}^{N}\left[n_j \ln\left(\frac{X_j}{n_j}\right) + n_j\right]. \tag{9}$$

The constraint is introduced by the method of Lagrange multipliers, i.e.,

$$\frac{\partial}{\partial n_i}\left[\ln F\{n_j\} + b\sum_j n_j\hbar\omega_j\right]_{n_i{}^0} = 0 \qquad \text{for all } i = 1, 2, \cdots, N, \tag{10}$$

where b is the Lagrange multiplier.

This leads to

$$-\ln n_i{}^0 + \ln X_i + b\hbar\omega_i = 0 \qquad \text{or} \qquad n_i{}^0 = X_i e^{b\hbar\omega_i} \qquad \text{for all } i, \tag{11}$$

and hence

$$F_{\max}(\Delta E, \{n_j{}^0\}) = \prod_{j=1}^{N} \frac{X_j^{n_j{}^0}}{\Gamma(n_j{}^0 + 1)}. \tag{12}$$

The value of the Lagrange multiplier b is determined by the equation

$$\Delta E = \sum_{j=1}^{N} n_j{}^0 \hbar\omega_j = \sum_{j=1}^{N} \hbar\omega_j X_j e^{b\hbar\omega_j}. \tag{13}$$

Stirling's approximation can also be applied to the gamma function

$$\Gamma(n_j{}^0 + 1) \cong (n_j{}^0)^{n_j{}^0} e^{-n_j{}^0}. \tag{14}$$

In order to make the result more transparent, note that $n_j{}^0$ depends exponentially on ω_j. Assume, therefore, that one frequency ω_M dominates the rate. Then

$$\ln F = n_M{}^0(+b\hbar\omega_M) - n_M{}^0. \tag{15}$$

But recalling

$$b\hbar\omega_M = \ln \frac{\Delta E}{X_M \hbar\omega_M} \qquad \text{and} \qquad n_M{}^0 = X_M e^{b\hbar\omega_M} = \frac{\Delta E}{\hbar\omega_M},$$

we obtain

$$\ln F = \frac{\Delta E}{\hbar\omega_M}\left(\ln \frac{\Delta E}{X_M \hbar\omega_M} - 1\right). \tag{16}$$

Introducing

$$\gamma = \ln \frac{\Delta E}{X_M \hbar\omega_M} - 1,$$

we finally obtain

$$F = e^{-\gamma\, \Delta E/\hbar\omega_M}. \tag{17}$$

Hence, the nonradiative rate k_{nr} falls off exponentially with the "energy gap" ΔE between the s and l surfaces, but is favored by high frequency modes ω_M. Some liberties have been taken even with this simplified derivation. Freed's review[18] gives other, more elegant derivations, and transcends some of the simplifications assumed in the above derivation. He also describes some of the experimental observations supporting the "energy gap law" from studies of neutral aromatic hydrocarbons and rare-earth ions in crystals.

We conclude this derivation by noting that the energy gap equation is a consequence of introducing the Born–Oppenheimer approximation and finding an approximation for the resulting Franck–Condon factors.

Appendix 3. Classical Approximation and Dirichlet Integral

The energy levels of a single harmonic oscillator, as obtained in quantum mechanics, are given by

$$\varepsilon_n = (n + \tfrac{1}{2})h\nu. \tag{18}$$

If we take a $\Delta\varepsilon_n$ which is large compared to $h\nu$, the number of eigenstates encompassed by $\Delta\varepsilon_n$ is $\Delta\varepsilon_n/h\nu$, or $g(\Delta\varepsilon) = (1/h\nu)\,\Delta\varepsilon$.

If we have many weakly coupled oscillators N, for each of which the above expression holds and for which the total energy is E, then

$$G(E) = \sum\sum\cdots\sum \frac{\Delta\varepsilon_1}{h\nu}\frac{\Delta\varepsilon_2}{h\nu}\cdots\frac{\Delta\varepsilon_N}{h\nu}, \tag{19}$$

where the various summations indicate that each $\Delta\varepsilon_n$ can have all or part of the total energy, restricted only by the overall requirement that

$$\varepsilon_1 + \varepsilon_2 + \cdots + \varepsilon_N \leq E. \tag{20}$$

In the limit of small $\Delta\varepsilon_n$, but always with the restriction that $\Delta\varepsilon_n > h\nu$ (which corresponds to a classical limit), we can write

$$G(E) = \left(\frac{1}{h\nu}\right)^N \int\cdots\int d\varepsilon_1\, d\varepsilon_2 \cdots d\varepsilon_N. \tag{21}$$

This happens to be of the form of a definite integral, studied by Dirichlet, which can be stated

$$\iint\cdots\int \varepsilon_1^{i_1-1} + \varepsilon_2^{i_2-1}\cdots\varepsilon_N^{i_N-1}\, d\varepsilon_1\, d\varepsilon_2\cdots d\varepsilon_N$$

$$= \frac{\Gamma(i_1)\Gamma(i_2)\cdots\Gamma(i_N)}{\Gamma(i_1 + i_2 + \cdots + i_N + 1)} E^{i_1+i_2+\cdots+i_N}, \tag{22}$$

where $\varepsilon_1 + \varepsilon_2 + \cdots + \varepsilon_N \leq E$, ($\varepsilon_1, \varepsilon_2, \ldots, \varepsilon_N$ are positive, $i_1, i_2, \ldots, i_N$ are any set of numbers greater than zero).

In our case, $i_1 = i_2 = \cdots = i_N = 1$, and hence

$$G(E) = \left(\frac{1}{h\nu}\right)^N \frac{1}{\Gamma(N+1)} E^N. \tag{23}$$

Within the limitations of the classical approximation indicated above, this expression tells us the total number of states in the energy range between 0 and E. If we wish to know the number of states in a small band of that entire energy gap, we differentiate and obtain

$$\frac{dG(E)}{dE} = \left(\frac{1}{h\nu}\right)^N \frac{N}{\Gamma(N+1)} E^{N-1}. \tag{24}$$

If the oscillators do not have identifical frequencies, then $(1/h\nu)^N$ becomes

$$\left(\frac{1}{h\nu_1}\right)\left(\frac{1}{h\nu_2}\right)\cdots\left(\frac{1}{h\nu}\right)_N = \prod_{n=1}^{N}\left(\frac{1}{h\nu_n}\right) \quad \text{and} \quad [\text{definition of } \Gamma(N+1) = N!]$$

$$\frac{N}{\Gamma(N+1)} = \frac{N}{N!} = \frac{1}{(N-1)!}.$$

The expression now reads

$$\frac{dG(E)}{dE} = \prod_{n=1}^{N}\left(\frac{1}{h\nu_n}\right)\frac{E^{N-1}}{(N-1)!}. \tag{25}$$

Marcus and Rice[23] find that if they identify E with an energy that includes the zero point energy, the classical approximation departs less from the exact state density.

Appendix 4. Metastable Ions in Magnetic Mass Spectrometers

1. Initial Acceleration

$$\tfrac{1}{2}mv^2 = Ve. \tag{26}$$

2. Magnetic Momentum Selection

$$\frac{mv^2}{r} = e\mathbf{v} \times \mathbf{B} = evB\sin\theta. \tag{27}$$

If $\mathbf{v}$ is $\perp B$, then evB.

In centimeter-gram-second units,

$$\frac{mv^2}{r} = \frac{Hev}{c}, \qquad v = \frac{r}{m}\cdot\frac{He}{c}, \tag{28}$$

where H is in Oersteds, e in electrostatic units, v in centimeters per second and the force in dynes.

Eliminating v,

$$\frac{2V}{r} = \frac{Hv}{c} = \left(\frac{H}{c}\right)\frac{r}{m}\left(\frac{He}{c}\right) \tag{29}$$

$$\frac{m}{e} = \frac{H^2}{c^2}\frac{r^2}{2V} \quad \text{or} \quad r^2 = \frac{2Vmc^2}{eH^2}. \tag{30}$$

3. Metastable—Starts out as m_p, spontaneously dissociates between initial acceleration and magnetic bending to $m_f + n$ ($n =$ neutral). Then

$$\tfrac{1}{2}m_p v^2 = Ve$$

$$\frac{m_f v}{r} = \frac{He}{c} \tag{31}$$

Again, eliminating v,

$$v = \frac{reH}{m_f c} \tag{32}$$

$$\frac{1}{2}m_p\left(\frac{reH}{m_f c}\right)^2 = Ve$$

$$r^2 = \frac{2Vc^2}{eH^2}\left(\frac{m_f{}^2}{m_p}\right). \tag{33}$$

See, for example, previous

$$r^2 = \frac{2Vc^2}{eH^2}(m). \tag{34}$$

Therefore, a peak will appear at an apparent mass $m_f{}^2/m_p$, in general nonintegral—called a metastable peak.

REFERENCES

1. K. Siegbahn, D. A. Allison, and J. H. Allison, *in* "Handbook of Spectroscopy" (J. W. Robinson, ed.), Vol. 1, pp. 258–749. CRC Press, Cleveland, Ohio, 1974.

1a. C. R. Brundle and A. D. Baker, eds., "Electron Spectroscopy: Theory, Techniques and Applications." Academic Press, New York, 1977.

1b. A. J. Lorquet and J. C. Lorquet, *Chem. Phys.* **4**, 353 (1974).

2. H. M. Rosenstock, *Adv. Mass Spectrom.* **4**, 523 (1968).

3. A. L. Wahrhaftig, *in* "Theory of Mass Spectra" (A. Maccoll, ed.), MTP International Review of Science, Physical Chemistry Series I, Vol. 5, p. 8. Butterworth, London, 1972.

4. R. C. Ehlert, and R. A. Mattson *J. Chem. Phys.* **48**, 5465 (1968).

4a. J. Nordgren, H. Ågren, L. O. Werme, C. Nordling, and K. Siegbahn, *J. Phys. B* **9**, 295 (1976).

5. B. Rosen, ed., "Spectroscopic Data Relative to Diatomic Molecules." Pergamon, Oxford, 1970.

6. G. Herzberg, *Q. Rev., Chem. Soc.* **25**, 201 (1971).

7. S. Leach, *J. Chim. Phys.* Spec. No., pp. 74–83 (1970).

8. J. H. D. Eland, M. Devoret, and S. Leach, *Chem. Phys. Lett.* **43**, 97 (1976).

8a. M. Allan and J. P. Maier, *Chem. Phys. Lett.* **41**, 231 (1976).

9. H. Lew and I. Heiber, *J. Chem. Phys.* **58**, 1246 (1973).

9a. P. A. Wehinger, S. Wyckoff, G. H. Herbig, G. Herzberg, and H. Lew, *Astrophys. J.* **190**, L43 (1974); H. Lew, *Can. J. Phys.* **54**, 2028 (1976).

10. R. N. Dixon, G. Duxbury, M. Horani, and J. Rostas, *Mol. Phys.* **22**, 977 (1971); G. Duxbury, M. Horani, and J. Rostas, *Proc. R. Soc. London, Ser. A* **331**, 109 (1972).
10a. K. T. Wu and A. J. Yencha, *Can. J. Phys.* **55**, 767 (1977).
11. M. Allan, E. Kloster-Jensen, and J. P. Maier, *J. Chem. Soc. Faraday Trans. II* **73**, 1406, 1417 (1977).
12. M. Allan, E. Kloster-Jensen, and J. P. Maier, *Chem. Phys.* **7**, 11 (1976).
12a. M. Allan, E. Kloster-Jensen, J. P. Maier, and O. Marthaler, *J. Electr. Spectrosc.* **14**, 359 (1978).
13. M. Allan and J. P. Maier, *Chem. Phys. Lett.* **34**, 442 (1975).
13a. M. Allan, J. P. Maier, and O. Marthaler, *Chem. Phys.* **26**, 131 (1977).
13b. J. P. Maier and O. Marthaler, *Chem. Phys.*, in press.
14. M. Allan and J. P. Maier, *Chem. Phys. Lett.* **43**, 94 (1976).
14a. M. Allan, J. P. Maier, O. Marthaler, and E. Kloster-Jensen, *Chem. Phys.* **29**, 331 (1978).
15. See, e.g., W. Rothman, F. Hirayama, and S. Lipsky, *J. Chem. Phys.* **58**, 1300 (1973).
16. K. F. Freed, *Top. Curr. Chem.* **31**, 105 (1972).
17. A. E. Douglas, *J. Chem. Phys.* **45**, 1007 (1966).
18. K. Freed, *Top. Appl. Phys.* **15**, 23 (1976).
19. M. Nakamura, M. Sasanuma, S. Sato, M. Watanabe, H. Yamashita, Y. Iguchi, A. Ejiri, S. Nakai, S. Yamaguichi, T. Sagawa, Y. Nakai, and T. Oshio, *Phys. Rev.* **178**, 80 (1969).
20. L. S. Kassel, "The Kinetics of Homogeneous Gas Reactions." Chem. Catalog Co., New York, 1932.
21. H. M. Rosenstock, *J. Chem. Phys.* **34**, 2182 (1961).
22. H. M. Rosenstock, M. B. Wallenstein, A. L. Wahrhaftig, and H. Eyring, *Proc. Natl. Acad. Sci. U.S.A.* **38**, 667 (1952).
23. R. Marcus and O. K. Rice, *J. Phys. Colloid Chem.* **55**, 896, footnote (1951).
24. See, e.g., W. A. Chupka and J. Berkowitz, *J. Chem. Phys.* **32**, 1546 (1960).
25. W. Forst, *Chem. Rev.* **71**, 339 (1971); see also P. J. Robinson and K. A. Holbrook, "Unimolecular Reactions." Cornell Univ. Press, Ithaca, New York, 1959.
25a. M. R. Hoare and P. Pal, *Mol. Phys.* **20**, 695 (1971).
26. N. B. Slater, "Theory of Unimolecular Reactions." Cornell Univ. Press, Ithaca, New York, 1959.
27. J. L. Lebowitz and O. Penrose, *Phys. Today* Feb., p. 23 (1973).
27a. N. Bloemberger and E. Yablonovitch, *Phys. Today* May, pp. 23–30 (1978).
28. W. L. Hase, *J. Chem. Phys.* **64**, 2442 (1976).
29. C. E. Klots, *J. Phys. Chem.* **75**, 1526 (1973).
30. C. E. Moore, "Atomic Energy Levels." *Natl. Bur. Stand.* (*U.S.*), *Circ.* No. 467 (1949); now available as *Natl. Stand. Ref. Data Ser.*, *Natl. Bur. Stand.* **NSRDS-NBS 35** (1971).
31. R. P. Madden and K. Codling, *Astrophys. J.* **141**, 364 (1965).
32. M. O. Krause and F. Wuilleumier, *J. Phys. B* **5**, L143 (1972).
33. V. L. Jacobs and P. G. Burke, *J. Phys. B* **4**, L67 (1972).
33a. F. Wuilleumier, M. Y. Adam, N. Sandner, V. Schmidt, and W. Mehlhorn, *J. Phys.* (*Paris*) (to be published).
34. V. Schmidt, N. Sandner, H. Kuntzemuller, P. Dhez, F. Wuilleumier, and E. Kallne, *Phys. Rev. A* **13**, 1748 (1976).
34a. D. M. P. Holland, K. Codling, J. B. West, and G. V. Marr, "Multiple Photoionization in the Rare Gases from Threshold to 280 eV," Daresbury Lab. Rept. DL/SRF/P 159, Daresbury, England (1978); submitted to *J. Phys. B*.
35. K. Codling, R. P. Madden, and D. L. Ederer, *Phys. Rev.* **155**, 26 (1967).
36. K. Radler and J. Berkowitz, *J. Chem. Phys.* (to be published) (1978).
37. J. A. R. Samson and R. B. Cairns, *Phys. Rev.* **173**, 80 (1968).
38. J. A. R. Samson, J. L. Gardner, and A. F. Starace, *Phys. Rev. A* **12**, 1459 (1975).

39. F. Wuilleumier and M. O. Krause, *Phys. Rev. A* **10**, 342 (1974).
40. F. Wuilleumier, *Adv. X-Ray Anal.* p. 63 (1973).
41. T. A. Carlson, M. O. Krause, and W. E. Moddeman, *J. Phys.* (*Paris*) **32** (C4), 76 (1971).
42. M. O. Krause, *J. Phys.* (*Paris*) **32**(C4), 67 (1971).
43. K. Siegbahn *et al.*, "ESCA Applied to Free Molecules," p. 31. North-Holland Publ., Amsterdam, 1969.
44. T. A. Carlson and M. O. Krause, *Phys. Rev.* **140** A1057 (1965).
45. T. A. Carlson and M. O. Krause, *Phys. Rev. Lett.* **14**. 390 (1965).
46. R. P. Madden and K. Codling, *Phys. Rev. Lett.* **10**, 516 (1963).
47. J. A. R. Samson and J. L. Gardner, *Phys. Rev. Lett.* **33**, 671 (1974).
48. R. G. Houlgate, J. B. West, K. Codling, and G. V. Marr, *J. Electron. Spectrosc. Relat. Phenom.* **9**, 205 (1976).
48a. M. Y. Adam, F. Wuilleumier, S. Krummacher, V. Schmidt, and W. Mehlhorn, *J. Electr. Spectr.* (to be published).
49. K. Seigbahn *et al.*, "ESCA II," p. 40. (1969).
50. M. O. Krause, T. A. Carlson, and R. D. Dismukes, *Phys. Rev.* **170**, 37 (1968).
51. F. Wuilleumier, *J. Phys.* (*Paris*) **26**, 776 (1965).
52. T. A. Carlson and M. O. Krause, *Phys. Rev.* **137**, A1655 (1965).
53. T. E. H. Walker, J. Berkowitz, J. L. Dehmer, and J. T. Waber, *Phys. Rev. Lett.* **31**, 678 (1973).
54. R. P. Madden and K. Codling, *J. Opt. Soc. Am.* **54**, 268 (1964).
55. J. A. R. Samson and G. N. Haddad, *Phys. Rev. Lett.* **33**, 875 (1974).
56. K. Codling and R. P. Madden, *Phys. Rev. Lett.* **12**, 106 (1964).
57. T. M. El-Sherbini and M. J. Van der Wiel, *Physica* (*Utrecht*) **62**, 119 (1972).
58. M. O. Krause, *Phys. Rev.* **177**, 151 (1969).
59. D. J. Kennedy and S. T. Manson, *Phys. Rev. A* **5**, 227 (1972).
60. Ref. 43, p. 43.
61. M. O. Krause and T. A. Carlson, *Phys. Rev.* **149**, 52 (1967); see also ref. 57.
62. M. O. Krause and T. A. Carlson, *Phys. Rev.* **158**, 18 (1967).
63. T. E. H. Walker and J. T. Waber, *J. Phys. B* **7**, 674 (1974).
63a. F. Wuilleumier, M. Y. Adam, N. Sandner, V. Schmidt, and W. Mechlhorn, *Phys. Rev.* **16**, 646 (1977).
64. F. Wuilleumier, M. Y. Adam, N. Sandner, V. Schmidt, W. Mechlhorn, and J. P. Desclaux, *Abstr. Int. Conf. VUV Radiat. Phys. 5th, Montpellier* **1**, 41 (1977).
64a. P. C. Kemeny, J. A. R. Samson, and A. F. Starace, *J. Phys. B* **10**, L 201 (1977).
64b. M. Y. Adam, F. Wuilleumier, N. Sandner, V. Schmidt, and W. Mehlhorn (to be published).
65. R. B. Cairns, H. Harrison, and R. I. Schoen, *Phys. Rev.* **183**, 52 (1969); see also ref. 55.
66. K. Codling and R. P. Madden, *Phys. Rev. Lett.* **12**, 106 (1964); *Appl. Opt.* **4**, 143 (1965).
66a. J. P. Connerade, *Abstr. Int. Conf. VUV Radiat. Phys., 5th, Montpellier* **1**, 70 (1977).
67. J. B. West, P. R. Woodruff, K. Codling, and R. G. Houlgak, *J. Phys. B* **9**, 407 (1976).
67a. M. Ya. Amusia, L. V. Chernysheva, and V. K. Ivanova, *Phys. Lett.* **43A**, 243 (1973); M. Ya. Amusia, V. K. Ivanova, N. A. Cherepkov, and L. V. Chernysheva, *Soviet Phys.-J.E.T.P.* **39**, 752 (1974).
68. Ref. 43, p. 49.
69. T. A. Carlson, W. E. Hunt, and M. O. Krause, *Phys. Rev.* **151**, 41 (1966).
70. J. L. Dehmer and J. Berkowitz, *Phys. Rev. A* **10**, 484 (1974).
70a. K. Codling, D. M. P. Holland, S. P. Shannon, and J. B. West, *Abstr. Int. Conf. VUV Radiat. Phys., 5th, Montpellier* **1**, 38 (1977).
71. R. Nilsson, R. Nyholm, A. Berndtsson, J. Hedman, and C. Nordling, *J. Electron Spectrosc. Relat. Phenom.* **9**, 337 (1976).

72. S. Svensson, N. Martensson, E. Basilier, P. A. Malmqvist, U. Gelius, and K. Siegbahn, *J. Electron Spectrosc. Relat. Phenom.* **9**, 51 (1976).
73. R. B. Cairns, H. Harrison, and R. I. Schoen, *J. Chem. Phys.* **53**, 96 (1970).
74. D. W. Turner, *Proc. R. Soc. London, Ser. A* **307**, 15 (1968).
75. D. C. Frost, C. A. McDowell, and D. A. Vroom, *Proc. R. Soc. London, Ser. A* **296**, 566 (1967).
76. L. Åsbrink, *Chem. Phys. Lett.* **7**, 549 (1970).
77. B. Brehm and R. Frey, *Z. Naturforsch., Teil A* **26**, 523 (1971).
78. J. Berkowitz and R. Spohr, *J. Electron Spectrosc. Relat. Phenom.* **2**, 143 (1973).
79. W. Kolos and L. Wohniewicz, *Rev. Mod. Phys.* **35**, 473 (1963); W. Kolos, *Acta Phys. Acad Sci. Hung.* **27**, 241 (1969).
80. J. M. Peek, personal communication (1971); see also ref. 78.
81. T. A. Carlson and A. E. Jonas, *J. Chem. Phys.* **55**, 4913 (1971).
82. R. Morgenstern, A. Nichaus, and M. W. Ruf, *Abstr. Int. Conf. Phys. Electron. At. Collisions, VII, Amsterdam*, p. 167 (1971).
83. Y. Itikawa, *J. Electron Spectrosc. Relat. Phenom.* **2**, 125 (1973).
84. Recent experimental results [J. L. Gardner and J. A. R. Samson, *J. Electron Spectrosc. Relat. Phenom.* **8**, 123 (1976)] confirm the decline in transition probability to vibrational levels of H_2^+ above $v' \sim 7$, compared to Itikawa's calculation.
85. J. Berkowitz and P. M. Guyon, *Int. J. Mass Spectrom. Ion Phys.* **6**, 302 (1971).
85a. J. A. R. Samson, *Phys. Repts.-Phys. Lett.* **28C**, 303–354 (1976).
85b. J. Berkowitz, unpublished (1971).
85c. J. Berkowitz, Physics Division Annual Review, April 1, 1977–March 31, 1978, ANL-78-66, p. 205. Argonne National Laboratory, Argonne, Illinois, 1978.
86. J. M. Sichel, *Mol. Phys.* **18**, 95 (1970).
87. A. Niehaus and M. W. Ruf, *Chem. Phys. Lett.* **11**, 55 (1971).
88. J. A. R. Samson, *Chem. Phys. Lett.* **12**, 625 (1972).
89. V. H. Dibeler, R. M. Reese, and M. Krauss, *J. Chem. Phys.* **42**, 2045 (1965).
90. F. J. Comes and H. O. Wellern, *Z. Naturforsch., Teil A* **23**, 881 (1968).
91. W. A. Chupka and J. Berkowitz, *J. Chem. Phys.* **51**, 4244 (1969).
92. G. R. Cook and P. H. Metzger, *J. Opt. Soc. Am.* **54**, 968 (1964).
93. Weak continua do exist. See A. Dalgarno and A. C. Allison, *J. Geophys. Res.* **74**, 4178 (1969); J. E. Mentall and E. P. Gentieu, *J. Chem. Phys.* **52**, 5641 (1970).
94. F. F. Rieke and W. Prepejchal, *Phys. Rev. A* **6**, 1507 (1972).
95. G. B. Shaw and R. S. Berry, *J. Chem. Phys.* **56**, 5808 (1972).
96. R. S. Berry, *J. Chem. Phys.* **45**, 1228 (1966).
97. J. N. Bardsley, *Chem. Phys. Lett.* **9**, 229 (1967).
98. R. S. Berry and S. E. Nielsen, *J. Chem. Phys.* **49**, 116 (1968).
99. S. E. Nielsen and R. S. Berry, *Chem. Phys. Lett.* **2**, 503 (1968).
100. R. S. Berry and S. E. Nielsen, *Phys. Rev. A* **1**, 383 (1970).
101. R. S. Berry and S. E. Nielsen, *Phys. Rev. A* **1**, 395 (1970).
102. J. T. Roncin, H. Damany, and C. Jungen, *Proc. Int. Conf. VUV Radiat. Phys., 4th, Hamburg*. Braunschweig p. 52. (1974).
103. W. Sroka, *Phys. Lett. A* **28**, 784 (1969).
103a. J. Breton, M. Glass-Maujean, and P. M. Guyon, *Abstr. Int. Conf. VUV Radiat. Phys., 5th, Montpelier, Sept. 5–9, 1977*. Extended abstracts **1**, 160 (1977).
104. P. M. Dehmer and W. A. Chupka, *J. Chem. Phys.* **65**, 2243 (1976) have recently reported photoionization spectra which purport to have a resolution ~2.5 times better than shown in Fig. VI. 22, but natural line widths are not in general obtained, nor is there any discussion of line shapes.
105. J. Berkowitz and W. A. Chupka, *J. Chem. Phys.* **51**, 2341 (1969).

106. H. Beutler and H. O. Jünger, *Z. Phys.* **100**, 80 (1936).
107. Y. Tanaka, *Sci. Pap. Inst. Phys. Chem. Res.* (*Jpn.*) **42**, 49 (1944).
108. G. Herzberg, "Molecular Spectra and Molecular Structure. I. Spectra of Diatomic Molecules," pp. 237, 416, 419. Van Nostrand, Princeton, New Jersey, 1950.
109. S. Takezawa, *J. Chem. Phys.* **52**, 2575 (1970).
110. G. Herzberg, *Phys. Rev. Lett.* **23**, 1081 (1969).
111. G. Herzberg and C. Jungen, *J. Mol. Spectrosc.* **41**, 425 (1972).
112. B. Jeziorski and W. Kolos, *Chem. Phys. Lett.* **3**, 677 (1969) (with correction incorporated).
113. U. Fano, *Phys. Rev. A* **2**, 353 (1970).
114. U. Fano, *Phys. Rev.* **124**, 1866 (1961).
115. G. Herzberg, *in* "Topics in Modern Physics—A Tribute to E. U. Condon" (W. E. Brittin and H. Odabasi, eds.), p. 191. Colorado Assoc. Univ. Press, Boulder, Colorado, 1971.
116. F. J. Comes, G. Schumpe, and U. Wenning, *Abstr. Int. Conf. VUV Radiat. Phys., 3rd, Tokyo* Pap. 1aA1-2 (1971).
117. J. H. de Boer, "Electron Emission and Adsorption Phenomena," pp. 178–180. Cambrdige Univ. Press, London, 1934.
118. C. J. Bakker, *Proc. R. Acad. Sci. Amsterdam* **36**, 589 (1933).
119. C. L. Pekeris, *Phys. Rev.* **126**, 1470 (1962).
120. J. Berkowitz and W. A. Chupka, *Abstracts Int. Conf. Phys. Electron. At. Collisions, 5th, Leningrad*, p. 608 (1967).
121. K. E. McCulloh and J. A. Walker, *Chem. Phys. Lett.* **25**, 439 (1974).
122. W. A. Chupka, P. M. Dehmer, and W. T. Jivery, *J. Chem. Phys.* **63**, 3929 (1975).
123. J. Berkowitz and P. L. Kronebusch, *Abstr. Annu. Conf. Mass Spectrom. Allied Top., 23rd, Houston, Tex.* p. 117 (1975).
123a. J. Durup, *J. de Phys.* **39**, 941 (1978).
124. R. Browning and J. Fryar, *J. Phys. B* **6**, 364 (1973).
125. S. Strathdee and R. Browning, *J. Phys. B* **9**, L505 (1976).
126. S. Takezawa and Y. Tanaka, *J. Chem. Phys.* **56**, 6125 (1972).
127. I. Dabrowski and G. Herzberg, *Can J. Phys.* **54**, 525 (1976).
128. I. Dabrowski and G. Herzberg, *Can J. Phys.* **52**, 1110 (1974).
129. S. Takezawa and Y. Tanaka, *J. Mol. Spectrosc.* **54**, 379 (1975).
129a. W. Kolos and L. Wolniewicz, *J. Chem. Phys.* **49**, 404 (1968).
129b. G. Herzberg, *J. Mol. Spectrosc.* **33**, 147 (1970).
129c. G. Herzberg, *Commentat. Pontif. Acad. Sci.* **2**(49), 1 (1972).
130. D. W. Turner, C. Baker, A. D. Baker, and C. R. Brundle, "Molecular Photoelectron Spectroscopy." Wiley (Interscience), New York, 1970.
131. J. A. R. Samson, G. N. Haddad, and J. L. Gardner, *J. Phys. B* **10**, 1749 (1977).
132. E. W. Plummer, T. Gustafsson, W. Gudat, and D. E. Eastman, *Phys. Rev. A* **15**, 2339 (1977).
132a. P. R. Woodruff and G. V. Marr, *Proc. Roy. Soc.* (*London*) **A358**, 87 (1977).
133. M. S. Banna and D. A. Shirley, *J. Electron Spectrosc. Relat. Phenom.* **8**, 255 (1976).
134. D. A. Allison and R. G. Cavell, *J. Chem. Soc., Faraday Trans. II* **72**, 118 (1976).
135. Ref. 43, p. 62.
136. A. Hamnett, W. Stoll, and C. E. Brion, *J. Electron Spectrosc. Relat. Phenom.* **8**, 367 (1976).
137. G. R. Wight, M. J. Van der Wiel, and C. E. Brion, *J. Phys. B* **9**, 675 (1976).
138. W. C. Price, A. W. Potts, and D. G. Streets, *in* "Electron Spectroscopy" (D. A. Shirley, ed.), p. 187. North-Holland Publ., Amsterdam, 1972.
139. V. L. Carter, *J. Chem. Phys.* **56**, 4195 (1972).
140. J. Berkowitz, unpublished observations (1977).
140a. P. Gürtler, V. Saile, and E. E. Koch, *Chem. Phys. Lett.* **48**, 245 (1977).
141. V. L. Carter and J. Berkowitz, *J. Chem. Phys.* **59**, 4573 (1973).
142. J. A. R. Samson and R. B. Cairns, *J. Geophys. Rev.* **69**, 4583 (1964).

143. M. Ogawa and Y. Tanaka, *Can. J. Phys.* **40**, 1604 (1962).
144. J. L. Bahr, A. J. Blake, J. H. Carver, J. L. Gardner, and V. Kumar, *J. Quant. Spectrosc. Radiat. Transfer* **11**, 1839 (1971).
144a. G. V. Marr and P. R. Woodruff, *J. Phys. B* **9**, L 377 (1976).
145. T. A. Carlson, *Chem. Phys. Lett.* **9**, 23 (1970).
146. A. W. Potts and T. A. Williams, *J. Electron Spectrosc. Relat. Phenom.* **3**, 3 (1974).
147. P. L. Kronebusch and J. Berkowitz, *Int. J. Mass Spectrom. Ion Phys.* **22**, 283 (1976).
148. P. Erman, *Phys. Scr.* **14**, 51 (1976).
149. R. Browning, J. Fryar, and R. Cunningham, *Adv. Mass Spectrom.* **6**, 933 (1974).
150. T. A. Carlson and M. O. Krause, *J. Chem. Phys.* **56**, 3206 (1972).
151. O. Edqvist, E. Lindholm, L. E. Selin, and L. Åsbrink, *Phys. Scr.* **1**, 25 (1970).
152. P. M. Guyon and J. Berkowitz, *J. Chem. Phys.* **54**, 1814 (1971).
153. J. A. R. Samson, J. L. Gardner, and G. N. Haddad, *J. Electron Spectrosc. Relat. Phenom.* **12**, 281 (1977).
154. A. J. Blake and J. H. Carver, *J. Chem. Phys.* **47**, 1038 (1967).
155. Ref. 43, pp. 70, 72.
156. F. M. Matsunaga and K. Watanabe, *Sci. Light* (*Tokyo*) **16**, 31, 191 (1967).
157. J. L. Bahr, A. J. Blake, J. H. Carver, J. L. Gardner, and V. Kumar, *J. Quant. Spectrosc. Radiat. Transfer* **11**, 1853 (1971).
158. K. Tanaka and I. Tanaka, *J. Chem. Phys.* **59**, 5042 (1973).
159. B. deB. Darwent, "Bond Dissociation Energies in Simple Molecules." *Natl. Stand. Ref. Data Ser., Natl. Bur. Stand.* **NSRDS-NBS 31** (1970).
160. C. E. Moore, "Ionization Potentials and Ionization Limits Derived from the Analyses of Optical Spectra." *Natl. Stand. Ref. Data Ser., Natl. Bur. Stand.* **NSRDS-NBS 34** (1970).
161. R. S. Berry, *Chem. Rev.* **69**, 533 (1969).
162. F. A. Elder, D. Villarejo, and M. G. Inghram, *J. Chem. Phys.* **43**, 758 (1966).
163. V. H. Dibeler and J. A. Walker, *J. Opt. Soc. Am.* **57**, 1007 (1967).
164. P. M. Dehmer and W. A. Chupka, *J. Chem. Phys.* **62**, 4525 (1975).
164a. P. M. Guyon, T. Baer, L. F. A. Ferreira, I. Nenner, A. Tabché-Fouhailé, R. Botter, and T. R. Govers, *J. Phys. B* **11**, L141 (1978).
165. P. H. Doolittle, R. I. Schoen, and K. E. Schubert, *J. Chem. Phys.* **49**, 5108 (1968).
166. J. H. D. Eland, *Int. J. Mass Spectrom. Ion Phys.* **9**, 400 (1972).
167. F. J. LeBlanc, *J. Chem. Phys.* **28**, 487 (1963).
168. R. J. Van Brunt, F. W. Powell, R. G. Hirsch, and W. D. Whitehead, *J. Chem. Phys.* **57**, 3120 (1972).
169. Ref. 130, p. 49.
170. Ref. 108, p. 522.
171. J. Berkowitz and J. L. Dehmer, *J. Chem. Phys.* **57**, 3194 (1972).
172. J. L. Bahr, A. J. Blake, J. H. Carver, J. L. Gardner, and V. Kumar, *J. Quant. Spectrosc. Radiat. Transfer* **12**, 59 (1972).
173. R. I. Schoen, *J. Chem. Phys.* **40**, 1830 (1964).
174. J. A. R. Samson and J. L. Gardner, *J. Electron Spectrosc. Relat. Phenom.* **8**, 35 (1976).
175. Ref. 43, pp. 76–77.
176. "JANAF Thermochemical Tables." Dow Chemical Co., Midland, Michigan, (1977).
176a. R. Locht, *Chem. Phys.* **22**, 13 (1977).
176b. H. Baumgärtel, private communication (1978).
177. L. Åsbrink, C. Fridh, E. Lindholm, and K. Codling, *Phys. Scr.* **10**, 183 (1974).
177a. P. S. Bagus and E.-K. Viinikka, *Phys. Rev. A* **15**, 1486 (1977).
178. J. L. Gardner and J. A. R. Samson, *J. Chem. Phys.* **62**, 1447 (1975).
178a. M. J. Van der Wiel, R. B. Kay, and P. E. Van der Leeuw, *Abstr. Int. Conf. VUV Radiat. Phys., 5th, Montpellier* **1**, 142 (1977).
178b. R. B. Kay, Ph. E. Van der Leeuw, and M. J. Van der Wiel, *J. Phys. B* **10**, 2521 (1977).

179. Ref. 130, p. 53.
180. Ref. 5, pp. 277–281.
181. O. Edqvist, E. Lindholm, L. E. Selin, H. Sjogren, and L. Åsbrink, *Ark. Eys.* **40**, 439 (1970).
182. O. Edqvist, L. Åsbrink, and E. Lindholm, *Z. Naturforsch., Teil A* **26**, 1407 (1971).
183. E. W. Thulstrup and Y. Ohrn, *J. Chem. Phys.* **57**, 3716 (1972).
184. J. A. R. Samson, *Phys. Lett. A* **28**, 391 (1968).
185. Ref. 43, p. 74.
186. K. Watanabe, *J. Chem. Phys.* **22**, 1564 (1954).
187. P. C. Killgoar, Jr., G. E. Leroi, J. Berkowitz, and W. A. Chupka, *J. Chem. Phys.* **58**, 803 (1973).
187a. C. Y. Ng, B. H. Mahan, and Y. T. Lee, *J. Chem. Phys.* **65**, 1965 (1976); see also E. Miescher, Y. T. Lee, and P. Gürtler, *J. Chem. Phys.* **68**, 2753 (1978).
188. See, e.g., H. Hurzeler, M. G. Inghram, and J. D. Morrison, *J. Chem. Phys.* **28**, 76 (1958).
188a. See also, e.g., A. J. C. Nicholson, *J. Chem. Phys.* **43**, 1171 (1965); R. M. Reese and H. M. Rosenstock, *J. Chem. Phys.* **44**, 2007 (1966).
189. V. I. Kleimenov, Y. V. Chizhov, and F. I. Vilesov, *Opt. Spektrosk.* **32**, 702 (1972); Engl. transl., *Opt. Spectrosc. (USSR)* **32**, 371 (1972).
190. J. L. Gardner and J. A. R. Samson, *J. Electron Spectrosc. Relat. Phenom.* **2**, 153 (1973).
190a. G. Caprace, J. Delwiche, P. Natalis, and J. E. Collin, *Chem. Phys.* **13**, 43 (1976).
190b. P. Natalis, J. Delwiche, J. E. Collin, G. Caprace, and M.-T. Praet, *Chem. Phys. Lett.* **49**, 177 (1977).
190c. H. Hertz, H. W. Jochims, H. Schenk, and W. Sroka, *Chem. Phys. Lett.* **29**, 572 (1974).
191. G. L. Weissler, J. A. R. Samson, M. Ogawa, and G. R. Cook, *J. Opt. Soc. Am.* **49**, 338 (1959).
192. T. R. Govers and J. Schopman, *Chem. Phys. Lett.* **12**, 414 (1971).
193. Ref. 43, p. 84.
194. Ref. 130, p. 77.
195. A. W. Potts and W. C. Price, *Proc. R. Soc. London, Ser. A* **326**, 165 (1972).
196. M. S. Banna and D. A. Shirley, *J. Chem. Phys.* **63**, 4759 (1975).
197. L. Karlsson, L. Mattson, R. Jadrny, R. G. Albridge, S. Pinchas, T. Bergmark, and K. Siegbahn, *J. Chem. Phys.* **62**, 4745 (1975).
197a. R. N. Dixon, G. Duxbury, J. W. Rabalais, and L. Åsbrink, *Mol. Phys.* **31**, 423 (1976).
198. L. Åsbrink and J. W. Rabalais, *Chem. Phys. Lett.* **12**, 182 (1971).
199. J. W. C. Johns, *Can. J. Phys.* **41**, 209 (1963).
200. C. R. Brundle and D. W. Turner, *Proc. R. Soc. London Ser. A* **307**, 27 (1968).
201. H. Lew, *Can J. Phys.* **54**, 2028 (1976).
202. K. H. Tan, C. E. Brion, P. Van der Leeuw, and M. J. Van der Wiel, *Chem. Phys.* **29**, 299 (1978).
202a. G. R. Branton and C. E. Brion, *J. Electron Spectrosc. Relat. Phenom.* **3**, 129 (1974).
203. Ref. 43, pp. 83–84.
204. K. Siegbahn, *J. Electron Spectrosc. Relat. Phenom.* **5**, 65 (1974).
205. A. R. P. Rau and U. Fano, *Phys. Rev.* **162**, 68 (1967).
206. J. H. D. Eland, *Chem. Phys.* **11**, 41 (1975).
207. V. H. Dibeler, J. A. Walker, and H. M. Rosenstock, *J. Res. Natl. Bur. Stand., Sect. A* **70**, 459 (1966).
208. R. B. Cairns, H. Harrison, and R. I. Schoen, *J. Chem. Phys.* **55**, 4886 (1971).
209. J. Nordgren, L. O. Werme, H. Ågren, C. Nordling, and K. Siegbahn, *J. Phys. B* **8**, L18 (1975); see also K. Siegbahn, *J. Electron Spectrosc. Relat. Phenom.* **5**, 66 (1974).
210. W. E. Moddeman, T. A. Carlson, M. O. Krause, B. P. Pullen, W. E. Bull, and G. K. Schweitzer, *J. Chem. Phys.* **55**, 2317 (1971).
211. D. L. Walters and C. P. Bhalla, *Phys. Rev. A* **3**, 1919 (1971).

212. See, e.g., T. A. Carlson and R. M. White, *J. Chem. Phys.* **44**, 4510 (1966); T. A. Carlson and M. O. Krause, *J. Chem. Phys.* **56**, 3206 (1972).
213. A. D. McLean and M. Yoshimine, "Tables of Linear Molecule Wave Functions." IBM Corp., San Jose, California, 1967.
214. S. D. Peyerimhoff, R. J. Buenker, and J. L. Whitten, *J. Chem. Phys.* **46**, 1707 (1967).
215. C. J. Allan, U. Gelius, D. A. Allison, G. Johansson, H. Siegbahn, and K. Siegbahn, *J. Electron Spectrosc. Relat. Phenom.* **1**, 131 (1972).
216. J. A. R. Samson and J. L. Gardner, *J. Geophys. Res.* **78**, 3663 (1973).
216a. L. S. Cederbaum, J. Schirmer, W. Domcke, and W. von Niessen, *J. Phys. B* **10**, L549 (1977).
217. G. Herzberg, "Molecular Spectra & Molecular Structure, III. Electronic Spectra & Electronic Structure of Polyatomic Molecules." Van Nostrand, Princeton, New Jersey, 1966.
218. Ref. 130, pp. 66–74, 103.
219. L. C. Lee and D. L. Judge, *J. Chem. Phys.* **57**, 4443 (1972).
219a. Y. Tanaka, A. S. Jursa, and F. J. LeBlanc, *J. Chem. Phys.* **32**, 1199, 1205 (1960).
220. J. A. R. Samson, J. L. Gardner, and J. E. Mentall, *J. Geophys. Res.* **77**, 5560 (1972); J. L. Gardner and J. A. R. Samson, *J. Electron Spectrosc. Relat. Phenom.* **8**, 469 (1976).
220a. T. Gustafsson, E. W. Plummer, D. E. Eastman, and W. Gudat, *Phys. Rev. A* **17**, 175 (1978).
221. C. J. Allan and K. Siegbahn, "Electron Spectroscopy for Chemical Application," U.U.I.P. No. 754. Uppsala Univ. Inst. Phys., Uppsala, Nov. (1971).
222. B. Henke, R. L. Elgin, R. E. Lent, and R. B. Ledingham *Norelco Rep.* **14**, 112 (1967).
223. D. J. Villarejo, R. Stockbauer, and M. G. Inghram, *J. Chem. Phys.* **48**, 3442 (1968).
224. S. Mrozowski, *Phys. Rev.* **60**, 730 (1941); **62**, 270 (1942); **72**, 682, 691 (1947).
225. J. W. C. Johns, *Can. J. Phys.* **42**, 1004 (1964).
226. Y. Tanaka and M. Ogawa, *Can. J. Phys.* **40**, 879 (1962).
227. K. E. McCulloh, *J. Chem. Phys.* **59**, 4250 (1973).
228. J. H. D. Eland, *Int. J. Mass Spectrom. Ion Phys.* **9**, 397 (1972).
229. J. H. D. Eland and J. Berkowitz, *J. Chem. Phys.* **67**, 5034 (1977).
230. J. Fryar and R. Browning, *Planet. Space Sci.* **21**, 709 (1972).
231. R. W. Carlson, D. L. Judge, and M. Ogawa, *J. Geophys. Res.* **78**, 3194 (1973).
232. For recent fluorescence studies and references to earlier work, see D. Gauyacq, M. Horani, S. Leach, and J. Rostas, *Can. J. Phys.* **53**, 2040 (1975).
232a. S. Leach, M. Devoret, and J. H. D. Eland, *Chem. Phys.* **33**, 113 (1978).
233. J. A. D. Stockdale, B. P. Pullen, and A. E. Carter, *Int. J. Mass Spectrom. Ion Phys.* **17**, 241 (1975).
233a. J. A. R. Samson, P. C. Kemeny, and G. N. Haddad, *Chem. Phys. Lett.* **51**, 75 (1977).
234. Ref. 130, p. 89.
235. U. Gelius, *J. Electron Spectrosc. Relat. Phenom.* **5**, 1042 (1974).
236. G. R. Cook, P. H. Metzger, and M. Ogawa, *J. Opt. Soc. Am.* **58**, 129 (1968).
237. R. E. Rebbert and P. Ausloos, *J. Res. Natl. Bur. Stand., Sect. A* **75**, 481 (1971).
238. S. W. Bennett, J. B. Tellinghuisen, and L. F. Phillips, *J. Phys. Chem.* **75**, 719 (1971).
239. J. Berkowitz and J. Ehrhardt, *Phys. Lett.* **21**, 531 (1966).
240. J. Berkowitz and J. H. D. Eland, *J. Chem. Phys.* **67**, 2740 (1977).
241. D. G. Hopper, *Chem. Phys. Lett.* **31**, 446 (1975).
242. J. H. Callomon and F. Creutzberg, *Philos. Trans. R. Soc. London, Ser. A* **227**, 157 (1974).
243. J. H. D. Eland, *Int. J. Mass Spectrom. Ion Phys.* **12**, 389 (1973).
244. R. J. Coleman, J. S. Delderfield, and B. G. Reuben, *Int. J. Mass Spectrom. Ion Phys.* **2**, 25 (1969).
245. A. S. Newton and A. F. Sciamanna, *J. Chem. Phys.* **52**, 327 (1970).

246. R. G. Orth and R. C. Dunbar, *J. Chem. Phys.* **66**, 1616 (1977).
247. H. Hertz, H. W. Jochims, and W. Sroka, *J. Phys. B* **7**, L548 (1974).
248. S. Gardner, student thesis, Argonne Natl. Lab., Argonne, Illinois, 1975.
249. U. Gelius, *J. Electron Spectrosc. Relat. Phenom.* **5**, 1006 (1974).
250. J. Nordgren, H. Ågren, L. O. Werme, C. Nordling, and Y. K. Siegbahn, *J. Phys. B* **9**, 295 (1976).
251. R. G. Hirsch, R. J. Van Brunt, and W. D. Whitehead, *J. Chem. Phys.* **59**, 3863 (1973).
252. U. Kaldor and I. Shavitt, *J. Chem. Phys.* **45**, 888 (1966).
253. A. W. Potts and W. C. Price, *Proc. R. Soc. London, Ser. A* **326**, 181 (1972).
254. G. R. Branton, D. C. Frost, F. G. Herring, C. A. McDowell, and I. A. Stenhouse, *Chem. Phys. Lett.* **3**, 581 (1969).
255. M. J. Weiss and G. M. Lawrence, *J. Chem. Phys.* **53**, 214 (1970).
256. J. W. Rabalais, L. Karlsson, L. O. Werme, T. Bergmark, and K. Siegbahn, *J. Chem. Phys.* **58**, 3370 (1973).
257. M. J. Van der Wiel and C. E. Brion, *J. Electron Spectrosc. Relat. Phenom.* **1**, 443 (1972).
257a. C. E. Brion, A. Hamnett, G. R. Wight, and M. J. Van der Wiel, *J. Electron Spectrosc. Relat. Phenom.* **12**, 323 (1977).
258. K. Watanabe and S. P. Sood, *Sci. Light (Tokyo)* **14**, 36 (1965).
259. K. E. McCulloh, *Int. J. Mass Spectrom. Ion Phys.* **21**, 333 (1976).
260. B. Brehm and E. von Puttkamer, *Adv. Mass Spectrom.* **4**, 591 (1968).
261. G. Herzberg and J. W. C. Johns, *J. Chem. Phys.* **54**, 2276 (1971).
262. A. Schweig and W. Thiel, *J. Electron Spectrosc. Relat. Phenom.* **3**, 27 (1974).
262a. G. R. Wight, M. J. Van der Wiel, and C. E. Brion, *J. Phys. B* **10**, 1863 (1977).
263. R. H. Shaw, Jr., J. S. Jen, and T. D. Thomas, *J. Electron Spectrosc. Relat. Phenom.* **11**, 91 (1977).
264. M. T. Økland, K. Faegri, Jr., and R. Manne, *Chem. Phys. Lett.* **40**, 185 (1976).
265. C. A. Coulson and H. L. Strauss, *Proc. R. Soc. London, Ser. A* **269**, 443 (1962).
266. G. Favini, G. Majorino, and M. Simonetta, *Atti Accad. Naz. Lincei, Cl. Sci. Fis., Mat. Nat., Rend.* **38**, 775 (1965).
267. R. G. Pearson, *J. Am. Chem. Soc.* **91**, 4947 (1969).
268. J. Arentz and L. C. Allen, *J. Chem. Phys.* **53**, 73 (1970).
269. G. S. Handler and H. W. Joy, *Int. J. Quantum Chem., Symp.* **3**, 529 (1970).
270. F. A. Grimm and J. Godoy, *Chem. Phys. Lett.* **6**, 336 (1970).
271. R. N. Dixon, *Mol. Phys.* **20**, 113 (1971).
272. J. W. Rabalais, T. Bergmark, L. O. Werme, L. Karlsson, and K. Siegbahn, *Phys. Scr.* **3**, 13 (1971).
272a. A. W. Potts and W. C. Price, *Proc. R. Soc. London, Ser. A* **326**, 165 (1972).
273. C. R. Brundle, M. B. Robin, and H. Basch, *J. Chem. Phys.* **53**, 2196 (1970).
274. K. Hamrin, G. Johansson, U. Gelius, A. Fahlman, C. Nordling, and K. Siegbahn, *Chem. Phys. Lett.* **1**, 613 (1968).
275. U. Gelius, *J. Electron Spectrosc. Relat. Phenom.* **5**, 1011 (1974).
276. C. Backx and M. J. Van der Wiel, *J. Phys. B* **8**, 3020 (1975).
277. M. J. Van der Wiel, W. Stoll, A. Hamnett, and C. E. Brion, *Chem. Phys. Lett.* **37**, 240 (1976).
278. S. Nishikawa and T. Watanabe, *Chem. Phys. Lett.* **22**, 590 (1973).
279. W. A. Chupka and J. Berkowitz, *J. Chem. Phys.* **54**, 4256 (1971).
280. V. H. Dibeler, M. Krauss, R. M. Reese, and F. N. Harlee, *J. Chem. Phys.* **42**, 3791 (1965).
281. W. A. Chupka, *J. Chem. Phys.* **48**, 2337 (1968).
282. R. Stockbauer, *J. Chem. Phys.* **58**, 3800 (1973).
283. M. L. Vestal, *in* "Fundamental Processes in Radiation Chemistry" (P. Ausloos, ed.), pp. 59–118. Wiley (Interscience), New York, 1968.

284. H. von Koch, *Ark. Fys.* **28**, 529 (1964).
284a. K. E. McCulloh and V. H. Dibeler, *J. Chem. Phys.* **64**, 4445 (1976).
284b. B. H. Solka, J. H. Beynon, and R. G. Cooks, *J. Phys. Chem.* **79**, 859 (1975).
284c. J. R. Flamme, H. Wankenne, R. Locht, J. Momigny, P. J. C. M. Nowak, and J. Los, *Chem. Phys.* **27**, 45 (1978).
284d. J. P. Flamme and J. Momigny, Theoretical study of the metastable peaks corresponding to the rotational predissociation of the CH_4^+ and CD_4^+ ions, *Chem. Phys.* **34**, 303 (1978).
284e. C. E. Klots, *Z. Naturforsch., Teil A* **27**, 553 (1972).
284f. R. Stockbauer, *Int. J. Mass Spectrom. Ion Phys.* **25**, 401 (1977).
284g. R. Browning, S. Strathdee, and R. Cunningham, *Abstr. Int. Conf. Phys. Electron. At. Collisions, 10th, Paris* p. 98 (1977).
284h. L. W. Sieck and R. Gordon, *Chem. Phys. Lett.* **19**, 509 (1973).
285. J. Appell, Ph.D. Thesis, Univ. Paris-Sud, Orsay (1972).
286. W. Meyer, *J. Chem. Phys.* **58**, 1017 (1973).
287. R. Spohr, T. Bergmark, N. Magnusson, L. O. Werme, C. Nordling, and K. Siegbahn, *Phys. Scr.* **2**, 31 (1970).
288. L. C. Lee, E. Phillips, and D. L. Judge, *J. Chem. Phys.* **67**, 1237 (1977).
289. J. Berkowitz, *in* Physics Division Annual Review, ANL-77-60, Argonne National Laboratory, p. 199. Argonne, Illinois, 1977.
290. V. H. Dibeler and J. A. Walker, *Int. J. Mass Spectrom. Ion Phys.* **11**, 49 (1973).
291. V. H. Dibeler and R. M. Reese, *J. Res. Natl. Bur. Stand., Sect. A* **68**, 409 (1964).
292. R. Botter, V. H. Dibeler, J. A. Walker, and H. M. Rosenstock, *J. Chem. Phys.* **44**, 1271 (1966).
293. P. H. Metzger and G. R. Cook, *J. Chem. Phys.* **41**, 642 (1964).
294. Ref. 130, pp. 170, 193.
295. V. H. Dibeler, J. A. Walker, and K. E. McCulloh, *J. Chem. Phys.* **59**, 2264 (1973).
296. D. G. Streets and A. W. Potts, *J. Chem. Soc., Faraday Trans. II* **70**, 1507 (1974).
297. R. G. Cavell and D. A. Allison, *Chem. Phys. Lett.* **36**, 514 (1975).
298. W. Thiel and A. Schweig, *Chem. Phys. Lett.* **16**, 409 (1972).
299. R. G. Cavell, *J. Electron Spectrosc. Relat. Phenom.* **6**, 281 (1975).
300. P. C. Haarhoff, *Mol. Phys.* **8**, 49 (1964).
301. M. L. Vestal, *in* "Fundamental Processes in Radiation Chemistry" (P. Ausloos, ed.), p. 102. Wiley (Interscience), New York, 1968.
302. R. G. Cooks, J. H. Beynon, R. M. Caprioli, and G. R. Lester, "Metastable Ions," p. 82, Fig. 44. Elsevier, Amsterdam, 1973.
303. E. von Puttkamer, *Z. Naturforsch., Teil A* **25**, 1062 (1970).
304. J. R. Wyatt and F. E. Stafford, *J. Phys. Chem.* **76**, 1913 (1972).
305. H. Okabe and V. H. Dibeler, *J. Chem. Phys.* **59**, 2430 (1973).
306. S. I. Miller and J. Berkowitz, unpublished observations (1971).
307. W. B. Maier, *J. Chem. Phys.* **42**, 1790 (1965).
308. F. Fiquet-Fayard, *J. Chim. Phys. Phys.-Chim. Biol.* **64**, 320 (1967).
309. P. M. Larson, undergraduate thesis, Argonne Natl. Lab., Argonne, Illinois, 1975.
310. H. M. Rosenstock, K. Draxl, B. Steiner, and J. T. Herron, *J. Phys. Chem. Ref. Data* **6**, *Suppl. 1*, 1–783 (1977).
311. Ref. 130, pp. 166, 167.
312. A. Berndtsson, E. Basilier, U. Gelius, J. Hedman, M. Klasson, R. Nilsson, C. Nordling, and S. Svensson, *Phys. Scr.* **12**, 235 (1975).
313. G. R. Branton, D. C. Frost, T. Makita, C. A. McDowell, and I. A. Stenhouse, *J. Chem. Phys.* **52**, 802 (1970).

314. C. R. Brundle and D. B. Brown, *Spectrochim. Acta, Part A* **27**, 2491 (1971).
315. R. M. White, T. A. Carlson, and D. P. Spears, *J. Electron Spectrosc. Relat. Phenom.* **3**, 59 (1974).
316. E. Lindholm, C. Fridh, and L. Åsbrink, *Faraday Discuss. Chem. Soc.* **54**, 129 (1972).
317. R. Stockbauer and M. G. Inghram, *J. Electron Spectrosc. Relat. Phenom* **7**, 492 (1975).
318. C. Benoit and J. A. Horsley, *Mol. Phys.* **30**, 557 (1975).
319. R. Stockbauer and M. G. Inghram, *J. Chem. Phys.* **62**, 4862 (1975).
320. M. Vestal and G. Lerner, Aerosp. Res. Lab. Rep. 67-0114 (1967) (available from Natl. Techn. Inf. Serv., Springfield, Virginia).
321. I. Nenner, H. Nguyen Nghi, and R. Botter, *Adv. Mass Spectrom.* **6**, 885 (1974).
322. R. Botter, V. H. Dibeler, J. A. Walker, and H. M. Rosenstock, *J. Chem. Phys.* **45**, 1298 (1966).
323. W. A. Lathan, W. J. Hehre, and J. A. Pople, *Mol. Phys.* **22**, 1081 (1971).
324. J. N. Murrell and W. Schmidt, *J. Chem. Soc., Faraday Trans. II* **68**, 1709 (1972).
325. J. W. Rabalais and A. Katrib, *Mol. Phys.* **27**, 923 (1974).
326. S. Dey, A. J. Dixon, I. E. McCarthy, and E. Weigold, *J. Electron Spectrosc. Relat. Phenom.* **9**, 397 (1976).
326a. A. Richartz, R. J. Buenker, P. J. Bruna, and S. I. Peyerimhoff, *Mol. Phys.* **33**, 1345 (1977).
327. A. W. Potts, T. A. Williams, and W. C. Price, *Faraday Discuss. Chem. Soc.* **54**, 105 (1972).
328. Ref. 130, p. 175.
329. W. A. Chupka and J. Berkowitz, *J. Chem. Phys.* **47**, 2921 (1967).
330. J. Berkowitz, unpublished observations, (1976), (1978).
331. Stockbauer notes that in C_2D_6 the dissociative ionization threshold is at 14.8 eV, whereas the $(1e_u)^{-1}$ adiabatic onset is at 14.45 eV. However, neither direct dissociation nor statistically controlled dissociation can be expected to have an onset below the thermochemical threshold.
332. Z. Prasil and W. Forst, *J. Phys. Chem.* **71**, 3166 (1967).
333. I. G. Simm, C. J. Danby, and J. H. D. Eland, *J. Chem. Soc., Chem. Commun.* p. 832 (1973); I. G. Simm, C. J. Danby, and H. D. Eland, *Int. J. Mass Spectrom. Ion Phys.* **14**, 285 (1974).
333a. K. Watanabe, T. Nakayama, and J. Mottl, *J. Quant. Spectrosc. Radiat. Transfer* **2**, 369 (1962).
333b. D. R. Stull and H. Prophet, JANAF Thermochemical Tables, 2nd ed., *Natl. Stand. Ref. Data Ser., Natl. Bur. Stand.* **NSRDS-NBS 37** (1971).
333c. M. W. Chase, J. L. Cornutt, A. T. Hu, H. Prophet, A. N. Syverud, and L. C. Walker, JANAF Thermochemical Tables, 1974 Supplement, *J. Phys. Chem. Ref. Data* **3**, 311 (1974); also M. W. Chase, J. L. Cornutt, H. Prophet, R. A. McDonald, and A. N. Syverud, JANAF Thermochemical Tables, 1975 Supplement, *J. Phys. Chem. Ref. Data*, **4**, 1 (1975).
333d. P. M. Guyon, W. A. Chupa, and J. Berkowitz, *J. Chem. Phys.* **64**, 1419 (1976).
333e. P. H. Krupenie, *Natl. Stand. Ref. Data Ser., Natl. Bur. Stand.* **NSRDS-NBS 5** (1966).
334. M. B. Robin and N. A. Kuebler, *J. Electron Spectrosc. Relat. Phenom.* **1**, 13 (1971).
335. A. W. Potts, T. A. Williams, and W. C. Price, *Faraday Discuss. Chem. Soc.* **54**, 107 (1972).
336. See also the spectrum recorded by L. Åsbrink, shown by B. O. Jonsson and J. Lind, *J. Chem. Soc., Faraday Trans. II* **72**, 906 (1976).
337. Ref. 43, pp. 98–100.
338. J. Berkowitz, *J. Chem. Phys.* **69**, 3044 (1978).
339. The threshold region (10.8–12.5 eV) is taken from K. M. A. Refaey and W. A. Chupka, *J. Chem. Phys.* **48**, 5205 (1968).
340. V. K. Potapov and V. V. Sorokin, *Khim. Vys. Energ.* **4**, 355 (1970); Engl transl., *High Energy Chem.* (*USSR*) **4**, 310 (1970).
341. B. Brehm, V. Fuchs, and P. Kebarle, *Int. J. Mass Spectrom. Ion Phys.* **6**, 279 (1971).

342. Courtesy of W. C. Price, reproduced in J. H. D. Eland, "Photoelectron Spectroscopy," p. 22. Wiley, New York, 1974.
343. S. Ikuta, K. Yoshihara, and T. Shiokawa, *Jpn. J. Mass Spectrosc.* **22**, 233 (1974).
344. B. O. Jonsson and J. Lind, *J. Chem. Soc., Faraday Trans. II* **72**, 906 (1976).
344a. P. Wilmenius and E. Lindholm, *J. Chem. Soc., Faraday Trans. II* **72**, 906 (1976).
344b. P. Warneck, *Z. Naturforsch., Teil A* **26**, 2047 (1971).
344c. J. C. Callomon, E. Hirota, K. Kuchitsu, W. J. Lafferty, A. G. Maki, and C. S. Pote, "Structure Data of Free Polyatomic Molecules," Landolt–Bornstein Tables, New Series, Group II. Springer-Verlag, Berlin and New York, 1976.
344d. T. H. Dunning, Jr., cited in ref. 338.
344e. H. B. Jansen and P. Ros, *Chem. Phys. Lett.* **3**, 140 (1969).
344f. S. Forsen and B. Roos, *Chem. Phys. Lett.* **6**, 128 (1970).
344g. L. Åsbrink, E. Lindholm, and O. Edqvist, *Chem. Phys. Lett.* **5**, 609 (1970).
345. L. Åsbrink, O. Edqvist, E. Lindholm, and L. E. Selin, *Chem. Phys. Lett.* **5**, 192 (1970).
346. Ref. 130, p. 264.
347. J. Momigny, C. Foffart, and L. D'Or, *Int. J. Mass Spectrom. Ion Phys.* **1**, 53 (1968).
348. P. Natalis, J. E. Collin, and J. Momigny, *Int. J. Mass Spectrom. Ion Phys.* **1**, 327 (1968).
349. J. A. R. Samson, *Chem. Phys. Lett.* **4**, 257 (1969).
350. B. O. Jonsson and E. Lindholm, *Ark. Fys.* **39**, 65 (1969).
351. A. W. Potts, W. C. Price, D. G. Streets, and T. A. Williams, *Faraday Discuss. Chem. Soc.* **54**, 168 (1972).
352. U. Gelius, *J. Electron Spectrosc Relat. Phenom.* **5**, 985 (1974).
353. J. M. Schulman and J. W. Moscowitz, *J. Chem. Phys.* **43**, 3287 (1965); **47**, 3491 (1967).
354. R. Hoffman, *J. Chem. Phys.* **39**, 1397 (1963).
355. D. A. Clark and J. R. Ragle, *J. Chem. Phys.* **46**, 4235 (1967).
356. M. D. Newton, F. P. Boer and W. N. Lipscomb, *J. Am. Chem. Soc.* **88**, 2367 (1966).
357. T. A. Carlson and C. P. Anderson, *Chem. Phys. Lett.* **10**, 561 (1971).
358. H. M. Rosenstock, J. T. Larkin and J. A. Walker, *Int. J. Mass Spectrom. Ion Phys.* **11**, 309 (1973).
359. H. M. Rosenstock and M. Krauss, *Adv. Mass Spectrom.* **2**, 251 (1963).
360. M. L. Vestal, *in* "Fundamental Processes in Radiation Chemistry" (P. Ausloos, ed.), p. 105. Wiley (Interscience), New York, 1968.
361. C. Ottinger, *Z. Naturforsch, Teil A* **20**, 1229 (1965).
362. O. Osberghaus and C. Ottinger, *Phys. Lett.* **16**, 121 (1965).
363. B. Andlauer and C. Ottinger, *J. Chem. Phys.* **55**, 1471 (1971).
364. B. Andlauer and C. Ottinger, *Z. Naturforsch., Teil A* **27**, 293 (1972).
365. J. H. D. Eland and H. Schulte, *J. Chem. Phys.* **62**, 3835 (1975).
366. C. E. Klots, *Z. Naturforsch., Teil A* **27**, 553 (1972).
367. J. H. D. Eland, J. Berkowitz, H. Schulte, and R. Frey, *Int. J. Mass Spectrom Ion Phys.* **28**, 297 (1978), have recently obtained a slightly lower value for $\Delta H^0_{f_0}$ ($C_4H_4{}^+$) than that estimated by Rosenstock *et al.*, ref. 358, by studying the reaction $C_5H_5N + h\nu \rightarrow C_4H_4{}^+ + HCN$.
367a. H. Schulte, Diplomarbeit, Univ. Freiburg, Freiburg, West Germany, 1975.
367b. H. Rindfleisch, Diplomarbeit, Univ. Freiburg, Freiburg, West Germany, 1974.
368. Y. V. Chizhov, V. I. Kleimenov, G. S. Medynskii, and F. I. Vilesov, *Opt. Spektrosk* **33**, 661 (1972); Engl. transl., *Opt. Spectrosc.* (*USSR*) **33**, 365 (1972).
369. B. Brehm, *Z. Naturforsch., Teil A* **21**, 196 (1966).
370. B. S. Freiser and J. L. Beauchamp, *Chem. Phys. Lett.* **35**, 35 (1975).
371. F. H. Dorman and J. D. Morrison, *J. Chem. Phys.* **35**, 575 (1961).
372. L. O. Werme, B. Grennberg, J. Nordgren, C. Nordling, and K. Siegbahn, *J. Electron Spectrosc. Relat. Phenom.* **2**, 435 (1973).

373. Ref. 43, pp. 95, 97.
374. A. W. Potts, H. J. Lempka, D. G. Streets, and W. C. Price, *Philos. Trans. R. Soc. London, Ser. A* **268**, (1970).
375. J. L. Dehmer, *J. Chem. Phys.* **56**, 4496 (1972).
376. J. W. D. Connolly and K. H. Johnson, *Chem. Phys. Lett.* **10**, 616 (1971).
377. U. Gelius, B. Roos, and P. Siegbahn, *Chem. Phys. Lett.* **4**, 471 (1970).
377a. W. von Niessen, L. S. Cederbaum, G. H. F. Diercksen, and G. Hohlneicher, *Chem. Phys.* **11**, 399 (1975).
378. U. Gelius, *J. Electron Spectrosc. Relat. Phenom.* **5**, 1046 (1974).
378a. T. Gustafsson. *Phys. Rev. A* **18**, 1481 (1978).
379. M. Sasanuma, E. Ishiguro, Y. Morioka, and M. Nakamura, *Int. Conf. VUV Radiat. Phys., 3rd, Tokyo* Pap. 1pA2-3 (1971).
380. D. C. Frost, C. A. McDowell, J. S. Sandhu, and D. A. Vroom, *J. Chem. Phys.* **46**, 2008 (1967).
381. J. Delwiche, *Bull. Cl. Sci., Acad. R. Belg.* **55**, 215 (1969).
382. U. Gelius, *J. Electr. Spectr.* **5**, 985 (1974).
383. V. H. Dibeler and J. A. Walker, *J. Chem. Phys.* **44**, 4405 (1966).
384. J. Berkowitz, W. A. Chupka, P. M. Guyon, J. H. Holloway, and R. Spohr, *J. Phys. Chem.* **75**, 1461 (1971).
385. T. A. Walter, C. Lifshitz, W. A. Chupka, and J. Berkowitz, *J. Chem. Phys.* **51**, 3531 (1969).
386. Ref. 130, p. 237.
387. J. Berkowitz and W. A. Chupka, *J. Chem. Phys.* **29**, 653 (1958).
388. J. Berkowitz and J. R. Marquart, *J. Chem. Phys.* **37**, 1853 (1962).
389. I. G. Simm, C. J. Danby, J. H. D. Eland, and P. J. Mansell, *J. Chem. Soc., Faraday Trans. II* **72**, 426 (1976).
390. B. P. Pullen and J. A. D. Stockdale, *Int. J. Mass Spectrom. Ion Phys.* **19**, 35 (1976).
391. C. E. Klots, *J. Chem. Phys.* **41**, 117 (1964).
391a. A. P. Hitchcock, C. E. Brion, and M. J. Van der Wiel, *J. Phys. B* **11**, 3245 (1978).
392. R. E. LaVilla, *J. Chem. Phys.* **57**, 899 (1972).
393. P. M. Dehmer, J. Berkowitz, and W. A. Chupka, *J. Chem. Phys.* **59**, 5777 (1973).
394. P. M. Dehmer, J. Berkowitz, and W. A. Chupka, *J. Chem. Phys.* **60**, 2676 (1974).
395. R. B. Cairns, H. Harrison, and R. I. Schoen, *J. Chem. Phys.* **51**, 5440 (1969).
396. H. Harrison, R. I. Schoen, R. B. Cairns, and K. E. Schubert, *J. Chem. Phys.* **50**, 3930 (1969).
397. J. Berkowitz and C. Lifshitz, *J. Phys. B. Ser. 2*, **1**, 438 (1968).
398. J. E. Hansen, B. Ziegenbein, R. Lineke, and H. P. Kelly, *J. Phys. B* **10**, 37 (1977).
399. K. Tegeder and R. Lincke, *Z. Phys.* **247**, 51 (1971).
400. A. C. Parr, *J. Chem. Phys.* **54**, 3161 (1971).
401. A. C. Parr and F. A. Elder, *J. Chem. Phys.* **49**, 2665 (1968).
402. A. C. Parr and M. G. Inghram, *J. Chem. Phys.* **52**, 4916 (1970).
403. J. Berkowitz, W. A. Chupka, P. M. Guyon, J. H. Holloway, and K. Spohr, *J. Chem. Phys.* **54**, 5165 (1971).
404. V. H. Dibeler, J. A. Walker, and K. E. McCulloh, *J. Chem. Phys.* **51**, 4230 (1969).
405. M. Krauss, J. A. Walker, and V. H. Dibeler, *J. Res. Natl. Bur. Stand., Sect. A* **72**, 281 (1968).
406. J. D. Morrison, H. Hurzeler, M. G. Inghram, and H. E. Stanton, *J. Chem. Phys.* **33**, 821 (1960).
407. V. H. Dibeler, J. A. Walter, and K. E. McCulloh, *J. Chem. Phys.* **50**, 4592 (1969).
408. V. H. Dibeler, J. A. Walker, K. E. McCulloh, and H. M. Rosenstock, *Int. J. Mass Spectrom. Ion Phys.* **7**, 209 (1971).

409. M. E. Akopyan, F. I. Vilesov, and Y. L. Sergeev, *Opt. Spektrosk.* **35**, 812 (1973); Engl. transl., *Opt. Spectrosc.* (*USSR*) **35**, 472 (1973).
410. V. H. Dibeler, J. A. Walker, and K. E. McCulloh, *J. Chem. Phys.* **53**, 4414 (1970).
411. J. Berkowitz and W. A. Chupka, *J. Chem. Phys.* **45**, 1287 (1966).
412. J. Berkowitz, *Adv. High Temp. Chem.* **3**, 123 (1971).
413. J. Berkowitz, *J. Chem. Phys.* **50**, 3503 (1969).
414. J. Berkowitz and T. A. Walter, *J. Chem. Phys.* **49**, 1184 (1968). *J. Phys. B* **1**, 438 (1968).
415. J. Berkowitz and C. Lifshitz, *J. Chem. Phys.* **48**, 4346 (1968).
416. J. Berkowitz and W. A. Chupka, *J. Chem. Phys.* **50**, 4245 (1968).
417. V. H. Dibeler and S. K. Liston, *J. Chem. Phys.* **49**, 482 (1968).
418. J. Berkowitz, E. H. Appelman, and W. A. Chupka, *J. Chem. Phys.* **58**, 1950 (1973).
419. J. Berkowitz, W. A. Chupka, and T. A. Walter, *J. Chem. Phys.* **50**, 1497 (1969).
420. V. H. Dibeler and S. K. Liston, *J. Chem. Phys.* **48**, 4765 (1968).
421. K. Radler and J. Berkowitz, *J. Chem. Phys.* **66**, 2176 (1977).
422. M. J. Weiss, J. Berkowitz, and E. H. Appelman, *J. Chem. Phys.* **66**, 2049 (1977).
423. G. R. Cook, *in* "Recent Developments in Mass Spectroscopy" (K. Ogata and T. Hayakawa, eds.), p. 761. Univ. Park Press, Baltimore, Maryland, 1970.
424. J. Berkowitz, J. H. D. Eland, and E. H. Appelman, *J. Chem. Phys.* **66**, 2183 (1977).
425. P. C. Killgoar, Jr., G. E. Leroi, W. A. Chupka, and J. Berkowitz, *J. Chem. Phys.* **59**, 1370 (1973).
426. V. H. Dibeler, J. A. Walker, and S. K. Liston, *J. Res. Natl. Bur. Stand., Sect. A* **71**, 371 (1967).
427. D. C. Frost, D. Mak, and C. A. McDowell, *Can. J. Chem.* **40**, 1064 (1962).
428. J. Berkowitz, P. M. Dehmer, and W. A. Chupka, *J. Chem. Phys.* **59**, 925 (1973).
429. J. Berkowitz and J. H. Holloway, *J. Chem. Soc. Faraday Trans. II* **11**, 2077 (1978).
430. J. D. Morrison, A. J. C. Nicholson, and T. A. O'Donnell, *J. Chem. Phys.* **49**, 959 (1968).
431. V. H. Dibeler and S. K. Liston, *J. Chem. Phys.* **47**, 4548 (1967).
432. W. A. Chupka and C. Lifshitz, *J. Chem. Phys.* **48**, 1109 (1968).
433. F. A. Elder, C. F. Giese, B. Steiner, and M. G. Inghram, *J. Chem. Phys.* **36**, 3292 (1962).
434. L. Sergeev, M. E. Akopyan, and F. I. Vilesov, *Opt. Spektrosk.* **30**, 230 (1971); Engl. transl., *Opt. Spectrosc.* (*USSR*) **30**, 121 (1971).
435. M. E. Akopyan, F. I. Vilesov, and S. N. Lopatin, *Khim. Vys. Energ.* **6**, 110 (1972); Engl. transl., *High Energy Chem.* **6**, 96 (1972).
436. F. A. Elder and A. C. Parr, *J. Chem. Phys.* **50**, 1027 (1969).
437. C. Lifshitz and W. A. Chupka, *J. Chem. Phys.* **47**, 3439 (1967).
438. M. M. Lipei, V. K. Potapov, and R. G. Kostyanovskii, *Khim. Vys. Energ.* **8**, 496 (1974); Engl. transl., *High Energy Chem.* **8**, 426 (1974).
439. B. Steiner, C. F. Giese, and M. G. Ingram, *J. Chem. Phys.* **34**, 189 (1961).
440. R. I. Schoen, *J. Chem. Phys.* **37**, 2032 (1962).
441. R. Krassig, D. Reinke, and H. Baumgärtel, *Ber. Bunsenges. Phys. Chem.* **78**, 425 (1974).
442. Y. L. Sergeev, M. E. Akopyan, F. I. Vilesov, and Y. V. Chizhov, *Khim. Vys. Energ.* **7**, 418 (1973); Engl. transl., *High Energy Chem.* **7**, 369 (1973).
443. A. S. Werner, B. P. Tsai, and T. Baer, *J. Chem. Phys.* **60**, 3650 (1974).
444. M. E. Akopyan, F. I. Vilesov, and Y. L. Sergeev, *Teor. Eksp. Khim.* **7**, 271 (1971); Engl. transl., *Theor. Exp. Chem.* **7**, 299 (1971).
445. M. E. Akopyan, Y. L. Sergeev, and F. I. Vilesov, *Khim. Vys. Energ.* **4**, 213 (1970); Engl. transl., *High Energy Chem.* **4**, 186 (1970).
446. A. D. Williamson, P. R. LeBreton, and J. L. Beauchamp, *J. Am. Chem. Soc.* **98**, 2705 (1976).
447. A. C. Parr and F. A. Elder, *J. Chem. Phys.* **49**, 2659 (1968).

448. D. Reinke, R. Krassig, and Baumgärtel, *Z. Naturforsch., Teil A* **28**, 1021 (1973).
449. W. Lohr, H. W. Jochims, and H. Baumgartel, *Ber. Bunsenges. Phys. Chem.* **79**, 901 (1975).
450. I. Omura, T. Kaneko, Y. Yamada, and K. Tanaka, *J. Phys. Soc. Jpn.* **27**, 178 (1969).
451. H. Hurzeler, M. G. Inghram, and J. D. Morrison, *J. Chem. Phys.* **27**, 313 (1957).
452. V. K. Potapov and L. I. Iskakov, *Khim. Vys. Energ.* **4**, 354 (1970); Engl. transl., *High Energy Chem.* **4**, 309 (1970).
453. L. I. Iskakov and V. K. Potapov, *Khim. Vys. Energ.* **4**, 381 (1970); Engl. transl., *High Energy Chem.* **4**, 345 (1970).
454. L. I. Iskakov and V. K. Potapov, *Khim. Vys. Energ.* **5**, 534 (1971); Engl. transl., *High Energy Chem.* **5**, 474 (1971).
455. M. E. Akopyan, F. I. Vilesov and A. N. Terenin, *Dokl. Akad. Nauk SSSR* **140**, 1037 (1961); Engl. transl., *Sov. Phys.—Dokl.* **6**, 890 (1962).
456. M. E. Akopyan, F. I. Vilesov, and Y. V. Loginov, *Khim. Vys. Energ.* **9**, 322 (1975); Engl. transl., *High Energy Chem.* **9**, 281 (1975).
457. M. E. Akopyan, F. I. Vilesov, and Y. V. Loginov, *Khim. Vys. Energ.* **9**, 327 (1975); Engl. transl., *High Energy Chem.* **9**, 286 (1975).
458. M. L. Foster, A. D. Williamson, and J. L. Beauchamp, *Int. J. Mass Spectrom. Ion. Phys.* **15**, 429 (1974).
459. M. M. Lipei and V. K. Potapov, *Khim. Vys. Energ.* **9**, 332 (1975); Engl. transl., *High Energy Chem.* **9**, 290 (1975).
460. V. H. Dibeler and S. K. Liston, *Inorg. Chem.* **8**, 1728 (1969).
461. M. E. Akopyan and F. I. Vilesov, *Kinet. Katal.* **4**, 39 (1963); Engl. transl., *Kinet. Catal. (USSR)* **4**, 32 (1963).
462. V. K. Potapov and V. V. Sorokin, *Dokl. Akad. Nauk SSSR* **195**, 616 (1970); Engl. transl., *Akad. Nauk SSSR. Proc. Section Chemistry* **195**, 848 (1970).
463. V. K. Potapov, M. M. Lipei, A. G. Kaz'min, and V. V. Sorokin, *Khim. Vys. Energ.* **8**, (1974); Engl. transl., *Higher Energy Chem.* **8**, 241 (1974).
464. V. K. Potapov and B. A. Bazhenov, *Khim. Vys. Energ.* **4**, 553 (1970); Engl. transl., *High Energy Chem.* **4**, 505 (1970).
465. V. K. Potapov and O. A. Yuzhakova, *Dokl. Akad. Nauk SSSR* **192**, 131 (1970); Engl. transl., *Akad. Nauk SSSR. Proc. Section Phys. Chem.* **192**, 365 (1970).
466. V. K. Potapov, A. D. Filyugina, D. N. Shigorin, and G. A. Ozerova, *Dokl. Akad. Nauk SSSR* **180**, 398 (1968); Engl. transl., *Dokl. Phys. Chem.* **180**, 352 (1968).
467. E. Murad and M. G. Inghram, *J. Chem. Phys.* **40**, 3263 (1964).
468. E. Murad and M. G. Inghram, *J. Chem. Phys.* **41**, 404 (1964).
468a. W. M. Trott, N. C. Blais, and E. A. Walters, *J. Chem. Phys.* **69**, 3150 (1978).
469. M. E. Akopyan, Y. L. Sergeev, and F. I. Vilesov, *Khim. Vys. Energ.* **4**, 305 (1970); Engl. transl., *High Energy Chem.* **4**, 265 (1970).
470. M. E. Akopyan, Y. L. Sergeev, and F. I. Vilesov, *Khim. Vys. Energ.* **4**, 351 (1970); Engl. transl., *High Energy Chem.* **4**, 306 (1970).
471. M. E. Akopyan, Y. L. Sergeev, and F. I. Vilesov, *Khim. Vys. Energ.* **4**, 351 (1970); Engl. transl., *High Energy Chem.* **4**, 307 (1970).
472. M. E. Akopyan, Y. L. Sergeev, and F. I. Vilesov, *Khim. Vys. Energ.* **4**, 353 (1970); Engl. transl., *High Energy Chem.* **5**, 308 (1970).
473. W. L. Stebbings and J. W. Yaylor, *Int. J. Mass Spectrom. Ion Phys.* **6**, 471 (1971).
474. M. E. Akopyan and F. I. Vilesov, *Khim. Vys. Energ.* **2**, 107 (1968); Engl. transl., *High Energy Chem.* **2**, 89 (1968).
475. M. E. Akopyan and F. I. Vilesov, *Zh. Fiz. Khim.* **40**, 115 (1966); Engl. transl., *Russ. J. Phys. Chem.* **40**, 63 (1966).
476. Y. L. Sergeev, M. E. Akopyan, F. I. Vilesov, and V. I. Kleimenov, *Opt. Spektrosk.* **29**, 119 (1970); Engl. transl., *Opt. Spectrosc. (USSR)* **29**, 63 (1970).

477. Y. Y. Villem, M. E. Akopyan, and F. I. Vilesov, *Khim. Vys. Energ.* **9**, 407 (1975); Engl. transl., *High Energy Chem.* **9**, 356 (1975).
478. Y. Y. Villem and M. E. Akopyan, *Zh. Fiz. Khim.* **50**, 674 (1976); Engl. transl., *Russ. J. Phys. Chem.* **50**, 394 (1976).
479. V. H. Dibeler and S. K. Liston, *Inorg. Chem.* **7**, 1741 (1968).
480. V. H. Dibeler and S. K. Liston, *Inorg. Chem.* **8**, 50 (1969).
481. G. Distefano and V. H. Dibeler, *Int. J. Mass Spectrom Ion Phys.* **4**, 59 (1970).
482. G. Distefano, *Inorg. Chem.* **9**, 1919 (1970).
483. V. K. Potapov and V. V. Sorokin, *Khim. Vys. Energ.* **5**, 487 (1971); Engl. transl., *High Energy Chem.* **5**, 435 (1971).
484. G. Distefano, *J. Res. Natl. Bur. Stand., Sect. A* **74**, 233 (1970).
485. C. Y. Ng, D. J. Trevor, B. H. Mahan, and Y. T. Lee, *J. Chem. Phys.* **66**, 446 (1977).
486. C. Y. Ng, D. J. Trevor, B. H. Mahan, and Y. T. Lee, *J. Chem. Phys.* **65**, 4327 (1976).
487. C. Y. Ng, P. W. Tiedemann, B. H. Mahan, and Y. T. Lee, *J. Chem. Phys.* **66**, 5737 (1977).
488. C. Y. Ng, D. J. Trevor, P. W. Tiedemann, S. T. Ceyer, P. L. Kronebusch, B. H. Mahan, and Y. T. Lee, *J. Chem. Phys.* **67**, 4235 (1977).
489. C. Y. Ng, P. W. Tiedemann, B. H. Mahan, and Y. T. Lee, *J. Chem. Phys.* **66**, 3985 (1977).
490. G. G. Jones and J. W. Taylor, *J. Chem. Phys.* **68**, 1768 (1978).

Chapter VII

THE ANGULAR DISTRIBUTION OF PHOTOELECTRONS

A. General Form of the Equation

1. Symmetry Considerations—Early History

Although Einstein gave a quantal description of the photoelectric effect in 1905,[1] there seems to have been little interest in the angular distribution of photoelectrons during the period when the Bohr theory held sway. In a rather later paper, Auger and Perrin[2] show that the Bohr picture (an electron in a fixed orbit around a nucleus, with velocity vector normal to the radial direction, and photon impulse in propagation direction) leads to a maximum in the photon propagation direction. In the early 1920s there were already experimental indications that this was the incorrect result, and Auger and Perrin, in fact, devote their paper to a symmetry-based consideration of the wave mechanical picture, giving more nearly correct results. They purport to prove the following points.

(a) For polarized light, the azimuthal angular distribution should have the form $A \cos \phi + B$ (ϕ measured from the polarization direction) and for completely unpolarized light, the azimuthal distribution must be isotropic.

(b) For $h\nu > \text{IP}$, but $h\nu \ll mc^2$, they deduce a longitudinal distribution $a \cos^2 \theta + b$ (θ measured from the electric vector toward the propagation direction in that plane). The constant term is attributed to the original velocity of the electron in its orbit and to deviations during its departure from the atom. The implication seems strong in their article that there exists a preference for photoejection along the electric vector.

(c) For $h\nu \gg$ IP, $h\nu \sim mc^2$ or not much less, they deduce a θ_m which equipartitions the number of photoelectrons forward and backward of this angle. This is the phenomenon that has been called retardation in the subsequent literature. It derives from the magnetic vector potential of the electromagnetic field, and provides forward momentum to the photoelectron that destroys the forward–backward symmetry.

(d) They recognize that K-shell electrons have $l = 0$, whereas L, M,... shells have an initial angular momentum greater than zero, and as a consequence, a dispersion results which can be important when $h\nu$ is only slightly > IP.

Some modern papers have credited Auger and Perrin with deducing correctly, for the first time, the general form of the angular distribution for dipole excitation, but it seems from their analysis that there would never be a minimum along the electric vector. Indeed, the early literature seemed convinced of the preference for ejection along the electric vector, and more attention was directed to the retardation effect, which exhibited some dynamic changes. This was in large part due to the nature of the experimental studies which were in the hard x-ray range.

2. Hydrogenic Approximation

Wentzel[3] first presented a mathematical wave mechanical description of the process. He showed that using hydrogenlike bound-state functions and outgoing spherical waves, the electric field component of light wave interaction with an atom (for $h\nu >$ IP) leads to a pure $\cos^2\theta$ dependence (θ between electric field vector and outgoing electron). He obtained this result not only for ejection from an s state, but showed that it is true for initial $l > 0$. This result really would have made the entire subject trivial. It is incorrect, and probably a consequence of the form he took for the outgoing wave, since Bethe and Salpeter[4] show that the lowest order Born approximation result, using plane waves, gives just this result, which is incorrect for $l > 0$.

Wentzel[5] subsequently also addressed himself to the magnetic vector potential interaction and obtained a formula for the retardation effect, but this also appears to be quantitatively incorrect. Sommerfeld and Schur,[6] using hydrogenic continuum functions and K-shell photoejection, obtained the correct result

$$\frac{d\sigma}{d\Omega} \propto (1 + \gamma\cos\theta)\sin^2\theta, \tag{1}$$

where θ is measured from the propagation direction and $\gamma = 4v/c$. (They point out that this value for γ is twice that expected on classical momentum transfer considerations.) Bethe[7] has shown that this is the first term of a rapidly converging series

$$\frac{d\sigma}{d\Omega} \propto \frac{\sin^2\theta}{[1-(v/c)\cos\theta]^4}. \tag{2}$$

(Rather recently, Cooper and Manson[8] have shown that for low energies and outer subshells, where screened hydrogenic momentum wave functions provide a better approximation, the retardation portion of the angular distribution takes the form

$$\frac{d\sigma}{d\Omega} \propto 1 - \frac{v}{c}\cos\theta^{-4-l} \tag{3}$$

This is of course equivalent to the previous expression for s-shell ejection.)

Schur[9] examined L-shell photoejection using hydrogenic continuum functions. For the L_I component (2s ejection), he obtained a form similar to K-shell ejection, but for the $L_{II,III}$ component he derived

$$\frac{d\sigma_{L_{II,III}}}{d\Omega} \propto 1 + \frac{8I_L}{h\nu}\sin^2\theta, \tag{4}$$

where I_L = IP for the L shell. This was the first fairly accurate wave mechanical calculation that demonstrated an isotropic, in addition to a $\sin^2\theta$ component, for the angular distribution. At large $h\nu$, this equation predicts isotropic behavior. We shall see later that the behavior of 2p ejection in atomic hydrogen is not typical in its asymptotic behavior.

3. Symmetry Considerations—Modern History

In 1948, Yang[10] perceived that numerous calculations of angular distributions in nuclear reactions had many terms cancel out after a laborious computation. He suspected underlying symmetry reasons and proved the following theorem: "If only incoming waves of orbital angular momentum L contribute appreciably to the reaction, the angular distribution of the outgoing particles in the center-of-mass system is an even polynomial of $\cos\theta$ with maximum exponent not higher than $2L$." Yang also treated the relativistic case and noted that odd powers of $\cos\theta$ can enter in this case. However, "the odd powers of $\cos\theta$ in the angular correlation have coefficients smaller than the even powers by a factor of v/c."

Translated into the electromagnetic domain for electric-dipole interactions, $L = 1$ and hence only terms in $(\cos\theta)^0$ and $(\cos\theta)^2$ are permitted in the nonrelativistic case, whereas the relativistic $\cos\theta$ term must have a coefficient of order v/c, as noted in specific calculations.

Peshkin[11] specifically examined the terms permitted by symmetry for certain types of electromagnetic interactions, i.e., electric-dipole ($E1$), magnetic-dipole ($M1$), and interference terms $E1 \cdot M1$ and $E1 \cdot E2$. He also explored the possibility of additional information being extracted by use of linearly polarized and circularly polarized light and concluded that polarization of the light gains no qualitatively new information, but plane polarization doubles the relative strength of the $\cos^2\theta$ (or $\sin^2\theta$) term.

Ritchie[12] has recently found a special case that can produce an odd power of $\cos\theta$ in $E1$ interactions, which manages to wend its way through the tight symmetry restrictions imposed by the above theorems. This special case involves the use of circularly polarized light on chiral molecules (molecules exhibiting optical rotatory dispersion). For these special examples, Ritchie finds

$$\frac{d\sigma}{d\Omega} \propto A \pm B\cos\theta + C\cos^2\theta. \tag{5}$$

In a model calculation, Ritchie finds $A = 9.77$, $B = -0.646$, and $C = -4.88$, so the effect may be measurable.

For oriented targets, i.e., aligned molecules or atoms in a nonstatistical distribution of m states, angular distributions having more terms than P_0 and P_2 are possible. This is also true for multiphoton processes. It is conceivable that molecules adsorbed on surfaces may be studied as oriented species, and there are experiments under way examining the angular distribution in two-photon ionization, but in the spirit of this monograph we exclude both of these areas from the present consideration.

4. Many-Electron Atom in LS Coupling

(Total orbital and spin angular momenta are considered to be well-defined quantities for both initial and final states.) Cooper and Zare[13] conclude that the angular distribution of photoelectrons from an $l' \to l$ electron jump in a many-electron atom whose states are described by LS coupling is *equivalent* to the angular distribution for an $l' \to l$ electron jump in a one-electron atom. However, if the effects of configuration interaction are included, this simple picture is no longer so.

Instead, it is necessary to superimpose different one-electron angular distributions computed by the square of the CI mixing coefficients. Example: Mg.

Primary configuration: $(1s)^2(2s)^2(2p)^6(3s)^2\ {}^1S_0$. Ejection of 3s electron leads to "pure" $\cos^2\theta$ dependence. However, including configurations which are symmetry allowed, such as

$$(1s)^2(2s)^2(2p)^6(3p)^2\ {}^1S_0,$$
$$(1s)^2(2s)^2(2p)^6(3d)^2\ {}^1S_0,$$

permits of an angular distribution no longer pure $\cos^2\theta$, depending upon the extent of configuration mixing.

The Cooper–Zare formula for β where β is defined by $I(\theta) \propto P_0 + \beta P_2 \cos\theta$, is

$$\beta = \frac{l(l-1)R_{l-1}^2 + (l+1)(l+2)R_{l+1}^2 - 6l(l+1)R_{l+1}R_{l-1}\cos(\delta_{l+1} - \delta_{l-1})}{(2l+1)[lR_{l-1}^2 + (l+1)R_{l+1}^2]}, \tag{6}$$

where

$$R_{l\pm1}(\varepsilon) = \int_0^\infty P_{nl}(r) r P_{\varepsilon,l\pm1}(r)\,dr \tag{7}$$

and $\delta_{l\pm1}$ are the phase shifts of the respective scattered waves.

This formula is implicitly given in Bethe's 1933 Handbuch.[7] Using hydrogenic functions, Bethe gives the angular distribution in the form

$$I(\theta) = A + B\cos^2\theta, \tag{8}$$

where

$$\begin{aligned} A &= (a - 2b + c)l(l+1), \\ B &= a(l)(l-1) + 6b(l)(l+1) + c(l+1)(l+2). \end{aligned} \tag{9}$$

To compare this expression with the Cooper–Zare formula, we first note that

$$\beta = \frac{2B}{3A + B}. \tag{10}$$

Then

$$\beta = \frac{2\{al(l-1) + 6bl(l+1) + c(l+1)(l+2)}{3l(l+1)(a - 2b + c) + al(l-1) + 6bl(l+1) + c(l+1)(l+2)}, \tag{11}$$

which reduces to

$$\beta = \frac{al(l-1) + 6bl(l+1) + c(l+1)(l+2)}{(2l+1)[la + (l+1)c]}. \tag{12}$$

The quantities a, b, and c are given by Bethe in terms of radial integrals. Thus,

$$a = (R_{w,\,l-1}^{nl})^2, \qquad c = (R_{w,\,l+1}^{nl})^2, \tag{13}$$

and

$$b = \frac{(n')^2 - l(l+1)R_{w,\,l-1}^{nl}R_{w,\,l+1}^{nl}}{\sqrt{[(l+1)^2 + (n')^2][l^2 + (n')^2]}}. \tag{14}$$

Bethe also gives, for the radial integrals,

$$R_{w,l} = \left(\frac{2}{\pi k}\right)^{1/2} \frac{1}{r} \exp\left[i(kr + n' \ln 2kr - \frac{\pi}{2}(l-1) + \sigma_l)\right] \tag{15}$$

and

$$R_{w,\,l+1} = -R_{w,\,l-1} \frac{(l+1-in')(l-in')}{\{[(l+1)^2 + (n')^2][l^2 + (n')^2]\}^{1/2}}. \tag{16}$$

From (15),

$$R_{w,\,l+1} = \left(\frac{2}{\pi k}\right)^{1/2} \frac{1}{r} \exp\left[i(kr + n' \ln 2kr - \frac{\pi}{2} l + \sigma_{l+1})\right], \tag{17}$$

$$R_{w,\,l-1} = \left(\frac{2}{\pi k}\right)^{1/2} \frac{1}{r} \exp\left[i(kr + n' \ln 2kr - \frac{\pi}{2}(l-2) + \sigma_{l-1})\right], \tag{18}$$

$$\begin{aligned} \frac{R_{w,\,l+1}}{R_{w,\,l-1}} &= \exp\left[i\left(-\frac{\pi}{2} l + \sigma_{l+1} + \frac{\pi}{2}(l-2) - \sigma_{l-1}\right)\right] \\ &= \exp(i\pi) \exp[i(\sigma_{l+1} - \sigma_{l-1})] \\ &= -[\cos(\sigma_{l+1} - \sigma_{l-1}) - i \sin(\sigma_{l+1} - \sigma_{l-1})]. \end{aligned} \tag{19}$$

Equating this last expression to (16) in transposed form,

$$-[\cos(\sigma_{l+1} - \sigma_{l-1}) - i \sin(\sigma_{l+1} - \sigma_{l-1})] = -\frac{l^2 + l - (n')^2 - 2in'l - in'}{\sqrt{[(l+1)^2 + (n')^2][l^2 + (n')^2]}}. \tag{20}$$

Comparing the real parts of these expressions,

$$-\cos(\sigma_{l+1} - \sigma_{l-1}) = \frac{(n')^2 - l(l+1)}{\{[(l+1)^2 + (n')^2][l^2 + (n')^2]\}^{1/2}}. \tag{21}$$

In this form, we can immediately see that Bethe's b is

$$b = -\cos(\sigma_{l+1} - \sigma_{l-1}) R^{nl}_{w,\,l-1} R^{nl}_{w,\,l+1}. \tag{22}$$

Substituting the values of a, b, and c into Eq. (12), we finally obtain

$$\begin{aligned} \beta = {} & [l(l-1)(R^{nl}_{W,\,l-1})^2 - 6l(l+1) R^{nl}_{W,\,l-1} R^{nl}_{W,\,l+1} \cos(\sigma_{l+1} - \sigma_{l-1}) \\ & + (R^{nl}_{W,\,l+1})^2 (l+1)(l+2)] \\ & \div (2l+1)[l(R^{nl}_{W,\,l-1})^2 + (l+1)(R^{nl}_{W,\,l+1})^2] \end{aligned} \tag{23}$$

which is the form of the Cooper–Zare formula.

Bethe derived his expression for a hydrogenic system, and hence the σ_l's are Coulomb phase shifts. If they are generalized to include non-Coulomb phase shifts as well, the two formulas are identical. (In Bethe and Salpeter,[4] the many-electron atom is treated, but the algebra is not carried as far as in the Handbuch article.[7]) Cooper and Zare also show that the same form results for a (nonspherical) molecule if an average is made over all (random) orientations of the initial molecular state. Tully *et al.*[14] achieved the same result at about the same time.

Summary to this point

(1) In electric-dipole approximation, the selection rules are $\Delta l = \pm 1$ for systems where l is a good quantum number. For photoejection from an s orbital, only p waves are permitted. This means $R_{l-1} = 0$ in Cooper–Zare (Bethe) formula, and hence $\beta = 2$. Therefore, such distributions will be "pure" $\cos^2 \theta$, as initially found by Wentzel,[3] and Sommerfeld and Schur.[6]

(2) For photoejection from orbital $l > 0$, there will be two outgoing waves, $l + 1$ and $l - 1$, which can interfere with one another. This interference is a function of photoelectron energy, as manifested by the difference in phase shifts $\cos(\delta_{l+1} - \delta_{l-1})$ in the formula for β. It is only in an interference experiment, such as photoelectron angular distribution, that one gets direct information on phase shifts, and on the sign of the dipole matrix element.

(3) It might be anticipated that photoelectron ejection would show preference along the electric vector direction of the photon. Indeed, Bethe's Handbuch article[7] (which is restricted to hydrogenic systems) comments (p. 484) that ". . . neglecting retardation, the angular distribution of photoelectrons from p-, d-, . . . shells has its maximum in the polarization direction." In the language of the Cooper–Zare formula, this means $\beta > 0$.

In 1967, Berkowitz *et al.*[15] observed β values for O_2, NO, and N_2O which were negative. Shortly thereafter, Hall and Siegel[16] reported a similar observation for O^-, C^- which was supported by calculations on these systems by Cooper and Zare.[17] This means that under some circumstances electron ejection can be a *minimum* along the electric vector direction.

The realizable range of β in the Cooper–Zare formula (to ensure a nonnegative differential cross section) can extend from $\beta = +2$ (pure $\cos^2 \theta$, preference along electric vector) to $\beta = 0$ (isotropic distribution) to $\beta = -1$ (pure $\sin^2 \theta$, preference perpendicular to electric vector).

(4) If unpolarized light is used instead of linearly polarized light, then the Cooper–Zare formula becomes (see experimental section)

$$\frac{d\sigma}{d\Omega} = \frac{\sigma}{4\pi}\left(1 - \frac{\beta}{2} P_2(\cos\phi)\right),$$

where ϕ is now the angle between the photon's *propagation direction* and the photoelectron direction. The same β appears in both formulas, and hence no new information is obtained by using polarized light.

(5) The above conclusions apply to electric dipole processes. The lowest order correction to β goes as $4v/c$, where v is the photoelectron velocity and c the velocity of light. This is the retardation term already considered by Schur and Sommerfeld. It introduced a forward–backward asymmetry in the angular distribution, relative to the photon propagation direction. For total cross sections, this term integrates out to zero, and the lowest order correction then goes as $(v/c)^2$. However, for calculations of β at modest x-ray energies (say 1 keV photoelectrons), the retardation term is more significant, and may amount to a 15% correction.

(6) In the dipole approximation, the angular variation is given by

$$\sigma_i(\theta) = (\sigma_i/4\pi)[1 + \beta P_2(\cos\theta)], \text{ where } P_2(\cos\theta) = (\tfrac{3}{2}\cos^2\theta - \tfrac{1}{2}).$$

If photoelectron spectroscopic studies are performed on several ionic states, and the relative intensities σ_{i1}/σ_{i2} are to be meaningful, then either one must know the asymmetry parameter β for each transition, or collect electrons ejected at all angles, or measure each transition at the "magic angle." where $\cos^2\theta = \frac{1}{3}$, $\theta = 54.73561°$.

5. Beyond the Central Field Approximation

Photoelectron emission can be considered[18,19] a two-step process: (1) absorption of a photon by a bound-state electron with the accompanying transfer of angular momentum to the photoelectron and (2) the escape of the photoelectron. During the escape, angular momentum can be exchanged between photoelectron and ion core. This second part is neglected in the central field approximation. For closed shell ($^1S_0$) systems, angular momentum conservation salvages the central field approach. For open shell systems, anisotropic interactions can influence the second step. [Even for closed shell systems, ejection of an electron leaves an open shell, but this interaction is a spin–orbit one which (see below, under relativistic treatment) turns out to be quite small in most cases.]

Dill has recently introduced a formalism for calculating β which lends itself to the calculation of these anistropic interactions, and which reduces to the Bethe–Cooper–Zare formula when these are absent. Photoelectron ejection by dipole interaction is represented schematically as

$$X(J_0\pi_0) + h\nu(j_\gamma = 1, \pi_\gamma = -1) \rightarrow X^+(J_c\pi_c) + e^-[lsj, \pi_e = (-1)^l]. \quad (24)$$

Here, X and X^+ represent the atom and resultant ion, the Js, angular momenta, and the πs, parity. In an experimental situation (typical), where neither the spin of the photoelectron nor the orientation of the residual ion is determined, the photoelectric process can be described by

$$\mathbf{j}_t = \mathbf{j}_\gamma - \mathbf{l} = \mathbf{J}_c + \mathbf{s} - \mathbf{J}_0. \quad (25)$$

In this expression, $\mathbf{j}_t$ is the angular momentum transferred between the total angular momentum of atom $X(J_0)$ and the combined angular momentum of the ion and photoelectron $(\mathbf{J}_c + \mathbf{s})$.

Conservation of angular momentum (**J**) and parity (π) determine the possible values of $\mathbf{j}_t$

$$\mathbf{J} = \mathbf{J}_0 + \mathbf{j}_\gamma = \mathbf{J}_c + \mathbf{s} + \mathbf{l}, \tag{26}$$

$$\pi = \pi_\gamma \pi_0 = -\pi_0 = \pi_c(-1)^l. \tag{27}$$

Each value of j_t is characterized as being parity favored or parity unfavored. Parity-favored transitions occur for $l = j_t \pm 1$, parity unfavored for $l = j_t$. No other values are possible. From angular momentum and parity considerations, the asymmetry parameter β for parity-unfavored transitions is -1, independent of dynamical considerations, which appear only in the parity-favored transitions.

The asymmetry parameter appropriate to an experimental observation is given as the weighted average of the various $\beta(j_t)$.

$$\beta = \sum_{J_{cs}} \frac{\left(\sum_{j_t}^{\text{fav}} \sigma(j_t)_{\text{fav}} \beta(j_t)_{\text{fav}} - \sum_{j_t}^{\text{unf}} \sigma(j_t)_{\text{unf}}\right)}{\sigma}, \tag{28}$$

where the σs are partial cross sections and $J_{cs} = J_c + s$. The detailed form of the $\sigma(j_t)$s and $\beta(j_t)_{\text{fav}}$'s are given[18] in terms of scattering amplitudes, which are related to dipole matrix elements and continuum phase shifts.

We are now in a position to examine the effect of departures from the central field approximation. In the photon absorption act (part 1), the final orbital angular momentum is

$$\mathbf{l}' = \mathbf{l}_0 + \mathbf{j}_\gamma .$$

From its definition, the angular momentum transfer j_t' for this first part is

$$\mathbf{j}_t' = \mathbf{j}_\gamma - \mathbf{l}' = -\mathbf{l}_0 .$$

From parity conservation, $l' = l_0 \pm 1$ and hence $j_t' = l_0 = l' \mp 1$ is parity favored. In the central field approximation, this is the only process since exchange between ion core and escaping photoelectron is not considered. Hence, parity-unfavored angular momentum transfers are excluded, as are values of j_t other than l_0.

The influence of these anisotropic interactions has been examined for the photoionization of atomic oxygen[20] and atomic sulfur,[21] both open shell systems with ^{3}P ground states which can produce ^{4}S, ^{2}D, and ^{2}P ionic states by removal of a single valence p electron. Within the Cooper–Zare framework, all three final states would have the same values of β.

Dill *et al.*[21] found that angular momentum transfers $j_t \neq l_0$ contribute only a small amount to β in the case of sulfur, and a still smaller effect was calculated (see Starace *et al.*[20]) for oxygen. Hence, while there is a manifest deviation from the Bethe–Cooper–Zare formula, it is not very dramatic.

6. Photoelectron Angular Distributions through Resonances

From Eq. (28) it is transparently obvious why β should experience oscillations through a resonance. Since β is a function of the contributory $\sigma(j_t)$ which are experiencing major changes through a resonance, this should also be reflected in β. In the words of Dill,[18] "... σ and β are merely two sides of the same coin, alternative manifestations of the resonant behavior of ionization amplitudes that is characteristic of autoionization." However, "... whereas the integrated cross section depends only on the squared moduli of the ionization amplitudes, the asymmetry parameter depends as well on the moduli and phases of these amplitudes through the interference terms in the $\beta(j_t)_{\text{fav}}$. Therefore, the resonance in β can be an even more pronounced spectral feature than the resonance in σ."

Dill[18] has calculated the β variation through the autoionization resonances between the $^2P_{3/2}$ and $^2P_{1/2}$ states of Xe^+, obtaining the data for the ionization amplitudes and phases from interaction parameters derived from multichannel quantum defect theory (MQDT) which, in turn, parameterizes experimental photoabsorption data. An experiment was subsequently performed by Samson and Gardner[22] on two resonances in this system. A comparison of the calculated β variation and the experimental values is shown in Fig. 1. The slight shift in the minima is attributed to a small error in the MQDT parameterization. The overall agreement is very good, and thus the experimental results provide strong corroboration for the theory. The experiment is significantly simplified, in this case, by the fact that electron energy analysis is unnecessary, since the photoelectrons between the $^2\Pi_{3/2}$ and $^2\Pi_{1/2}$ states can have only one energy at a given incident photon energy.

Kabachnik and collaborators[22a] have independently made an analysis of the β variation through resonances and its relationship to the parameters characteristic of the absorption process. They have also extended the analysis to consider the variation of polarization of the ejected electron through resonances. Specific calculations were performed on the $(1s)^2(2s)^2(2p)^6 \rightarrow (1s)^2 2s(2p)^6 np$ resonances in neon and the $(1s)^2(2s)^2(2p)^6(3s)^2(3p^6) \rightarrow (1s)^2(2s)^2(2p)^6(3s)(3p)^6 np$ resonances in argon. In their scheme, these resonances can be characterized by six parameters: three can be obtained from the photoabsorption profiles, the other three either from model calculations or from a measurement of the β variation through the resonance. Conversely, such a measurement can be utilized to yield information about the relation between the transitions into different channels and the ratio of partial decay widths.

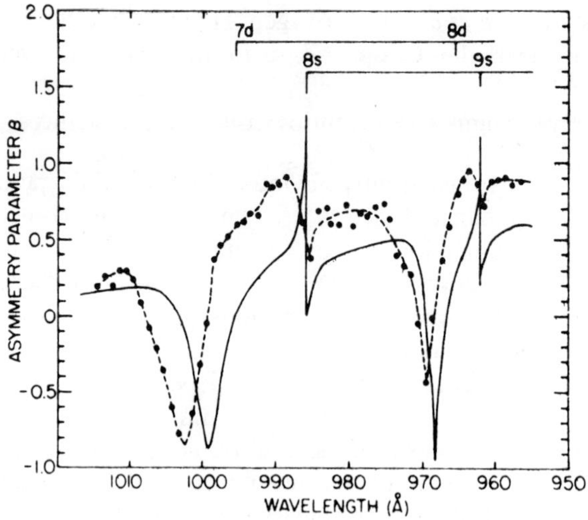

Fig. 1 The variation of the angular distribution asymmetry parameter β through two of the autoionization resonances between the $^2P_{3/2}$ and $^2P_{1/2}$ states of Xe^+. ——— Calculation using multichannel quantum defect theory performed by D. Dill, *Phys. Rev. A* **7**, 1976 (1973). ●-●- Experimental data obtained by J. A. R. Samson and J. L. Gardner, *Phys. Rev. Lett.* **31**, 1327 (1973). [From Samson and Gardner, cited here.]

7. Relativistic Treatment

A general formulation of β appropriate for relativistic wave functions has been derived by Walker and Waber.[23] The treatment has been carried out for multipole transitions (with circularly polarized light), but when specialized to electric-dipole transitions, it takes the form

$$\beta = \left\{ \frac{(2j-3)(2j-1)}{48j^2} R_{j-1}^2 - \frac{(2j-1)(2j+3)}{48j^2(j+1)^2} R_j^{\,2} + \frac{(2j+3)(2j+5)}{48(j+1)^2} R_{j+1}^2 + \frac{(2j-1)}{8j^2(j+1)} |R_j R_{j-1}^*| + \frac{2j+3}{8j(j+1)^2} |R_j R_{j+1}^*| + \frac{(2j-1)(2j+3)}{8j(j+1)} |R_{j-1} R_{j+1}^*| \right\} \div \left\{ \frac{2j-1}{12j} R_{j-1}^2 + \frac{1}{12j(j+1)} R_j^{\,2} + \frac{(2j+3)}{12(j+1)} R_{j+1}^2 \right\}, \qquad (29)$$

where the Rs are radial matrix elements using relativistic wave functions.

This expression reduces to the Bethe–Cooper–Zare formula when nonrelativistic wave functions are used, and the R_js are translated to R_ls.

In principle, this expression allows for different values of β for different spin–orbit components. A surprisingly large effect can be seen for ionization from s orbitals. Substituting $j = \frac{1}{2}$ into the above equation,

$$\beta = \frac{2R_{\mathrm{p}}^{\;2} + 4|R_{\mathrm{p}}R_{\mathrm{p}}^{*}|}{R_{\bar{\mathrm{p}}}^{\;2} + 2R_{\mathrm{p}}^{\;2}}, \tag{30}$$

where $\bar{\mathrm{p}}$ refers to $\mathrm{p}_{1/2}$, and p to $\mathrm{p}_{3/2}$. When $R_{\mathrm{p}} = R_{\bar{\mathrm{p}}}$ and the phase shifts are identical for the two continuum orbitals, $\beta = 2$, which is the nonrelativistic value we had previously noted.

However, when $R_{\mathrm{p}} = 0$ and $R_{\bar{\mathrm{p}}} \neq 0$, $\beta = 0$; alternatively, when $R_{\bar{\mathrm{p}}} = 0$ and $R_{\mathrm{p}} \neq 0$, $\beta = 1$. Accordingly, a rapid variation of β should be observable in the vicinity of a Cooper minimum for ionization from an s orbital, which would be a major deviation from nonrelativistic theory.

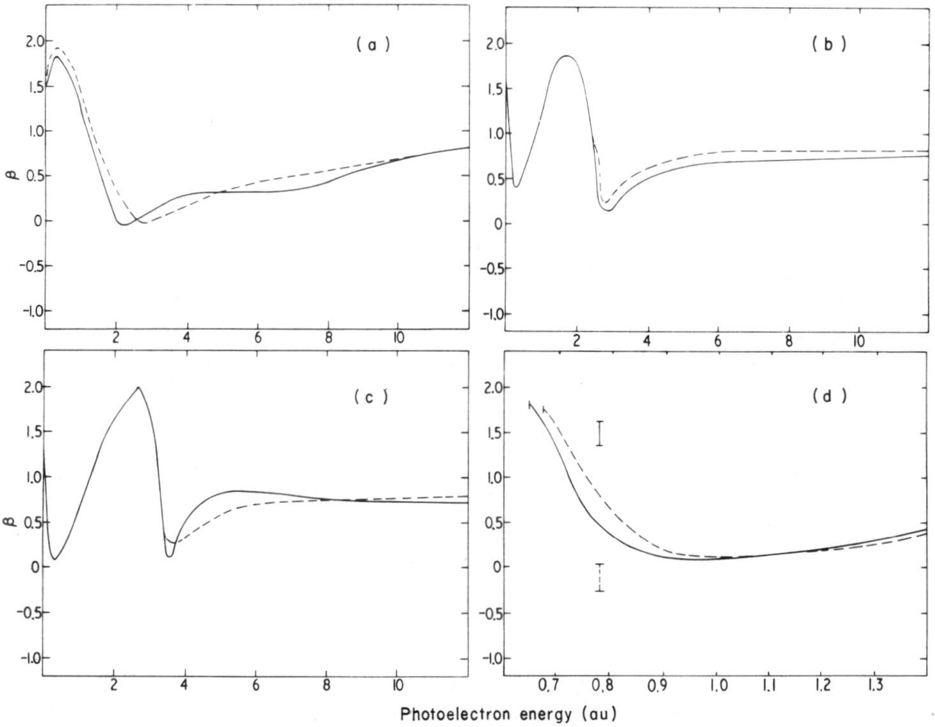

Fig. 2 Calculated variation of the asymmetry parameter β with photoelectron energy for ionization from (a) the 5p orbital of xenon: ——— $5\mathrm{p}_{3/2}$, ---- $5\mathrm{p}_{1/2}$; (b) the 4d orbital of xenon: ——— $4\mathrm{d}_{5/2}$, ---- $4\mathrm{d}_{3/2}$; (c) the 4d orbital of cadmium: ——— $4\mathrm{d}_{5/2}$, ---- $4\mathrm{d}_{3/2}$; (d) the 4d orbital of cadmium near threshold: ——— $4\mathrm{d}_{5/2}$, ---- $4\mathrm{d}_{3/2}$. Experimental values of Harrison (1970) are shown with error bars. [From T. E. H. Walker and J. T. Waber, *J. Phys. B* 7, 674 (1974).]

Experiments by Niehaus and Ruf[24] on ionization from the 6s orbital of Hg have obtained $\beta = 1.68$. However, this deviation from the nonrelativistic value of 2.0 may also be due to configuration interaction effects, since the initial state has significant p^2 and d^2 coefficients. The Niehaus and Ruf experiment was performed at 21.2 eV, about 3 eV beyond the Cooper minimum for s-orbital ionization.

When calculations are made on spin–orbit components removed from the Cooper minimum, very slight departures from the nonrelativistic values are obtained. Such calculations have been performed by Walker and Waber[25] on ionization from the 5p and 4d orbitals of xenon and the 4d orbital of cadmium (see Fig. 2). The experimental result of Harrison[26] on cadmium, performed with 21.2 eV radiation, obtained $\beta_{5/2} = 1.49$, $\beta_{3/2} = -0.12$, which is seen to be completely at odds with the relativistic or nonrelativistic calculation, nor is there evidence for autoionization at this wavelength in cadmium which could influence the result (see above on β through autoionizations).

By contrast, the relativistic and nonrelativistic (Manson[27]) calculations and experiment (Niehaus–Ruf) are in rather good agreement for Hg, as seen in Fig. 3. Since β varies with energy, and experiments performed with fixed photon energy will produce spin–orbit partners whose respective photoelectrons have different kinetic energies, one can anticipate that the experimental βs will be different when measured under these conditions. The conclusions of the Walker–Waber calculations are that this is the major difference of β for spin–orbit components, i.e., if one corrects for the kinetic energy dependence of β, the spin–orbit components would have nearly identical asymmetry parameters. This deduction had been made on other grounds by Niehaus and Ruf and by Manson.[27]

8. Molecular Effects—Rotation

The theory of the angular distribution of photoelectrons for molecules is more complex than for atoms for the following reasons:

(1) The molecular orbitals are not eigenfunctions of the orbital angular momentum operator l^2, due to the absence of spherical symmetry.

(2) The bound orbital, from which the photoelectron is ejected, is quantized on the molecular axis, while the continuum function of the ejected electron is more readily referred to space-fixed axes.

(3) Individual vibrational (and in a few cases, rotational) states can be resolved in photoelectron spectra. Hence, it is necessary not only to calculate an average β for a transition to a given electronic state of the ion, but to individual vibrational and rotational states as well.

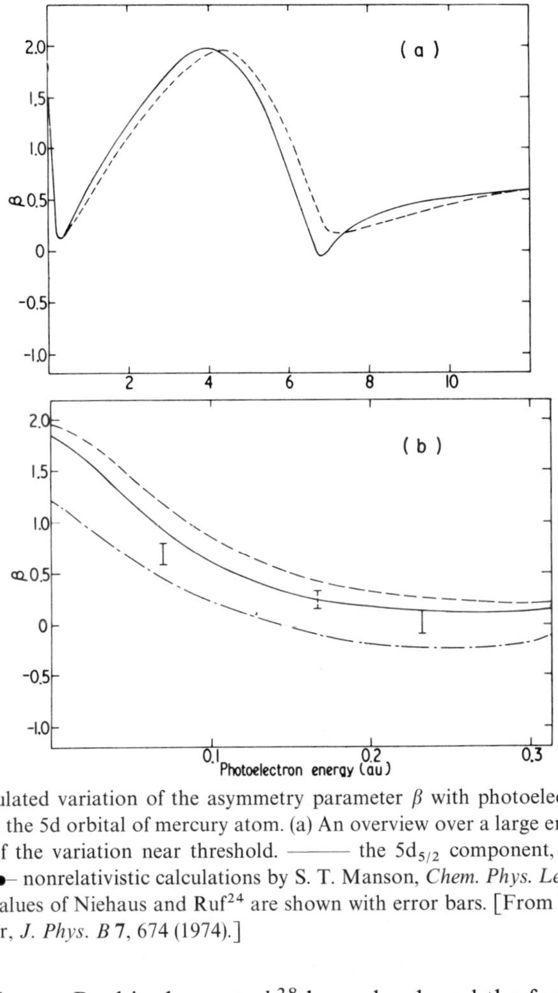

Fig. 3 Calculated variation of the asymmetry parameter β with photoelectron energy for ionization from the 5d orbital of mercury atom. (a) An overview over a large energy range, and (b) a closeup of the variation near threshold. ——— the $5d_{5/2}$ component, ---- the $5d_{3/2}$ component, •–•– nonrelativistic calculations by S. T. Manson, *Chem. Phys. Lett.* **19**, 76 (1973). Experimental values of Niehaus and Ruf[24] are shown with error bars. [From T. E. H. Walker and J. T. Waber, *J. Phys. B* 7, 674 (1974).]

a. Rotations. Buckingham *et al.*[28] have developed the formalism for β in molecules, when the photoelectron spectrum can resolve individual rotational levels. These authors obtain

$$\beta = \Bigg[\sum_{l=|\Delta\Lambda|}^{\infty} |C_{nl}|^2 Q(l)(2l+1)^{-2}[l(l-1)|F_{nl}^{E,\,l-1}|^2 + (l+1)(l+2)|F_{nl}^{E,\,l+1}|^2 - 6l(l+1)R_e(F_{nl}^{E,\,l-1}F_{nl}^{E,\,l+1})]\Bigg] \div \sum_{l=|\Delta\Lambda|}^{\infty} |C_{nl}|^2 Q(l)(2l+1)^{-1}[l|F_{nl}^{E,\,l-1}|^2 + (l+1)|F_{nl}^{E,\,l+1}|^2]. \quad (31)$$

The relationship between this equation and the Bethe–Cooper–Zare formula is immediately apparent if only a single value of l is considered, and hence the summations are reduced to a single term. We recognize that the F_{nl} are the radial integrals previously defined as

$$R_{l\pm 1}(\varepsilon) = \int_0^\infty P_{nl}(r) r P_{\varepsilon, l\pm 1}(r)\, dr$$

and currently defined as

$$F_{nl}^{E\lambda} = \int_0^\infty r^2\, dr\, R_\lambda^*(r) r R_{nl}(r) \tag{32}$$

and that the radial wave functions $R_{nl}(r)$ in the molecular case are obtained from an expansion of the bound molecular orbital from which the photoelectron is ejected, in terms of *central field functions*, i.e.,

$$u_{n,\,-\Delta\Lambda} = \sum_{l=|\Delta\Lambda|}^{\infty} C_{nl} R_{nl}(r) Y_{l,\,-\Delta\Lambda}(\theta', \varphi'). \tag{33}$$

In the limit of a single l being dominant, the Bethe–Cooper–Zare form is recovered, with a multiplicative factor $Q(l)$, which contains the specific dependence on molecular rotation. Buckingham *et al.* obtain the form of $Q(l)$ for Hund's cases (a) and (b).

In Hund's case (b), the spin is weakly coupled to the resultant of the orbital and rotational angular momenta, and the spin is quantized along a space-fixed (laboratory) axis. This description is suitable for $^2\Sigma \leftarrow {}^1\Sigma$ ionization. In this coupling scheme,

$$Q(l) = (2K'+1)\begin{pmatrix} K' & l & K'' \\ -\Lambda' & \Delta\Lambda & \Lambda'' \end{pmatrix}^2, \tag{34}$$

where the term in large parentheses is a $3j$ symbol. Here, Λ is the component of orbital angular momentum along the molecular axis, K the resultant of orbital and rotational angular momenta $\Delta\Lambda = \Lambda' - \Lambda''$, and l is a dummy variable ranging from $|\lambda - 1|$ to $\lambda + 1$. The λ designate the partial waves of the outgoing electron.

In Hund's case (a), the spin is strongly coupled to the molecular axis. This description is more appropriate for $^2\Pi \leftarrow {}^1\Sigma$ ionization. For this coupling scheme, Buckingham *et al.* obtain

$$Q(l) = (2J'+1)(2S''+1)\begin{pmatrix} S' & 1/2 & S'' \\ \Sigma' & -\Delta\Sigma & -\Sigma'' \end{pmatrix}^2$$

$$\times \sum_{\chi} (2\chi+1)\begin{pmatrix} l & 1/2 & \chi \\ \Delta\Lambda & \Delta\Sigma & -\Delta\Omega \end{pmatrix}^2 \begin{pmatrix} J' & \chi & J'' \\ -\Omega' & \Delta\Omega & \Omega'' \end{pmatrix}^2, \tag{35}$$

where $\Delta\Sigma = \Sigma' - \Sigma''$, $\Delta\Omega = \Omega' - \Omega''$, Σ is the component of spin along the molecular axis and $\Omega = \Lambda + \Sigma$. The dummy variable χ has the values $l \pm \frac{1}{2}$.

Still another form of $Q(l)$ is found when Hund's case (a) is appropriate to one of the states involved in the transition and Hund's case (b) to the other, as may be necessary for NO and O_2.

When rotational fine structure is not resolved in the experiment, the appropriate theoretical expression for comparison with experiment must average the differential cross section over the rotational levels of the molecule, and sum over the levels of the ion. Under these circumstances, Buckingham *et al.* obtain for Hund's case (b), $Q(l) = 1$. If only a single value of l were involved, Eq. (31) reduces to the Bethe–Cooper–Zare formula. For Hund's case (a), the averaging and summation yield

$$Q(l) = (2S'' + 1)\begin{pmatrix} S' & 1/2 & S'' \\ \Sigma' & -\Delta\Sigma & -\Sigma'' \end{pmatrix}^2. \tag{36}$$

As for Hund's case (b), the $Q(l)$ is independent of l, the l dependence being contained in the familiar terms of the Bethe–Cooper–Zare formula.

Buckingham *et al.*[28] also give the form of the total cross section with and without resolution of rotational fine structure, which would be applicable to the analysis of photoelectron line profiles. However, they have not evaluated the matrix elements for any particular case, and hence it is not clear from their work how different the βs can be for alternative rotational states involved in the ionization.

Some general considerations have been given by Sichel[29] for applying the above formalism to experimental situations. He points out that two experimental observables, the cross section for formation of a particular ionic state σ_i and the asymmetry parameter for that process β can be written for atoms as

$$\sigma_i = \text{const} \cdot \nu \cdot x_l, \tag{37}$$

$$\beta = y_l/x_l, \tag{38}$$

where

$$x_l = (2l+1)^{-1}[l|F_{nl}^{E,\,l-1}|^2 + (l+1)|F_{nl}^{E,\,l+1}|^2] \tag{39}$$

$$y_l = (2l+1)^{-2}[l(l-1)|F_{nl}^{E,\,l-1}|^2 + (l+1)(l+2)|F_{nl}^{E,\,l+1}|^2 - 6l(l+1)|F_{nl}^{E,\,l-1}|\,|F_{nl}^{E,\,l+1}|\cos(\delta_{l+1} - \delta_{l-1})]. \tag{40}$$

The asymmetry parameter is seen to be a ratio of radial matrix elements. There are three parameters to be determined, $F_{nl}^{E,\,l-1}$, $F_{nl}^{E,\,l+1}$, and the difference in phase shifts, $\delta_{l+1} - \delta_{l-1}$, but only two experimental values σ_i and β, so that the theoretical parameters cannot be directly determined by experiment. It is necessary to use some theoretical model for the continuum orbitals in order to calculate transition integrals and phase shifts which yield σ_i and β to be tested against experiment.

For molecules, these expressions take the form

$$\sigma_{\mathrm{i}} = \mathrm{const} \cdot v \cdot \sum_{l=|\Delta\Lambda|}^{\infty} Q(l) C_l^2 x_l \tag{41}$$

$$\beta = \sum_{l=|\Delta\Lambda|}^{\infty} Q(l) C_l^2 y_l \Bigg/ \sum_{l=|\Delta\Lambda|}^{\infty} Q(l) C_l^2 x_l, \tag{42}$$

where x_l, y_l have their previous meanings and the possible influence of alternative vibrational states has been suppressed.

If rotational energy levels are resolved in the molecular ion, then measurements of σ_{i} and β for a particular rotational transition would provide the values $C_l^2 x_l$ and $C_l^2 y_l$, respectively, for each transition. These two experimental parameters for each rotational transition can be related to four theoretical parameters, i.e., the three previously mentioned ones and the expansion coefficients C_l.

The expansion coefficients C_l can independently be obtained from appropriate ab initio calculations which employ a single-center basis set. The two experimental quantities σ_{i} and β can then be used to test continuum models, as for atoms.

Alternatively, if other evidence exists which can be used to define continuum parameters, such as quantum defect analysis of the discrete spectrum, then comparison with experiment can provide information about the relative importance of the expansion coefficients C_l.

b. Application to H_2. The electronic ground state of H_2 is $^1\Sigma_g^+$. It can be expanded in even single-center functions s, d, g, ... A one-center numerical Hartree–Fock calculation[30] has shown that the convergence is so rapid that only s and d terms contribute significantly. This conclusion implies that in the photoionization transition $\Delta K = 0, \pm 2$ only would be observable, where K is the resultant of orbital and rotational angular momenta, as before. In a high-resolution photoelectron spectrum of H_2 in which rotational levels of H_2^+ were resolved, Åsbrink[31] did indeed observe $\Delta N = 0, \pm 2$ transitions, the ± 2 transitions being about one order of magnitude weaker than $\Delta N = 0$. (ΔN is the change in rotational angular momentum between H_2 and H_2^+.)

Dill[32] has applied his angular momentum transfer approach to H_2 photoionization. For $\Delta N = \pm 2$, $j_t = 2$, while for $\Delta N = 0$, $j_t = 0$ or 2, but is dominated by $j_t = 0$. When individual rotational levels can be resolved, Dill deduces $\beta = \frac{1}{5}$ for $\Delta N = \pm 2$ and

$$\beta = \frac{2|C(0)|^2 + \frac{1}{5} f(N)|C(2)|^2}{|C(0)|^2 + f(N)|C(2)|^2} \tag{43}$$

for $\Delta N = 0$, with $f(N) = N(N+1)/(2N-1)(2N+3)$. The $C(0)$ and $C(2)$ are given by (see, for example, H_2 partial cross section discussion of rotational autoionization in Chapter VI)

$$C(0) = \sqrt{1/3}(d_\sigma e^{i\pi\mu_\sigma} + 2d_\pi e^{i\pi\mu_\pi}) \tag{44}$$

and

$$C(2) = \sqrt{2/3}(-d_\sigma e^{i\pi\mu_\sigma} + d_\pi e^{i\pi\mu_\pi}). \tag{45}$$

Dill also gives expressions for the corresponding partial cross sections σ_i, as well as the rotationally averaged $\overline{\beta}$:

$$\overline{\beta} = \sum_{\Delta N} \frac{\sigma_{\Delta N}\beta_{\Delta N}}{\sigma_{\Delta N}} = \frac{2(2d_\sigma{}^2 + 7d_\pi{}^2 + 6d_\sigma d_\pi \cos \pi\delta)}{5(d_\sigma{}^2 + 2d_\pi{}^2)}. \tag{46}$$

From analyses of the absorption spectrum of H_2, Herzberg and Jungen[33] (see also partial cross section H_2) deduce $\delta = 0.285$ and $\rho = d_\sigma/d_\pi = 1 \pm 0.3$. Substituting these quantites, we find

$$\overline{\beta} = \tfrac{2}{5}(3 + 2\cos(0.285\pi)) = 1.70. \tag{47}$$

Values of β from rotationally unresolved transitions in H_2 have been reported by Carlson and Jonas[34] who obtain $\beta = 1.75$ and by Niehaus and Ruf[35] who obtain $\beta = 1.93 \pm 0.03$. Large differences are predicted when rotational levels of $H_2{}^+$ are resolved. As already noted, $\beta_{\Delta N=\pm 2} = \frac{1}{5}$, whereas Dill calculates $\beta_{\Delta N=0} = 2$, 1.87, or 1.91, when the transitions are 0–0, 1–1, and 2–2, respectively. Niehaus and Ruf[35] have reported βs for partially resolved rotational transitions. For $\beta_{\Delta N=0}$ (unresolved further), they obtain 1.95 ± 0.03, in good agreement with Dill's prediction. For $\beta_{\Delta N=+2}$, they observe 0.85 ± 0.14, which is significantly different from $\beta_{\Delta N=0}$ and in the direction predicted by Dill, but rather far removed from the theoretical expectation $\beta = 0.2$. Dill indicates that the difference could be due to photoelectron f waves (which he has not included in his calculations), but it may also be due[35a] to incomplete separation of $\Delta N = 0$ and $\Delta N = +2$ transitions in the experimental studies of Niehaus and Ruf.

Dill[32] has also calculated the variation of β through the rotational autoionization resonances in H_2.

9. Molecular Effects

Vibrations. The neutral molecule and molecular ion wave functions are usually expressed (following the Born–Oppenheimer approximation) as products of electronic Φ and vibrational χ wave functions, i.e.,

$$\psi_m = \Phi_n(r_1, r_2, \ldots, r_n, R)\chi_m(R) \tag{48}$$

and

$$\psi_i = \Phi_i(r_1, r_2, \ldots, r_{n-1}, R)\chi_i(R), \tag{49}$$

where the r_js are electron coordinates and the Rs nuclear coordinates. The transition moments appearing in the expression for β will be of the form

$$\int \Phi_i^* \chi_i^* \varphi_e \left(\sum_j \mathscr{E} \cdot \mathbf{r}_j \right) \Phi_m \chi_m \, dr_1 \, dr_2 \cdots dr_n dR. \tag{50}$$

Integration over the electronic coordinates does not involve the vibrational functions χ. Often, the electronic matrix elements are slowly varying functions of R, and the transition moment is factored into

$$\int \chi_i^* \chi_m \, dR \left\langle \Phi_i^* \varphi_e \middle| \left(\sum_j \mathscr{E} \cdot \mathbf{r}_j \right) \middle| \varphi_m \right\rangle. \tag{51}$$

In this expression, the first integral is a vibrational overlap whose square is the Franck–Condon factor, and the expression in brackets contains the $R_{l\pm 1}$ (or alternatively, $F_{nl}^{E,\, l\pm 1}$) appearing in the expression for β.

In the zero-order approximation, different vibrational members formed in ionization to a particular ionic state would have the same β value, as determined by the electronic matrix element. Several corrections must be considered, however.

(1) The radial integrals are functions of the photoelectron energy, which will be different for the various vibrational states when a fixed incident photon energy is used.

(2) The electronic matrix elements are R dependent, and the vibrational overlap for different vibrational states encompasses different R regions. Hence, the electronic matrix element will vary somewhat with v', the vibrational quantum number of the ionic state.

(3) The use of a product function $\psi = \Phi\chi$ may introduce inaccuracies, i.e., there may occur a breakdown in the Born–Oppenheimer approximation. This was seen to be the explanation for autoionization in H_2.

(4) Transitions proceeding through an intermediate state (i.e., autoionization resonances, either electronically or vibrationally induced) will introduce intermediate vibrational wave functions and a different set of vibrational overlap integrals will enter.[36]

In general, when there is no evidence of autoionization resonances, the electronic matrix elements are weakly varying over the energy range of an electronic band, and the β values of the various members of the band are not expected to change dramatically. The effect of the R (internuclear distance) dependence is nonnegligible, however. Itikawa[36a] has made an explicit calculation for photoionization of H_2 at 21.2 eV, and found a monotonic variation in β from 1.675 for $v' = 0$ to 1.944 for $v' = 18$. Although Carlson and Jonas observed no significant variation of β between $v' = 0, 1, \ldots, 7$, other experiments[36b] do indicate such a trend. The effect is similar to that previously noted for the relative vibrational intensities which are not given by the corresponding Franck–Condon factors, but must be corrected for the R and photoelectron energy dependence of the electronic matrix elements.[37,38]

Similar observations (i.e., little dependence of β on v) have been reported on most other molecules with two significant exceptions, the $v' = 0, 1$ states in $N_2{}^+, X^2\Sigma_g{}^+ \leftarrow N_2, X^1\Sigma_g{}^+$ and $CO^+, X^2\Sigma \leftarrow CO, X^1\Sigma$ studied with 21.2 eV incident photon energy.[34] For N_2, $\beta_{v'=0}$ was found to be $+0.5$ and for $\beta_{v'=1}$, $+1.4$. For CO, Carlson and Jonas reported $\beta_{v'=0} = +0.8$, $\beta_{v'=1} = +0.4$. Duzy and Berry[39] have recently tried to explain the N_2 result in terms of the R dependence on the electronic transition moment. They obtain $\beta_{v'=0} = 1.01$ and a slight departure from this value for $v' = 1$, presented in the form of a graph of $I(v' = 1)/I(v' = 0)$ which is much too weak a variation to explain this persistent anomaly. (Morgenstern *et al.*[40] have verified the experimental observation of Carlson and Jonas.[34])

Kalman[40a] has presented a more extensive mathematical exercise of the factors influencing a dependence of β upon the vibrational state of the ion. He specifically treats H_2, N_2, and CO, and is unable to account for the aforementioned large variations in the cases of N_2 and CO.

10. Comparison of Plane Wave, Orthogonalized Plane Wave, Hydrogenic Coulomb Wave, and Coulomb Wave from a Central Field Potential

A major problem in the computation of β in the Bethe–Cooper–Zare formula is the determination of the continuum wave function which enters not only the $R_{l\pm 1}$, but also (and especially) the calculation of phase shifts.

The continuum wave function should be a superposition of plane wave plus incoming spherical wave (see Bethe and Salpeter,[4] p. 296)

$$u_\Omega(r) = \sqrt{\left(\frac{k}{(2\pi)^3}\right)^{1/2}}\,[e^{i\mathbf{k}\cdot\mathbf{r}} + v(\mathbf{r})], \tag{52}$$

where $v(\mathbf{r})$ is the "scattered part" of this continuum eigenstate.

In the photon energy domain where $h\nu \gg I_1$ (the ionization potential) but $h\nu \ll mc^2 = 0.5$ MeV, $\hbar k > p_0$, the Bohr momentum. In this case, $v(\mathbf{r}) < e^{i\mathbf{k}\cdot\mathbf{r}}$. However, $e^{i\mathbf{k}\cdot\mathbf{r}}$ is a plane wave state of the electron in the absence of any Coulomb potential. For this condition, Bethe and Salpeter[4] show that ionization from an s state ($l = 0$) leads to

$$\frac{d\sigma}{d\Omega} = \frac{2^4}{\pi n^3}\left(\frac{e^2}{mc}\right)\frac{(I_1)^{5/2}}{\nu^{7/2}}\cos^2\theta. \tag{53}$$

Hence, the angular distribution from an s state in the plane wave approximation is "pure" $\cos^2\theta$, as it is for a hydrogenic system[3] and for a more general calculation to which the Bethe–Cooper–Zare formula is applicable.

However, this is not true for initial states $l > 0$. For p-orbital ionization, Bethe and Salpeter[4] give

$$\frac{d\sigma}{d\Omega} = \frac{2^6(n^2-1)}{9\pi n^5}\left(\frac{e^2}{mc}\right)\frac{I_1^{7/2}}{\nu^{9/2}}\cos^2\theta, \tag{54}$$

and in general, for $l > 0$, the plane wave Born approximation gives

$$\frac{d\sigma}{d\Omega} \propto \frac{\nu^{-l-7/2}}{n^3}\cos^2\theta. \tag{55}$$

If one now takes into account the $v(\mathbf{r})$ term, as is done in the first-order Born approximation, one finds for p-orbital ionization

$$\frac{d\sigma}{d\Omega} = \frac{2^4}{9\pi}\frac{n^2-1}{n^5}\left(\frac{e^2}{mc}\right)\frac{I_1^{7/2}}{\nu^{9/2}}. \tag{56}$$

This equation has no θ dependence, i.e., the angular distribution is predicted to be *asymptotically* isotropic. It will be recalled that Schur's explicit calculation[9] for the angular distribution from 2p orbitals of hydrogenic systems gave

$$\frac{d\sigma}{d\Omega} \propto 1 + \frac{2(Z-s_2)^2}{\nu}\,\mathrm{Ryd}\cos^2\theta, \tag{57}$$

which, for $h\nu \gg I \approx Z^2$ Ryd, also predicts an isotropic behavior.

The plane wave result is thus clearly in error asymptotically, when compared to the corrected plane wave (first-order Born approximation) or the exact solution for atomic hydrogen.

Bethe and Salpeter[4] point out why the result for s-shell ionization still manages to give the correct result in the plane wave calculation. For $l > 0$, transitions are possible to $l' = l \pm 1$. For $l' = l - 1$, the first-order Born wave function is still sufficiently large at higher k so that the contribution from $v(\mathbf{r})$ is of the same order as $e^{i\mathbf{k}\cdot\mathbf{r}}$. For $l' = l + 1$, the first-order Born wave function is so small for higher k that the contribution from $v(\mathbf{r})$ is negligible compared to $e^{i\mathbf{k}\cdot\mathbf{r}}$. In ionization from s states, there can only be $l' = l + 1$ contributions and $e^{i\mathbf{k}\cdot\mathbf{r}}$ is dominant, but for $l > 0$, the first-order Born correction is very important for angular distributions. Hence, the plane wave calculation is probably worse for angular distributions than for partial cross sections, which are generally not reliable.

Ritchie[41] has recently reexamined the limiting behavior at high energy for β as determined by (a) the plane wave length form (PWL), (b) the plane wave velocity from (PWV), and (c) the Coulomb length form (CL). (See, e.g., Manson[19] for a discussion of the dipole length, dipole velocity, and dipole acceleration forms.) Ritchie finds that PWL is in disagreement with PWV and CL at all energies, except for photoejection from the 1s orbital. For $l > 0$, use of the first-order Born correction to PWV makes it agree with CL.

Orthogonalization of the plane wave to other bound orbitals of the system gives a result which will agree with CL and corrected PWV only for particular values of the screening parameters of these orbitals, demonstrating that the orthogonalized plane wave calculation *cannot* lead to correct asymptotic forms except by accident. This is an important result, since a number of calculations[42–45] have recently been undertaken using plane wave or orthogonalized plane wave continuum wave functions. The justification for these calculations has been that the plane wave approach, while admittedly inaccurate in the low-energy region where the details of the atomic or molecular potential are most pronounced, should become appropriate at higher photoelectron energy. It is now seen that the asymptotic values of β computed with plane wave or orthogonalized plane wave continuum functions cannot be correct except by accident.

Paradoxically, the first-order Born and hydrogenic approximations also seem to give incorrect asymptotic behavior for $l > 0$. The hydrogenic functions will not exhibit a non-Coulomb phase shift which probably is responsible for their incorrect asymptotic behavior compared to continuum waves derived from a more realistic central field potential. The Hartree–Fock and Hartree–Slater calculations of Manson and associates (see Section B below) on photoelectron angular distributions from a large number of atoms strongly suggest that $\beta_{\varepsilon\to\infty} \sim 1.5$ for ionization from p orbitals and for d orbitals, $1 \leq \beta \leq 1.5$. Ritchie[46] has derived an asymptotic lim $\beta_{\varepsilon\to\infty} = 1.4545$ for ejection from p orbitals and $\beta_{\varepsilon\to\infty} = 1.18182$ for ejection from d orbitals. The experimental results of Krause[47] which cover a range from 20 to 80 Ryd of photoelectron energy in ionization of Kr 3p tend to confirm this asymptotic behavior.

B. Calculations of Angular Distributions

1. Atoms Using Central Field Approximations

In the Bethe–Cooper–Zare expression for β, there exist radial integrals $R_{l\pm 1}$ and a difference in the cosines of phase shifts for the $l + 1$ and $l - 1$ waves. To compute these quantities, one requires radial wave functions for bound and continuum states. For bound states of atoms, accurate Hartree–Fock (HF) solutions of both nonrelativistic (see Froese-Fischer[48] and Roetti and Clementi[49]) and relativistic (see Desclaux[50]) types are available. The major problem is the evaluation of the corresponding continuum wave function.

It is desirable to use Hartree–Fock functions for the neutral atom and the residual ion core, in order to include the effects of core relaxation. Such a scheme yields initial and final states in different basis sets, introducing complexity. In addition, wave functions of hole states of ions are not generally available. Hence, it is convenient to use the same discrete orbitals in both initial and final states. Hartree–Fock discrete orbitals are obtained by setting up an antisymmetric product of one-electron functions

$$\psi \sim \sum_{n,l} r^{-1} P_{nl}(r) Y_{lm}(\theta, \varphi),$$

constructing the energy functional $\langle\psi|H|\psi\rangle$, and applying the variation principle. This results in a set of self-consistent coupled integrodifferential equations for the $P_{nl}(r)$ which can be solved, yielding the HF wave functions for the given state.

The continuum wave function is more difficult since the Hartree–Fock problem is not defined for wave functions containing continuum orbitals. This results because the HF method solves for each orbital in the field generated by the charge distribution of the other orbitals. The charge distribution for a continuum orbital is not defined since continuum orbitals are not normalized in the same way as bound orbitals.

The methodology in this case is to perform a HF calculation for the residual ion core (recall the previously mentioned problem with hole states), freeze the core orbitals and carry out the calculation of the final state (ion core plus photoelectron), solving for the radial part of the continuum orbital $P_{\varepsilon l}(r)$. This procedure yields a single integrodifferential equation for $P_{\varepsilon l}(r)$ known as the continuum HF equation. The effects of core relaxation are embodied in an overlap integral S:

$$S = \prod_{\substack{\text{passive} \\ \text{electrons}}} \left| \int_0^\infty P_{nl}^{\mathrm{i}}(r) P_{nl}^{\mathrm{f}}(r)\, dr \right|^2, \tag{58}$$

where i and f refer to orbitals from initial and final states and the dipole matrix elements for the active orbital are

$$R_{l\pm 1} = \int_0^\infty P_{nl}^{\mathrm{i}}(r) r P_{\varepsilon,\, l\pm 1}^{\mathrm{f}}(r)\, dr.$$

If one uses the same discrete orbitals in initial and final states, $S = 1$ and $R_{l\pm 1}$ will embody slightly different functions $P_{\varepsilon,\, l\pm 1}(r)$.

The HF wave function is the most accurate independent particle wave function possible, since it is obtained by employing the variation principle.

By its construction, it implicitly includes exchange effects which can be important for the evaluation of the $R_{l\pm 1}$ and phase shifts. However, solution of the HF equations is rather tedious, and often a simpler type of wave function has been used in the calculation of β.

This approach, referred to as Hartree–Slater or as Hermann–Skillman (HS in either case) is based on a central field *approximation* to the exact Hamiltonian, i.e.,

$$H_0 = \sum_j \left[\frac{p_j^2}{2m} + u(r_j) \right] \tag{59}$$

for both the initial and final states of the atom. The central potential seen by each electron $u(r_j)$ is a function of *scalar* r_j only. The radial parts of the one-electron functions are solutions to the one-body Schrödinger equation

$$\left[\frac{d^2}{dr^2} - \frac{l(l+1)}{r^2} - u(r) + E \right] P(r) = 0. \tag{60}$$

In using these central-field wave functions, only the one electron is permitted to change quantum numbers in the photoionization process, or else the matrix element vanishes. Since the initial and final states are solutions to the Schrödinger equation in the same potential $u(r)$, the equivalent S integral for passive electrons is unity, and hence core-relaxation effects are eliminated in the treatment, as with the corresponding HF procedure. The HS wave functions generated by the above prescription do not implicitly include exchange, which makes them less complete than HF functions. Nevertheless, it can be shown (see, e.g., Manson[19]) that the length and velocity forms of the dipole matrix element are identical in this framework, although they are not in the HF framework. Hence, although identical results from length and velocity formulations of the dipole matrix element represent a necessary condition for exact wave functions, it is not a sufficient condition.

One can approximate the effects of exchange in the central potential scheme by using the Slater scheme[51] which amounts to using the form of the exchange potential to be that of a free electron gas. Thus,

$$V^{\mathrm{ex}}(r) = -6\left[\frac{3}{8\pi} |\rho(r)| \right]^{1/3}, \tag{61}$$

where ρ is the total charge density. If this approximate exchange potential is explicitly introduced into the central field problem, one obtains the Hartree–Slater (HS) central-field wave functions and potential.

Manson and associates[8,27,52–56] have made numerous calculations of $\beta(\varepsilon)$ for many atoms from various subshells, using HS and (in fewer instances) HF wave functions. Several generalizations are possible.

(a) For ionization from 2p orbitals (which have no node, and hence no Cooper minimum), the HS and HF wave functions give almost identical results for $\beta(\varepsilon)$. (See Fig. 4.)

(b) There is a similarity in the behavior of $\beta(\varepsilon)$ from 2p subshells for all elements, displaying rapid change in the near threshold region, then a rather slow, generally monotonic rise to an asymptotic value for $\beta_{\varepsilon \to \infty} \sim 1.5$. The asymptotic behavior has been discussed previously. The threshold variation can be understood by the following simple analysis. The Bethe–Cooper–Zare formula for ionization from 2p orbitals can be written

$$\beta(\varepsilon) = \frac{1 - 2\rho(\varepsilon)\cos\Delta(\varepsilon)}{1 + \frac{1}{2}[\rho(\varepsilon)]^2}, \tag{62}$$

where

$$\rho(\varepsilon) = R_{l-1}(\varepsilon)/R_{l+1}(\varepsilon). \tag{63}$$

The contribution of $R_{l-1}(\varepsilon)$ [p → s] is generally small compared to $R_{l+1}(\varepsilon)$ [p → d] for photoionization, and hence $\beta(\varepsilon)$ can be approximated as

$$\beta(\varepsilon) = 1 - 2\rho(\varepsilon)\cos\Delta(\varepsilon). \tag{64}$$

The phase difference $\Delta(\varepsilon)$ can be written as the sum of a Coulomb phase

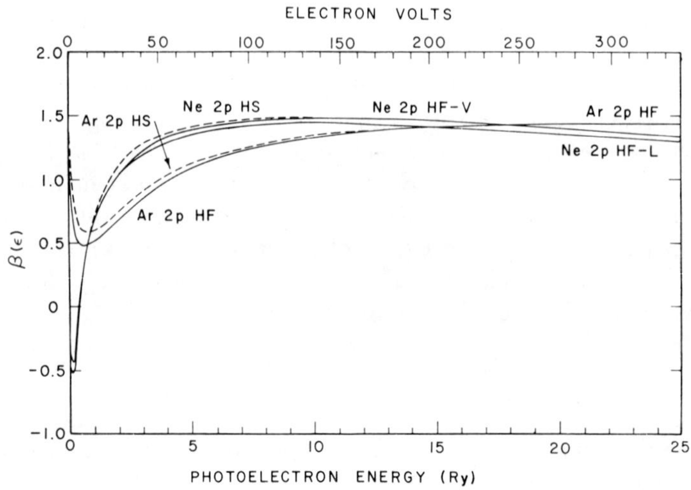

Fig. 4 Calculated variation of asymmetry parameter β with photoelectron energy for ionization from 2p orbitals of neon and argon, in the Hartree–Fock length (HF–L), Hartree–Fock velocity (HF–V), and Hartree–Slater (HS) approximations. The three calculations give nearly identical results for ionization from 2p orbitals. [From S. T. Manson, *J. Electr. Spectr.* **1**, 413 (1973).]

shift difference $\sigma_{l+1}(\varepsilon) - \sigma_{l-1}(\varepsilon)$ and a non-Coulomb phase shift difference $\eta_{l+1}(\varepsilon) - \eta_{l-1}(\varepsilon)$. The non-Coulomb phase shift difference generally varies slowly over a small energy interval, whereas the Coulomb phase shift difference can vary rapidly at low ε. Explicitly, the Coulomb phase shift difference may be written[55]

$$\sigma_{l+1}(\varepsilon) - \sigma_{l-1}(\varepsilon) = \tan^{-1}\left[\frac{-(2l+1)}{\varepsilon^{1/2}l(l+1) - \varepsilon^{-1/2}}\right]. \tag{65}$$

For $l = 1$ (ε in Ryd), the phase shift difference varies from $-\pi$ at $\varepsilon = 0$ to $-\pi/2$ at $\varepsilon = 0.5$ Ryd.

(c) For ionization from 3p (and rather generally, np, where $n \geq 3$) orbitals, there will be a Cooper minimum in this partial cross section, where $R_{l+1} \to 0$. In the energy region where this occurs, we can rewrite the Bethe–Cooper–Zare formula, defining $\delta(\varepsilon) = R_{\mathrm{d}}(\varepsilon)/R_{\mathrm{s}}(\varepsilon)$:

$$\beta_{\mathrm{p}}(\varepsilon) = \frac{2\delta(\varepsilon)\{\delta(\varepsilon) - 2\cos\Delta(\varepsilon)\}}{1 + 2(\delta(\varepsilon))^2}. \tag{66}$$

For $\delta(\varepsilon) \ll 1$, this becomes

$$\beta_{\mathrm{p}}(\varepsilon) = -4\delta(\varepsilon)\cos\Delta(\varepsilon). \tag{67}$$

It can readily be seen that at the Cooper minimum, β goes to zero; furthermore, in the vicinity of the Cooper minimum, $\beta(\varepsilon)$ will vary rapidly. A characteristic difference in photoionization calculations when using HS and HF wave functions is that HS calculations result in Cooper minima (and related maxima) at smaller values of ε than HF functions. As a consequence, the zero value of β will occur at smaller values of ε with HS than with HF functions. Apart from this systematic difference between HS and HF calculations, there is again a similarity in the energy dependence of β for ionization from np orbitals, displaying a rapid change near the threshold (see above for behavior of 2p orbitals), a decline to $\beta = 0$ at the Cooper minimum, then a recovery toward higher energy, asymptotically approaching $\beta \sim 1.5$. A typical variation of $\beta(\varepsilon)$ for a 3p orbital is shown in Fig. 5.

(d) Ionization from d orbitals has not been studied as extensively as from p orbitals, but similar generalizations can be made. In the near threshold region, making the approximation $R_{l-1}/R_{l+1} = \rho \ll 1$, one obtains

$$\beta_{\mathrm{d}}(\varepsilon) = \frac{4 - 12\rho(\varepsilon)\cos\Delta(\varepsilon)}{15}. \tag{68}$$

Kennedy and Manson[54] have calculated β for Kr 3d and Xe 3d, Xe 4d and Manson[27] has calculated β for Hg 5d. The rapid change in β near threshold

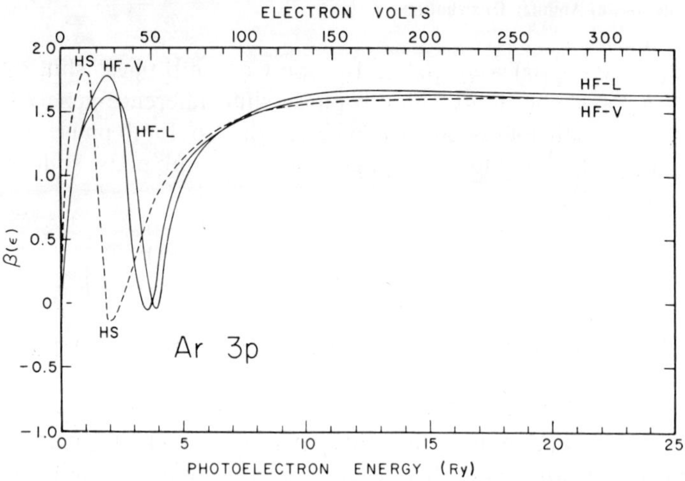

Fig. 5 Calculated variation of asymmetry parameter β with photoelectron energy for ionization from the 3p orbital of argon in the Hartree–Fock length (HF–L), Hartree–Fock velocity (HF–V), and Hartree–Slater (HS) approximations. The HS differs significantly from the HF calculations in the energy of the Cooper minimum. [From S. T. Manson, *J. Electr. Spectr.* **1**, 413 (1973).]

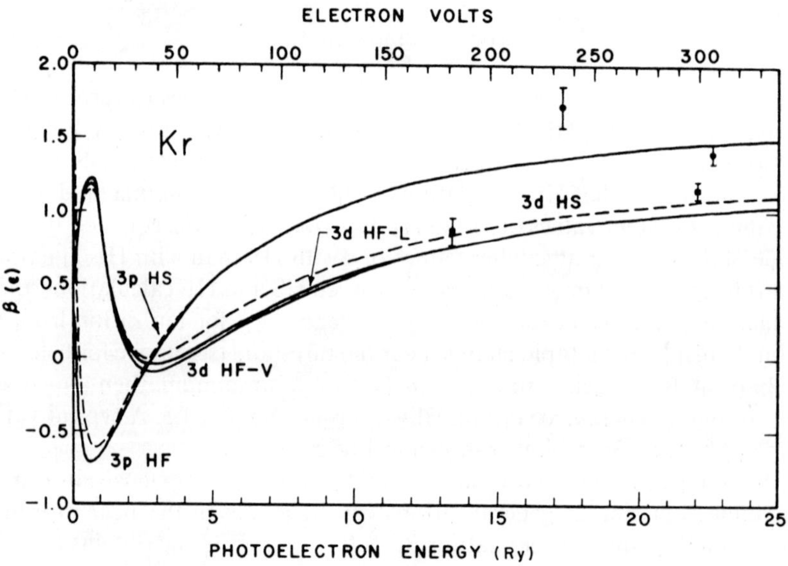

Fig. 6 Calculated variation of asymmetry parameter β with photoelectron energy for ionization from 3p and 3d orbitals of krypton in the Hartree–Fock length (HF–L), Hartree–Fock velocity (HF–V), and Hartree–Slater (HS) approximations. The 3d wave function has no node, and hence agreement between HS and HF is seen to be good. Note that the asymptotic behavior of β (3d) appears to be $1.0 < \beta < 1.5$. Experimental values of Krause[47] are shown with error bars. [From D. J. Kennedy and S. T. Manson, *Phys. Rev. A* **5**, 227 (1972).]

is due principally to the rapid variation in the Coulomb phase shift, as for np orbitals. The change in β for the d orbitals is somewhat more rapid than for p orbitals, a behavior which is attributable to the fact that the ratio of dipole matrix elements varies somewhat more rapidly in d-orbital ionization.

The 3d orbitals have no node and hence no Cooper minimum. As with 2p orbitals, agreement between HS and HF calculations is very good, but the high-energy asymptotic value of β approaches a value slightly greater than 1, but apparently less than 1.5 (see Fig. 6).

For ionization from nd orbitals with $n \geq 4$, there should be a Cooper minimum. In the vicinity of the Cooper minimum, the Bethe–Cooper–Zare formula may be written

$$\beta_{\mathrm{d}} = \frac{1 - 18\delta(\varepsilon)\cos\Delta(\varepsilon)}{10}. \tag{69}$$

where as before, $\delta(\varepsilon) = R_{l+1}(\varepsilon)/R_{l-1}(\varepsilon)$. Unlike the corresponding p-orbital ionization, β_{d} is not identically zero at the Cooper minimum. In the calculation of Kennedy and Manson[54] (see Fig. 7), β_{d} for Xe 4d, in fact, has a pronounced negative value in the vicinity of the Cooper minimum. Similar

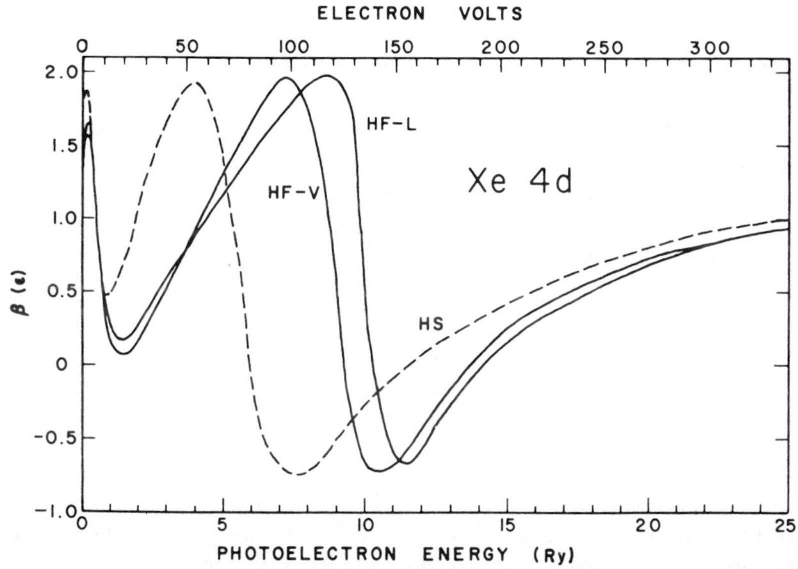

Fig. 7 Calculated variation of asymmetry parameter β with photoelectron energy for ionization from the 4d orbital of xenon in the Hartree–Fock length (HF–L), Hartree–Fock velocity (HF–V), and Hartree–Slater (HS) approximations. The 4d wave function has a node, and hence HS and HF differ significantly in the energy of the Cooper minimum. [From D. J. Kennedy and S. T. Manson, *Phys. Rev. A* **5**, 227 (1972).]

values are obtained in both HS and HF approximations, although the minima are displaced by 3–4 Ryd. The asymptotic value of $\beta(\varepsilon)$ again seems to be somewhat greater than 1 but less than 1.5.

(e) Ionization from the 4f orbital of mercury has been investigated by Shyu and Manson.[56] This is again a case where the bound orbital has no nodes, and hence there should not exist a Cooper minimum. The HS and HF calculations should give similar results and apparently do. The Bethe–Cooper–Zare formula becomes

$$\beta_{\mathrm{f}} = \frac{6R_{\mathrm{d}}^{2} + 20R_{\mathrm{g}}^{2} - 72R_{\mathrm{d}}R_{\mathrm{g}}\cos\Delta(\varepsilon)}{7[3R_{\mathrm{d}}^{2} + 4R_{\mathrm{g}}^{2}]}. \tag{70}$$

This time, the g wave is severely retarded near threshold because of the large centrifugal barrier ($l(l+1)/r^2$) in the potential. Hence, $\delta(\varepsilon) \equiv R_{\mathrm{g}}/R_{\mathrm{d}} \ll 1$ near threshold and

$$\beta_{\mathrm{f}} = \frac{6 + 20\delta(\varepsilon)^2 - 72\delta(\varepsilon)\cos\Delta(\varepsilon)}{7[3 + 4\delta(\varepsilon)^2]}, \tag{71}$$

which, for $\delta(\varepsilon) \ll 1$, reduces to

$$\beta_{\mathrm{f}} = \tfrac{2}{7}[1 - 12\delta(\varepsilon)\cos\Delta(\varepsilon)]. \tag{72}$$

At threshold, $\delta(\varepsilon)$ should be smallest since the g wave has no kinetic energy with which to overcome the centrifugal barrier. Hence, one would predict $\beta_{\mathrm{f}} = \frac{2}{7}$ at threshold, which is just about the value given by Shyu and Manson (Fig. 8).

Just above threshold, $\delta(\varepsilon)$ increases and $\cos\Delta(\varepsilon)$ is negative; as a result, Eq. (72) predicts an increase in β_{f}. In an intermediate energy range, R_{d} and R_{g} are comparable. When they are equal, we obtain from Eq. (70)

$$\beta = \frac{26 - 72\cos\Delta(\varepsilon)}{49}. \tag{73}$$

If the matrix elements R_{d} and R_{g} are slowly varying and $\cos\Delta(\varepsilon)$ can achieve its maximum negative value (-1) in this region, β_{f} should go through a maximum. Shyu and Manson[56] calculate that $\cos\Delta(\varepsilon) = -1$ at $\varepsilon \cong 3$ Ryd, which is approximately where β_{f} does go through a maximum. If R_{d} and R_{g} were identical at this energy, β_{f} would be 2. The calculated value is ~1.7, indicating a slight departure from this equality of R_{d} and R_{g}.

At higher energy, β_{f} declines and slowly approaches some asymptotic value as $R_{\mathrm{d}}/R_{\mathrm{g}}$ becomes $\ll 1$.

(f) In many-body perturbation theory, Amusia *et al.*[57] have included many-electron correlations in calculation of photoionization cross sections.

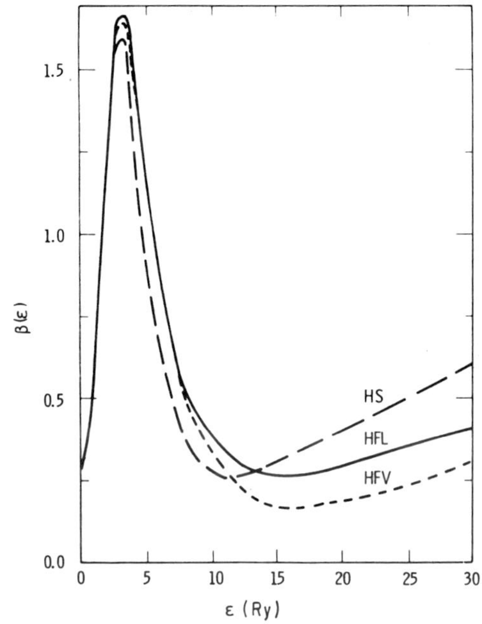

Fig. 8 Calculated variation of asymmetry parameter β with photoelectron energy for ionization from the 4f orbital of mercury in the HFL, HFV, and HS approximations. The 4f wave function is nodeless, and hence the three calculations are in reasonably good agreement. Other features of this β variation are discussed in the text. [From J. S. Shyu and S. T. Manson, *Phys. Rev. A* **11**, 166 (1975).]

In doing so, the dipole matrix elements become complex:

$$R_{nl,\,\varepsilon l\pm 1} = R'_{l\pm 1} + iR''_{l\pm 1}, \tag{74}$$

which changes the Bethe–Cooper–Zare formula slightly.[57a]

$$\begin{aligned}\beta(\varepsilon) = \{&l(l-1)|R_{l-1}|^2 + (l+1)(l+2)|R_{l+1}|^2 - 6l(l+1)[(R'_{l-1}R'_{l+1} \\ &+ R''_{l-1}R''_{l+1})\cos(\delta_{l+1} - \delta_{l-1}) + (R'_{l-1}R''_{l+1} \\ &- R'_{l+1}R''_{l-1})\sin(\delta_{l+1} - \delta_{l-1})]\}/(2l+1)[l|R_{l-1}|^2 + (l+1)|R_{l+1}|^2].\end{aligned} \tag{75}$$

The dipole matrix elements have been computed by the random phase approximation with exchange (RPAE). Both length and velocity forms of the operator have been used. With the inclusion of correlations, these two forms of calculating β gave identical results, whereas they do not in the Hartree–Fock scheme. Amusia *et al.* give β variations with photoelectron energy from the outer p electrons for Ne, Ar, Kr, and Xe from threshold to ~88 eV. Of note at this point is the asymptotic behavior, which seems to be approaching

$\beta \sim 1.5$ for Ne and Ar, with Kr and Xe still in the Cooper minimum region at the upper limit of the calculations. The major disagreement with Hartree–Fock calculations is in the Cooper minimum region. We postpone discussion of this point to the experimental section.

Another many-body formalism, the *R*-matrix method, has recently been applied[57b] to the computation of β for outer p-shell ionization of neon and argon. The results are very similar to those obtained by the RPAE method. In addition, Taylor[57b] has applied the *R*-matrix calculational scheme to the resonant regions of these atoms, corresponding to the excitation of the outermost s electron to Rydberg levels. This calculation provides the three missing parameters in the formulation of Kabachnik and Sazhina[22a] and this enables him to predict the β variation through those resonances.

2. Molecules

On the basis of arguments presented in Section A.10 of this chapter, we shall not present details of the plane wave and orthogonalized plane wave calculations used by several groups recently to compute β for molecular photoionization. Although such an approach has become popular because of its relative simplicity, it has been demonstrated to be fundamentally in error at all photoelectron energies except by accident.

We have already noted (Section A.8, this chapter) that molecular orbitals are not eigenfunctions of l^2; also, that for diatomic molecules the bound orbital is quantized on the molecular axis, while the continuum function is referred to the laboratory frame. The bound state molecular orbital is usually available in the multicenter "LCAO" form. Such orbitals can be transformed into a single-center expansion, where the natural origin is the center of *mass*, since this is the only point in the molecule which remains fixed in space as the molecule rotates. The continuum function, however, is determined by the potential of the residual ion; for heteronuclear molecules this is most closely approximated by an energy eigenfunction for a central potential when the origin is at the center of *charge* of the ion. Thus far, calculational attempts for molecules have apparently not advanced to the point where this is a problem.

Ritchie has made a number of model calculations, primarily applicable to molecules having a center of symmetry, or more particularly, to homonuclear diatomic molecules. He[46] expands the molecular bound state in spherical functions about the molecular center, e.g.,

$$\psi_g \sim \varphi_0(\zeta_0 r) + \lambda_2 \varphi_2(\zeta_2 r) P_2(\cos\theta) + \cdots \tag{76}$$

$$\psi_u \sim \varphi_1(\zeta_1 r) P_1(\cos\theta) + \lambda_3 \varphi_3(\zeta_3 r) P_3(\cos\theta) + \cdots, \tag{77}$$

where g, u denote gerade and ungerade symmetry, $\varphi_{l_0}(\zeta_{l_0} r)$ is the l_0th radial component, the λs are expansion coefficients, and ζ is an effective charge parameter, taking into account screening.

In one such study,[46] he examines the asymptotic behavior of β_g and β_u at sufficiently high energy, which he argues occurs when $(v_{\text{ejected electron}}/v_{\text{orbital}}) \geq 2$. Under this condition, the potential is dominated by the spherically symmetric monopole term, which simplifies the construction of the continuum wave. In this way he obtains Coulomb waves calculated in a potential of strength Z, where

$$k^2 < Z^2, \qquad \zeta \leq Z,$$

and k is the wave vector of the departing electron. This assumes that the ejected electron tends to move in the Coulomb potential of strength equal to the full nuclear charge Z in the region of space in which most of the photoelectric conversion occurs. He then calculates the necessary radial matrix elements connecting bound and continuum states and then the rotationally averaged βs. In the large k limit, he finds ionization from an ungerade orbital to be dominated by ionization from a p-type orbital (the first term in the σ_u expansion) and he analytically deduces

$$\lim_{k\to\infty} \beta_u = 1.4545.$$

Correspondingly, under most conditions, the expansion of σ_g is dominated by the leading s-type orbital and $\beta_g \to 2$ as $k \to \infty$. For some values of the charge parameter ζ, however, the s $\to$ p matrix element in the expression for β_g becomes very small or vanishes due to nodal structure, and the dominant terms derive from the d-like nature of the bound state. This is particularly true when $\zeta = Z/2$. In the special case when $\zeta = 1$ and $Z = 2$, he analytically deduces $\text{limit}_{k\to\infty}\ \beta_g = 1.18182$, which appears similar to the asymptotic behavior of ionization from d orbitals calculated by Manson and associates (see Section B.1(d) of this chapter). In summary, ionization from a σ_g-type orbital is predicted to generally have $\beta \to 2$ as $k \to \infty$, that from a σ_u-type orbital to have $\beta \to 1.4545$ as $k \to \infty$, and for some values of the charge parameter (which can be transposed to some values of the internuclear distance R of the molecule) β can be expected to vary rapidly. This latter behavior can introduce a dependence of β on the vibrational members of the same photoelectron transition if the rapid variation occurs in the vicinity of $R = R_e$, the equilibrium internuclear distance of the molecule.

Ritchie has also made a model calculation[58] of β for similar systems in the low-energy region (1–50 eV). This model consists of a two-electron molecule, one of which is a "core" electron, the other the photoelectron, both from the same type of molecular orbital.

This prototype calculation neglects (1) screening of other electrons and (2) orthogonalization of the continuum orbital to all bound orbitals of the same symmetry. Ritchie argues that this neglect is not too serious. He considers three cases: $1s\sigma_g$, $2p\sigma_g$, and $2p\sigma_u$. He constructs an antisymmetrized product of the two electrons, and to this extent his model calculation includes "exchange."

On the assumption that a $1s\sigma_g$ orbital has a characteristic behavior to photoejection which can be parameterized for internuclear distance and photoelectron energy, he calculates $\beta(1s\sigma_g)$ at $R = 1.4$ au (atomic units) (appropriate for H_2) and a photoelectron energy typical for 21.2 eV photons. He obtains $\beta = 1.97$ with inclusion of the f-wave, $\beta = 1.93$ for the p-wave contribution only.

Carlson and Jonas[34] have reported $\beta = 1.75$ from an experimental investigation, but Niehaus and Ruf[35] obtained $\beta = 1.93 \pm 0.03$. It will be recalled that Dill[32] deduced $\beta = 1.71$ from a multichannel quantum defect analysis [see Section B.8(a)]. Ritchie thinks this may be too low because he did not include f waves, which provide constructive interference in the numerator of β. For ionization from the $2p\sigma_g$ orbital, Ritchie calculates a rather strong variation of β with R, in contrast to $1s\sigma_g$. This behavior of $2p\sigma_g$ has also been noted by Duzy and Berry[39] in a calculation of photoionization in N_2.

At $R = 2.0$ au (close to R_e for N_2) and a photoelectron energy close to that expected with 21.2 eV photons on N_2, Ritchie[58] calculates $\beta = 0.448$, very close to the value of the $v' = 0$ transition observed by Carlson and Jonas[34] ($\beta = 0.5 \pm 0.1$). It will be recalled that Duzy and Berry[39] obtained $\beta = 1.01$ in their calculation of this transition.

For ionization from $2p\sigma_u$ with R and ε parameters appropriate to N_2, Ritchie calculates $\beta = 1.31$, whereas Carlson and Jonas report $\beta = 1.25$ for formation of $^2\Sigma_u$ of $N_2{}^+$, but this state results from $(2s\sigma_u)^{-1}$, not $(2p\sigma_u)^{-1}$. Hence, Ritchie would like to conclude that it is the molecular symmetry of the molecular orbital, not its LCAO composition, that primarily determines β. If his surmise is correct, it would enable a simplification to be made for the energy dependence of β in molecular photoionization similar to that described for atoms in Section B.1(b)–(e).

C. Experimental Considerations

We begin with the symmetry-based equation

$$\frac{d\sigma}{d\Omega} = \frac{\sigma}{4\pi}[1 + \beta P_2(\cos\theta)] \tag{78}$$

which assumes linearly polarized light with θ measured from the polarization direction.

With partially polarized light[59] we can immediately write

$$\frac{d\sigma}{d\Omega} = \frac{I_x}{I_0}\frac{\sigma}{4\pi}[1 + \beta P_2(\cos\theta_x)] + \frac{I_y}{I_0}\cdot\frac{\sigma}{4\pi}[1 + \beta P_2(\cos\theta_y)], \tag{79}$$

where (I_x/I_0) and (I_y/I_0) describe the fraction of light intensity polarized along the x and y axes and $I_x + I_y = I_0$. We define the polarization of the incident light beam,

$$p \equiv \frac{I_y - I_x}{I_y + I_x}.$$

Then, using the direction cosine relation

$$\cos^2\theta_x + \cos^2\theta_y + \cos^2\theta_z = 1 \tag{80}$$

(with z the photon propagation direction) and some algebra

$$\begin{aligned}
&\frac{I_x}{I_0}\left[1 + \beta\left(\frac{3}{2}\cos^2\theta_x - \frac{1}{2}\right)\right] + \frac{I_y}{I_0}\left[1 + \beta\left(\frac{3}{2}\cos^2\theta_y - \frac{1}{2}\right)\right]\\
&\quad = 1 - \frac{\beta}{2} + \frac{3}{2}\frac{\beta}{I_0}(I_x\cos^2\theta_x + I_y\cos^2\theta_y)\\
&\quad = 1 + \frac{\beta}{4} - \frac{3\beta}{4}\left(\frac{I_x + I_y}{I_0}\right)(\cos^2\theta_x + \cos^2\theta_y + \cos^2\theta_z)\\
&\qquad + \frac{3}{2}\frac{\beta}{I_0}(I_x\cos^2\theta_x + I_y\cos^2\theta_y)\\
&\quad = 1 + \frac{\beta}{4} - \frac{3\beta}{4}\cos^2\theta_z + \frac{3\beta}{4I_0}(I_y\cos^2\theta_y - I_x\cos^2\theta_y - I_y\cos^2\theta_x + I_x\cos^2\theta_x)\\
&\quad = 1 - \frac{\beta}{2}(P_2(\cos\theta_z)) + \frac{3\beta}{4I_0}(I_y - I_x)(\cos^2\theta_y - \cos^2\theta_x)\\
&\quad = 1 - \frac{\beta}{2}\left[(P_2(\cos\theta_z)) - \frac{3}{2}p(\cos^2\theta_y - \cos^2\theta_x)\right].
\end{aligned} \tag{81}$$

We obtain

$$\frac{d\sigma}{d\Omega} = \frac{\sigma}{4\pi}\left\{1 - \frac{\beta}{2}\left[P_2(\cos\theta_z) - \frac{3}{2}p(\cos^2\theta_y - \cos^2\theta_x)\right]\right\}. \tag{82}$$

For unpolarized light, $p = 0$ and

$$\frac{d\sigma}{d\Omega} = \frac{\sigma}{4\pi}\left[1 - \frac{\beta}{2}P_2(\cos\theta_z)\right]. \tag{83}$$

Thus, the differential cross section for unpolarized light has a form similar to that for polarized light, but the angle θ_z is measured from the propagation direction, rather than the polarization direction.

1. Experiments with Polarized Radiation

If one has a source of linearly polarized light and can rotate the plane of polarization conveniently, the experimental arrangement simplifies considerably since light source and detector can remain fixed. This minimizes the effect of geometry changes and stray fields. Such experiments[16] have been performed in the visible spectrum, using an argon-ion laser operating at 4880 or 5145 Å and Brewster-angle polarizing optics to determine β in the photodetachment of O^- and C^-. Earlier, this type of experiment had been done in the ultraviolet at $\sim$2400 Å, using a pile of quartz plates as a polarizer in the photoionization of potassium.[60] Recently, Hancock and Samson[61] have used a three-mirror polarizer at 584 Å to obtain β values. In those instances when one has linearly polarized light but cannot rotate the plane of polarization, the greater sensitivity to β in the basic equation can sometimes make the experiment possible or more accurate. The bremsstrahlung from x-ray tubes is partially polarized. This property was exploited by Bubb[62] and Kirchner[63] in determining angular distributions in the x-ray region of such gases as N_2, CO_2, and Ar using the Wilson cloud chamber to observe the electron paths. Pietenpol[64] used an angle-defining orifice and Geiger counter for such measurements. No energy analysis accompanied these experiments, which were probably emphasizing K-electron ejection since they always observed a $\cos^2 \theta$ distribution about the plane of polarization.

Recently, a particular feature of synchrotron radiation has been utilized to advantage for β measurements. In the orbital plane of the synchrotron, the component of radiation polarized in the vertical direction is zero. In an initial experiment, Mitchell and Codling[65] used a grazing incidence monochromator to disperse the radiation in the vertical plane (thus preserving the polarization) and a spherical retarding field photoelectron detector fixed at right angles to the photon beam. The normalized photoelectron signal as a function of photon energy was compared with the photoionization cross section (equivalent to collecting all photoelectron angles) and thus a β variation with $h\nu$ was inferred. This was put on an absolute scale by fixing $\beta \sim 0.85$ at $h\nu = 584$ Å for argon, a value obtained in previous experiments and Hartree–Fock calculations. Neon,[66] krypton, and xenon[67] were subsequently studied in the same manner. In all four examples, very good agreement was obtained with the Hartree–Fock calculations in a region of photoelectron energy $\sim$0–20 eV above threshold. The agreement with Hartree–Slater calculations was not as good; the deviation between the predictions of many-body perturbation theory[57] and the Hartree–Fock results occur in the vicinity of Cooper minima, which were just beyond the range of these experiments.

More recently, Houlgate *et al.*[68,68a] have dispersed radiation from the NINA synchrotron in the horizontal plane, with both a normal incidence[68] and grazing incidence[68a] spectrometer. This dispersion does not preserve polarization as well—they deduced 82% polarization at normal incidence, and 93% polarization at grazing incidence. The photoelectron detector was a 127° electrostatic analyzer, which accepts a line source (see Chapter VIII). The horizontal entrance slit of the analyzer was aligned along the photon propagation direction and the analyzer rotated in a plane perpendicular to the photon propagation direction, the latter being the axis of rotation. In this way, the dependence upon other sources of data for determination of β was eliminated. The variation of β with photoelectron energy was studied in the range 0–80 eV above threshold for ejection of 3p electrons from argon. These data now extend well beyond the region of the Cooper minimum, where calculations differ. The experimental results (Fig. 9 is an earlier version extending to 35 eV; ref. 68a provides a figure out to 80 eV) display a distinct

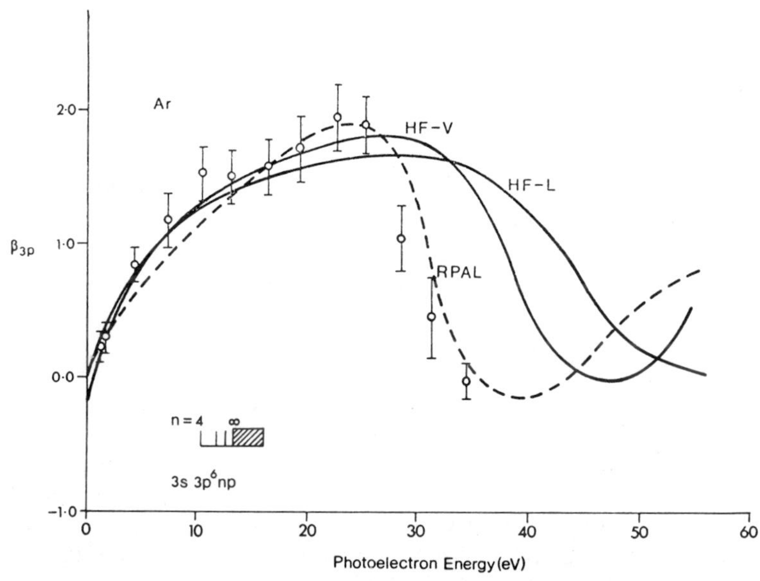

Fig. 9 Variation of asymmetry parameter β with photoelectron energy for ionization from the 3p orbital of argon. Ī Experimental data of Houlgate *et al.*, *J. Phys. B* **7**, L470 (1974). --- Many-body theory with random phase approximation and exchange (RPAE) by M. Ya. Amusia *et al.*, *Phys. Lett.* **40A** 15 (1972). ——— Hartree–Fock length and velocity calculations of D. J. Kennedy and S. T. Manson, *Phys. Rev. A* **5**, 227 (1972). The experimental data support the many-body theory. [From Houlgate *et al.* cited here.]

preference for the many-body RPAE method, rather than the Hartree–Fock calculations.

This experimental technique has now been applied to the determination of $\beta(E)$ for $(2p)^{-1}$ ionization from neon,[68b] where HF and RPAE agree (it will be recalled that ionization from 2p does not result in a Cooper minimum) and the experimental results are comfortingly close to the calculated ones. A subsequent experiment with xenon[68c] has focused on $\beta(E)$ for $(4d)^{-1}$ and $(5p)^{-1}$ ionization. While the $(4d)^{-1}$ experimental results were in reasonable agreement with HF calculations, the $(5p)^{-1}$ values of $\beta(E)$ were not; there was also discrepancy in this latter case with RPAE calculations. The authors concluded that the difference is probably attributable to the effect of intershell correlation with 4d on photoemission from the 5p subshell. The corresponding study on the $(4p)^{-1}$ ionization in krypton has also been performed,[68d] but both experimental data and RPAE calculations extend barely beyond the Cooper minimum and hence do not permit of any conclusions regarding intershell correlation between 3d and 4p subshells.

A still simpler device for utilization of synchrotron radiation in favorable cases has recently been described by Watson and Stewart.[69] They allow the undispersed synchrotron radiation to enter the ionization region and use a simple retarding grid analyzer to detect the photoelectrons. The analyzer can sweep the entire 360° range in the plane perpendicular to the photon direction. Maximum sensitivity to β is achieved by measuring at $\theta = 0°$ and $\theta = 90°$. The technique was applied to argon and krypton, where photoejection is largely limited to the outer p shell, so that photoelectron energy analysis (without photon monochromatization) suffices to specify the process. Their results in the range 0–20 eV electron energy above threshold are in good agreement with previous experiments and calculations, but do not extend to the more controversial Cooper minimum region.

2. Experiments with Unpolarized Radiation

In experiments with unpolarized radiation, the reference axis is the photon propagation direction (Eq. (83)). If the radiation is not completely unpolarized, however, Eq. (82) is the proper one. Most practitioners, to date, who work with assumedly unpolarized radiation do not test for possible partial polarization. The data are analyzed in the form $I(\theta) = A + B \sin^2 \theta_z$, and by comparison with Eq. (83), yield β in the form

$$\beta = \frac{4B}{3A + 2B}. \tag{84}$$

If the radiation happens to be partially polarized, analysis of the experimental data in the form $A + B\sin^2\theta_z$ and comparison with Eq. (82) would yield

$$\beta = \frac{4B}{3A + 2B - 3pB(\cos^2\theta_y - \cos^2\theta_x)}. \tag{85}$$

For $B = 0 = \beta$, there is no error incurred, but for positive or negative B there is a possibility for significant error.

Another source of possible confusion in measurements, in which the photoelectron detector rotates about the reaction center, is the nature of the reaction zone. If it is a molecular beam uniformly bathed by the radiation source, no angular correction should be required. If, on the other hand, the photon source is a narrow cylinder that intersects (at right angles) a much wider cylinder describing the molecular beam, then a larger ionization zone may be viewed by the electron detector along the photon direction than perpendicular to it, if the electron detector has a small angle of acceptance (usually required). The idealized correction, in this latter instance, is to multiply the measured $I(\theta)$ by $\sin\theta$, to compensate for the larger reaction zone seen by the detector at small θ. Some of the discrepancy among various workers can perhaps be traced to the use or neglect of this $\sin\theta$ correction and its accuracy in particular cases (see Table 1). Use of the $\sin\theta$ correction will make β more positive than otherwise.

If the pressure outside the reaction zone is not sufficiently low, electron scattering will occur which will tend to diminish any anisotropy in the angular distribution.

Stray electric and/or magnetic fields can give rise to forward–backward anisotropies which should not be present until v/c becomes appreciable.

TABLE 1 *Some β Measurements for the Noble Gases, with Varying Conditions, Using Nominally Unpolarized Radiation*

Target gas	λ_i, Å	Berkowitz *et al.* (1966, 1967)[a]	Morgenstern *et al.* (1971)[b]	Niehaus and Ruf (1971)[c]	Carlson and Jonas (1971)[d]	Dehmer *et al.* (1975)[e]
Ar	584	0.55	1.47	0.95	0.85	0.89
Ar	736		1.09	0.27	0.3	0.08
Kr	584	0.7	1.60	1.37	1.20	1.27
Kr	736		1.14	0.8	0.59	0.99
Xe	584	1.11	1.73	1.68	1.40	1.70
Xe	736		1.56	1.20–1.45	1.22	1.31

[a] Berkowitz *et al.*[15] and Berkowitz and Ehrhardt.[70] Retarding field analyzer, rotated detector, no $\sin\theta$ correction.

[b] Morgenstern *et al.*[40] 127° cylindrical analyzer, rotated detector, used $\sin\theta$ correction.

[c] Niehaus and Ruf.[24] 127° cylindrical analyzer, rotated detector, used $\sin\theta$ correction.

[d] Carlson and Jonas.[34] Spherical sector analyzer, rotated lamp, used $\sin\theta$ correction.

[e] Dehmer *et al.*[71] Hemispherical analyzer, rotated detector, no $\sin\theta$ correction.

Table 1 compares the results of several groups on Ar, Kr, and Xe. It is of interest to note the convergence of the β values as a function of time, and the fact that some groups have employed the full $\sin\theta$ correction, others none at all. Note also the different experimental arrangements. A full $\sin\theta$ correction applied to the early work of Berkowitz *et al.*[15,70] would have overcorrected, compared to the recent results, but a partial $\sin\theta$ correction may have been necessary. A comparison of the data of Berkowitz *et al.* with the values of Carlson and co-workers, not only on the rare gases, but also N_2, CO, O_2, NO, and N_2O (Table 2), is revealing in that an increase in β of ~ 0.4 in the earlier data, in almost all cases, brings the two results into good agreement. This is the kind of deviation which may be attributable to a partial $\sin\theta$ correction (or possibly a partial polarization) but not a pressure scattering effect.

Even the most recent data on the noble gases reveal discrepancies of ~ 0.3 in β in some cases (Table 1). Before leaving the noble gases, we take note of

TABLE 2 *Comparison of β Measurements for Some Common Molecules, Using Nominally Unpolarized 584 Å Radiation*

Target gas	State formed	Berkowitz *et al.* (1967)[a]	Carlson *et al.* (1971, 1972)
N_2	$X^2\Sigma_g^+$	0	0.5 ± 0.1[b]
	$A^2\Pi_u$	0	0.3 ± 0.1[b]
CO	$X^2\Sigma^+$	+0.36	0.8 ± 0.1[b]
	$A^2\Pi$	0	0.3 ± 0.1[b]
O_2	$X^2\Pi_g$	−0.615	-0.3 ± 0.1[c]
	$a^4\Pi_u$		0.4[c]
	$A^2\Pi_u$		0.3[c]
	$b^4\Sigma_g^-$		0.7[c]
NO	$X^1\Sigma^+$	−0.45	$+0.32 \pm 0.7$[c]
	$^3\Sigma^+$		+0.4[c]
	$^3\Pi$		0.7[c]
	$^3\Delta$		0.35[c]
	$^3\Sigma^-$		0.0[c]
NO	$^1\Pi$		+0.45[c]
	$^1\Sigma^-$		0.0[c]
	$^1\Delta$		−0.1[c]
N_2O	$X^2\Pi$	−0.4	0.0 ± 0.1[d]
	$A^2\Sigma^+$		0.35 ± 0.1[d]
	$B^2\Pi$		-0.2 ± 0.2[d]

[a] Berkowitz *et al.*[15]
[b] Carlson and Jonas.[34]
[c] Carlson *et al.*[79]
[d] Carlson and McGuire.[80]

some higher energy studies on Kr by Krause[47] (already alluded to and displayed in Fig. 6) which are in very good agreement with Hartree–Fock calculations and experiments on the 2p shell of neon[71] (Fig. 10). In this latter example, which has no Cooper minimum, there is good agreement at low and high energies, and no significant discrepancy exists between Hartree–Fock and many-body calculations.

One surprising feature that appeared first in the data of Berkowitz *et al.*[15] was the observation of negative β values, implying a tendency for electrons to be ejected preferentially in a direction opposite to the electric vector. In the earlier hydrogenic theories and the early experiments with x rays, which explored the K and L shells of atomic systems, either pure $\sin^2\theta$ behavior or an isotropic component superimposed upon $\sin^2\theta$ was predicted and observed. Much of this early experimental work has been reviewed by Bothe.[72] Some of these early workers (among them Bothe[73] and Auger[74]) directed an x-ray beam across the diameter of a gas chamber and used a rotatable Geiger counter with a narrow window as the photoelectron detector. They found a maximum in photoelectron intensity around 90° with respect to the photon propagation direction but "pushed" in the forward direction. Others[75] used the Wilson cloud chamber method, sending a collimated x-ray beam through a cloud chamber containing the gas of interest and photographing tracks from two perpendicular sites, then counting the paths. The isotropic term in L-shell photoejection became more apparent in these studies, as did the retardation effect.

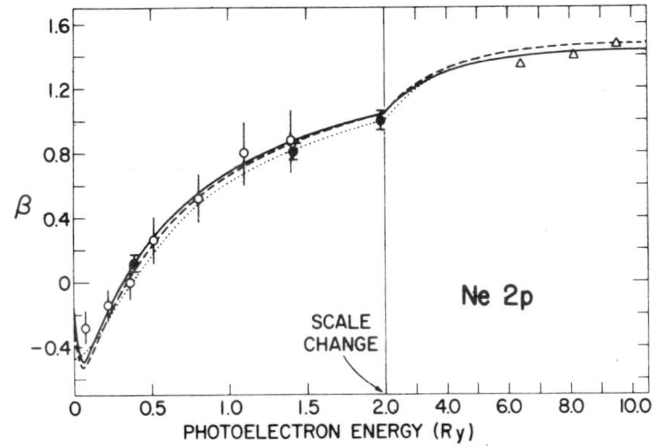

Fig. 10 Variation of asymmetry parameter β with photoelectron energy for ionization from the 2p orbital of neon. ● Experimental data of J. L. Dehmer *et al.*, *Phys. Rev. A* **2**, 1966 (1975). ○ Experimental data of M. J. Lynch *et al.*, *Phys. Lett.* **40A**, 349 (1972). △ Experimental data of F. Wuilleumier and M. O. Krause, *Phys. Rev. A* **10**, 242 (1974). ——— HFL and --- HFV calculations of D. J. Kennedy and S. T. Manson, *Phys. Rev. A* **5**, 227 (1972) •••• RPAE calculations of M. Ya. Amusia *et al.*, *Phys. Lett.* **40A**, 15 (1972). [From Dehmer *et al.*, cited here.]

The incorporation of energy analysis with angular studies was initiated by Watson and Van den Akker.[76] They used a collimated x-ray source striking a metal foil and two magnetic energy analyzers, both rotatable. The relative intensities of photoelectrons were determined photographically at several angles.

Using an AgCl target, they found:

(1) The isotropic component occurred only for AgL_{II} and AgL_{III}.

(2) The L_{II} and L_{III} ejected electrons had their maxima at smaller angles than K and L_I. [This would be in conformity with the refined retardation effect predictions of Cooper and Manson.[8] See Eq. (3).]

(3) The difference in behavior of M_I, relative to M_{II} and M_{III}, was similar to that already noted for the L sublevels.

In 1931, Van den Akker and Watson[77] substituted a Geiger–Mueller counter for the photographic plate, thus arriving at a direct analog to modern studies, with electrostatic analyzer substituted for magnetic analyzer and electron multiplier for G–M tube.

As recently as 1936, Hall[78] noted in his review article "... Bethe has found that the angular distribution for any (n, l) shell may be written $\alpha + \beta \sin^2 \theta \cos^2 \varphi$, where α and β are given in terms of radial integrals corresponding to the transition $(n, l \rightarrow E, l \pm 1)$. *In the above treatment β is positive. The direction of maximum emission is consequently in the direction of the electric vector.*"

Hall's review paper in 1936 (and the Bethe[7] and Bothe[72] Handbuch articles) summarized a period of intense activity on this subject in the late 1920s and early 1930s. The predominant portion of the experimental work was with x rays, and only a few of the later experiments involved energy analysis of the photoelectrons. Since the photoelectrons produced by x rays are largely from deep inner shells (mostly K and L_I) which are well-described by hydrogenic functions, the early wave mechanical theory was able to account for the observed phenomena.

The study of photoelectron angular distributions was quiescent for over 30 years, until the advent of ultraviolet photoelectron spectroscopy which focused attention on photoejection from valence shells. In 1966, Berkowitz and Ehrhardt[70] reopened the subject for investigation utilizing the new technology of photoelectron spectroscopy. Subsequently, Berkowitz *et al.*[15] observed *negative* values for β in the formation of ground states of O_2^+, NO^+, and N_2O^+ with 21.2 eV photon energy. In each of these instances the electrons were ejected from π_g or π_g-like orbitals. The O_2^+ behavior has been verified by Carlson *et al.*[79] (see Table 2) who also observed this behavior in the π_g ejection from CO_2. Subsequently, Hall and Siegel[16] observed negative values for β in the photodetachment of O^- and C^-, and the current status of theory readily accommodates negative β values.

Another surprising (and as yet unsatisfactorily explained) phenomenon resulting from studies of β in molecules using ultraviolet radiation was the observation by Carlson[81] and Carlson and Jonas[34] of vibrational components of the same photoelectron band having very different β values (see Section A.9 of this chapter). Although eight vibrational levels in H_2^+ studied by Carlson and Jonas showed no significant systematic variation of β, the $v' = 0$ and $v' = 1$ levels of N_2^+, $X^2\Sigma_g^+$ were found to have $\beta = +0.5$ and $+1.4$, respectively. This result was verified by Morgenstern *et al.*[40] who obtained $\beta_{v'=0} = +0.69 \pm 0.03$ and $\beta_{v'=1} = +1.40 \pm 0.03$. Carlson and Jonas[34] noted a similar, though less dramatic, behavior for the corresponding transition in the isoelectronic CO, where $\beta_{v'=0} = +0.8$ and $\beta_{v'=1} = +0.4$. Subsequently, Carlson and McGuire[80] observed such behavior in the fourth bands of CO_2^+, COS^+, and CS_2^+ ($C^2\Sigma^+$ states). One vibrational component in each of these bands was found to have an anomalous value of β. This behavior can be readily explained if autoionization and direct ionization are simultaneously giving rise to photoelectrons.

Thus, Carlson[81] has measured values of $\beta = -0.6$ for some vibrational levels in the photoionization of O_2 with NeI resonance radiation, and others with $\beta = 0$, the latter resulting from a known autoionization in O_2 and apparently producing an isotropic distribution. However, the other examples cited cannot readily be correlated with known autoionization behavior.

The dependence of β upon rotational states has been examined in an experimental study on H_2 by Niehaus and Ruf[35] and already discussed in Section A.8 of this chapter.

D. Summary of Other Recent Measurements of β

1. Atoms

Apart from the noble gases, the only other atomic systems for which angular distributions have been measured in the past decade are Zn, Cd, and Hg. Table 3, taken from Niehaus and Ruf[24], summarizes their data and those of Harrison[26] on Hg. The departure of β from its expected value of two for ionization from an s orbital has been discussed in the theoretical section under relativistic effects. The value at 1066 Å is influenced by autoionization. (Dill[18] has explored the β variation through these resonances theoretically.) The three data points corresponding to ejection from the d shell fall on a smooth plot of β versus photoelectron energy, indicating that this variation in β is due to the variation of matrix elements and phase shifts

TABLE 3 *Some β Measurements for Hg Using Nominally Unpolarized Radiation*

State formed	λ, Å	Niehaus and Ruf (1971)[a]	Harrison (1970)[b]
$^2S_{1/2}$	584	1.68 ± 0.1	—
	1048	2.13 ± 0.1	—
	1066	1.25 ± 0.1	—
$^2D_{5/2}$	584	0.03 ± 0.1	-0.09 ± 0.01
	736	0.68 ± 0.1	
$^2D_{3/2}$	584	0.25 ± 0.1	0.14 ± 0.02

[a] Niehaus and Ruf.[24]
[b] Harrison.[26]

with electron energy, but that for a given electron energy the β values for $^2D_{5/2}$ and $^2D_{3/2}$ would be very nearly the same. This behavior has generally been found for the noble gases as well and is supported by theory. (See Section A.7 in this chapter.)

Harrison's[26] results for $Cd^+(^2S_{1/2}) \leftarrow Cd(^1S_0)$ and $Zn^+(^2S_{1/2}) \leftarrow Zn(^1S_0)$ at 1048 Å are $\beta = 1.97 \pm 0.17$ and $\beta = 2.02 \pm 0.48$, respectively, both well in the range expected for ejection of an s-electron. However, for formation of $^2D_{3/2}$ and $^2D_{5/2}$ of Cd^+ with 584 Å radiation, he obtains $\beta = -0.12 \pm 0.03$ and $\beta = +1.49 \pm 0.13$, respectively, which is dramatically different from the anticipated similarity in behavior of spin–orbit partners as noted above for Hg and the noble gases. No adequate explanation has been given for this peculiar result in cadmium, nor has the experimental result been tested by other investigators.

2. Molecules

The molecular determinations of β beyond those already discussed are not distinguished by any unusual properties and consist primarily of data obtained at a single wavelength of 584 Å. Hence, β variations, which would be desirable to test any forthcoming theory for β in molecules, constitute an important project for future research. Carlson and Anderson[82] have made practical utilization of the dependence of β upon the molecular orbital from which the photoelectron is ejected to sort out some overlapping bands in the 21.2 eV photoelectron spectrum of benzene. Their results are displayed in Fig. 95 in Chapter VI. Jonas *et al.*[83] have made similar, but less dramatic, attempts in the 21.2 eV photoelectron spectra of the tetramethyl compounds of C, Si, Ge, Sn, and Pb.

White *et al.*[84] have examined C_2H_4, C_3H_6, *cis*- and *trans*-butene, 1-butene, and 1,3-butadiene. They find relatively high values of β from π

orbitals, lower ones from σ orbitals. This result is in direct contrast to Lohr's[85] calculation based on the plane wave calculation for C_2H_4.

Carlson and White[86] have reported βs for CH_3F, CH_2F_2, CHF_3, CF_4, CH_3Cl, CH_2Cl_2, $CHCl_3$, CCl_4, CH_3Br, and CH_3I. Very little correlation is found. Most of the βs are positive and <1. When Jahn–Teller and spin–orbit split states occur, they yield the same β. (The spin–orbit behavior has been discussed in connection with atoms.) Some overlapping of bands in the photoelectron spectra of CH_2F_2 and CF_4 was separable by β analysis.

The results of Carlson and co-workers on some simple molecules not previously covered are summarized in Table 4.

Vilesov and collaborators[87,88] have reported values of β for (a) Ar, Xe, H_2, O_2, CH_4, C_6H_6 (21.2 eV) and (b) Ar, Xe, H_2, N_2, O_2, H_2O, NO (21.2 and 16.8 eV). In Vilesov and Lopatin,[87] all values of β reported seemed remarkably close to $+1$ (especially surprising for O_2), and there was a large forward–backward asymmetry.

The Milyaev and Vilesov[88] paper gave results more in line with other data, but the $O_2 b^4\Sigma^-$ curve gave a distribution not conforming to $A + B\sin^2\theta$, and more in line with the more complicated distributions reported

TABLE 4 *Some Values of for Simple Molecules Obtained with 21.2 eV Photons*

Ion state	CO_2	COS^a	CS_2
	(A) From Carlson and McGuire.[80]		
$X^2\Pi_g$	-0.1 ± 0.1	$+1.1 \pm 0.1$	$+1.3 \pm 0.1$
$A^2\Pi_u$	$+0.8$	$+0.6$	$+1.2$
$B^2\Sigma_u$	-0.4	$+0.35$	$+0.4$
$C^2\Sigma_g^+$	$+1.3$	$+0.55$	$+1.2 \pm 0.2$
$D^2\Sigma_u$			$+1.5 \pm 0.2$
Ion state	H_2O	H_2S	
	(B) From Carlson and McGuire.[80]		
$^2B_\Delta$	$+1.0 \pm 0.1$	$+1.6 \pm 0.1$	
$^2A_\Delta$	$+0.3$	$+1.1$	
$^2B_2$	-0.1 ± 0.2	$+0.9$	
Ion state	HCl		
	(C) From Carlson *et al.*[79]		
$^2\Pi$	$+1.40 \pm 0.05$		
$^2\Sigma$	$+0.75 \pm 0.1$		

^{a} The g or u designation is not appropriate for this molecule.

by McGowan *et al.*,[89] which are not permissible on symmetry grounds, such as Yang's theorem.

Note added in proof: Sell *et al.* have reported β values with 21.2 eV incident photon energy for benzene and the monohalobenzenes,[89a] sulfur hexafluoride,[89b] and ethylene, benzene, and their fluorinated derivatives.[89c] The first measurements of β over an extended energy range for molecules are beginning to appear. McCoy *et al.*[89d] have studied

$$O_2(X^3\Sigma_g^-) + h\nu \rightarrow O_2^+(X^2\Pi_g) + e$$

between 20 and 45 eV photon energy. Near threshold the value of β is negative (as found by previous workers) but it rises to $+1$ at ~ 30 eV and appears to reach a plateau between 30 and 45 eV. Marr *et al.*[89e] have obtained β values for the first three ionic states in the photoionization of N_2 and CO, also in the 20–45 eV photon energy range. In these latter examples, the β values cluster around $+1$, with rather mild variations.

3. Pseudophotoelectron Spectroscopy

When an electron impact ionization experiment is performed with high incident energy and the inelastically scattered electron is observed very near to the initial electron direction (i.e., the momentum transfer $\rightarrow 0$) and the electron ejected in the ionization act is detected in coincidence, the situation is equivalent to photoelectron ejection with plane polarized radiation.[90] When the momentum transfer is randomly oriented, coincidence studies approximate photoejection with unpolarized radiation. Such experiments have recently been performed on Ne,[91] Ar, and Kr.[92] For Ne the results are in good concordance with the aforementioned theories and experiments, except for some anomalous behavior in a region of autoionization. For Ar, and particularly Kr, the values of β above 20 eV fall significantly below even the many-body RPAE theory in the approach to the Cooper minimum. It should be noted that these determinations of β are relative measurements, normalized at low energy to the calculated β values of Kennedy and Manson.[54]

Appendix

A substantial subset of the literature on photoelectron angular distributions concerns truly relativistic effects. Hall[78] summarizes the successive complications as follows:

Neglecting

(1) momentum of the light quantum,
(2) change of photoelectron mass with velocity,

(3) effects due to deep binding of the initial electron (large Z) and restricting ourselves for the moment to K-shell electrons

$$\frac{d\sigma_K}{d\Omega} \sim \sin^2\theta \qquad (\theta \text{ from propagation direction}).$$

(a) If (1) is taken into account, but not (2) and (3), we obtain the retardation formula

$$\frac{d\sigma_K}{d\Omega} \sim \sin^2\theta\left(1 + 4\frac{v}{c}\cos\theta\right).$$

All other relativistic effects enter only through integer powers of $(v/c)^2$.

(b) If now (1) is disregarded, but the Dirac formulation (rather than Schrödinger) is used, one gets the first $(v/c)^2$ correction:

$$\frac{d\sigma_K}{d\Omega} \sim \sin^2\theta[1 + (0.280\cos^2\theta - 0.056)\alpha^2],$$

where $\alpha = \alpha_0 Z = (2\pi e^2/hc)Z = (1/137.3)Z$. Note that (i) $\cos\theta$ enters only in $\cos^2\theta$ form, and hence the angle of maximum emission remains at 90°. (ii) For $\cos^2\theta = 0.056/0.280 = \frac{1}{5}(\theta = 67°)$, correction term $= 0$.

(c) With (1) and the Dirac formulation, the effects are additive.

$$\frac{d\sigma_K}{d\Omega} \sim \sin^2\theta\left[1 + (0.280\cos^2\theta - 0.056)\alpha^2 + \frac{4v}{c}\cos\theta\right].$$

When $\alpha^2 \to 0$, the simple retardation form is obtained.

(d) When $(v/c)^2$ is not to be neglected, but $\alpha = Z/137 \ll 1$, the Sauter[93] formula is obtained.

$$\frac{d\sigma_K}{d\Omega} \sim \frac{\sin^2\theta}{(1-(v/c)\cos\theta)^4}\left[\cos^2\varphi\{(1-(v/c)^2)^{1/2} - \frac{1}{2}(1-(1-(v/c)^2)^{1/2})(1-(v/c)\cos\theta)\} + \frac{1}{4}\frac{(1-(1-(v/c)^2)^{1/2})^2}{1-(v/c)^2}(1-(v/c)\cos\theta)\right].$$

When $(v/c)^2 \ll 1$, this reduces to

$$\frac{d\sigma_K}{d\Omega} \sim \frac{\sin^2\theta}{(1-(v/c)\cos\theta)^4}\cos^2\varphi,$$

which is the simple retardation formula ($\cos^2\varphi$ being the azimuthal dependence with plane polarized light).

At extremely relativistic energies,

$$\frac{h\nu}{mc^2} \gg 1$$

and

$$\left(1 - \left(\frac{v}{c}\right)^2\right)^{-1/2} \approx 1 + \frac{h\nu}{mc^2} \gg 1,$$

the last term becomes the dominant one. Setting $\gamma = (1 - (v/c)^2)^{-1/2}$ and noting that the distribution will be peaked strongly in the forward direction so that $\sin^2 \theta \to \theta^2$, Bethe and Salpeter[4] give

$$\frac{d\sigma_K}{d\Omega} \propto \frac{\theta^2}{(1 + \gamma^2\theta^2)^3}.$$

Upon differentiation,

$$\theta_{\max} = \frac{1}{\sqrt{2}} \cdot \frac{1}{\gamma} = \sqrt{\frac{1 - (v/c)^2}{2}}.$$

Note, however, that the intensity is predicted to vanish at $\theta = 0$.

Gavrila[94] has treated the L shell relativistically, and obtains a similar behavior for L_I and L_{III} as for K, but for L_{II} the intensity is maximum at $\theta = 0$, decreasing rapidly therefrom.

In 1954, Hedgran and Hultberg[95] performed experiments with 1.33 MeV photons on Pb foil and 411 keV photons on Au foil, and found that the intensity of photoelectrons from the K shell did not vanish at $\theta = 0$.

Although the approximate formula given above doesn't explain this effect, Sauter and Wüster[96] point out that Sauter's original formula without approximations[97] can explain this behavior, but the formula is very cumbersome.

Sauter and Wüster provide a descriptive explanation for this behavior. In the Dirac equation, only j and m_j are good quantum numbers, not l. Hence, $s_{1/2}$ and $p_{1/2}$ combine; $p_{3/2}$ and $d_{3/2}$ combine. The $s_{1/2}$ and $p_{1/2}$ are spherically symmetric; $p_{3/2}$ and $d_{3/2}$ are not. The allowed photoeffect transitions are $s_{1/2} \to p_{1/2}$ and $p_{3/2}$ ($\Delta m_j = \pm 1$). In the nonrelativistic domain, $s \to p$ leads to vanishing intensity at $\theta = 0$. In the relativistic domain, $s_{1/2} \to p_{3/2}$ has similar behavior, but $s_{1/2} \to p_{1/2}$ provides a spherically symmetric component, and hence a finite intensity at $\theta = 0$.

REFERENCES

1. A. Einstein, *Ann. Phys. (Leipzig)* **17**, 132 (1905).
2. P. Auger and F. Perrin, *J. Phys. Radium* **8**, 93 (1927).

3. G. Wentzel, *Z. Phys.* **40**, 574 (1926).
4. H. A. Bethe and E. E. Salpeter, "Quantum Mechanics of One- and Two-Electron Atoms." Springer-Verlag, Berlin and New York, 1957.
5. G. Wentzel, *Z. Phys.* **41**, 828 (1927).
6. A. Sommerfeld and G. Schur, *Ann. Phys.* (*Leipzig*) **4**, 410 (1930).
7. H. A. Bethe, *Handb. Phys.* **24**, 475–487 (1933).
8. J. W. Cooper and S. T. Manson, *Phys. Rev.* **177**, 157 (1969).
9. G. Schur, *Ann. Phys.* (*Leipzig*) **4**, 433 (1930).
10. C. N. Yang, *Phys. Rev.* **74**, 764 (1948).
11. M. Peshkin, *Adv. Chem. Phys.* **18**, 1 (1970).
12. B. Ritchie, *Phys. Rev. A* **13**, 1411 (1976).
13. J. Cooper and R. N. Zare, *Lect. Theor. Phys.* **11C**, 323 (1969).
14. J. C. Tully, R. S. Berry, and B. J. Dalton, *Phys. Rev.* **176**, 95 (1968).
15. J. Berkowitz, H. Ehrhardt, and T. Tekaat, *Z. Phys.* **200**, 69 (1967).
16. J. L. Hall and M. W. Siegel, *J. Chem. Phys.* **48**, 943 (1968).
17. J. Cooper and R. N. Zare, *J. Chem. Phys.* **48**, 942 (1968).
18. D. Dill, *Phys. Rev. A* **7**, 1976 (1973).
19. S. T. Manson, *Adv. Electron. Electron Phys.* **41**, 73 (1976).
20. A. F. Starace, S. T. Manson, and D. J. Kennedy, *Phys. Rev. A* **9**, 2453 (1974).
21. D. Dill, A. F. Starace, and S. T. Manson, *Phys. Rev. A* **11**, 1596 (1975).
22. J. A. R. Samson and J. L. Gardner, *Phys. Rev. Lett.* **31**, 1327 (1973).

22a. N. M. Kabachnik and I. P. Sazhina, *J. Phys. B* **9**, 1681 (1976); V. V. Balashov, N. M. Kabachnik, and I, P. Sazhina, *Proc. Intl. Conf. Physics Electronic Atomic Collisions, 8th Belgrade*, p. 527. (Belgrade: Institute of Physics) (1973).

23. T. E. H. Walker and J. T. Waber, *J. Phys. B* **6**, 1165 (1973).
24. A. Niehaus and M. W. Ruf, *Z. Phys.* **252**, 84 (1971).
25. T. E. H. Walker and J. T. Waber, *J. Phys. B* **7**, 674 (1974).
26. H. Harrison, *J. Chem. Phys.* **52**, 901 (1970).
27. S. T. Manson, *Chem. Phys. Lett.* **19**, 76 (1973).
28. A. D. Buckingham, B. J. Orr, and J. M. Sichel, *Philos. Trans. R. Soc. London, Ser. A* **268**, 147 (1970).
29. J. M. Sichel, *Mol. Phys.* **18**, 95 (1970).
30. R. L. Wilkins and H. S. Taylor, *J. Chem. Phys.* **48**, 4505 (1968).
31. L. Åsbrink, *Chem. Phys. Lett.* **7**, 549 (1970).
32. D. Dill, *Phys. Rev. A* **6**, 160 (1972).
33. G. Herzberg and C. Jungen, *J. Mol. Spectrosc.* **41**, 425 (1972).
34. T. A. Carlson and A. E. Jonas, *J. Chem. Phys.* **55**, 4913 (1971).
35. A. Niehaus and M. W. Ruf, *Chem. Phys. Lett.* **11**, 55 (1971).

35a. E. S. Chang, *J. Phys. B* **11**, L69 (1978).

36. J. Berkowitz and W. A. Chupka, *J. Chem. Phys.* **51**, 2341 (1969).

35a. Y. Itikawa, *Chem. Phys.* **30**, 109 (1978).

36b. M. Ruf, Diplomarbeit, Fak. Phys. Univ. Freiburg, Freiburg, West Germany, 1971.

37. J. Berkowitz and R. Spohr, *J. Electron Spectrosc. Relat. Phenom.* **2**, 143 (1973).
38. Y. Itikawa, *J. Electron Spectrosc. Relat. Phenom.* **2**, 125 (1973).
39. C. Duzy and R. S. Berry, *J. Chem. Phys.* **64**, 2421 (1976).
40. R. Morgenstern, A. Niehaus, and M. W. Ruf, *Abstr. Int. Conf. Phys. Electron. A. Collisions, 7th, Amsterdam*, p. 167 (1971).

40a. O. F. Kalman, *Mol. Phys.* **34**, 397 (1977).

41. B. Ritchie, *J. Chem. Phys.* **64**, 3050 (1976).
42. L. L. Lohr and M. B. Robin, *J. Am. Chem. Soc.* **92**, 7241 (1970).
43. G. Kaplan and A. P. Markin, *Opt. Spectrosc.* (*USSR*) **24**, 475 (1968); **25**, 275 (1968).

44. W. Thiel and A. Schweig, *Chem. Phys. Lett.* **12**, 49 (1971); **16**, 409 (1972); **21**, 541 (1973); *J. Chem. Phys.* **60**, 951 (1974).
45. F. O. Ellison, *J. Chem. Phys.* **61**, 507 (1974); J. W. Rabalais, T. P. Debies, J. L. Berkosky, J. T. J. Huang, and F. O. Ellison, *J. Chem. Phys.* **61**, 516 (1974); **61**, 529 (1974); T. P. Debies and J. W. Rabalais, *J. Am. Chem. Soc.* **97**, 487 (1975).
46. B. Ritchie, *J. Chem. Phys.* **61**, 3291 (1974).
47. M. O. Krause, *Phys. Rev.* **177**, 151 (1969).
48. C. Froese-Fischer, *At. Data* **4**, 301 (1972); **12**, 87 (1973).
49. C. Roetti and E. Clementi, *J. Chem. Phys.* **60**, 3342 (1974).
50. J. P. Desclaux, *At. Data* **12**, 311 (1973).
51. J. C. Slater, *Phys. Rev.* **81**, 385 (1951).
52. S. T. Manson and J. W. Cooper, *Phys. Rev. A* **2**, 2170 (1970).
53. S. T. Manson and D. J. Kennedy, *Chem. Phys. Lett.* **7**, 387 (1970).
54. D. J. Kennedy and S. T. Manson, *Phys. Rev. A* **5**, 227 (1972).
55. S. T. Manson, *J. Electron Spectrosc. Relat. Phenom.* **1**, 413 (1972).
56. J. S. Shyu and S. T. Manson, *Phys. Rev. A* **11**, 166 (1975).
57. M. Y. Amusia, N. A. Cherepkov, and L. V. Chernysheva, *Phys. Lett. A* **40**, 15 (1972).
57a. The reader who may be contemplating such calculations should note that Amusia has reproduced this equation elsewhere, with different notation, and apparently with a different sign of the sine term in the phase shift. See M. Y. Amusia, *Proc. Int. Conf. VUV Radiat. Phys., 4th, Hamburg* p. 217, eq. 10 (1974).
57b. K. T. Taylor, *J. Phys. B* **10**, L 699 (1977).
58. B. Ritchie, *J. Chem. Phys.* **63**, 1351 (1975).
59. J. A. R. Samson and A. F. Starace, *J. Phys. B* **8**, 1806 (1975).
60. E. O. Lawrence and M. A. Chaffee, *Phys. Rev.* **36**, 1099 (1930); M. A. Chaffee, *Phys. Rev.* **37**, 1233 (1931), A. Kraus, *Ann. Phys.* (*Liepzig*) **14**, 103 (1932).
61. W. H. Hancock and J. A. R. Samson, *J. Electr. Spectr.* **9**, 211 (1976).
62. F. W. Bubb, *Phys. Rev.* **23**, 137 (1924).
63. F. Kirchner, *Ann. Phys.* (*Leipzig*) **83**, 521 (1927); **84**, 899 (1927); *Phys. Z.* **27**, 385, 800 (1926).
64. C. J. Pietenpol, *Phys. Rev.* **32**, 564 (1928).
65. P. Mitchell and K. Codling, *Phys. Lett. A* **38**, 31 (1972).
66. M. J. Lynch, A. B. Gardner, and K. Codling, *Phys. Lett. A* **40**, 349 (1972).
67. M. J. Lynch, K. Codling, and A. B. Gardner, *Phys. Lett. A* **43**, 213 (1973).
68. R. G. Houlgate, J. B. West, K. Codling, and G. V. Marr, *J. Phys. B* **7**, L470 (1974).
68a. R. G. Houlgate, J. B. West, K. Codling, and G. V. Marr, *J. Electron Spectrosc. Relat. Phenom.* **9**, 205 (1976).
68b. K. Codling, R. G. Houlgate, J. B. West, and P. R. Woodruff, *J. Phys. B* **9**, L83 (1976).
68c. L. Torop, J. Morton, and J. B. West, *J. Phys. B* **9**, 2035 (1976).
68d. D. L. Miller, J. D. Dow, R. G. Houlgate, G. V. Marr, and J. B. West, *J. Phys. B* **10**, 3205 (1977).
69. W. S. Watson and D. T. Stewart, *J. Phys. B* **7**, L466 (1974).
70. J. Berkowitz and H. Ehrhardt, *Phys. Lett.* **21**, 531 (1966).
71. J. L. Dehmer, W. A. Chupka, J. Berkowitz, and W. T. Jivery, *Phys. Rev. A* **12**, 1966 (1975).
72. W. Bothe, *Handb. Phys.* **23** (2), 57 (1933).
73. W. Bothe, *Z. Phys.* **26**, 59 (1924).
74. P. Auger, *J. Phys. Radium* **9**, 225 (1928).
75. W. Bothe, *Z. Phys.* **17**, 148 (1923); E. J. Williams, *Nature* (*London*) **121**, 134 (1928); E. J. Williams, J. M. Nuttall, and H. S. Barlow, *Proc. R. Soc. London, Ser. A* **121**, 611 (1928);

C. D. Anderson, *Phys. Rev.* **34**, 547 (1929); **35**, 1139 (1930); P. Auger, *C. R. Acad. Sci.* **186**, 758 (1928); **187**, 1141 (1928); **188**, 447 (1929).

76. E. C. Watson, *Phys. Rev.* **30**, 479 (1927); E. C. Watson and J. A. Van den Akker, *Proc. R. Soc. London, Ser. A* **126**, 138 (1930).
77. J. A. Van den Akker and E. C. Watson, *Phys. Rev.* **37**, 1631 (1931).
78. H. Hall, *Rev. Mod. Phys.* **8**, 358 (1936).
79. T. A. Carlson, G. E. McGuire, A. E. Jonas, K. L. Cheng, C. P. Anderson, C. C. Lu, and B. P. Pullen, *in* "Electron Spectroscopy" (D. A. Shirley, ed.), p. 207. North-Holland Publ., Amsterdam. 1972.
80. T. A. Carlson and G. E. McGuire, *J. Electron Spectrosc. Relat. Phenom.* **1**, 209 (1972).
81. T. A. Carlson, *Chem. Phys. Lett.* **9**, 23 (1971).
82. T. A. Carlson and C. P. Anderson, *Chem. Phys. Lett.* **10**, 561 (1971).
83. A. E. Jonas, G. K. Schweitzer, F. A. Grimm, and T. A. Carlson, *J. Electron Spectrosc. Relat. Phenom.* **1**, 29 (1972).
84. R. M. White, T. A. Carlson, and D. P. Spears, *J. Electron Spectrosc. Relat. Phenom.* **3**, 59 (1974).
85. L. Lohr, *in* "Electron Spectroscopy" (D. A. Shirley, ed.), p. 245. North-Holland, Amsterdam, 1972.
86. T. A. Carlson and R. M. White, *Faraday Discuss. Chem. Soc.* **54**, 285 (1972).
87. F. I. Vilesov and S. N. Lopatin, *Vestn. Leningr. Univ.* No. 4, p. 64 (1970).
88. V. B. Milyaev and F. I. Vilesov, *Khim. Vys. Energ.* **6**, 354 (1972).
89. J. W. McGowan, D. A. Vroom, and A. R. Comeaux, *J. Chem. Phys.* **51**, 5626 (1969).

89a. J. A. Sell and A. Kuppermann, *Chem. Phys.* **33**, 367 (1978).

89b. J. A. Sell and A. Kuppermann, *Chem. Phys.* **33**, 379 (1978).

89c. J. A. Sell, D. M. Mintz, and A. Kuppermann, *Chem. Phys. Lett.* **58**, 601 (1978).

89d. D. G. McCoy, J. M. Morton, and G. V. Marr, *J. Phys. B* **11**, L 547 (1978).

89e. G. V. Marr, J. M. Morton, R. M. Holmes, and D. G. McCoy, Daresbury Lab. Rept. DL/SRF/P142, 1978. Submitted to *J. Phys. B.*

90. Y.-K. Kim, *Phys. Rev. A* **6**, 666 (1972).
91. M. J. van der Wiel and C. E. Brion, *J. Electron Spectrosc. Relat. Phenom.* **1**, 439 (1972).
92. G. R. Branton and C. E. Brion, *J. Electron Spectrosc. Relat. Phenom.* **3**, 123 (1974).
93. F. Sauter, *Ann. Phys.* (*Liepzig*) **9**, 217 (1931).
94. M. Gavrila, *Phys. Rev.* **124**, 1132 (1961).
95. A. Hedgran and S. Hultberg, *Phys. Rev.* **94**, 498 (1954); *Ark. Fys.* **9**, 245 (1955).
96. F. Sauter and H. O. Wüster, *Z. Phys.* **141**, 83 (1955).
97. F. Sauter, *Ann. Phys.* (*Paris*) **11**, 454 (1931).

Chapter VIII

INSTRUMENTS AND METHODS

A. Techniques for Measuring Photoabsorption Cross Sections and Ionization Yields

The experimental strategy for these measurements has been thoroughly covered by Samson[1] and hence an overview will suffice. Both single and double ion chambers are used. An example of the latter is displayed in Fig. 1. In this figure, a gas with number density n fills the chamber. At the measuring wavelength, its absorption cross section is σ and its ionization yield (ions/photons absorbed) is η. It is assumed that appropriate potentials can be applied[1] so that i_1 is a measure of all the ions produced in the first d gap and

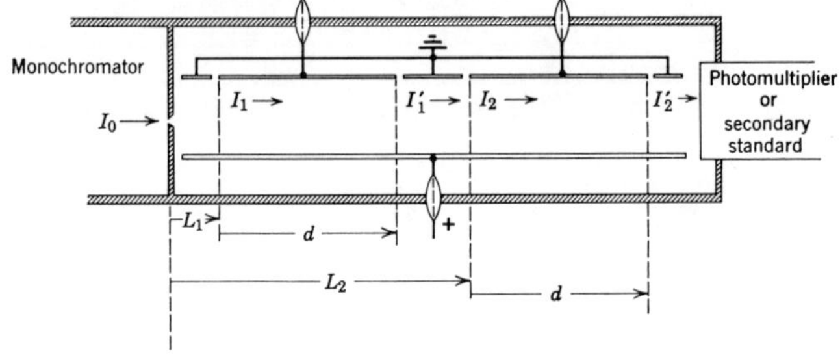

Fig. 1 An illustration of the double ion chamber used to measure the absolute photoabsorption cross section (σ) and the quantum yield of ionization (η). [From J. A. R. Samson, "Techniques of Vacuum Ultraviolet Spectroscopy," p. 267. Wiley, New York, 1967.]

i_2 the corresponding ionization in the second d gap. The light entering the first d gap is $I_0 e^{-\sigma n L_1}$, that leaving the first d region is $I_0 e^{-\sigma n L_1}(1 - e^{-\sigma n d})$. The difference is the light absorbed. From a measurement of the total ionization in this region and the definition of ionization yield (η),

$$\eta = \frac{i_1/q}{I_0 e^{-\sigma n L_1}(1 - e^{-\sigma n d})}. \tag{1}$$

Similarly, for the second gap having the identical length d,

$$\eta = \frac{i_2/q}{I_0 e^{-\sigma n L_2}(1 - e^{-\sigma n d})}. \tag{2}$$

From the ratio of these two measurements, one can deduce

$$\sigma = \frac{\ln(i_1/i_2)}{n(L_2 - L_1)}. \tag{3}$$

Hence, a measurement of the pressure, two ion currents and the dimensions of the apparatus is sufficient for the measurement of an absorption cross section. If the ionization yield η is known, then insertion of the inferred value of σ from (3) enables one to determine I_0. If one wishes to measure η, it is necessary to make an auxiliary measurement with a gas whose value of η is known. The noble gases are ideal for this purpose since they are monatomic and their ionization yields have been verified to be unity from ionization threshold to the domain of multiple ionization. This method can be used from the ionization potential of xenon (12.13 eV = 1022 Å) down to ~300 Å. For the energy range below 12.13 eV, nitric oxide has been used as the calibrating gas, with experimental values for η(NO) (see Chapter V, Section B.4).

Samson and Haddad[2] have recently extended the lower λ limit of this approach by examining some of the errors introduced with higher energy radiation. In addition to double and triple ionization (see Chapter VI, Sections B.1–B.5), the single ionization events produce energetic photoelectrons which result in additional ionization, exacerbated by the collecting fields present. They propose two methods to make measurements in the higher energy domain, both using single ion chambers. In one version, the collecting field is low and measurements at several pressures are extrapolated to zero pressure. In the other, measurements are made at high field and at high pressure.

They note that the measurement of η for *molecules* at short wavelength is best made by using the high-field method. This is because dissociative ionization becomes important and the kinetic energies of the ion fragments can be significant. With the low-field ion chamber, fragment ions with a few electron volts of kinetic energy can escape the collector, giving an erroneous ion yield.

To date, very few measurements of ion yield have been made below ~300 Å. By contrast, photoabsorption cross section measurements have been made from 300 Å on into the x-ray region. In this domain, the gas is isolated between thin windows (e.g., zapon or celluloid) and the photon detector is a Geiger counter.[3,4]

B. Photoionization Mass Spectrometry

A basic difference between the requirements in photoionization mass spectrometry and photoabsorption is that the latter is essentially a measurement of the attenuation of light. Hence, the magnitude of the signal I is of the same order as that of the incident light intensity I_0. By contrast, in photoionization mass spectrometry one is directly measuring the product of the reaction i_m, whose magnitude is typically 2–4 orders weaker than I_0. This is partly due to the fact that the column length in photoabsorption can be adjusted until I/I_0 is in the range 0.01–1.0, whereas in photoionization mass spectrometry one generally arranges for the detection of i_m at conditions where the attenuation of the incident light is negligible. Hence, high light source intensity, high collection efficiency, and very low background must be stressed in order to achieve a large dynamic range of measurement.

We can subdivide the technology of photoionization mass spectrometry into four basic components: light source, dispersive instrument, light detector, and ion detector.

1. Light Source

In the range of photon energies from ionization threshold (~10 eV) to about 20 eV, three light sources are most commonly used: the many-lined molecular hydrogen emission and the continua generated by argon and helium. Typical spectral distributions of the three sources are shown in Fig. 2a–d. In addition, resonance line sources obtained by low pressure capillary discharges through hydrogen, argon, neon, and helium are sometimes used to obtain mass spectra at selected wavelengths. Electron synchrotrons and storage rings are also becoming useful as light sources for these studies. We shall include a recent direct comparison of these sources with laboratory light sources.

(a) The many-lined hydrogen emission consists of the Werner and Lyman bands of H_2. It can readily be obtained from a dc capillary discharge through hydrogen at ~1 torr pressure. The emitted light intensity is roughly propor-

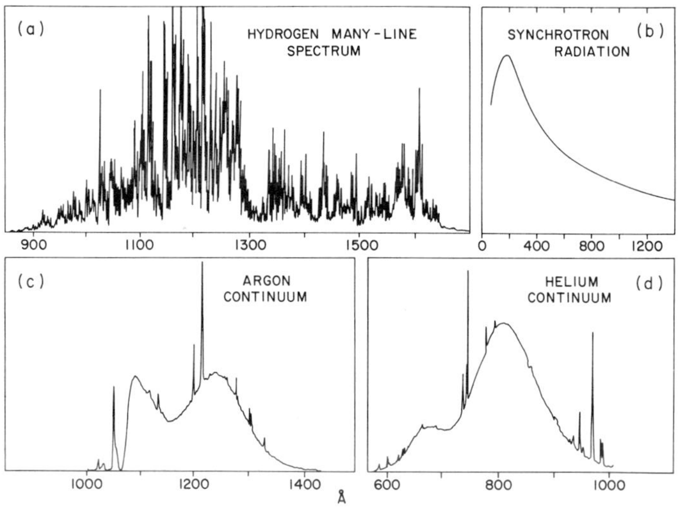

Fig. 2 Spectral distribution of four light sources commonly used in photoionization mass spectrometry. The hydrogen, argon, and helium spectra are measured light intensities at the exit of a normal incidence monochromator, whereas the synchrotron spectrum is that calculated for the 240 MeV storage ring in Wisconsin. See Fig. 7 for a measured dispersed spectrum utilizing this light source. [From J. Berkowitz, *Adv. in High Temp. Chem.* **3**, 123 (1971).]

tional to current. Hence, for the highest intensities, it is necessary to water cool the anode and cathode, as well as the discharge tube. Care must be taken to provide sufficient electrical insulation between the water-cooled high voltage cathode and the grounded water source. To minimize stray ground currents, it often helps to maintain the anode ground separate from the common apparatus ground.

To minimize electrical noise, the lamp pressure often requires critical adjustment. For high current operation, the discharge tube should be constructed of quartz, with annular water cooling. The lamp design used at Argonne, and commonly operated with 0.5–1 A current, is illustrated in Fig. 3. Needless to say, the power supply for this discharge should be capable of handling this current, should provide at least 5 kV for initiating the discharge, and should include some current-limiting elements, such as a resistor bank or a saturable reactor.

(b) The continua generated by argon and helium (and other noble gases) have been know for some time[5] but the technology for generating these continua has improved over the last decade, providing substantially more intensity. These continua require about 100 torr of the particular noble gas. Hence, massive differential pumping (typically, two stages before the monochromator entrance slit) is required to reduce the monochromator pressure

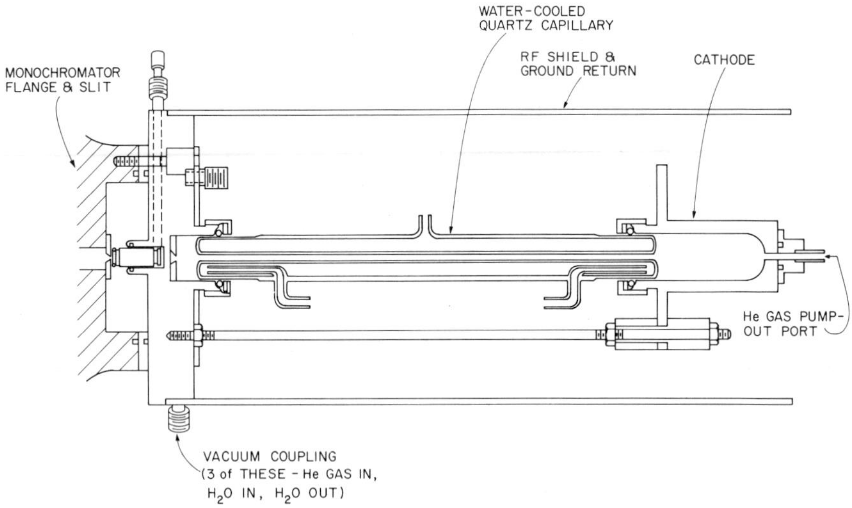

Fig. 3 Schematic illustration of the water-cooled capillary discharge lamp used at the author's laboratory to generate the low pressure hydrogen many-lined spectrum and the high pressure continua in helium and argon.

to $\sim 10^{-5}$ torr. It is believed that the mechanism responsible for these continua involves excitation of an atom by electron impact, collisional formation of a weakly bound, excited diatomic molecule, and emission from this weakly bound state to the repulsive ground state. Support for this mechanism derives from the observation that the continua possess high-energy cutoffs at the first resonance lines—584 Å for He, 1066 and 1048 Å for Ar.

These continua cannot be generated by a dc discharge. The Ar, Kr, and Xe continua can be produced by microwave discharge, but the He continuum requires repetitive high power pulses. In early versions, a condenser was permitted to charge up and then discharge through the high pressure gas. The repetition rate was irregular and determined by the properties of the lamp (length, pressure) as well as the external circuit.

More control, and substantially more intensity, was achieved by triggering the discharge with a thyratron. This approach was introduced by Huffman *et al.*[6] The pulse repetition rate could be controlled in the range 0–10 kc (kilocycles). A schematic of the electronic circuit is given in Fig. 4. Analysis of the time structure of the electronic pulse and light pulse reveals that the former persists for ~ 1–$2\ \mu$sec, whereas the light pulse commences toward the end of the electrical pulse and has a decay width of $\sim 10\ \mu$sec. With 10 kc repetition frequency, there remains a dead time of about 100 μsec before the start of the next pulse. Since light intensity is roughly proportional to pulse

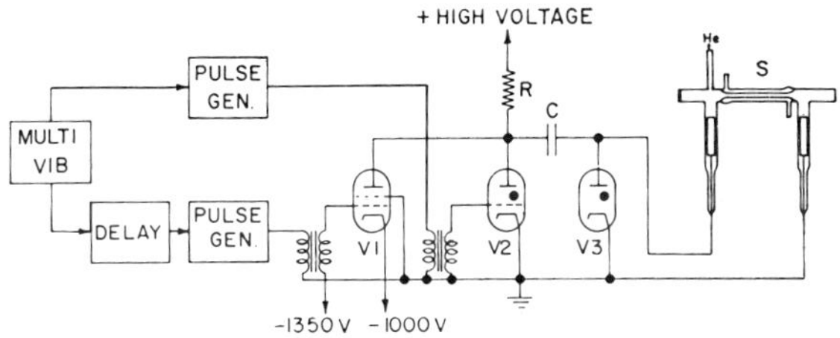

Fig. 4 Diagram of the thyratron-triggered circuit used to generate the helium and argon continua. *V*1, 4PR 1000 A; *V*2, E.G.G. Type 7322; *V*3, E.G.G. Type HR 1; *C*, energy storage capacitor; *R*, current limiting resistor; *S*, quartz discharge tube. [From R. E. Huffman *et al.*, *Appl. Opt.* **4**, 1145 (1965).]

frequency, it appeared possible to improve upon this design with a high power pulser capable of achieving 100 kc repetition rate. To overcome the limitation of thyratron triggering, a high power pulser circuit was designed at Argonne for achieving this goal. It consists essentially of a commercial high frequency pulsing circuit and two large pulse amplifier tubes, operated in parallel. This circuit is schematically shown in Fig. 5. The lamp-circuit system exhibits a voltage rise time of 60 nsec and a width of 0.15 μsec (FWHM), thereby allowing the use of the higher repetition rate.

In practice, the optimum performance occurs at $\sim$88 kc. A direct comparison between this device and the thyratron triggered one has indicated an

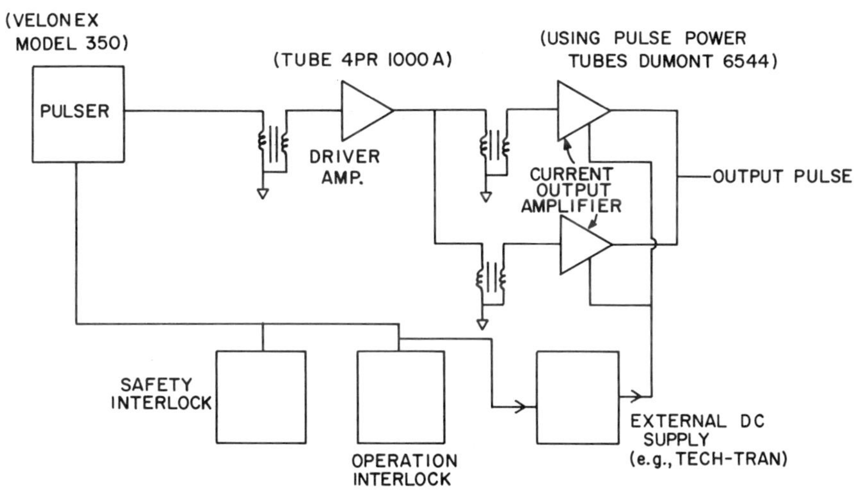

Fig. 5 Schematic diagram of the high power, high repetition rate pulser used at the author's laboratory to generate the helium and argon continua. (Circuit developed by J. R. Haumann, Argonne National Laboratory.)

improvement in light intensity by a factor four to five. The cost of the thyratron-triggered discharge is ~\$10,000, the high-power pulser, \$15,000. It is also possible to employ a commercial 100 kc pulser without the high power amplification. The cost (~\$5,000) is substantially less, as is the light intensity. Light output also increases approximately with the length of the lamp since the emitted light is not self-absorbed. Practical limitations exist here in initiating the discharge and supporting the lamp structure. A convenient lamp length of ~50 cm is employed at Argonne. The purity of helium can significantly influence the light intensity achieved with this lamp, presumably because of collisions of the second kind which transfer the excitation energy of helium to the impurities. One must often select tanks of low impurity content from among the commercial grades of helium, or use ultrapure helium. Purification through a liquid-nitrogen cooled zeolite or charcoal trap is generally helpful.

Clogging of inlet slits, which may occur rapidly when using narrow slits, can be retarded by introducing gas near the front of the lamp and pumping at the rear, thereby maintaining a counterflow that removes sputtered and entrained particles from the vicinity of the slits.

(c) Electron synchrotrons and storage rings have been used as light sources for a number of years. In the investigation of gas phase systems, they have been used for photoabsorption studies at the outset. Recently, as synchrotron intensities have increased, some investigators have begun to employ these devices as light sources for photoionization mass spectrometry.[7] It is obviously of considerable interest to compare the useable light intensity from a synchrotron with that of modern laboratory light sources, so that one can judge whether transporting one's apparatus to a synchrotron site is worth the candle.

The absolute light intensity, spectral distribution, and polarization of synchrotron radiation are calculable from the Schwinger equation.[8] Several review articles have described experimental tests of this equation and have provided simplified forms for different spectral domains.[9] In the final analysis, however, one needs to know the useable wavelength-selected radiation emanating from the exist slit of a monochromator and how this compares with that attainable from a laboratory light source. An experiment of this type was recently performed by us,[10] comparing the high power pulsed helium continuum with the effective light output of the 240 MeV storage ring at Stoughton, Wisconsin. The apparatus used for this experiment (3 m normal-incidence monochromator, experimental chamber, and aluminum photocathode) is displayed in Fig. 6. Note that substantial differential pumping is required in the storage ring experiment when gas samples are introduced in order to maintain the requisite high vacuum (10^{-9} torr) in the storage ring. Diffusion pumps are forbidden for this purpose, and one must employ turbomolecular and cryogenic pumping. Apart from these

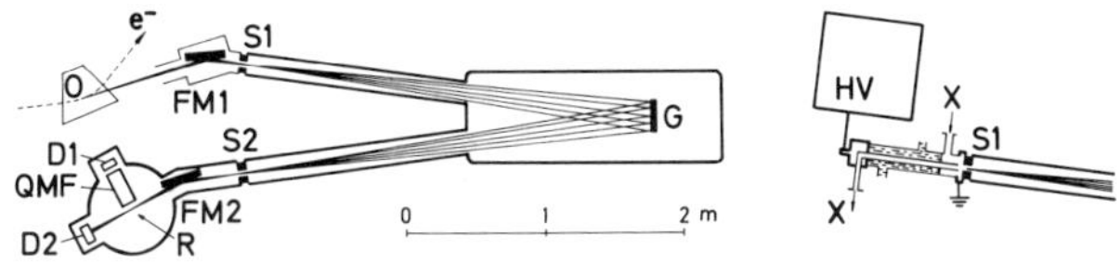

Fig. 6 Schematic diagram of apparatus used with synchrotron and pulsed discharge light sources. e^-, electron orbit of synchrotron; O, bending magnet, origin of synchrotron radiation; *FM* 1, focusing mirror; *S* 1, entrance slit of 3 m monochromator; *G*, diffraction grating; *S* 2, exit slit; *FM* 2, focusing mirror; *R*, photon-gas interaction zone; *QMF*, quadrupole mass filter; *D* 1, ion detector; *D* 2, photon detector; *HV*, high voltage pulsed power supply; *X*, purified rare gas source. [From K. Radler and J. Berkowitz, *J. Opt. Soc. Am.* **68**, 1181 (1978).]

pumping features, the identical apparatus and arrangement was used for both experiments.

The results of the comparison are summarized in Fig. 7. Curve a represents the calculated spectral output of the Stoughton storage ring at 240 MeV

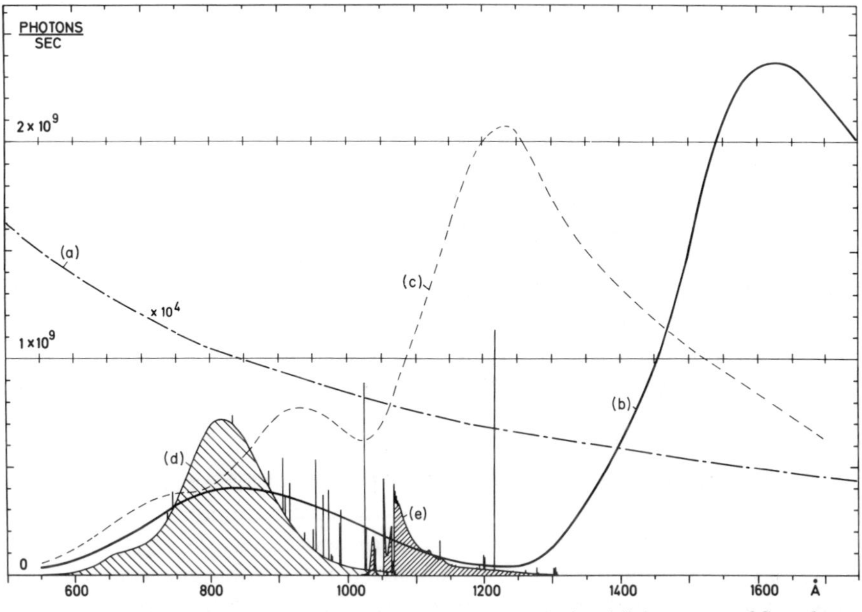

Fig. 7 Comparison of spectral intensities. Curve (a). Calculated light output of Stoughton storage ring at 240 MeV and 30 mA electron current, with 50 mrad horizontal acceptance and total vertical acceptance. Curve (b). Measured light output of Stoughton storage ring after passing through mirrors and grating described in Fig. 6. Conditions: Orbiting electron energy, 230 MeV; electron current, 30 mA; monochromator slits both 100 μ. Curve (c). Measured light output of DORIS storage ring after passing through a 3 m monochromator with 100 μ slits. Curve (d). Measured light output from helium continuum lamp. The physical arrangement was identical to that used in obtaining curve (b). Curve (e). Same as curve (d), except that argon continuum was generated, [From K. Radler and J. Berkowitz, *J. Opt. Soc. Am.* **68**, 1181 (1978).]

and 30 mA electron current, with 50 mrad horizontal acceptance and total vertical acceptance. Curve b is the observed spectral output after traversing the optics of our apparatus, as shown in Fig. 6. The photoelectric current at the light detector has been corrected for the spectral dependence on the photoelectric yield. Above 1100 Å, the light intensity was measured with a sodium salicylate-coated multiplier. The curves were matched at 1000 Å. Curve c displays the measured output from the DORIS ring through a similar apparatus when the ring is operated at 1.8 GeV and 200 mA current. Curve d is the measured output of our helium continuum under conditions similar to b, i.e., same grating, detector, and slit widths (100 μ). Curve e is similar to d, but with the argon continuum.

In comparing curves b and d, the following conclusions can be drawn:

(1) At ~800 Å, the two sources are approximately equivalent. The synchrotron output (curve b) is shown weaker here, but the experiment was performed with 30 mA circulating current. At the time of our experiments, the injection current was typically ~100 mA, but with a half-life of ~90 min.

(2) Below 600 Å, the synchrotron source is distinctly preferable. Below 584 Å, the helium continuum produces almost no detectable light, whereas the synchrotron source provides useable light to 500 Å with the grating available for this experiment, and to still shorter wavelengths with gratings blazed for shorter wavelengths.

The grating used for these experiments was a 1200 1/mm replica blazed for 1360 Å, coated with aluminum, and overcoated with MgF_2. It has less than optimum reflectivity below 600 Å and also appeared to have a large absorption "hole" around 1200–1250 Å. Hence, the synchrotron radiation also has advantages on the long wavelength side of the helium continuum between 950–1100 Å, where it comes into the range of the argon continuum. Above 1300 Å, there is substantial intensity from the synchrotron source.

Several qualifying statements must be kept in mind.

(1) For photoionization measurements, the final focusing mirror can be eliminated. This would increase the ordinate scale by about a factor of three for both helium continuum and synchrotron source.

(2) The experiments described employed horizontal reflection and diffraction optics. The horizontally polarized synchrotron radiation has substantially higher reflectivity with vertical reflection and diffraction optics. The effective spot size of the synchrotron source, and its concentration in the horizontal plane, also favor imaging through a slit long in the horizontal dimension. Hence, if one can utilize a vertical diffraction monochromator and corresponding reflection components, one can gain perhaps a factor four.

(3) The undispersed radiation emanating from the 240 MeV storage ring has a calculated spectral distribution of radiation shown in curve a of Fig. 7 (see also Fig. 26). The maximum occurs at ~200 Å. Hence, it is unavoidable that some second-order radiation enters the experimental chamber when a single dispersing element is used. It is particularly difficult if one wishes to work in second or higher order with 600–1000 Å radiation from the storage ring since the first-order radiation at 1200–2000 Å swamps the second-order radiation. No such problem exists with the helium continuum source.

Can we rationalize the scale of undispersed radiation intensity (curve a of Fig. 7) with the intensity observed after passing through the dispersing and reflecting optics (curve b of Fig. 7)?

At 800 Å, the calculated gross intensity emitted by the Wisconsin storage ring is

$$\frac{6.8 \times 10^9 \text{ photons}}{\text{sec} - \text{Å} - \text{mA} - \text{mrad}}.$$

In our experiment, the bandwidth was 0.28 Å, the acceptance angle 16.6 mrad, and the storage ring current was 50 mA, which should provide 0.95×10^{12} photons/sec. The estimated loss factors are premirror: 2.5, ratio of image spot size to entrance slit width: 6, grating inefficiency: 50, and refocusing mirror: 3. With this total loss factor of 2250, the calculated net intensity is 4.2×10^8 photons/sec at 800 Å, compared to the observed value of $\sim 3.8 \times 10^8$ photons/sec. In view of the uncertainties in some of the factors entering into the calculation, the agreement must be considered satisfactory.

We are now in a position to compare with other synchrotron and/or storage ring sources. Figure 8 displays calculated spectral outputs of some

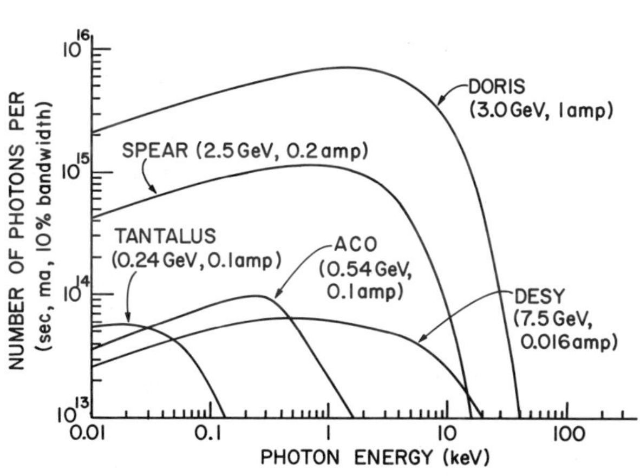

Fig. 8 Calculated spectral outputs of some active storage ring sources and one synchrotron at conditions not yet necessarily realized. [From K. Radler and J. Berkowitz, *J. Opt. Soc. Am.* **68**, 1181 (1978).]

active storage ring sources and one synchrotron (DESY). We note that in the normal incidence VUV region and up to ~50 eV, the Wisconsin storage ring (Tantalus), DESY, and ACO are quite comparable. The SPEAR ring is presently operating at $\frac{1}{10}$ of the current plotted, whereas the big DORIS ring is performing at about $\frac{1}{5}$ of the current shown. The higher currents are displayed to provide a feeling for the probable level of performance in the next few years. In Table 1, the significant parameters of other synchrotron and storage ring sources currently in operation are listed so that their spectral outputs can be roughly interpolated by reference to Fig. 8. Table II has similar information concerning such sources in construction as of June 1978, while Table III includes additional facilities proposed as of June 1978. The author is indebted to Dr. Herman Winick of the Stanford Synchrotron Radiation Facility for providing him with these compilations.

We have mentioned that a grating blazed to shorter wavelength can enable measurements to be made down to ~300–350 Å with a normal incidence monochromator. For studies at still shorter wavelengths, it is necessary to use a grazing incidence monochromator. Measurements with a grazing incidence device pose further problems, to be elaborated in the next section.

2. Monochromators

While perhaps not essential, it is most desirable that the entrance and exit slits of a monochromator intended for photoionization mass spectrometry remain fixed while the wavelength is scanned, and the image at the exit slit must remain in focus. Without this provision, relatively bulky apparatus must ride on a movable slit, maintained under vacuum. The constraints are inconvenient and can affect the ultimate resolution. The light emanating from the exit slit should also remain fixed in direction during wavelength variation.

In the normal incidence region, several designs exist which satisfy these requirements. The simplest is the Seya-Namioka solution. When the angular separation between entrance and exit beam is 70° 15′, a simple rotation of the concave grating varies wavelength and maintains the image in approximate focus. However, the image is astigmatic and curved. The length of the image is much larger than that of the object slit, resulting in a loss of intensity of perhaps a factor of three. The curvature of the image limits the resolution. Usually, Seya-Namioka monochromators are not used for high resolution (slit widths $<50\ \mu$) without corrective optics which require special features (curved slits, additional reflections).

For precise focusing, the slits and grating center must lie on the Rowland circle (ref. 1, p. 5) as the wavelength is changed. A somewhat more com-

TABLE 1 *Synchrotron Radiation Sources—Storage Rings in Operation—June 1978*[a]

Machine, location	E (GeV)	I (mA)	R (m)	ε_c (keV)	Remarks
VEPP-4 Novosibirsk, USSR	7 4.5	10	16.5 (18.6)	46.1 (10.9)	Initial operation at 4.5 GeV (from 8 KG wiggler-part of magnet lattice)
DORIS Hamburg, Germany	2.2 5	300 50	12.1	1.95 22.9	
SPEAR Stanford, USA	2.25 4.0	300(10) 100(35)	12.7	2.0 11.1	( ) indicates colliding beam current limits
VEPP-3 Novosibirsk, USSR	2.25	100	6.15 (2.14)	4.2 (11.8)	Operation to 3 GeV is possible (from 35 KG wiggler-planned)
DCI Orsay, France	1.8	500	4.0	3.63	
ADONE Frascati, Italy	1.5	60	5.0 (2.8)	1.5 (2.7)	Synchrotron Radiation lines in construction (from 18 KG wiggler-planned)
VEPP-2M Novosibirsk, USSR	.67	100	1.22	0.54	
ACO Orsay, France	0.54	100	1.1	0.32	Dedicated[b]
SOR Ring Tokyo, Japan	0.40	250	1.1	0.13	Dedicated[b]
TANTALUS I Wisconsin, USA	0.24	200	0.64	0.048	Dedicated[b]
SURF II Washington, DC, USA	0.25	25	0.84	0.036	Dedicated[b]
N-100 Karkhov, USSR	0.10	25	0.50	0.004	

[a] The author is indebted to Dr. H. Winick of the Stanford Synchrotron Radiation Laboratory for providing him with this compilation.

[b] Dedicated machines are used only for synchrotron radiation research. Other machines are used for both high energy physics research and synchrotron radiation research.

plicated motion, involving translation as well as rotation of the grating as the wavelength is scanned, can achieve this condition. The angle between entrance and exit arms is commonly 15°, reducing astigmatism to acceptable limits, and the image remains linear. Various mechanical designs exist for this motion, among them the McPherson cam[11] and the Robin mount.[12]

TABLE 2 *Synchrotron Radiation Sources—Storage Rings in Construction—June 1978*[a]

Machine, location expected completion date	E (GeV)	I (mA)	R (m)	ε_c (keV)	Remarks
PETRA[b]	18	18	192	67.4	
Hamburg, Germany	15	80		39.0	
1978	10	50		11.6	
PEP[b]	18	10	165.5	78	
Stanford, USA	15	55		45.2	
1979	10	35		13.4	
	12	45	(23.6)	(163)	(From 17 KG wiggler-part of PEP lattice)
CESR[b]	8	100	32.5	35	Synchrotron Radiation
Cornell, USA	4	50		4.4	facility planned
1979					
PHOTON FACTORY	2.5	500	8.33	4.16	Dedicated
Tsukuba, Japan			(1.67)	(20.5)	(for a 50 KG wiggler)
1982					
NSLS					
Brookhaven Natl. Lab	2.5	500	8.17	4.2	Dedicated
USA, 1981			1.67	20.5	(for a 50 KG wiggler)
SRS	2.0	500	5.55	3.2	Dedicated
Daresbury, UK			(1.33)	(13.3)	(from 50 KG
1979					wiggler-planned)
ALADDIN	1.0	500	2.08	1.70	Dedicated
Wisconsin, USA					
1980					
BESSY					
Berlin, W. Germany					
1982	0.80	500	1.83	0.62	Dedicated; industrial use planned
NSLS					
Brookhaven Natl. Lab	0.70	500	1.90	0.40	Dedicated
USA, 1981					

[a] The author is indebted to Dr. H. Winick of the Stanford Synchrotron Radiation Laboratory for providing him with this compilation.

[b] PETRA, PEP, and CESR will operate over a range of energies. Examples of expected characteristics are given.

With the commercial McPherson apparatus, one can maintain satisfactory resolution with 10 μ slits in a carefully aligned system. Thus, with a 1200 1/mm grating (1 m radius of curvature), the dispersion is 8.3 Å/mm. With 10 μ slits, the theoretical resolution (FWHM) is 0.083 Å, and 0.12 Å is attainable. Working in the third order, experiments[13] have achieved 0.04 Å (FWHM). The corresponding performance of a 3 m apparatus is claimed[14] to be 0.016 Å.

TABLE 3 *Synchrotron Radiation Sources—Storage Rings Proposed as of June 1978*[a]

Machine, location	E (GeV)	I (mA)	R (m)	ε_c (keV)	Remarks
ERSINE					
Erevan, USSR	2.5	500	5.21	6.67	Dedicated
PAMPUS					
Amsterdam, Netherlands	1.5	500	4.17	1.80	Dedicated (2 GeV operation possible)
IPP					
Moscow, USSR	1.35	100	2.5	2.2	Dedicated (2 GeV also considered)
CSRL					
Canada	1.2	100	3.1	1.33	Dedicated
Electrotechnical Lab Tokyo, Japan	0.6	100	2.0	0.24	Dedicated

[a] The author is indebted to Dr. H. Winick of the Stanford Synchroton Radiation Laboratory for providing him with this compilation.

In the grazing incidence region (< 500 Å), the desired conditions of fixed slits and exiting light direction, while retaining focus, are more difficult to achieve. Various designs have been constructed, primarily for use with synchrotrons, but, up to this time, have been mostly employed for absorption studies. Kunz[15] has reviewed the various designs of grazing incidence and crystal monochromators proposed or in use as of early 1973. Several of these warrant consideration here.

(a) The Miyake *et al.*[16] design used at the Tokyo synchrotron is one of the simplest. The synchrotron light passes through a slit onto a plane grating, from whence it is diffracted onto a spherical focusing mirror. The latter focuses the selected wavelengths on a fixed exit slit. Rotation of the grating varies wavelength, while maintaining a fixed *sum* of entrance and exit angles. This simple, convenient arrangement unfortunately permits a considerable quantity of stray light to emerge.

(b) The Kunz *et al.*[17] design used at DESY has the incident radiation strike a plane mirror, then a plane grating, and a paraboloidal mirror which focuses the selected wavelength onto a fixed exit slit. The motion of the premirror and grating is rather complex. Higher-order radiation is reported to be efficiently suppressed over the entire operating range. The instrument is arranged to always operate at the blaze maximum, independent of wavelength. It covers the energy range 15–280 eV with a resolution of $\frac{1}{400}$ to $\frac{1}{800}$.

(c) The design of Codling and Mitchell[18] currently being used at Daresbury, has the synchrotron light focused by a mirror onto an entrance slit–mirror combination. The radiation which is passed and reflected by the entrance slit is incident upon a concave grating, which travels along the Rowland circle. The diffracted light of selected wavelength passes through an exit slit–mirror combination and is reflected onto the target. The slits rotate, but do not translate, so that this requirement for photoionization studies is met.

(d) Madden and Ederer[19] have constructed an instrument tailored to the NBS synchrotron. The almost parallel light from the synchrotron strikes a single optical element, a grating ruled on a toroidal surface, the diffracted spectrum being selected by the exit slit. The slight rotation of the grating causes only second-order focusing errors, and the toroidal grating makes a nearly stigmatic focus possible. The design is not easily transferable since the grating is ruled for a particular geometry. Furthermore, since the grating is an original, one must pay particular attention to contamination or imperfection of the grating.

Two important devices not covered in the Kunz review are the so-called "grasshopper"[20] constructed at the SPEAR ring and the double grating[21] instrument used at the ACO ring at Orsay. In the grasshopper arrangement (see Fig. 9), a large (40 cm) horizontally deflecting mirror located six meters from orbit sends synchrotron radiation onto a second, vertically deflecting mirror which focuses this radiation onto the horizontal entrance slit S_1. Both

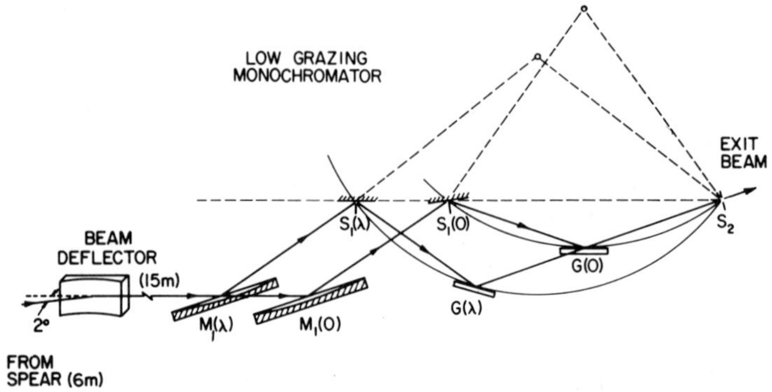

Fig. 9 Optical arrangement of the "grasshopper" grazing incidence monochromator used at the SPEAR storage ring. M_1, mirror; S_1, entrance slit; G, grating; S_2 exit slit. Elements are shown in two positions, corresponding to wavelength λ and to zero order. Angles are highly exaggerated. [From F. C. Brown *et al.*, "Vacuum Ultraviolet Radiation Physics," Proc. IV. Int. Conf. on VUV Radiation Physics, p. 785. Pergamon-Vieweg, Braunschweig, 1974.]

S_1 and the exit slit S_2 are mirror–slit combinations, as described by Codling and Mitchell. The rotating, but nontranslating, slits and the motion of the grating along the Rowland circle are also the conditions used by the Daresbury group, but the mechanical details differ. Ultrahigh vacuum technology is used throughout, the feed screw and drive system being external to the vacuum system. The first mirror is platinum coated copper which is cooled and serves to filter out hard x-ray radiation, thereby sparing the subsequent optical elements. The elements M_1, S_1, and G translate parallel to the incoming beam as a unit.

One possible disadvantage of this arrangement involves its four reflections and one diffraction, which can cost intensity in the vacuum ultraviolet.

The Jaegle[21] design is intended to remove higher-order components (see Fig. 10a). Radiation from the synchrotron is reflected by a focusing pre-mirror onto the entrance slit S_1. The light falling on the first grating G_1 is diffracted and strikes the intermediate mirror M. This mirror makes the various diffracted beams convergent on the midpoint of the second grating G_2, which is the optical image of the first grating's midpoint. Only a narrow band of the spectrum diffracted by G_1 is reflected by M onto G_2. All optical elements except the prefocusing mirror are mounted on a horizontal 50 cm radius Rowland circle. The exit slit S_2 is again a mirror–knife edge of the Codling type, so that the selected wavelength remains fixed in position and direction as one scans. The wavelength scan is accomplished by driving the mirror M and grating G_2, the exit slit rotating through $\frac{1}{4}$ of the angle traversed by G_2.

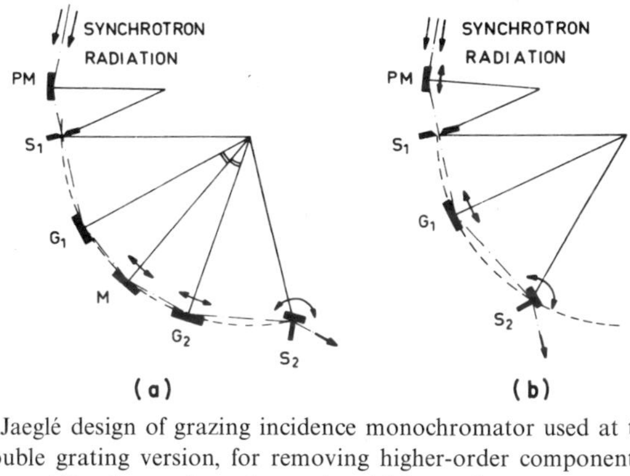

Fig. 10 The Jaeglé design of grazing incidence monochromator used at the ACO storage ring. (a) The double grating version, for removing higher-order components. (b) The single grating version, providing larger throughout when higher orders are not as troublesome, i.e., in the high energy range. *PM*, premirror; S_1, entrance slit; S_2 exit slit; G_1, G_2, gratings; *M*, mirror; ---- Rowland circle. In both (a) and (b), the direction of the monochromatized radiation is kept fixed by rotating the mirror exit slit when the wavelength range is scanned. [From V. Schmidt *et al.*, *Phys. Rev. A* **13**, 1748 (1976).]

This monochromator is very effective in removing higher orders, but at the price of three reflections and two diffractions. When higher orders are not as troublesome (high energy range), the one-grating mode (Fig. 10b) is used. In this case the wavelength is scanned by driving the grating G_1 along the Rowland circle, while simultaneously moving the prefocusing mirror.

Recently, the Jaegle-designed monochromator has been used with the ACO storage ring at Orsay to study double photoionization in He, Ne, and Ar, using magnetic mass analysis.[22] This is one of the few examples of photoionization mass spectrometry with dispersed radiation in the high (40–200 eV) photon energy range.

As of this writing, commercial grazing incidence monochromators with the required properties are not available.

Even when available, commercial monochromators must generally be modified for photoionization mass spectrometric purposes. The vacuum housing of the exit arm should be cut back to allow the exit slit to protrude so that it is in close proximity to the ionization region, which, in turn, is determined by the ion optics. If the ionization chamber is maintained at high voltage (as is common with magnetic mass spectrometers), the mounting of the exit slit must be appropriately insulated from ground potential. Since it is very difficult to retain bilaterally adjustable slits under these conditions, changes in slit width must either be made manually after venting, or a selection of fixed slits must be sequentially positioned along accurate slides which can be controlled through a vacuum seal.

To date, the gratings most frequently used in these monochromators are replicas of ruled gratings. The gratings are blazed (see Samson,[1] p. 26) to enhance the intensity of diffracted light in selected wavelength regions. Holographic gratings have recently become available. They are reported to have much lower scattered light than ruled replicas, and can, in principle, be made to compensate for aberrations in the diffraction optics. At the present time, there is sparse experience relating the relative performance of the two types of gratings. The diminution of scattered light attributable to holographic gratings may be partly illusory in practice. Monolayers of hydrocarbons or other impurities in a vacuum system, when bombarded with vacuum ultraviolet light, tend to form polymeric layers on gratings in the course of time, which may become the dominant cause of scattered light with both ruled and holographic gratings.

The usual photoionization mass spectrometric measurements involve long counting times, and some form of automation becomes necessary. A common practice is to use a stepping motor, either directly or through a gear system, preferably connected directly to the drive screw that varies the position of the grating. A timing device can be arranged to count ion and light signals for a preset time, after which a pulse is sent to an indexer which relays information to the stepping motor to advance a preset number of steps, corresponding to a wavelength increment. The wavelength scale for the normal incidence devices described above is very nearly linear, and hence a few calibration points suffice to determine the wavelength scale. We have found, however, that some stepping motors are notoriously uneven in step size at the limits of step resolution. When this is the case, it is most desirable to incorporate a shaft encoder directly coupled to the drive screw.

There has been little activity in photoionization mass spectrometry in wavelength regions shorter than ~ 50 Å (somewhat arbitrarily, the low energy cutoff of grazing incidence monochromators). (There have been some studies with fixed wavelength x-ray sources, obviating the need for monochromators.) Traditionally, crystals have been used as the dispersive elements in this high energy region. Recently, Bragg reflection from parallel surfaces of a channel-cut silicon crystal has been used[23] to monochromatize the synchrotron radiation at SPEAR out to 9 keV (see Fig. 11). The high degree of collimation (in the vertical plane) from the storage ring enables one to achieve an energy resolution of 1 eV at 9 keV, using the Si (220) planes. The output beam is parallel to the input beam, and is displaced vertically by $2d \cos \theta$, where θ is the Bragg angle and d the distance between the reflecting surfaces.

Although the resolution of these elements is not as high as that attainable from dispersive crystal monochromators, the simplicity of the arrangement and the small light loss appear to make this technique highly attractive for photoionization mass spectrometry in the largely unexplored x-ray region.

Another intriguing development in the 10–1000 eV domain is the proposed[24] use of transmission gratings, which (it is claimed) can provide resolution of $\frac{1}{2000}$ with higher transmission than multiple reflection monochromators.

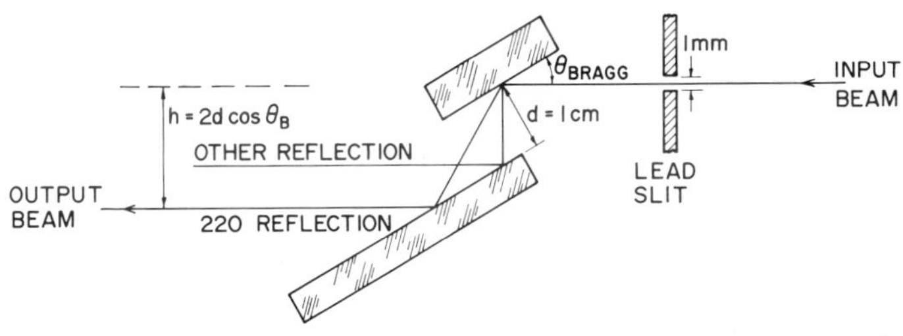

Fig. 11 Schematic diagram of channel-cut silicon crystal monochromator used to monochromatize x-rays at the SPEAR storage ring. [From the Stanford Synchrotron Radiation Project SSRP Rept. No. 76/100, p. 150, Aug. 1976.]

3. Light Detectors

Various techniques are available here, and the selection is to some extent a matter of taste. Direct photon counting is possible. One must use a multiplier at high voltage, and this implies either a limitation in the allowable pressure in the vacuum chamber, or additional differential pumping between the chamber surrounding the ionization region and the photodetector chamber. Care must be taken not to overload the counter (photon rates of 10^8–10^9 per second are not uncommon) and the wavelength dependence of the photocathode must be determined.

Alternatively, one can use a multiplier at significantly lower voltage to measure dc current, which is converted to voltage by allowing the multiplied photocurrent to pass across a large resistor. The resulting signal can be digitized for counting purposes with the aid of a voltage-to-frequency converter. We have found this technique a useful one for low level signals (high resolution studies). A simple metal plate with extracting field is adequate for higher level signals. In both instances, the spectral yield of the photocathode must be determined.

A rather different approach that seems attractive at first glance is to employ a phototransducer, such as sodium salicylate. A thin film of this material applied to a glass surface will fluoresce when vacuum ultraviolet radiation impinges upon it. Furthermore, the fluorescence spectrum is independent of the wavelength of the incident light, and the quantum yield is constant over a large wavelength range.[25] In principle, one can either coat the glass surface of a photomultiplier tube sensitive to the fluorescent radiation, or coat a glass surface that serves as the vacuum seal with sodium salicylate or another fluorescent material.[26] The fluorescent signal can then be monitored by photon counting or the dc method (with voltage-to-frequency conversion) can be employed. Advantages include isolation of the photodetector from the not-too-high vacuum in the ionization region, and a well-established wavelength dependence of the photodetector. This device is sensitive to stray light from ion gauges, heating or ionizing filaments, or room light. In practice, we have found that geometric and other losses and inherent noise in the photomultiplier (which can be minimized by cooling the photomultiplier) make the signal: noise response of this detector somewhat less satisfactory than the bare multiplier approach.

In all of the above methods, care must be taken to assure that either *all* of the light beam causing ionization, or a *representative sample* of this beam, reach the light detector. It has been found that the efficiency of different portions of a grating may differ. This can result in spurious structure when ion intensity is divided by light intensity as one passes through structure in the light source if the above precautions have not been taken.

4. Ion Detectors

For the study of permanent gases, it is advantageous to employ an enclosed ionization chamber to enhance the probability of photoionization. (One possible exception is the use of an aerodynamic nozzle to cool the gas, which would necessitate using a molecular beam source.) To minimize absorption corrections, it is convenient not to have more than, say, 0.1% of the incident light beam absorbed by the gas. A characteristically large absorption cross section would be 1000 Mb and a reasonable path length about 2 cm. Under such circumstances, the pressure of target gas in the ionization chamber should not exceed 10^{-3} torr. If one has reason to believe that the cross section will not exceed 100 Mb, or if one is content to permit 1% gas absorption, the target gas pressure can be raised to 10^{-2} torr. Already at this pressure, collisional effects (ion–molecule reactions, chemi-ionization, collisional ionization) are playing a role.

Another limiting factor is the pressure in the vacuum envelope surrounding the ionization chamber. We would like to extract and measure as large a fraction as possible of the photoions formed. For order-of-magnitude calculations, let us assume an ion extraction orifice of area 1 cm^2. With a pumping speed of 1000 liter/sec (on the good side of typical), the vacuum envelope would have a steady state pressure two orders of magnitude lower than that in the ionization chamber. To minimize scattering effects, the pressure in the vacuum envelope should not exceed 10^{-4} torr, and hence, the ionization chamber pressure should not exceed 10^{-2} torr. In more detailed considerations, there is a tradeoff between smaller orifice for ion extraction and higher ionization chamber pressure that depends upon the ion optics employed and the other limits on ionization chamber pressure mentioned previously. For operation with a molecular beam (either high temperature vapor, free radical source, or supersonic nozzle), the absorption of light by the beam is not likely to be a limiting factor, but the constraints due to pumping speed still exist for the latter two types of experiments.

The ion optics from ionization chamber through mass analyzer to particle detector depend to some extent on the type of mass analyzer to be used. Rather standard mass spectrometric practice is used, with emphasis upon high transmission. Magnetic deflection or quadrupole mass spectrometers are commonly used. Time-of-flight mass spectrometers have small duty cycles and result in poorer signal: noise ratios; they can be useful in some experiments that make use of their special features, such as coincidence mass spectrometry and measurement of ion kinetic energy (see Section D in this chapter).

The routine ion optics of a mass spectrometer includes an ion repeller plate located in the rear of an ionization chamber, typically a few volts positive with respect to the ionization chamber (negative for negative ions). Then, in sequence, one has a drawout plate, a second focusing plate, and a set of deflection plates (for directing the beam onto an entrance slit) or two sets of deflection plates at right angles (for directing the beam onto a circular orifice, as in a quadrupole mass spectrometer).

For obtaining high transmission through a magnetic sector mass spectrometer, we have found that an ion optical system using strong-focus principles is quite effective. The idea is simple, the calculation is straightforward, and it works. Visualize the ion exit orifice of the ionization chamber. For maximum transmission, our goal is to produce an ion beam which converges to a crossover at the entrance slit width of the mass spectrometer, while remaining parallel or slightly convergent along the slit length. The electrostatic quadrupole ion lens enables one to achieve this effect while gathering ions from a fairly large area.[27] In such a lens, the ions pass through a field which has the potential distribution

$$V = \pm V_0(r^2/R_0{}^2)\cos 2\theta. \tag{4}$$

The equipotentials for such a field are rectangular hyperbolas. This field can be adequately approximated by the use of circular cylinders (see Fig. 12) with much simpler construction.

A convenient matrix formulation has been devised[28] for the properties of such lenses, in which transfer matrices are given for a converging section, a diverging section, and for drift spaces. The transfer matrix for a combination of elements is then the matrix product of these, taken in the proper order.

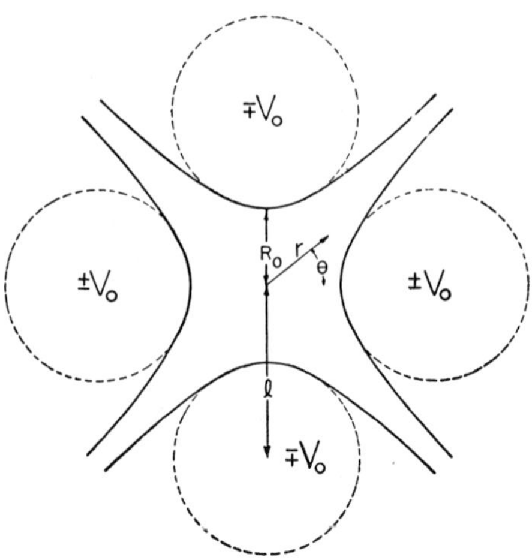

Fig. 12 Cross section of an electrostatic quadrupole lens. The equipotentials for the field should have $V = \pm V_0\,(r^2/R_0{}^2)\cos 2\theta$. The circular cylinders (dotted lines) are used to approximate this field. [From C. F. Giese, *Rev. Sci. Instr.* **30**, 260 (1959).]

Let us define the parameter

$$K = (1/R_0)(V_0/V_a)^{1/2}, \tag{5}$$

where V_0 and R_0 are defined in Fig. 12 and V_a is the ion energy. For D, the effective length of the lens element, we may take the physical length plus $\frac{1}{4}R_0$ at each end to allow for the fringing field. Then, from Teng,[28] the convergent matrix operating on the slit width dimension can be written

$$\begin{pmatrix} \cos KD & (1/K)\sin KD \\ -K\sin KD & \cos KD \end{pmatrix} \begin{pmatrix} y_0 \\ y_0' \end{pmatrix} = \begin{pmatrix} y_1 \\ y_1' \end{pmatrix}, \tag{6}$$

where y_0 and y_0' are the initial position and slope of the beam in the slit width direction, and y_1, y_1' are the resultant position and slope in that direction. By arranging for a uniform accelerating field from the ion exit orifice, y_0' can be made nearly zero. A gauze placed across the ion exit orifice can help to avoid lens action in this region. We recall here that in magnetic mass spectrometers, it is conventional to accelerate from high potential in the ionization chamber to ground potential at the magnetic field, which is immediately preceded by the quadrupole lenses.

While the lens is converging the beam along the slit width (y) direction, it is diverging the beam along the slit length (z) direction in a manner described by the matrix formula

$$\begin{pmatrix} \cosh KD & (1/K)\sinh DK \\ K\sinh KD & \cosh KD \end{pmatrix} \begin{pmatrix} z_0 \\ z_0' \end{pmatrix} = \begin{pmatrix} z_1 \\ z_1' \end{pmatrix}. \tag{7}$$

Hence, starting with a certain width and height of the ion orifice and specified lens parameters, we have calculated the resultant positions and slopes of the ion beam.

We now allow for a drift space (whose effective length is ρ) before entering the second quadrupole. The matrix notation for the drift space acting on the new position of the ion beam is

$$\begin{pmatrix} 1 & \rho \\ 0 & 1 \end{pmatrix} \begin{pmatrix} y_1 \\ y_1' \end{pmatrix} = \begin{pmatrix} y_2 \\ y_2' \end{pmatrix} \tag{8}$$

and

$$\begin{pmatrix} 1 & \rho \\ 0 & 1 \end{pmatrix} \begin{pmatrix} z_1 \\ z_1' \end{pmatrix} = \begin{pmatrix} z_2 \\ z_2' \end{pmatrix}. \tag{9}$$

The second quadrupole lens is made divergent in the y direction, convergent in the z direction. Thus,

$$\begin{pmatrix} \cosh KD & (1/K)\sinh KD \\ K\sinh KD & \cosh KD \end{pmatrix} \begin{pmatrix} y_2 \\ y_2' \end{pmatrix} = \begin{pmatrix} y_3 \\ y_3' \end{pmatrix} \tag{10}$$

and

$$\begin{pmatrix} \cos KD & (1/K)\sin KD \\ -K\sin KD & \cos KD \end{pmatrix}\begin{pmatrix} z_2 \\ z_2' \end{pmatrix} = \begin{pmatrix} z_3 \\ z_3' \end{pmatrix}. \tag{11}$$

Finally, we have a second drift space between the second quadrupole lens and the entrance slit to the mass spectrometer

$$\begin{pmatrix} 1 & \rho \\ 0 & 1 \end{pmatrix}\begin{pmatrix} y_3 \\ y_3' \end{pmatrix} = \begin{pmatrix} y_4 \\ y_4' \end{pmatrix}, \tag{12}$$

$$\begin{pmatrix} 1 & \rho \\ 0 & 1 \end{pmatrix}\begin{pmatrix} z_3 \\ z_3' \end{pmatrix} = \begin{pmatrix} z_4 \\ z_4' \end{pmatrix}. \tag{13}$$

Our constraints are that y_4 must not exceed the entrance slit width of the mass spectrometer and z_4' must be near zero. We prefer to choose z_4' so that the beam crosses over in this dimension midway in its trajectory through the magnetic sector.

A further simplification results if we choose to make the lenses symmetric, i.e., if KD remains the same for the two lenses. A set of solutions which we have found satisfactory, and which can serve as a starting point for other workers, is as follows:

Actual length of cylinders	1.329 in.
Effective length of cylinders	1.640 in.
Actual length of drift space	0.860 in.
Effective length of drift space	0.546 in.
R_0	0.625 in.
V_0 for first section	$V_a/16$
V_0 for second section	$V_a/20$
Cross section of entering beam (i.e., nearly 1 cm^2).	0.375×0.375 in.^{2}

This solution allows for crossover in the z direction in the middle of the magnetic field of a 60° sector, 12 in. radius magnetic mass spectrometer. The influence of the initial Maxwellian distribution in velocities of the neutral molecular source, which is reflected in the ion velocities at the ion orifice, can be readily incorporated into the transfer matrices. In practice, the lenses are tuned for maximum transmission in the approximate domain of hyperspace for which the parameters have been calculated.

With higher incident photon energies and smaller molecules, fragment ions may be formed with substantial kinetic energy. To achieve a collection efficiency more-or-less independent of initial ion kinetic energy, it is advantageous to apply a strong drawout field at the exit orifice of the ionization chamber.[29,30]

The ions emerging from a quadrupole mass spectrometer may have larger divergence than those from a magnetic mass spectrometer, and hence an additional large orifice plate may be helpful in converging these ions onto the first stage of a particle multiplier. Various ion multipliers can be employed, including box-type, Venetian blind, or channeltron multipliers. For the divergent rays of a quadrupole mass spectrometer, multipliers with large entrance areas may be preferred. With discrete dynode multipliers, 20 stages have been found useful.

The technology of quadrupole mass spectrometers and magnetic mass spectrometers is well documented, commercially available, and would unnecessarily expand this monograph. Similar comments apply to the mechanics of pulse counting.

Some general comments are in order regarding the relative merits of quadrupole and magnetic mass spectrometers. Quadrupole mass spectrometers are generally less expensive and more compact. They require different heads for different mass ranges, and the transmission varies with mass. Magnetic mass spectrometers have a transmission which is more-or-less constant with mass (if one scans magnetically) and they can be used to detect "metastable ions" which provide a useful probe in studying the decay of molecular ions (see Section D in this chapter). They also provide a less ambiguous separation of negative and positive ions since quadrupole mass spectrometers do not distinguish between sign, and the distinction must be made with electrostatic deflectors before or beyond the quadrupole mass spectrometer.

C. Photoelectron Spectroscopy

Although some studies of this type were performed in the late 1920s and early 1930s (see Chapter VII on angular distributions), the modern emphasis on photoelectron spectroscopy began, apparently independently, in England[31] and the USSR[32] in the early 1960s. Since that time, a large array of devices has been explored to optimize the energy resolution and sensitivity of electron energy analysis.

Let us try to systematize the various approaches. We can make a gross distinction between spatially dispersive and nondispersive instruments. A spatially dispersive apparatus is one in which electrons having different kinetic energies are deflected into different positions in real space. Among the nondispersive instruments we may include retarding field analyzers, time-of-flight (TOF) analyzers, and a rather special and restrictive type, zero energy electron detectors. The wide variety of analyzers used is indicative of the fact that each has merits and disadvantages, depending upon the source of photoelectrons and the information desired. We consider the nondispersive instruments first.

1. Retarding Field Analyzers

In this class of energy analyzers, the photoelectrons are formed at some well-defined potential V_0, and some distance away the electrons leaving the source encounter a repulsive potential V_R. In order to achieve high transmission, this retarding potential is generally imposed on a rather open grid structure. In principle, all those electrons having an initial kinetic energy larger than $e(V_R - V_0)$ will penetrate the retarding field grid, whereas the slower ones will be reflected and ultimately escape or encounter noncritical walls of the apparatus. In practice, the resolving capability depends upon how well defined the initial source is, the mesh size of the grid system, and the geometry.

There are three common geometrical arrangements—planar, cylindrical, and spherical.

a. Cylindrical Retarding Field. We consider this arrangement first for historical reasons. Both Vilesov *et al.*[32] and Turner and Al-Jobory[31] used this geometry in their initial studies of ultraviolet photoelectron spectroscopy. A beam of monochromatic photons is directed along the axis of a cylinder. An inner cylindrical grid surface should concentrically surround this photon beam and determine the potential V_0. An outer cylindrical grid is then given the potential V_R, and finally, a collector cylinder concentrically surrounds the entire system (see Fig. 13). The target gas permeates the entire apparatus. For sufficiently high values of V_R, no electron signal other than background should be detected. As V_R is decreased, a value is reached at which only the fastest group (a) of electrons can penetrate, giving rise to a step in the graph of count rate versus V_R. At a still lower value of V_R, a second fastest group (b), in addition to group (a), will penetrate, causing another abrupt rise in the graph (see Fig. 14). This is usually referred to as an integral photoelectron spectrum.

In practice, the steps are not sharp rectangles, but exhibit rounding and sloping due to the resolution limitations discussed above, and in this cylindrical geometry because the electron trajectories are not all in planes normal to the cylindrical axis. To minimize this latter problem, slots have been interposed to absorb those electrons deviating by some selected angle from the normal.

This geometric arrangement, though convenient for an extended source of photoelectrons, is rarely used today. Limitations in resolution and the desirability in many cases of having a differential spectrum rather than the integral one obtained here are probably the major reasons. For some ap-

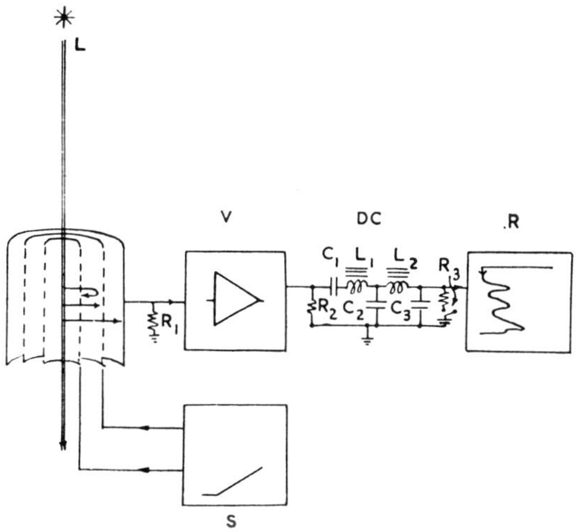

Fig. 13 General circuit arrangement of a cylindrical retarding field spectrometer. L symbolizes a monochromatic light beam along the cylindrical axis. Two concentric grids connected to the sweep generator (S) provide the retarding field. The photoelectron signal is measured by a vibrating reed electrometer (V). This signal is processed by a differentiating circuit (DC) to generate a differential spectrum, which is seen on the recorder (R). [From M. I. Al-Joboury and D. W. Turner, *J. Chem. Soc.* 5141 (1963).]

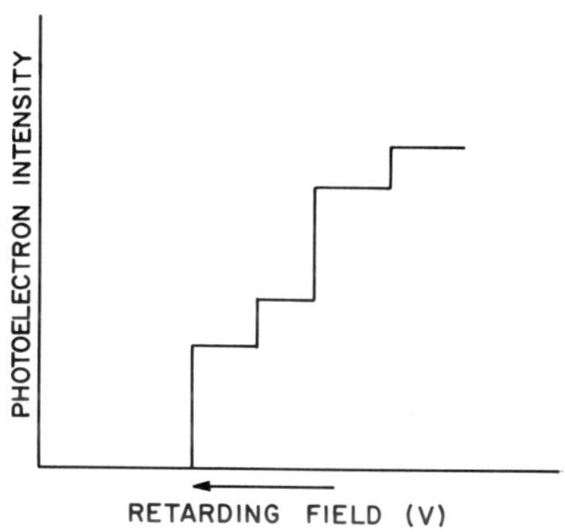

Fig. 14 Schematic illustration of a retarding field photoelectron spectrum.

plications one may actually prefer the integral spectrum, but even in such cases the angular effect of the cylindrical retarding field makes other geometries preferable. It is awkward to adapt counting techniques to this geometry and, up to this time, it has not been found practicable. A differential spectrum can be obtained by electronic or numerical techniques, but the signal-to-noise is not nearly as satisfactory as that obtained by a direct differential spectrum.

b. Planar Retarding Field. This arrangement is sometimes convenient for a quasi-point source of photoelectrons. The more-or-less isotropic source is made parallel by a lens (see Fig. 15 from von Puttkamer). The retarding field is imposed between two plane parallel grids normal to the now parallel electron beam. The photoelectrons that have penetrated the retarding grid can be refocused onto a small region, e.g., the cathode of a particle multiplier, and counting techniques can be conveniently employed. This device is compact and has high transmission. It has been found useful for weak sources (such as PES from photons selected by a monochromator[33] and for photoion–photoelectron coincidence studies (see Section D in this chapter). The resolution of this device is not only limited by the parameters discussed earlier, but also by the quality of the lens. The transmission has also been found to be a strong function of the electron energy for one practical design.[33]

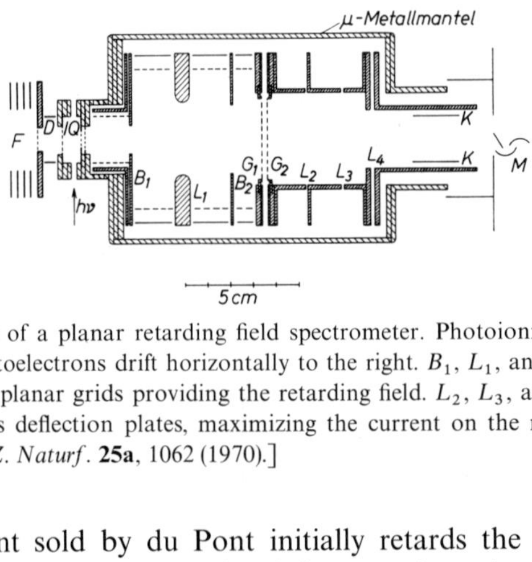

Fig. 15 Example of a planar retarding field spectrometer. Photoionization occurs at *IQ*, and the desired photoelectrons drift horizontally to the right. B_1, L_1, and B_2 act as an einzel lens. G_1 and G_2 are planar grids providing the retarding field. L_2, L_3, and L_4 are cylindrical lenses. *K* symbolizes deflection plates, maximizing the current on the multiplier *M*. [From K. von Puttkamer, *Z. Naturf.* **25a**, 1062 (1970).]

The instrument sold by du Pont initially retards the electrons with a planar field, then sends the transmitted electrons through a coarse deflecting field (see Fig. 16). These electrons are then reflected from a grid held at potential $V - \Delta V$ and attempt to penetrate a grid at the opposite end of the apparatus, held at $V + \Delta V$. The result is that only electrons with energy between $e(V + \Delta V)$ and $e(V - \Delta V)$ reach a particle multiplier.

c. Spherical Retarding Field. The geometry here consists of quasi-point source of photoelectrons whose potential V_0 is determined by a small spherical grid surrounding the ionization volume. A concentric spherical grid of much larger radius is maintained at the variable potential V_R, and finally a detector sphere surrounds the entire system (see Fig. 17). It has been shown[34] that the resolution width of such a device is approximately

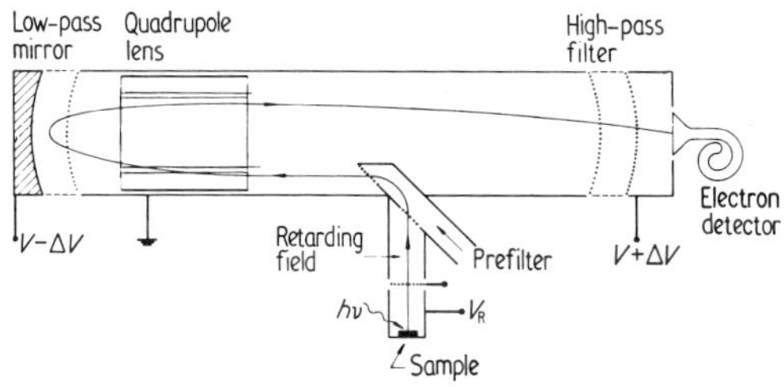

Fig. 16 Schematic view of the du Pont "double filter" instrument. [From Wannberg *et al.*, *J. Phys. E* **7**, 149 (1974).]

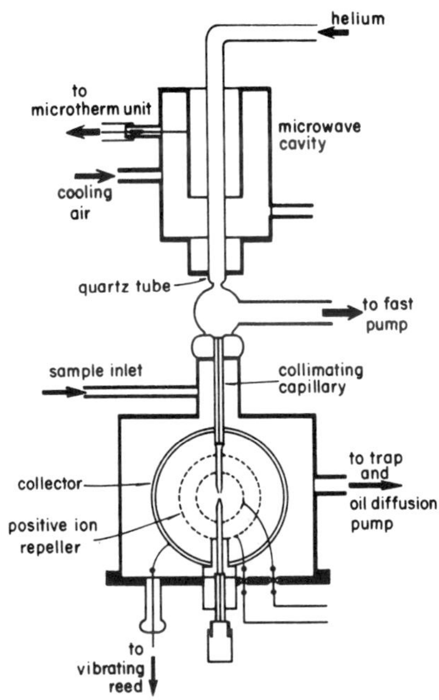

Fig. 17 Diagram of a spherical retarding field spectrometer. Also shown is a resonance line source generated by a microwave discharge. [From D. C. Frost *et al.*, *Phys. Rev. Lett.* **15**, 612 (1965).]

$\Delta E_{1/2} \approx 5 \times 10^{-2}\, Fa$, where F is the field strength between V_0 and V_R and a the distance between mesh wires. Hence, it behooves one to make the distance between inner and outer grids large and to keep the mesh size tight. Electrons formed within the inner grid, but off the center point, can traverse the retarding field at an angle and appear to have a lower energy because the radial component of their velocity is smaller. If the radius of the inner spherical grid is r and that of the outer one R, an approximation for the theoretical resolution is $\Delta E/E = (r/R)^2$.

There are several attractive features to this apparatus. It can accommodate a much larger surface area than a dispersive instrument of similar size, it has an almost ideal solid angle ~ two to three orders of magnitude larger than a dispersive instrument and (perhaps most important) relative partial cross sections can be directly measured from the integral spectrum, largely unaffected by angular effects or transmission corrections. The major disadvantages are those characteristic of nondispersive devices—limited resolution and the necessity of differentiating to obtain a differential spectrum with the attendant limitation in signal to noise. Again, it is difficult to use counting techniques with such a large solid angle. Some strategies have been proposed and used[34,35] to refocus the electrons after traversing the spherical retarding field.

With all analyzers, dispersive or nondispersive, stray magnetic fields must be minimized, either by using magnetic shielding or with Helmholtz coils. Nondispersive instruments are somewhat less sensitive to stray fields, in this author's experience. Another characteristic of these analyzers is sensitivity to varying surface potentials. The retarding grid devices are affected by such potentials as seriously as deflection instruments. For example, it has been reported[36] that a planar retarding field apparatus could achieve 5 meV resolution with argon, but with very few other target gases. It is common to coat sensitive surfaces (after careful cleaning) with a graphite film. This is done either by allowing the soot from a hydrocarbon flame with insufficient oxygen (benzene is often used) to deposit on such surfaces, or to use a colloidal dispersion of graphite in isopropanol, which can be painted or sprayed.

The larger the apparatus, the higher the potential resolving capability. However, the trajectory of electrons is longer, and hence one needs to pay more attention to the influence of stray fields and varying surface potentials.

2. The Electron Time-of-Flight Techniques

The drift time t of an electron from source to detector (length $= l$) is related to its energy E by $E = \frac{1}{2}m_e(l^2/t^2)$, where m_e is the electron's mass. Experimental limitations in the accuracy of length and time measurements lead to corresponding errors in the determination of energy, i.e.,

$$\frac{\Delta E}{E} \propto 2\frac{\Delta l}{l}$$

$$\frac{\Delta E}{E} \propto 2\frac{\Delta t}{t}.$$

Since only time is measured, length inaccuracies Δl appear as additional timing inaccuracies, the two being related by the drift velocity

$$v = \left(\frac{2E}{m_e}\right)^{1/2}.$$

Thus,

$$\Delta t' = \frac{\Delta l}{v},$$

where Δl is caused by the length difference between the various electron trajectories. Adding the two contributions in quadrature, one obtains

$$\frac{\Delta E}{E} = \left[\left(\frac{2\,\Delta t}{t}\right)^2 + \left(\frac{2\,\Delta t'}{t}\right)^2\right]^{1/2}.$$

Since $t \propto E^{-1/2}$, we deduce $\Delta E \propto E^{3/2}$. Hence, this approach becomes more attractive for electrons of low kinetic energy since for dispersive analyzers (Section C.4) $\Delta E \propto E$.

For illustrative purposes, let us consider an idealized situation in which $\Delta l = 0$. With a typical electronic timing uncertainty $\Delta t = 10^{-9}$ sec and a path length $l = 20$ cm, one readily calculates $\Delta E = 6$ meV for 1 eV electrons and $\Delta E = 189$ meV for 10 eV electrons. In practice, $\Delta l \neq 0$ and stray electric fields may introduce additional complication. Tsai *et al.*[37] have used an approach like this to analyze photoelectrons in the 0–1 eV range and obtain a resolution of ≤ 100 meV for electrons with "near zero kinetic energy." Brown *et al.*[38] have studied solid targets with such a device, using the monochromatized light and timing from the SPEAR storage ring, and deduce a system resolution of 0.2 eV. Detailed design criteria for an electron TOF spectrometer have been given by Baldwin and Friedman[39] and more recently by Wilden *et al.*[40]

3. The Threshold Photoelectron Technique

In order to achieve sufficiently strong signals, most photoelectron spectroscopic studies to date have used resonance line sources or x-ray lines (see below) and have avoided the use of monochromators to select photon band widths. With fixed *hv* having a magnitude considerably larger than several ionization energies of the molecule, the kinetic energy of the measured electrons can be directly correlated with states of the molecular ion. When the kinetic energy of photoelectrons is <1 eV, the transmission of dispersive analyzers frequently falls off. As shown above, TOF electron analysis can be useful in this region. For some purposes, one would like to measure just those photoelectrons within a few milli-electron-volts of zero. As an idealized example, let us hypothesize an electron analyzer that has a window only for 1 meV electrons. If our apparatus is provided with a monochromator and our signal to noise is satisfactory, then a scan of photon energy should reveal 1 meV spikes at those photon energies just sufficient to form states of the ion and concomitant 1 meV electrons. Hence, if photoelectron spectroscopy is restricted to determination of energy states, rather than relative intensities, angular distributions, etc., our idealized system could perform this function. Since the energy scale is determined by wavelength calibrations, some of the problems intrinsic to conventional PES (stray magnetic fields, surface potentials, varying calibrations during a run) are minimized or eliminated. Such an apparatus also has obvious attractiveness for photoelectron–photoion coincidence measurements, where the primary purpose of the photoelectron analyzer is to identify a state of the molecular ion.

Several techniques have been devised to approach the idealized performance described above. In one such scheme,[41] the near-zero photoelectrons are accelerated to a fixed energy and analyzed by a 127° dispersive analyzer. This method suffers from the typical low transmission characteristics of a dispersive analyzer, and the resolution width attained (~100 meV) was probably limited by the necessity of using somewhat larger slits to achieve satisfactory intensities.

An alternative approach makes use of the special properties of near-zero electrons in a homogeneous field. Consider a source of photoelectrons with a range of kinetic energies, which (for convenience) we assume to be isotropic. If we apply an electric field between two parallel plates surrounding this source and arrange for one of these plates (the positive one) to have parallel tubular apertures, then only those electrons in a narrow cone about the normal to these apertures will penetrate through them without hitting walls. The electrons with near-zero kinetic energy will have initial velocities correspondingly near zero. In the presence of the electric field, their velocities in the field direction will increase to

$$v_f = (2qV_f/m_e)^{1/2},$$

while the transverse velocities will remain near zero. Such electrons should penetrate the apertures, once having entered one of the orifices. Electrons with initial kinetic energies, and hence transverse velocities, will reach an orifice at some angle given by $\tan \gamma_f = v_{\text{transverse}}/v_f$, and hence may not penetrate the aperture if this angle is larger than that given by the length to diameter ratio of the aperture. Of course, some of the electrons with initial kinetic energy will already be moving along the field direction, and they too will be transmitted, but this fraction can be made very small relative to the

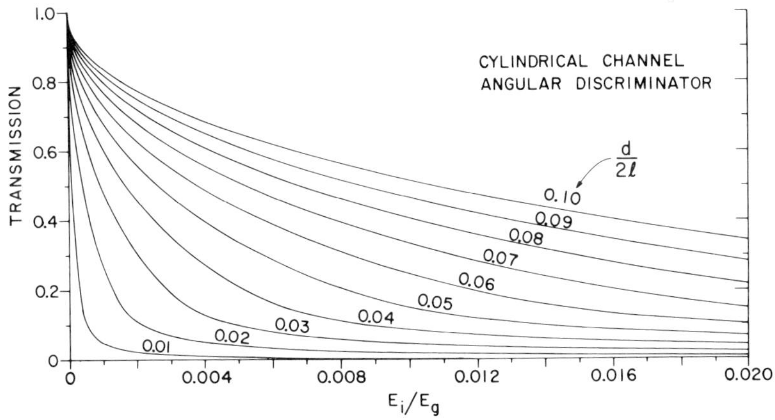

Fig. 18 Transmission function for a cylindrical channel angular discriminator with non-reflective channel walls, having a smooth angular discrimination function. The abscissa (E_i/E_g) is the ratio of the initial kinetic energy of the electron to energy gain in the extraction field. The cylindrical channels have diameter d and length l. One notes that for an aspect ratio ($d/2l$) of ~ 0.02 a very narrow band of electron energies near zero is transmitted. [From Spohr *et al.*, *Rev. Sci. Instr.* **42**, 1872 (1971).]

fraction of near-zero electrons which penetrates. The straightforward geometric analysis is given in a paper by Spohr *et al.*[42] The fraction of electrons transmitted, plotted as an inverse function of the applied potential V_f for different diameter to length ratios of the apertures, is shown in Fig. 18. An analysis of these curves shows that $\Delta E_{1/2} = (d/2l)^2 V_f$ for $(d/2l) \ll 1$. For a diameter to length ratio of $\frac{1}{40}$, and an applied potential of 1 eV, the calculated half-width would be 0.16 meV!

An example of a practical attempt to implement the design is shown in Fig. 19. The transmitted photoelectron spectrum (sometimes referred to as threshold photoelectron spectrum, TPES) obtained with this device, using krypton as a target gas, is given in Fig. 20 (top). The corresponding photoion spectrum appears in Fig. 20 (middle). Whereas a conventional 584 Å photoelectron spectrum of krypton reveals only 2 peaks, corresponding to formation of the $^2P_{3/2}$ and $^2P_{1/2}$ states of Kr, the TPES provides much additional structure, which (comparing the top and middle parts of the figure) is clearly due to autoionizing features in the ionization of krypton. These features correspond to electrons with nonzero kinetic energy that have managed to permeate the apertures.

Several techniques have been used to suppress these nonzero electrons. A numerical procedure having some theoretical basis is to divide the TPES by the photoionization spectrum. The resulting curve, Fig. 20 (bottom), has cleaned up most of the autoionizing features. (In this figure, approximately

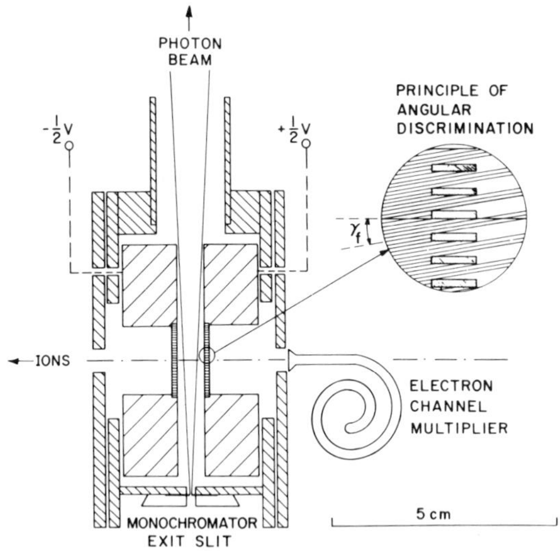

Fig. 19 Cross section of a threshold photoelectron detector. Photoelectrons generated at infinitesimal kinetic energies in the central region of the electrostatic condenser are made nearly parallel by the homogeneous electrostatic field, penetrate the channels at the right with high probability, and are detected by a channeltron multiplier. [From R. Spohr *et al.*, *Rev. Sci. Instr.* **42**, 1872 (1971).]

half of the bandwidth is attributable to the photon bandwidth.) Some experimental ploys have also been used to minimize the nonzero electrons. The approach used by Peatman *et al.*[43] is to arrange for the source of photoelectrons not to have direct line-of-sight access to the aperture(s). A particular choice of fields in the ionization box (see Fig. 21) then imposes those trajectories of near-zero electrons which approach near normal to the aperture, whereas the higher energy components have different trajectories. Peatman *et al.* calculate transmission functions for this device similar to that shown in Fig. 18, and display an argon spectrum with suppressed (though not entirely absent) autoionization features. Lohr *et al.*[44] impose a 90° electrostatic deflection "postmonochromator" to those electrons that have penetrated the apertures, thereby crudely separating the near-zero electrons from interlopers. They have not published a noble gas spectrum to display the efficacy of this solution.

The strategy of Baer and collaborators[44a] has been to combine the angular discrimination of TPES with time-of-flight electron discrimination. The latter effect need only serve to eliminate high energy components, as may be produced by autoionization. This approach obviously requires pulsed operation. Synchrotron radiation from an electron storage ring which has

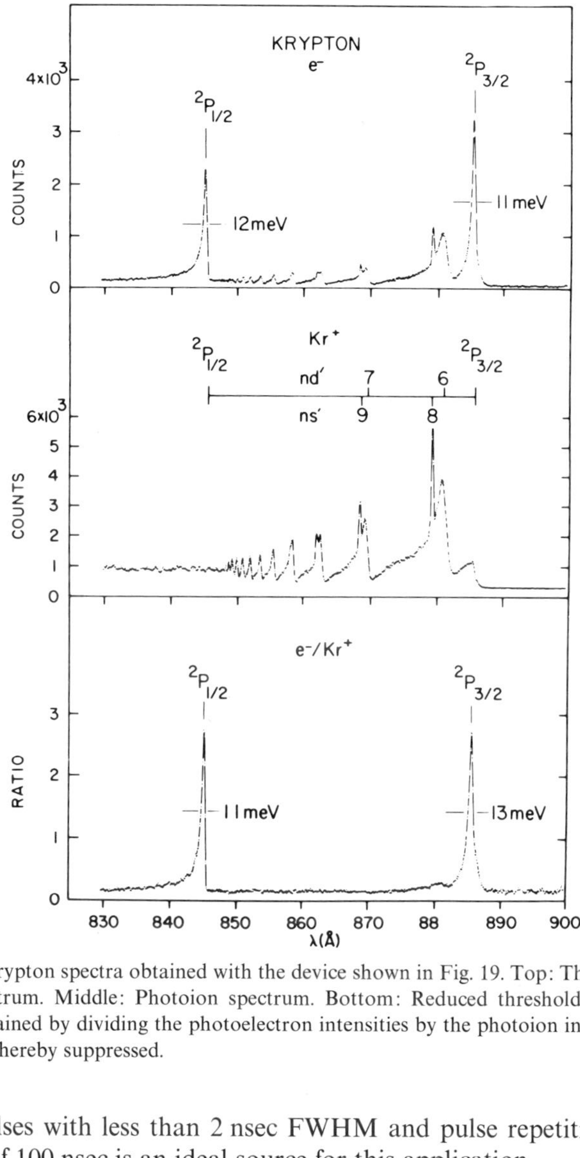

Fig. 20 Krypton spectra obtained with the device shown in Fig. 19. Top: Threshold photoelectron spectrum. Middle: Photoion spectrum. Bottom: Reduced threshold photoelectron spectrum obtained by dividing the photoelectron intensities by the photoion intensities. Autoionization is thereby suppressed.

photon pulses with less than 2 nsec FWHM and pulse repetition times of the order of 100 nsec is an ideal source for this application.

A rather different approach directed at the detection of near-zero energy electrons is to use another molecule (in addition to the target gas) as a scavenger for low energy electrons. SF_6 has a high cross section for electron attachment at near-zero energy, and the cross section falls rapidly with excess

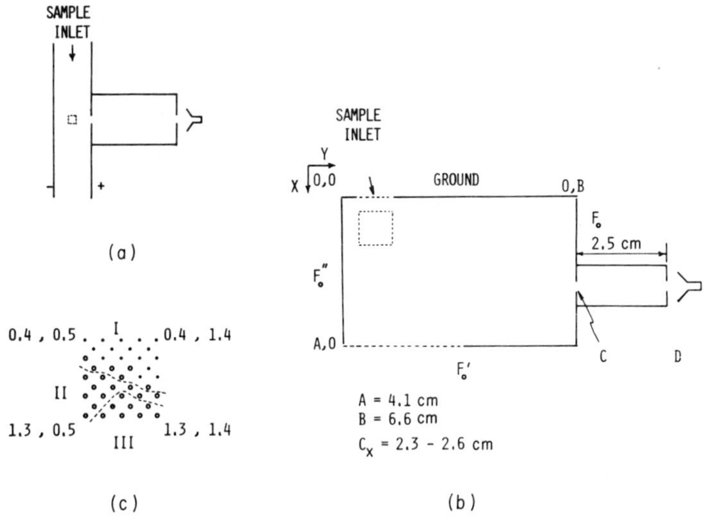

Fig. 21 Pictorial comparison of (a) a line-of-sight threshold photoelectron analyzer and (b) an indirect trajectory analyzer, with dimensions given. In (c), the photoionization region is shown in expanded scale. The indirect trajectory analyzer suppresses fast electrons in autoionization regions. [From W. B. Peatman *et al.*, *J. Electr. Spectr.* 7, 233 (1975).]

energy as a new channel (SF_5^-) opens. Hence, the combination of monochromator and mass spectrometer tuned to SF_6^- can serve as a near-zero energy electron detector.

The merits of these schemes include the ability to sample a large source area of photoelectrons at relatively high transmission and potentially very good resolution. The disadvantages relate to the interpretation of the results. The various schemes are still susceptible to transmission of electrons from autoionizing states (in the neighborhood of the first threshold) and electrons from collisional ionization and/or field ionization of high Rydberg states. (In the SF_6 scheme, the charge transfer reaction

$$M^{**} + SF_6 \rightarrow M^+ + SF_6^-$$

can occur below threshold.) As a consequence, meaningful relative intensities, line widths, and even precise energies of states are difficult to unravel.

4. Dispersive Analyzers

This class of instruments is the most widely used in photoelectron spectroscopy. As a consequence, a great deal of theory and technology has been devoted to development of these analyzers which we can only briefly summarize here. The interested reader can select from a number of references to be cited below.

In principle, a fraction of the photoelectrons from a source volume are allowed to enter the apparatus through a slit (or pseudo slit). From this origin, they are deflected (electrostatically or magnetically) to a different region in space, the deflection being a function of the kinetic energy of the electrons. The dispersion is the distribution in real space for a range of kinetic energies.

Magnetic dispersion, though generally used for the analysis of very energetic electrons in beta-ray spectroscopy, has come into disfavor in the lower energy domain. The iron-cored instruments are very difficult to use at low energies. Iron-free magnetic instruments are still used in x-ray photoelectron spectroscopy, but suffer from high sensitivity to stray magnetic fields.[45] An early effort to make ultraviolet photoelectron spectroscopic measurements with a magnetic instrument[46] has apparently been abandoned. Hence, we shall focus our attention here on electrostatic deflection devices and neglect other variations involving magnetic fields, such as Wien filters and trochoidal analyzers.

Numerous geometries for electrostatic deflection have been examined, including parallel plate and cylindrical condensers, concentric cylinders, concentric spheres and hemispheres, concentric toroids, and trochoidal paths. We shall try to sketch here the relative merits of many of these arrangements. It is useful to keep in mind that the energy resolution of these analyzers can be written rather generally as

$$\Delta E/E = (S_1 + S_2)/R + \text{higher-order terms},$$

where S_1 and S_2 are entrance and exit slit widths and R is some characteristic distance, such as the length of electron trajectory. On the other hand, the angular acceptance of the apparatus is proportional to the area of the entrance slit. For typical values, the ratio of acceptance solid angle to 4π is ~ 0.01–0.0001. Hence, photoelectron spectroscopy with such devices is correspondingly less sensitive than photoionization mass spectrometry. To compensate for this, stronger sources (particularly stronger light sources) are generally used, and will be briefly described at the conclusion of this section.

a. Parallel Plate Analyzers. Consider a parallel plate condenser with two parallel slits in the lower plate (Fig. 22) separated by a distance $2z_m$. Now, allow a beam of electrons to enter the condenser region through the left-hand slit, at an angle α to the horizontal, with velocity v_0. Then, with a

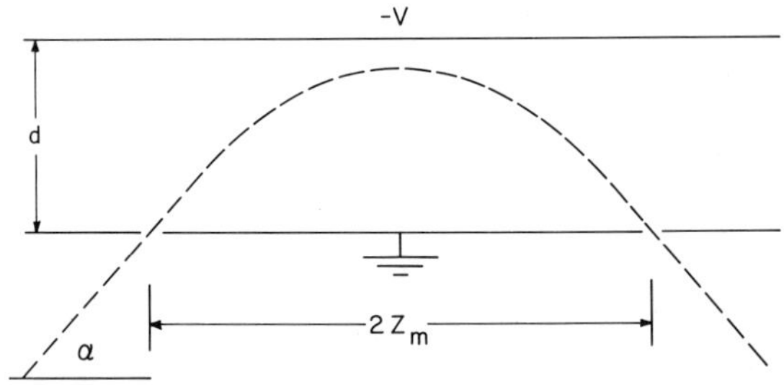

Fig. 22 Schematic diagram of the simple parallel plate dispersive analyzer.

potential $-V$ on the upper plate and a separation d between plates, the equations of motion may be written

$$\begin{aligned} F_y &= q\left(\frac{-V}{d}\right) = -qE_y = m_e \frac{d^2y}{dt^2} \\ F_z &= 0. \end{aligned} \tag{14}$$

Thus,

$$\begin{aligned} \frac{dz}{dt} &= v_0 \cos\alpha \\ \frac{dy}{dt} &= v_0 \sin\alpha - \frac{q}{m_e} E_y t. \end{aligned} \tag{15}$$

The electrons follow a parabolic trajectory. At the y maximum, $dy/dt = 0$ and

$$t = \frac{m_e v_0 \sin\alpha}{qE_y}.$$

The trajectory is symmetric on both sides of the maximum so that

$$2z_m = v_0 \cos\alpha(2t) = v_0 \cos\alpha\left(\frac{2m_e v_0 \sin\alpha}{qE_y}\right) \tag{16}$$

or

$$z_m = \frac{m_e}{q} \frac{v_0{}^2}{E_y} \frac{\sin 2\alpha}{2}.$$

In calculations of this type, it is common to expand the trajectory in a Taylor series as a function of various perturbations. In this simple case, we consider z_m as a function of $\Delta\alpha$

$$z_m = z_0 + \frac{\partial z}{\partial \alpha}(\Delta\alpha) + \frac{1}{2!}\left(\frac{\partial^2 z}{\partial \alpha^2}\right)(\Delta\alpha)^2 + \cdots. \tag{17}$$

The condition for first-order focus is that the coefficient of $\Delta\alpha$ be zero.

$$\left(\frac{\partial z_{\mathrm{m}}}{\partial \alpha}\right) = \frac{m_{\mathrm{e}}}{q}\frac{v_0}{E_y}\cos(2\alpha) = 0. \tag{18}$$

This occurs for $2\alpha = 90^\circ$ or $\alpha = 45^\circ$.

The second-order term (coefficient of $(\Delta\alpha)^2$ is

$$\left(\frac{\partial^2 z_{\mathrm{m}}}{\partial \alpha^2}\right) = -\frac{2m_{\mathrm{e}}}{q}\frac{v_0^{\,2}}{E_y}\sin(2\alpha). \tag{19}$$

This term has a maximum at $\alpha = 45^\circ$, and represents a severe limitation for this design. It limits the angular aperture to $\pm 1.5^\circ$ for satisfactory resolution, and this, in turn, severely curtails the acceptance angle and hence sensitivity. The 45° entry parallel plate analyzer does have some advantages. It is probably the easiest dispersion analyzer to construct, the slits can be made very long (and hence match a line source of photoelectrons) and are conveniently located at ground potential, obviating the need for fringing field correctors. Also, it does have an image plane, so that a multidetector array can be placed in the vicinity (and in place of) the exit slit to simultaneously scan a range of energies.

A simple variant[47] on the parallel plate analyzer described above allows for the image point to occur outside the condenser region, rather than at the exit slit, and will be seen to have superior focusing properties. With reference to Fig. 23, and using the same notation as before, consider a source of electrons at a distance l below the condenser, and $z = 0$. Electrons that enter the slit A while traversing the field-free region make an angle α with the z axis. The z displacement of the electron will then have the functional form

$$l\cot\alpha + \frac{m_{\mathrm{e}}}{q}\frac{v_0^{\,2}}{E_y}\sin(2\alpha) - y\cot\alpha$$

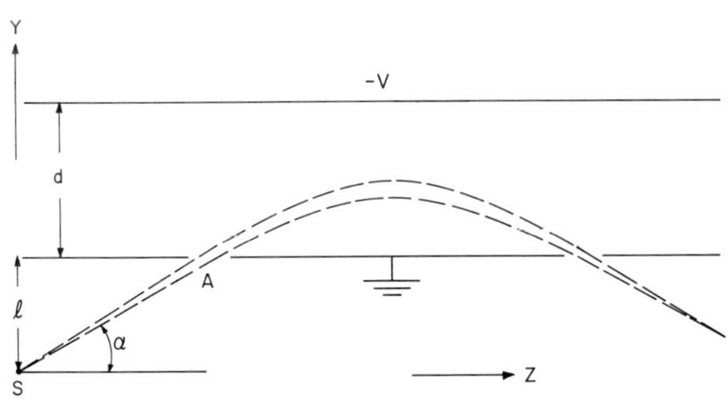

Fig. 23 Schematic diagram of the 30° parallel plate dispersive analyzer which has perfect second-order focusing properties.

or

$$z = (l - y)\cot\alpha + \frac{2(\mathrm{KE})}{E_y}\sin(2\alpha). \tag{20}$$

As discussed above, we explore the condition for first-order focus by equating $\partial z/\partial\alpha$ to zero

$$\frac{\partial z}{\partial\alpha} = -(l - y)\csc^2\alpha + \frac{4(\mathrm{KE})}{E_y}\cos(2\alpha)$$

and (21)

$$(l - y) = \frac{4(\mathrm{KE})}{E_y}\cos(2\alpha)\sin^2\alpha.$$

More freedom for first-order focus is evident from Eq. (21). A large range of angles α can be used and focus can be achieved at correspondingly different values of y (and therefore z). The locus of foci for different kinetic energies is found by substituting the value of (KE/E_y) from Eq. (21) into Eq. (20) Thus,

$$\begin{aligned} z &= (l - y)\cot\alpha + \frac{(l - y)}{2}\frac{\sin(2\alpha)}{\cos(2\alpha)\sin^2\alpha} \\ z &= (l - y)\cot\alpha[1 + \sec(2\alpha)]. \end{aligned} \tag{22}$$

The locus is seen to be a straight line, whose slope (angle with respect to plane of condenser plate) can be readily deduced by differentiation. The angle ζ of this line with respect to the z axis is

$$-\tan\zeta = \frac{\tan\alpha\cos(2\alpha)}{1 + \cos(2\alpha)}. \tag{23}$$

The criterion that the focal plane be at the lower condenser plate ($\zeta = 0$) is seen to lead to $\alpha = 45^\circ$, as initially obtained for the parallel plate analyzer. For positive ζ, we are dealing with a focal plane between the condenser plates, which must be rejected as a solution. This corresponds to $\alpha > 45^\circ$.

The attractiveness of this design becomes apparent when one examines the second-order focusing term. Proceeding as before,

$$\frac{\partial^2 z}{\partial\alpha^2} = 2(l - y)\csc^2\alpha\cot\alpha - 8\left(\frac{\mathrm{KE}}{E_y}\right)\sin(2\alpha) = 0$$

or (24)

$$\left(\frac{\mathrm{KE}}{E_y}\right) = \left(\frac{l - y}{4}\right)\frac{\csc^2\alpha\cot\alpha}{\sin(2\alpha)}.$$

But from Eq. (21), the condition for first-order focus is

$$\left(\frac{\mathrm{KE}}{E_y}\right) = \left(\frac{l-y}{4}\right)\frac{\csc^2\alpha}{\cos(2\alpha)}. \tag{25}$$

Equating these expressions, we obtain

$$\tan(2\alpha) = \cot\alpha$$

or (26)

$$\alpha = 30^\circ.$$

Hence, perfect second-order focusing is achieved with this modification, whereas the 45° parallel plate design was shown to have very poor second-order aberrations. With the 30° analyzer, it is possible to operate with an angular aperture $\Delta\alpha = \pm 3^\circ$ before third-order terms begin to contribute significantly.[47] From Eq. (23), one finds that the focal plane is tilted by

$$\zeta = \arctan\frac{1}{3\sqrt{3}} = 10.89^\circ. \tag{27}$$

With both parallel plate analyzers, as with the next to be described 127° cylindrical analyzer, trajectories in a plane non-normal to the slits cause additional aberrations which require a curved exit slit for correction. The 30° parallel plate would appear to be an attractive device for operating with a line source with good transmission at fairly high resolution, but (to this author's knowledge) only one group[48] has currently adopted this approach.

b. The 127° Cylindrical Analyzer. This design, first given a theoretical basis by Hughes and Rojansky[49] in 1929, has until recently produced most of the fairly high resolution photoelectron spectra in the UV region. Trajectories are not as simple as those in the parallel plate spectrometer. The development below is from Arnow and Jones.[50]

Referring to Fig. 24, we consider a pair of cylindrical surfaces having the radii of curvature r_1 and r_2 and maintained at the potentials V_1 and V_2. They are assumed to be of infinite length in the direction perpendicular to the page. The electron beam enters the analyzer at the median position r_0 with a velocity v_0. The electric field between the cylindrical surfaces is

$$\mathscr{E}(r) = \frac{V_1 - V_2}{r\ln(r_2/r_1)}. \tag{28}$$

For circular orbits, centrifugal and radial forces must be equal.

$$\frac{m_e v_0{}^2}{r_0} = q\mathscr{E}(r_0). \tag{29}$$

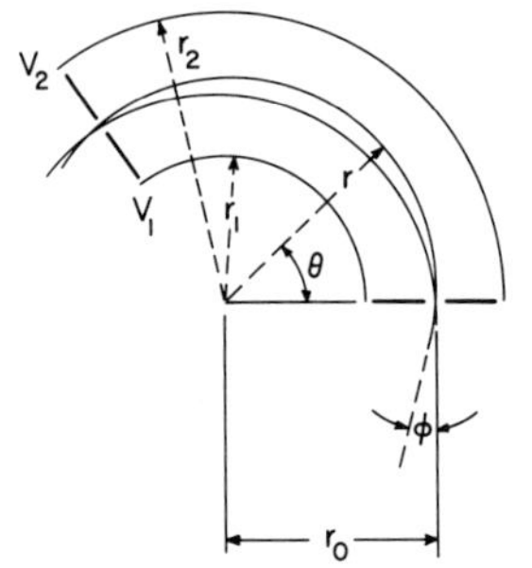

Fig. 24 Schematic diagram of the 127° cylindrical dispersive analyzer. [From M. Arnow and D. R. Jones, *Rev. Sci. Instr.* **43**, 72 (1972).]

The differential equation for the trajectory is given by

$$\frac{d^2u}{d\theta^2} + u = \frac{c^2}{u},$$

where

$$u = \frac{r_0}{r}, \qquad E = \frac{E_n}{E_0}, \qquad \text{and} \qquad c^2 = \frac{1}{E\cos^2\phi}. \tag{30}$$

The solution (accurate to first order) given by Arnow and Jones is

$$u_1 = \exp\left\{\frac{2(c^2-1)}{c^2+1}\sin^2\left[\frac{1}{2}(c^2+1)^{1/2}\theta\right] - \frac{\tan\phi}{(c^2+1)^{1/2}}\sin[(c^2+1)^{1/2}\theta]\right\}. \tag{31}$$

The approximate solution initially deduced by Hughes and Rojansky, which gives the correct resolving power, predicts first-order focusing at $\theta = \pi/\sqrt{2}$ (i.e., 127°) and the departure from perfect refocusing is

$$u_2 = c + (1-c)\cos(\theta\sqrt{2}) - \frac{\sqrt{2}}{2}\tan\phi\sin(\theta\sqrt{2}). \tag{32}$$

From the second term of this latter solution, one can readily see that a beam non-normal at entry aiming toward the outer cylinder will have its maximum deviation from the median ray at $\theta = \pi/2\sqrt{2}$ and intersect the path of the median ray at $\theta = \pi/\sqrt{2}$. If the trajectories are continued, the outer ray becomes an inner ray and crosses the median path again at $\theta = 2\pi/\sqrt{2}$, which is another first-order focus. The initial inner ray has the inverse behavior.

Turner[51] has introduced small perturbation potentials to inner and outer cylinders at $\theta = \pi/2\sqrt{2}$ and realizes an improvement in resolution, but no analysis is given.

The entrance and exit slits introduce additional perturbations. Turner[51] minimizes the distortion introduced by the external fringing field by placing the slits at the apices of cusps formed from two grounded half-cylinders, having axes coinciding with the ends of the capacitor electrodes. Alternatively, Herzog[52] correctors can be used, thus foreshortening the total deflection.

The second-order correction [coefficient of ϕ^2, or $(\Delta\alpha)^2$] for this analyzer is $\frac{4}{3}$. It shares with the parallel plate analyzers the lack of double focusing (although some "vertical focusing" has been reported[53]), relative ease of construction, and good matching with line sources (e.g., sources from polarized synchrotron radiation—see Chapter VII).

c. Spherical (or in General, Toroidal) Analyzers. In this type of analyzer, electrons traverse the field between sections of concentric spheres. The properties of this analyzer were first laid out in detail by Purcell.[54] This device is related to the 127° cylindrical apparatus in that the median ray travels along a constant potential path from source slit to detector slit in a circular orbit. It is superior in having double focusing capability. (For concentric hemispheres, rays emerging from the object point along the median path converge at the conjugate point like arcs of great circles.) With a pair of concentric spherical surfaces having radii r_1 and r_2, maintained at potentials V_1 and V_2, the potential of the median ray is given by[55]

$$V_0 = \frac{V_2 - V_1}{(r_2/r_1) - (r_1/r_2)}. \tag{33}$$

The field varies as $1/r^2$ and the path of the median ray is a circle. First-order focus occurs at 180°, and the coefficient of $(\Delta\alpha)^2$ is 2. However, solutions exist for other spherical segments.

In general, for spherical segments less than 180°, the object and image points and the center of curvature lie on a straight line (see Fig. 25), a property known as Barber's Rule.[56] This design enables sources and detector to be placed outside the field. A transfer matrix formalism has been devised for

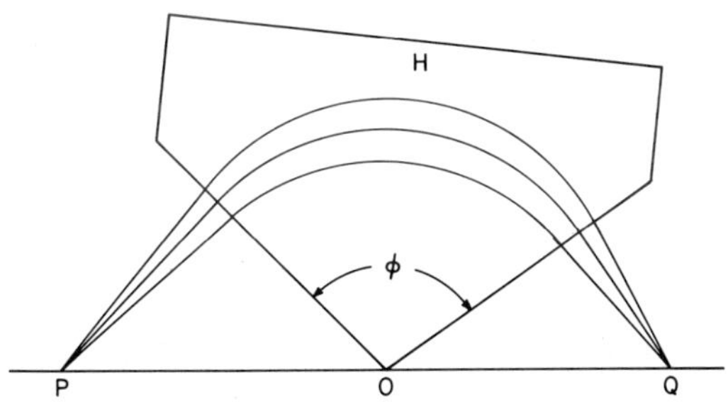

Fig. 25 Schematic illustration of Barber's rule. For spherical segments less than 180°, the object point (P), the image point (Q), and the center of curvature (O) lie on a straight line.

such systems[57] rather similar to that described previously for the design of a quadrupole lens. (See Section B.4 in this chapter.) Wannberg *et al.*[58] have obtained various second- and third-order aberration coefficients from computer solutions of this formalism. They have also explored more general toroidal fields (radii of curvature in orthogonal directions differing) and find superior properties for spherical segments of 218.9° and 211.4°, with ratios of radii of 1.6205 and 1.6288, respectively.

The spherical analyzer has the capability of accepting a large azimuthal angle, and hence its sensitivity for a quasi-point source can be very good. Alternatively, it can be adapted to a multidetector array for simultaneous measurement of a range of energies. Perhaps the highest resolution photoelectron spectra in both the ultraviolet[59] and x-ray regions[60] have been recorded with this type of analyzer.

d. The Cylindrical Mirror Analyzer. In Section C.4(b), a cylindrical analyzer was described in which the primary trajectory occurs in a plane *normal* to the cylinder axis. A configuration of concentric cylinders in which the primary path occurs in a plane *containing* the cylinder axis would appear to have a geometrical advantage since the entire azimuthal domain is accessible. It is therefore surprising that the properties of this analyzer were not described in the literature until 1966.[61,62] We follow here a development largely due to Zashkvara *et al.*[61] The field between the analyzer surfaces is, as with the 127° cylindrical analyzer,

$$\mathscr{E}(r) = \frac{V_1 - V_2}{r \ln(r_2/r_1)}$$

(see Fig. 26 for the geometrical parameters). In the field-free region from O to the entrance between analyzer cylinders.

$$\frac{dz}{dt} = v_0 \cos\alpha, \qquad \frac{dr}{dt} = v_0 \sin\alpha, \qquad \text{and} \qquad \frac{dz}{dr} = \cot\alpha. \tag{34}$$

In the field region, the electron's r component of velocity is initially retarded to zero, then accelerated to its initial velocity. The z component of velocity remains constant. The total (potential plus kinetic) energy is a constant of the motion.

$$\frac{1}{2} m_e v^2 + \frac{q(V_1 - V_2)\ln(r/r_1)}{\ln(r_2/r_1)} = \text{const.} \tag{35}$$

The constant is readily seen to be $m_e v_0{}^2/2$ by letting $r = r_1$. Now we express

$$\frac{1}{2} m_e v_0{}^2 = \frac{1}{2} m_e \left[\left(\frac{dz}{dt}\right)^2 + \left(\frac{dr}{dt}\right)^2 \right] \tag{36}$$

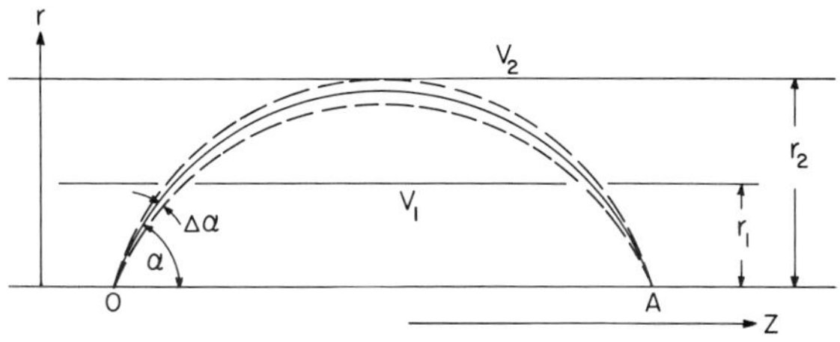

Fig. 26 Geometrical parameters employed in the analysis of the cylindrical mirror dispersive analyzer. There is cylindrical symmetry about the axis OA; The inner cylinder is at r_1 and the outer cylinder at r_2.

and note that $dz/dt = v_0 \cos\alpha$ in the field region also. Replacing dr/dt with $v_0 \cos\alpha \, dr/dz$ leads to

$$\frac{1}{2} m_e v_0{}^2 \cos^2\alpha \left[1 + \left(\frac{dr}{dz}\right)^2\right] + \frac{q(V_1 - V_2)\ln(r/r_1)}{\ln(r_2/r_1)} = \frac{1}{2} m_e v_0{}^2$$

or (37)

$$\frac{dr}{dz} = \frac{(2/m_e)^{1/2}}{v_0 \cos\alpha}\left[\frac{m_e v_0{}^2 \sin^2\alpha}{2} - \frac{q(V_1 - V_2)\ln(r/r_1)}{\ln(r_2/r_1)}\right]^{1/2}.$$

At the maximum (r_m) of the trajectory, $dr/dz = 0$ or

$$r_m = r_1 \exp\left(\frac{m_e v_0{}^2}{2} \frac{\ln[r_2/r_1]}{q[V_1 - V_2]} \sin^2\alpha\right). \tag{38}$$

The exponent of Eq. (38) is of the form $T_0 \sin^2\alpha/U_0$, where T_0 is the initial kinetic energy and $U_0 = q(V_1 - V_2/\ln(r_2/r_1)$. To determine how far the particle has travelled in the z direction when it has reached its apex, we integrate Eq. (37) from entry into the field (r_1) to the point of maximum trajectory (r_m)

$$\frac{dr}{dz} = \frac{(2/m_e)^{1/2}}{v_0 \cos\alpha}[T_0 \sin^2\alpha - U_0 \ln(r/r_1)]^{1/2} \tag{39}$$

$$\int_{r_1}^{r_m} \frac{dr}{[(T_0/U_0)\sin^2\alpha - \ln(r/r_1)]^{1/2}} = \frac{U_0^{1/2}(2/m_e)^{1/2}}{v_0 \cos\alpha} \int_0^{z_m} dz. \tag{40}$$

Changing now to the variable

$$y^2 = (T_0/U_0)\sin^2\alpha - \ln(r/r_1)$$

and

$$dr = 2r_1 y(\exp((T_0/U_0)\sin^2\alpha - y^2)\, dy$$

we have

$$2r_1 \exp\left[\left(\frac{T_0}{U_0}\right)\sin^2\alpha\right]\int_0^{(T_0U_0)^{1/2}\sin\alpha} e^{-y^2}\,dy = \frac{(2U_0/m_e)^{1/2}}{v_0\cos\alpha}\,z_m. \tag{41}$$

Noting that

$$\begin{aligned}\int_0^{(T_0/U_0)^{1/2}\sin\alpha} e^{-y^2}\,dy &= \frac{1}{\sqrt{2}}\int_0^{(2T_0/U_0)^{1/2}\sin\alpha} e^{-t_2/2}\,dt \\ &= \sqrt{\pi}\,\mathrm{ERF}((2T_0/U_0)^{1/2}\sin\alpha)\end{aligned} \tag{42}$$

the median slit-to-slit distance is

$$\begin{aligned}L_1 = 2z_m = 4r_1((T_0/U_0)^{1/2}&\exp[(T_0/U_0)\sin^2\alpha] \\ &\times \cos\alpha\sqrt{\pi}\,\mathrm{ERF}((2T_0/U_0)^{1/2}\sin\alpha)\end{aligned} \tag{43}$$

and the axial point-to-point distance is

$$L_2 = 2r_1\cot\alpha + 2z_m. \tag{44}$$

The condition for first-order focus in the slit-to-slit configuration is $\partial L_1/\partial\alpha = 0$, which leads to

$$\exp\left[\left(\frac{T_0}{U_0}\right)\sin^2\alpha\right]\left(\frac{T_0}{U_0}\right)^{1/2}\sin\alpha\sqrt{\pi}\,\mathrm{ERF}\left[1 - 2\left(\frac{T_0}{U_0}\right)\cos^2\alpha\right] = \frac{T_0}{U_0}\cos^2\alpha. \tag{45}$$

Sar-el[63] has plotted this function in the form $K_0 = T_0/U_0$ as a function of α, which is shown in Fig. 27.

The specific energy dispersion and the apparatus dimensions, as functions of α (also calculated by Sar-el), are shown in Figs. 28 and 29. The specific energy dispersion increases very rapidly with α (note the logarithmic scale) but the dimensions of the apparatus begin to increase precipitously beyond $\alpha = 60°$, setting practical limits.

The first-order focus for the point-to-point arrangement is given by

$$\frac{\partial L_2}{\partial\alpha} = 0$$

and leads to

$$2\left(\frac{T_0}{U_0}\right)^{1/2}\sin^3\alpha\exp\left[\left(\frac{T_0}{U_0}\right)\sin^2\alpha\right]\sqrt{\pi}\,\mathrm{ERF} = \frac{1 - 2(T_0/U_0)\sin^2\alpha\cos^2\alpha}{2(T_0/U_0)\cos^2\alpha - 1}. \tag{46}$$

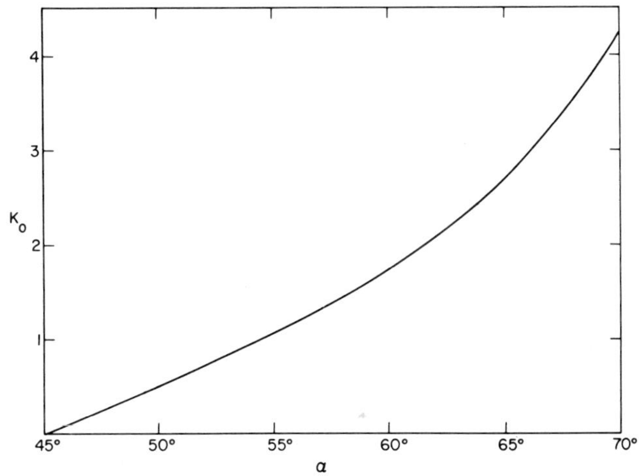

Fig. 27 Dependence of K_0 on α, according to the condition for first-order focusing of the slit-to-slit focusing cylindrical mirror analyzer. [From H. Z. Sar-el, *Rev. Sci. Instr.* **42**, 1601 (1971).]

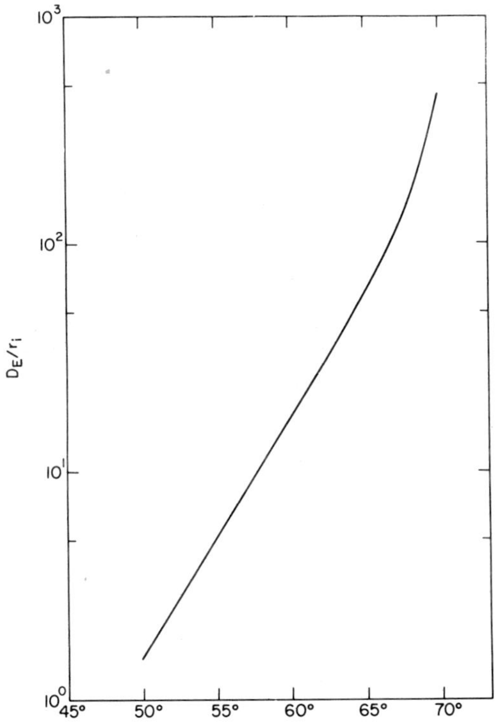

Fig. 28 Specific energy dispersion D_E, in units of r_1, as a function of launching angle α for the slit-to-slit focusing cylindrical mirror analyzer. Note the logarithmic scale of the ordinate, and hence the rapidly improving dispersion between 50° and 60°. [From H. Z. Sar-el, *Rev. Sci. Instr.* **42**, 1601 (1971).]

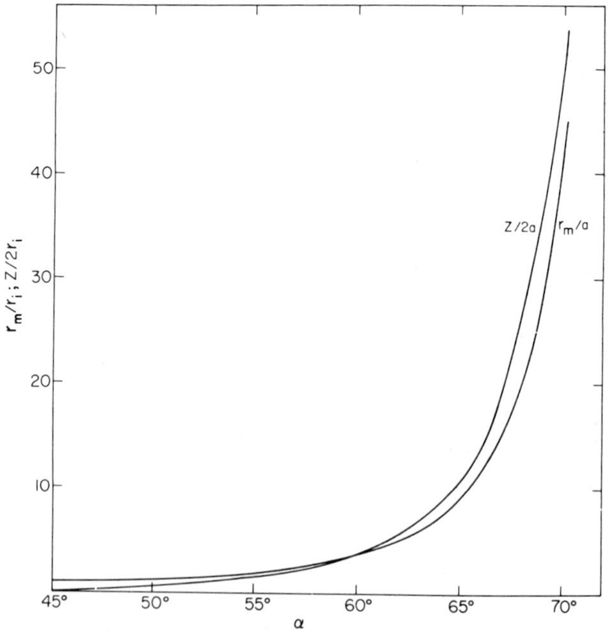

Fig. 29 Variation of the axial distance Z traversed by the particle (in units of $2r_i$) and the maximum height r_m (in units of r_i) as a function of the launching angle α for the slit-to-slit focusing cylindrical mirror analyzer. Although Fig. 28 favors ever larger launching angles, the dimensions of the apparatus rapidly become unwieldy beyond $\alpha \approx 60°–65°$, according to the curves given above. [From H. Z. Sar-el, *Rev. Sci. Instr.* **42**, 1601 (1971).]

A plot of this function (also given by Sar-el[62] in the form $K_0 = T_0/U_0$ versus α) is reproduced in Fig. 30. The surprising observation is that two energies can be brought to first-order focus for every angle in the domain $10° < \alpha < 80°$ except at $\alpha = 42°18.5'$. At this angle, it turns out that second-order focusing

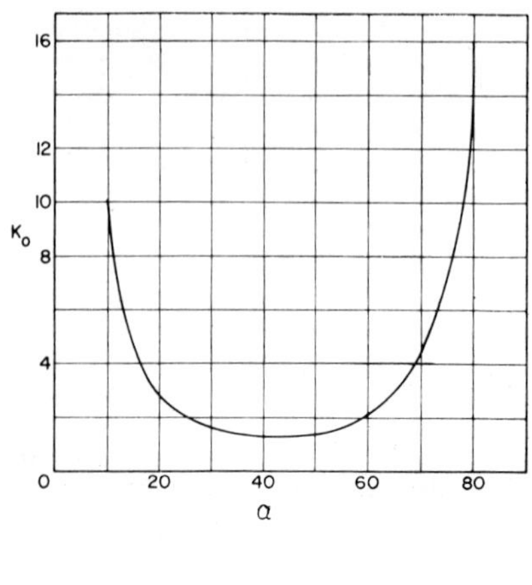

Fig. 30 Graphical representation of K_0 as a function of α for the axial focusing cylindrical mirror analyzer. Two electron energies can be brought to first-order focus for every angle between 10° and 80°, except at $\alpha = 42°\ 18'$. [From H. Z. Sar-el, *Rev. Sci. Instr.* **38**, 1210 (1967).]

is also achieved, i.e.,

$$\frac{\partial^2 L_2}{\partial \alpha^2} = 0$$

In Fig. 31 we plot the function $(1/r_1)(\partial^2 L_2/\partial \alpha^2)$ as a function of α (taken from Zashkvara *et al.*[61]). We had previously found second-order focus also for the 30° parallel plate analyzer. Risley[64] has shown that the coefficient of $(\Delta\alpha)^3$ is more favorable for the cylindrical mirror than the parallel plate, and that it also has higher energy dispersion by a factor ~1.5. The reduced energy dispersion for this configuration (see Fig. 32) is similar to that for the slit-to-slit case, though the vertical scale is somewhat larger due to the second-order focusing. It is nonetheless noteworthy that the reduced dispersion is substantially higher at ~60° than at 42°, negating to considerable extent the advantages of second-order focus. When other aberrations (particularly off-axis trajectories) are taken into account for a diffuse source, the second-order focus becomes still less attractive. The point-to-point cylindrical mirror analyzer has the highly attractive feature of having both double focusing and second-order focusing. Its disadvantages, in addition to limited dispersion at this angle, include its deviation from the nondiscriminatory angle (54° 44′, see Chapter VII), difficulty of access to ionization source and concomitant

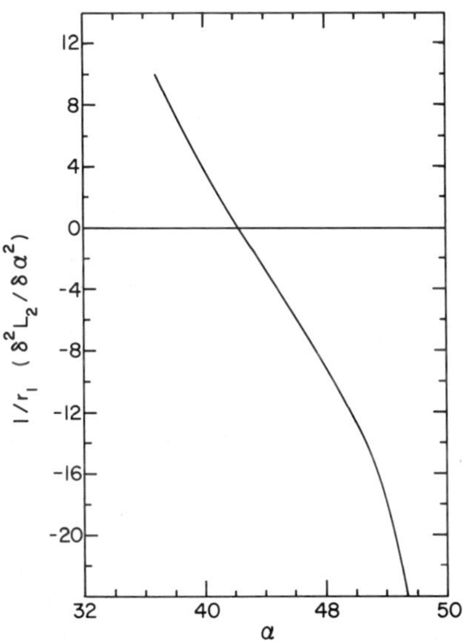

Fig. 31 The second-order focusing term as a function of the launching angle α for the axial focusing cylindrical mirror analyzer. This term vanishes, and hence perfect second-order focusing is achieved at $\alpha = 42°\ 18'$. The curve is plotted from calculations given by V. V. Zashkvara *et al.*, *Sov. Phys.–Tech. Phys.* **11**, 96 (1966).

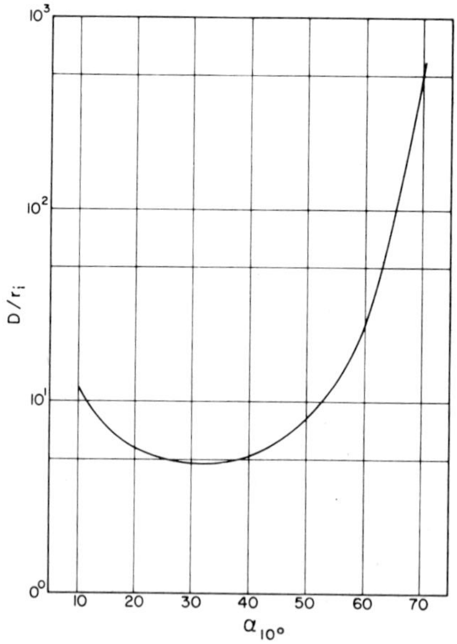

Fig. 32 The specific dispersion (D/r_i) as a function of launching angle α for the point-to-point (axial focusing) version of the cylindrical mirror analyzer. [From H. Z. Sar-el, *Rev. Sci. Instr.* **38**, 1210 (1967).]

problems in using retarding lenses, and the lack of a focal plane which would enable one to use a multidetector array.

The slit-to-slit modification, operated at $\alpha \sim 60^\circ$ in first-order focus, has been used by this author[65] for photoelectron analysis of molecular beam samples with some success. In particular, relative intensities appear to be quite reliable.[66–68] Gardner and Samson[69] have interposed retarding/accelerating slits within the inner cylinder in the slit-to-slit configuration, with $\alpha = 54^\circ\, 44'$, for UPS studies. Any retardation or lens system in these dispersive devices requires a transmission calibration if one is interested in relative cross sections.

Citrin *et al.*[70] and Maeda[71] have designed cylindrical mirrors for XPS studies, the former using $\alpha = 60^\circ$, the latter opting for the second-order focusing condition and interposting retarding slits.

e. The Pill Box, or Bessel Box. An energy analyzer featuring simple construction and no fringing fields outside the slits (as in the parallel plate analyzers) has been described by Allen *et al.*[72] It consists of a squat cylindrical cavity with grounded end plates containing circular slits (see Fig. 33). The design parameters are arrived at by computer analysis of trajectories

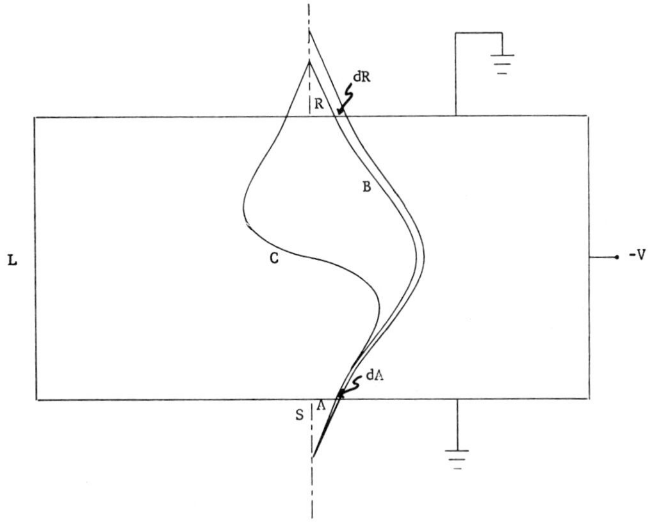

Fig. 33 Schematic diagram of the "pill box" photoelectron spectrometer. [From Allen *et al.*, *Int. J. Mass Spec. Ion Phys.* **8**, 81 (1972).]

since an analytical solution does not exist. A commercial device of this type is being marketed[73] which claims very good resolution (~ 5 meV) with high transmission.

5. Auxiliary Technical Considerations in Photoelectron Spectroscopy with Dispersive Instruments

a. Preretardation and Lenses. In general, the fractional energy resolution, expressed as $\Delta E/E$, is a function of slit widths, length of trajectory, and aberration terms such as those in $(\Delta\alpha)^2$. If electrons are retarded before entering an analyzer, then ΔE should decrease with E. Alternatively, if one maintains the same ΔE, then retardation should enhance intensity since slits can be opened for the lower fractional resolution required. As a consequence of Liouville's Theorem in the form of the Helmholtz–Lagrange relation, the product of position, divergence and $(\mathrm{KE})^{1/2}$ remains constant. It follows that the net gain in intensity by retardation is only $(E_1/E_2)^{1/2}$, the ratio of kinetic energies.[45]

Some practitioners of PES choose to scan photoelectron energies by varying the preretarding (or accelerating) voltage, while maintaining the analyzer at a fixed pass energy. In such circumstances, it is most desirable to have a lens system incorporated with the retarding field to maintain as constant a transmission as possible. The subject of "zoom" lenses is covered in a number of research articles, replete with graphs. Rather than attempting to review this intricate field, we refer the interested reader to a recent book[74] which covers theory, zoom lenses, cylinder and aperture triple lenses, and includes the necessary tables and graphs.

b. Single Channel Scans and Multidetector Arrays. A convenient method of scanning a spectrum with a two-slit dispersive analyzer is to employ a ramp generator, which supplies a linearly varying staircase voltage between critical plates. This voltage may be applied between inner and outer cylinders or spheres of the deflector, or it may be a preretarding (preaccelerating) voltage if the deflectors are set to transmit a fixed kinetic energy. With each step of the staircase voltage, an address advance pulse is applied to a multichannel analyzer, sending each new signal to the new channel.

If the analyzer has a focal plane (or other focal surface), the rate at which information can be accumulated can be increased by ~ 100 if the energy range spanned is 10–20% of the entire spectrum.[45] This effective gain in sensitivity with slight loss of resolution can be achieved by using a multidetector array, instead of exit slit plus detector. Ideally, the individual elements of the array should have widths $\leq$ entrance slit width.

The basic component for this purpose is a compact array of avalanche particle (or photon) multipliers, such as the commercially available[75] microchannel plate channeltron electron multiplier array, or a large number of parallel plate channel multipliers, as described by Nilsson *et al.*[76] In order to achieve the necessary gain (10^7) without ion feedback, two such arrays in tandem (chevron arrangement) are desirable. The spatial resolution of the individual detectors can be as small as 10 μ.

The amplified pulses from each detector element can be measured and stored in one of three ways—scanning with a TV camera, employing a self-scanning silicon photodiode array, or using a resistive anode image converter.

In the first scheme[77], the electron pulses emanating from each element in the channel plate are accelerated by 4 kV onto a phosphor screen, similar to that of a cathode ray tube. With the aid of a TV camera and associated electronics, the bright spots created on the phosphor screen are counted and assigned to a particular energy of the electron spectrum. Since the vidicon tube of the camera can "remember" light pulses long enough for the picture scan to reach the corresponding picture element, all information can be conveniently converted to serial form.

The self-scanning photodiode arrays[78] are monolithic silicon integrated circuits containing sensors, multiplex switches, and shift registers for scanning. They consist of a row of silicon photodiodes (with center-to-center spacing as small as 15 μ), each with an associated storage capacitor on which to integrate photocurrent and a multiplex switch for periodic readout via an integrated shift register scanning circuit. The switches are turned on and off in sequence by the shift register scanning circuit, thereby periodically recharging each cell to 5 V and storing a charge on its capacitance. The shift register is driven by a clock, with a periodic pulse being introduced to initiate each scan. The output obtained from each scan of an N-element array is a train of N charge pulses, each proportional to the light intensity on the corresponding photodiode.

The silicon photodiode array can be used directly for x-ray detection. For application to low energy electron detection, the output electron pulses from a multichannel plate must be accelerated to 4000–5000 V and impinge on a phosphor, which can be deposited on a fiberoptics plate. The fiberoptics plate is pressed directly against the photosensitive surface of the silicon diode array with only a thin light coupling film and a protective sheet of SiO_2 in between.[79]

Probably the simplest method of extracting the information from the multichannel plate is to back it with a resistive anode ($R \approx 10^5$–$10^6 \Omega$) which can be deposited on a substrate by various methods. The charge pulses from the multichannel plate are "proximity focused" onto the resistive anode plate which is equipped with pickup electrodes on its edges. Each individual event is located electronically by the ratios of the charges collected at the edges, or by the differences of the signals' rise times. A number of working models have been described in the literature.[80,81]

The spatial resolution of this system can be made small compared to the inherent limitation imposed by the size and spacing of the individual elements of the multichannel plate.

c. Light Sources in Photoelectron Spectroscopy. Although advances in technology promise to make photoelectron spectroscopy with arbitrarily selected wavelengths more feasible in the near future, the preponderance of investigations to date has been with line sources. Low pressure capillary discharges in the noble gases can be used to generate the corresponding resonance line (see Samson,[1] p. 135). Of these, the most commonly used are NeI (743.7195 Å = 16.6710 eV; 735.8962 Å = 16.8482 eV) and HeI (584.3340 Å = 21.2182 eV).

A hollow cathode discharge is particularly effective in generating He II (303.781/303.786 Å = 40.814/40.813 eV) and Ne II and Ne III lines. Some recent studies have utilized the soft x-ray lines of Y M_ζ (132.3 eV) and Zr M_ζ (151.4 eV). These lines, though inherently somewhat broad and weak, help to bridge the enormous gap between the vacuum ultraviolet and the x-ray domain. The latter commonly uses Mg $K_{\alpha 1,2}$ (1253.60 eV) and Al $K_{\alpha 1,2}$ (1486.70 and 1483.27 eV). The recent development at Uppsala[60] of a rotating anode x-ray tube and crystal monochromator for generating and monochromatizing the Al K_α line has enabled them to achieve unprecedented resolution in this energy range, permitting them to observe incipient vibrational structure in highly excited states of some small molecular ions.

D. Photoelectron–Photoion Coincidence Spectroscopy

The goal of these experiments is to select molecular ions with a predetermined internal energy and then to determine the relative probability of decomposition into various fragment channels. An additional useful piece of information that is obtained in these experiments is the kinetic energy of fragment ions, which aids in the elucidation of the dynamics and in testing theories of unimolecular decay.

At the present time, four groups (Freiburg,[82] Oxford,[83] Chicago,[84] and North Carolina[85]) have published their methods, and others are in preparation. The technologies have differed, though there is a tendency toward convergence in the recent publications. Two groups (Oxford and Freiburg) have generated photoions with the 584 Å resonance line and performed electron energy selection (and hence internal energy selection) by use of a dispersive analyzer. (The Oxford group has used a 45° parallel plate, the Freiburg group two cylindrical mirror analyzers in series). The other groups have employed broadbanded light sources and a monochromator to select a photon bandwidth and combined this with a threshold photoelectron detector. In this latter scheme, the internal energy of molecular ions is altered by varying the incident photon energy as selected by the monochromator. The North Carolina experiments[85] currently use a line-of-sight collimated hole structure as threshold electron selector (see Section C.3), whereas the Chicago group[84] accelerate near-zero energy electrons by 3 V through a narrow drift tube (so-called steradiancy analyzer) and postanalyze with a 127° section analyzer set to accept 3 eV electrons.

The 584 Å and threshold electron approaches, to some extent, have compensatory merits and disadvantages. The threshold electron technique has relatively high collection efficiency, but one must initially monochromatize the radiation with its attendant loss. In addition, it is difficult to completely reject autoionization electrons (which may have the wrong energy) in the threshold detector approach.

A critical zone in both schemes is the photoionization region. Positive ions must be accelerated by a rather strong field which concomitantly acts upon the photoelectrons and can degrade the energy resolution. In this respect, the threshold electron detection approach has an advantage. (The Oxford and Freiburg groups report electron energy resolutions of 150 and 200 meV, respectively, whereas the North Carolina and Chicago groups achieve 30–40 meV.) Some compromise is necessary between ion extraction voltage (which affects ion collection efficiency and mass resolution) and electron energy resolution. Typical values reported by the various groups range from 1–5 V/cm.

In all cases, the detected electron signal acts as the start pulse, either for one or more coincidence gates, or for a time-to-pulse height converter. In the former approach, a variable delay is chosen so that the flight time of a particular mass will be in coincidence with the electron pulse. With several such coincidence gates, it is possible to simultaneous monitor a corresponding number of ion masses. It is convenient to set one coincidence gate to monitor accidental coincidences which must be subtracted.

In the alternative technique, the output of the time-to-pulse height converter can be fed to a multichannel analyzer, which can then accumulate and display the number of coincidence events having a particular pulse height or time-of-flight (and hence mass). One strategy to monitor accidental coincidences here is to employ two additional single channel analyzers before the multichannel analyzer. The first is set to accept all relevant pulse heights corresponding to reasonable masses. The second is set to accept unreasonable pulse heights (coincidences). Both can be simultaneously monitored on the multichannel analyzer. The time-to-pulse height method appears to be gaining favor for these experiments.[84]

Mass separation is achieved by time-of-flight in each case except at Freiburg, where a magnetic sector mass spectrometer is additionally employed. The latter provides significantly better mass resolution and affords the opportunity to study metastable ions. The time-of-flight spectrum provides supplementary information about the kinetic energy distribution of the ions. High collection efficiency for both electrons and ions is vital for the success of these experiments.

A more detailed discussion of the experimental considerations and a summary of recent applications of this technique can be found in a current review article by Baer.[44a]

REFERENCES

1. J. A. R. Samson, "Techniques of Vacuum Ultraviolet Spectroscopy," Wiley, New York, 1967.
2. J. A. R. Samson and G. N. Haddad, *J. Opt. Soc. Am.* **64**, 47 (1974).
3. A. P. Lukirskii, M. A. Rumsh, and L. A. Smirnov, *Opt. Spektrosk.* **9**, 505 (1960); Engl. transl., *Opt. Spectrosc.* (*USSR*) **9**, 262 (1960).
4. D. L. Ederer and D. H. Tomboulian, *Appl. Opt.* **3**, 1073 (1964).
5. J. J. Hopfield, *Astrophys. J.* **72**, 1133 (1930); Y. Tanaka, *Sci. Pap. Inst. Phys. Chem. Res.* (*Jpn.*) **39**, 465 (1942); Y. Tanaka, A. S. Jursa, and F. J. LeBlanc, *J. Opt. Soc. Am.* **48**, 304 (1958).

6. R. E. Huffman, J. C. Larrabee, and D. Chambers, *Appl. Opt.* **4**, 1145 (1965); R. E. Huffman. J. C. Larrabee, Y. Tanaka, and D. Chambers, *J. Opt. Soc. Am.* **55**, 101 (1965); R. E. Huffman, J. C. Larrabee, and Y. Tanaka, *Appl. Opt.* **4**, 1581 (1965).
7. For normal incidence region, see D. Reinke, R. Kraessig, and H. Baumgartl, *Z. Naturforsch., Teil A* **28**, 1021 (1973); W. L. Stebbings and J. W. Taylor, *Int. J. Mass Spectrom. Ion Phys.* **9**, 471 (1972); G. R. Parr and J. W. Taylor, *Rev. Sci. Instrum.* **44**, 1578 (1973). For grazing incidence region, see V. Schmidt, N. Sendner, H. Kuntzemuller, P. Dhez, F. Wuilleumier, and E. Kallne, *Phys. Rev. A* **13**, 1748 (1976).
8. J. Schwinger, *Phys. Rev.* **75**, 1912 (1949).
9. See, e.g., R. P. Godwin, *Springer Tracts Mod. Phys.* **51**, 1 (1969); K. Codling, *Rep. Prog. Phys.* **36**, 541 (1973).
10. K. Radler and J. Berkowitz, *J. Dpt. Soc. Am.* **68**, 1181 (1978).
11. GCA—McPherson Model 225, available from Geophysics Corp. of America.
12. S. Robin, *J. Phys. Radium* **14**, 551 (1953).
13. W. A. Chupka and J. Berkowitz, *J. Chem. Phys.* **51**, 4244 (1969).
14. P. M. Dehmer and W. A. Chupka, *J. Chem. Phys.* **65**, 2243 (1976).
15. C. Kunz, *Proc. Int. Symp. Synchrotron Radiat. Users, Daresbury, England* DNPL/R26, p. 68 (1973); see also DESY Rep. SR-73/3 Hamburg, Germany, (1973).
16. K. P. Miyake, R. Kato, and H. Yamashita, *Sci. Light (Tokyo)* **18**, 39 (1969); R. Kato, H. Yamashita, M. Watanabe, T. Ishii, S. Sato, T. Iizuka, M. Sawada, and K. P. Miyake, *Int. Conf. VUV Radiat. Phys., 3rd, Tokyo* p. 31aA2-3 (1971).
17. C. Kunz, R. Haensel, and B. Sonntaf, *J. Ppt. Soc. Am.* **58**, 1415 (1968); H. Dietrich and C. Kunz. *Rev. Sci. Instrum.* **43**, 434 (1972).
18. K. Codling and P. Mitchell, *J. Phys. E.* **3**, 685 (1970).
19. R. P. Madden and D. L. Ederer, *J. Opt. Soc. Am.* **62**, 722 (1972).
20. F. C. Brown, R. Z. Bachrach, S. B. M. Hagstrom, N. Lien, and C. H. Pruett, *Proc. Int. Conf. VUV Radiat, Phys., 4th*, Hamburg, p. 785 (1974).
21. P. Jaegle, *J. Phys. (Paris)* **24**, 179 (1963); *C. R. Acad. Sci.* **259**, 533, 4556 (1964).
22. V. Schmidt, N. Sendner, H. Kuntzemuller, P. Dhez, F. Wuilleumier, and E. Kallne, *Phys. Rev. A* **13**, 1748 (1976).
23. Stanford Synchrotron Radiat. Proj. SSRP Rep. No. 76/100, p. 150 (1976).
24. H. W. Schnopper and L. Van Speybrosck, cited in ref. 23, pp. 26–27.
25. J. A. R. Samson and G. N. Haddad, *J. Opt. Soc. Am.* **64**, 1346 (1974).
26. P. T. Perdue and E. T. Arakawa, *Nucl. Instrum. Methods* **128**, 201 (1975).
27. C. F. Giese, *Rev. Sci. Instrum.* **30**, 260 (1959).
28. L. Teng, *Rev. Sci. Instrum.* **25**, 264 (1954).
29. T. A. Carlson and R. M. White, *J, Chem. Phys.* **44**, 4510 (1966).
30. R. Browning and J. Fryar, *J. Phys. B* **6**, 364 (1973).
31. D. W. Turner and M. I. Al-Jobory, *J. Chem. Phys.* **37**, 3007 (1962).
32. F. I. Vilesov, B. L. Kurbatov, and A. N. Terenin, *Dokl. Akad. Nauk SSSR* **138**, 1329 (1961); **140**, 797 (1961); Engl. transl., *Sov. Phys.—Dokl.* **6**, 490 (1961); **6**, 883 (1961).
33. J. Berkowitz and W. A. Chupka, *J. Chem. Phys.* **51**, 2342 (1969).
34. P. Staib, *J. Phys. E.* **5**, 484 (1972).
35. D. A. Hutchital and J. D. Rigden, *in* "Electron Spectroscopy" (D. A. Shirley, ed.), p. 79. North-Holland Publ. Amsterdam, 1972.
36. J. H. D. Eland, personal communication.
37. B. Tsai, T. Baer, and M. L. Horovitz, *Rev. Sci. Instrum.* **45**, 494 (1974).
38. Ref. 20, p. 795.
39. G. C. Baldwin and S. I. Friedman, *Rev. Sci. Instrum.* **38**, 519 (1967).
40. D. G. Wilden, P. J. Hicks, and J. Comer, *J. Phys. B* **9**, 1959 (1976).

41. D. Villarejo, R. R. Herm, and M. G. Inghram, *J. Chem. Phys.* **46**, 4995 (1967), and subsequent papers.
42. R. Spohr, P. M. Guyon, W. A. Chupka, and J. Berkowitz, *Rev. Sci. Instrum.* **42**, 1872 (1971).
43. W. B. Peatman, G. B. Kasting, and D. J. Wilson, *J. Electron Spectrosc. Relat. Phenom.* **7**, 233 (1975).
44. W. Lohr, H. W. Jochims, and H. Baumgartel, *Ber. Bunsenges. Phys. Chem.* **79**, 901 (1975).
44a. T. Baer, State Selection by Photoion–Photoelectron Coincidence, *in* "Gas Phase Ion Chemistry" (M. T. Bowers, ed.), Chap. 5. Academic Press, New York (1979).
45. B. Wannberg, U. Gelius, and K. Siegbahn, *J. Phys. E* **7**, 149 (1974).
46. D. W. Turner and D. P. May, *J. Chem. Phys.* **45**, 471 (1966).
47. T. S. Green and G. A. Proca, *Rev. Sci. Instrum.* **41**, 1409 (1970).
48. W. Schmitz and W. Mehlhorn, *J. Phys. E* **5**, 64 (1972).
49. A. L. Hughes and V. Rojansky, *Phys. Rev.* **34**, 284 (1929).
50. M. Arnow and D. R. Jones, *Rev. Sci. Instrum.* **43**, 72 (1972).
51. D. W. Turner, *Proc. R. Soc. London, Ser. A* **307**, 15 (1968).
52. R. Herzog, *Z. Phys.* **97**, 586 (1935).
53. J. J. Leventhal and G. R. North, *Rev. Sci. Instrum.* **42**, 120 (1971).
54. E. M. Purcell, *Phys. Rev.* **54**, 818 (1938).
55. C. E. Kuyatt and J. A. Simpson, *Rev. Sci. Instrum.* **38**, 103 (1967).
56. N. F. Barber, *Proc. Leeds Phil. Lit. Soc. Sci., Sect.* **2**, 427 (1933).
57. H. Matsuda, *Nucl. Instrum. Methods* **91**, 637 (1971); T. Matsuo, H. Matsuda, and H. Wollnik, *Nucl. Instrum. Methods* **103**, 515 (1972); H. Wollnik, T. Matsuo, and H. Matsuda, *Nucl. Instrum. Methods* **102**, 13 (1972).
58. B. Wannberg, G. Engdahl, and A. Skollermo, *J. Electron Spectrosc. Relat. Phenom.* **9**, 111(1976).
59. See, e.g., L. Åsbrink, *Chem. Phys. Lett.* **7**, 549 (1970); L. Karlsson, L. Mattsson, R. Jadrny, R. G. Albridge, S. Pinchas, T. Bergmark, and K. Siegbahn, *J. Chem. Phys.* **62**, 4745 (1975).
60. U. Gelius, E. Basilier, S. Svensson, T. Bergmark, and K. Siegbahn, *J. Electron Spectrosc. Relat. Phenom.* **2**, 405 (1974).
61. V. V. Zashkvara, M. I. Korsunskii, and O. S. Kosmachev, *Sov. Phys.—Tech. Phys.* **11**, 96 (1966).
62. H. Z. Sar-el, *Rev. Sci. Instrum.* **38**, 1210 (1967).
63. H. Z. Sar-el, *Rev. Sci. Instrum.* **42**, 1601 (1971).
64. J. S. Risley, *Rev. Sci. Instrum.* **43**, 95 (1972).
65. J. Berkowitz, *J. Chem. Phys.* **56**, 2766 (1972), and subsequent papers.
66. J. Berkowitz, *Chem. Phys. Lett.* **11**, 21 (1971).
67. J. Berkowitz and R. Spohr, *J. Electron Spectrosc. Relat. Phenom.* **2**, 143 (1973).
68. J. L. Dehmer and J. Berkowitz, *Phys. Rev. A* **10**, 484 (1974).
69. J. L. Gardner and J. A. R. Samson, *J. Electron Spectrosc. Relat. Phenom.* **2**, 267 (1973).
70. P. H. Citrin, R. W. Shaw, and T. D. Thomas, ref. 35, p. 105.
71. K. Maeda, ref. 35, p. 177.
72. J. D. Allen, J. P. Wolfe, and G. K. Schweitzer, *Int. J. Mass Spectrom. Ion Phys.* **8**, 81 (1972).
73. Available from Special Instruments Laboratory, Inc., P. O. Box 950, Knoxville, Tennessee 37901.
74. E. Harting and F. H. Read, "Electrostatic Lenses." Elsevier, Amsterdam, 1976.
75. Available from Galileo Electro-Optics Corp., Sturbridge, Massachusetts.
76. O. Nilsson, L. Hasselgren, K. Siegbahn, S. Berg, L. P. Anderson, and P. A. Tove, *Nucl. Instrum. Methods* **84**, 301 (1970).

77. U. Gelius, E. Basilier, S. Svensson, T. Bergmark, and K. Siegbahn, *J. Electron Spectrosc. Relat. Phenom.* **2**, 405 (1974).
78. Available from Reticon Corp., Sunnyvale, California.
79. G. R. Riegler and K. A. More, *IEEE Trans. Nucl. Sci.* **NS-20**, 102 (1973).
80. W. Parkes, K. D. Evans, and E. Mathieson, *Nucl. Instrum. Methods* **121**, 151 (1974).
81. M. Lampton and F. Paresce, *Rev. Sci. Instrum.* **45**, 1098 (1974).
82. Freiburg apparatus: B. Brehm, J. H. D. Eland, R. Frey, and A. Kustler, *Int. J. Mass Spectrom. Ion Phys.* **12**, 197 (1973); R. Frey, Diplomarbeit, Univ. Freiburg, Freiburg, West Germany, 1975.
83. Oxford apparatus: C. J. Danby and J. H. D. Eland, *Int. J. Mass Spectrom. Ion Phys.* **8**, 153 (1972).
84. Chicago apparatus: R. Stockbauer, *J. Chem. Phys.* **58**, 3800 (1973); R. Stockbauer and M. G. Inghram, *J. Chem. Phys.* **62**, 4862 (1975); R. Stockbauer and M. G. Inghram, *J. Chem. Phys.* **65**, 4081 (1976).
85. North Carolina apparatus: D. M. Mintz and T. Baer, *J. Chem. Phys.* **65**, 2407 (1976).

INDEX

Note: Appendix 1 of Chapter VI (pp. 329–336) has an extensive tabulation of molecules for which some photoionization mass spectrometric studies exist (including wavelength range, resolution, and references).

N

O

P

Q

R

S

T

U

V

W

X

Y

Z